Nutrition and Nutritive Soul in Aristotle and Aristotelianism

Topics in Ancient Philosophy/ Themen der antiken Philosophie

Edited by / Herausgegeben von
Ludger Jansen, Christoph Jedan, Christof Rapp

Volume 9

Nutrition and Nutritive Soul in Aristotle and Aristotelianism

Edited by
Giouli Korobili, Roberto Lo Presti

With the assistance of Dorothea Keller

DE GRUYTER

ISBN 978-3-11-110407-2
e-ISBN (PDF) 978-3-11-069055-2
e-ISBN (EPUB) 978-3-11-069056-9
ISSN 2198-3100

Library of Congress Control Number: 2020939771

Bibliographic information published by the Deutsche Nationalbibliothek
The Deutsche Nationalbibliothek lists this publication in the Deutsche Nationalbibliografie;
detailed bibliographic data are available on the Internet at http://dnb.dnb.de.

© 2022 Walter de Gruyter GmbH, Berlin/Boston
This volume is text- and page-identical with the hardback published in 2021.
Printing and binding: CPI books GmbH, Leck

www.degruyter.com

Table of Contents

Abbreviations —— VII

Introduction —— IX

Aristotle

James G. Lennox
'Most Natural Among the Functions of Living Things'
          Puzzles about Reproduction as a Nutritive Function —— 3

Mary Louise Gill
Method and Nutritive Soul in Aristotle's *De Anima* II,4 —— 21

R.A.H. King
Nutrition and Hylomorphism in Aristotle —— 43

Sophia M. Connell
The Female Contribution to Generation and Nutritive Soul in Aristotle's Embryology —— 63

Andrea Libero Carbone
Why do not Animals Grow on Without End?
          Aristotle on Nutrition and Form —— 85

David Lefebvre
Looking for the Formative Power in Aristotle's Nutritive Soul —— 101

Hynek Bartoš
Aristotle and his Medical Precursors on Digestion and Nutrition —— 127

Giouli Korobili
Aristotle on the Role of Heat in Plant Life —— 153

Aristotelianism

Robert Mayhew
Reading and Sleep in Pseudo-Aristotle, *Problemata* XVIII,7
On the Nutritive Soul's Influence on the Intellect, and *Vice Versa* —— 173

Gweltaz Guyomarc'h
Dividing an Apple
The Nutritive Soul and Soul Parts in Alexander of Aphrodisias —— 197

Tommaso Alpina
Is Nutrition a Sufficient Condition for Life?
Avicenna's Position Between Natural Philosophy and Medicine —— 221

Martin Klein
Digestive Problems
John Buridan on Human Nutrition —— 259

Christoph Sander
Magnetism and Nutrition
An Ancient Idea Fleshed out in Early Modern Natural Philosophy, Medicine, and Alchemy —— 285

Elisabeth Moreau
From Food to Elements and Humors
Digestion in Late Renaissance Galenism —— 319

Bernd Roling
Standstill or Death
Early modern Debates on the Hibernation of Animals —— 339

Andreas Blank
Antonio Ponce de Santacruz on Nutrition and the Question of Emergence —— 355

Index locorum —— 379

Index rerum —— 405

Abbreviations

Abbreviations for ancient Greek authors and works follow the Greek-English Dictionary by Liddell, Scott and Jones [LSJ]; we use the abbreviation *Juv.* to refer to the treatise *De Juventute et Senectute, De Vita et Morte, De Respiratione* as a whole.

Works by Galen and Hippocrates are being cited according to the CMG abbreviations as reported in Fichtner's bibliographies (http://cmg.bbaw.de/online-publications/hippokrates-und-galenbibliographie-fichtner [last visited April 2020]).

Citations of scholastic works use the following abbreviations:

arg.	argumentum	ep.	epistula
art.	articulus	ex.	exercitatio
cap.	capitulum	lect.	lectio
co.	corpus articuli	lib.	liber
contr.	contradictio	membr.	membrum
descr.	descriptio	n.	numerus
dist.	distinctio	q.	quaestio
doctr.	doctrina	sent.	sententia
dub.	dubium	tract.	tractatus

Introduction

1 Aristotle

According to the Oxford English Dictionary (online version), in contemporary English usage the word 'nutrition' carries four meanings:
1. The action or process of supplying, or of receiving, nourishment or food.
2. That which nourishes; food, nourishment.
3. The state or condition of being (well or badly) nourished; a person's state of health considered as a result or indicator of (good or bad) nourishment.
4. The branch of science that deals with nutrition (*sense 1*) and nutrients, esp. in humans; the study of food and diet.

In light of the above definitions, 'nutrition' seems an appropriate English rendering for the Greek words τροφή or τὸ τρέφειν/τρέφεσθαι, which are used by authors of the 5th and 4th centuries BCE to refer to processes, activities or functions related to nourishment, or even to kinds of food or nutriment that are able to nourish or procure sustenance. The term θρέψις is not attested before the 2nd century CE. In Galen θρέψις is acknowledged as one of the three main activities (ἐνέργειαι) of nature – the other two being growth and generation (*De facultatibus naturalibus* I,5, K. II,10).

Specifically, rendering 4 resonates with what in Hippocratic texts is sometimes referred to as dietetics, that is, that part of the medical art (and not science) which deals with diet (δίαιτα). The gradual development of dietetics into a cornerstone of medicine was set off by medical ideas of that time which are closely akin to rendering 3, most notably the belief that a person's state of health depends heavily on the food (s)he consumes. In one of its two main meanings (the other being 'rearing', 'bringing up'), τὸ τρέφειν, apart from the act of nourishing, refers to the substances that are able to nourish (so, rendering 2). Now concerning nutrition as a bodily function related to a set of individual physiological activities of certain body parts, Aristotle seems to deserve, at least to a certain degree, credit for being the first to explicitly make such a progress (cf. *de Partibus Animalium* II,3, 650a9; III,14, 674b10, 19). Of course the ancients did not talk about metabolism, in the strict biochemical sense of the word, nor did they reflect on matters related to the energy value of food. They did nonetheless speak of assimilation of food as a sort of change taking place due to mutual interaction, in some cases of opposite, while in others of like qualities or powers.

In the place of the brain, some of whose networks are nowadays considered to be associated with the control of feeding, the ancients put the soul.

Aristotle was the first to systematically describe a particular part of the soul as responsible for the physiological process of nutrition. This is the nutritive part, which is acknowledged as the necessary and sufficient condition for life, and is therefore held to be shared by all living beings, plants, animals and humans. Nevertheless, as will become clear from the contributions to this volume, nutrition is far from being the only act in which the nutritive soul, biologically speaking, manifests itself. Breathing, cooling, growth, reproduction and, to a certain degree, sleep and vigilance are directly connected to nourishment from food, which explains why they all fall within the nutritive soul's realm of responsibility. Aristotle did not, thus, single out, say, a 'breathing/cooling' or a 'formative' psychic part, but rather subsumed the respective functions, along with a variety of other functions, under the umbrella of the nutritive faculty. And he did so not only because these functions are common, as he repeatedly insists, to all living things, but also because, in order to be performed, there must be some form of direct interaction between them and phenomena occurring during the nutritive process.

Overarchingly, addressing the fundamental problems concerning the nutritive part of the soul as well as the variety of physical manifestations it directs lies at the core of this volume. Its principal aim is to highlight the much-neglected multifacetedness of the 'lowest' part of the soul and its physiological aspects, thus opening the way for further investigation of Aristotle's and his successors' views on the subject. Divided into two sections, 'Aristotle' and 'Aristotelianism', each made up of 8 fresh contributions, this volume lays no claim to an exhaustive coverage. The variety of digestive residues, the contribution of evaporation to the nutritive process as a whole, or even the role of heat and cooling in animals that do not respire are only a few examples of the many topics that are directly related to nutrition and nutritive soul and need further clarification.

The contributions to this volume centre around two crucial research topics which have greatly troubled thinkers since antiquity, and over which floods of ink have been poured: the relationship between body and soul, and the partition of the soul. It is true that the nutritive soul and its physical manifestations have not been discussed in the relevant literature as adequately as the other two parts of the soul, the perceptive and the rational, although it has been almost two decades since Richard King established the breadth of the subject area and highlighted its importance. In his monograph *Aristotle on Life and Death*, King explores the last part of Aristotle's *Parva Naturalia*, namely the part that deals with topics such as the length of life, youth, old age, life, death and respiration, which he sees as a continuation and completion of the discussion of the nutritive soul and its activities that began in *De Anima*. King analyses Aristotle's conception of life-cycle, stressing the indispensability of nutrition in passing through each stage of a living body's life-cycle, and the role of the nutritive soul as the efficient cause for growth and decay.

Nevertheless, reports of empirical observations about a wide variety of physical manifestations and states such as Aristotle's reasonably give rise to ontological questions regarding the 'identity' of the nutritive soul and its relation to the soul as a whole: What kind of entity is this nutritive soul? By means of what criteria did the ancients (or, perhaps *we* as interpreters?) distinguish this psychic part from other psychic parts, and why should this part be thought of as 'the lowest'? Should we speak of a part-whole relationship between the nutritive part and the rest of the *psychê* (cf. Perler 2015, p. 11–14)? Published in 2012, Thomas Johansen's *The Powers of Aristotle's Soul* offers a systematic analysis of *de An.* in which a chapter is devoted to nutrition and its importance in Aristotle's theory of the soul. In Johansen's own words, "Aristotle gives priority to nutritive soul in his account of the soul because nutrition serves as a paradigm of how the soul works as the nature of living beings. The nutritive soul thus has a special status among the capacities of the soul by illustrating how the soul works so as to bring about life" (p. 119). If indeed for Aristotle the nutritive soul holds a prominent place among the other psychic capacities, ought we not to reformulate our understanding of what it means for it to be 'the lowest' part of the soul? In any case, we hope that the collective effort undertaken for the present volume, oriented as it is towards investigating the subjects of nutrition and nutritive soul in Aristotle and Aristotelianism, will help readers explore more fully Aristotle's and his heirs' conception of the 'nature' of living things, and will give a new impetus to the study of Aristotelian psychology.

2 Philosophers and Physicians on Soul, Life and Nutrition

While the fact that the soul exists and somehow distinguishes a living thing from a non-living one has almost never been truly disputed, what the soul really is and how it activates the body have been hotly debated since antiquity. In the Homeric poems, a person's soul is often described as being the last breath that leaves his body at the moment of death. In Presocratic thought the soul, either immortal or mortal, is usually held to be of a material nature, being associated for example with air by Anaximenes and Diogenes of Apollonia, and with fire by Heraclitus and Democritus. Later, in Plato we find the view that the soul, though incorporeal, is imprisoned in and can be affected by the body as long as the latter is physically alive, whereas in Aristotle we learn that *nous* comes "from outside" (θύραθεν), even though the soul is a unified entity. Aristotle frequently attacks his predecessors for their materialistic accounts of the soul. If we also bear in mind that for Plato (a) the soul runs the danger of being affected by one's struggle to gratify bodily desires, which explains why one should rather abstain from the so-called pleasures connected with food and drink, from sexual pleasures, and from the pleasures of ornament (*Phaedo* 64d); and (b) the third part of the soul, the appetitive part, is better depicted as a wild beast that must remain tied up and be located far away at least from the rational part of the soul (*Timaeus* 70d–71a), then we can draw a quite clear picture of the intellectual obstacles confronting Aristotle as he undertook to rework and argue in favour of the 'lowest' part of the soul.

While putting to the test Plato's tripartite division of the soul (*Res publica* IV 434d–441c), Aristotle does not focus exclusively on reason, but introduces the nutritive capacity with a view to accounting for a variety of life activities (*de An.* III,9, 432b3–8; III,10, 433a21–6). For Aristotle, both 'being' and 'living' are better than 'not being' and 'non-living' respectively (*de Generatione Animalium* II,1, 731b29–30), but in order for a being to be alive it needs nourishment. In purging nourishment of its previous negative associations (see also Heraclitus' DK 22 B 117 and DK 22 B 118; Euripides *Ion* 1170; Claus 1981, p. 73–74) and highlighting it as a prerequisite of life, Aristotle initiates a shifting of philosophical interest towards knowledge of the body and its physiology. Nutrition, thus, comes to be regarded as a complex process with distinct stages, affected by a variety of both bodily (body heat, moisture and cavities, stage of growth) and extrabodily (environment, external air) factors – as a function, above all, that affects and supports the performance of other functions, such as cooling of the body, growth of individual body parts, and reproduction.

The above issues are addressed by many of Aristotle's predecessors in natural philosophy, but only sporadically, not systematically. Anaxagoras, for example, is said to have wondered how hair can come from what is not hair and flesh from what is not flesh, concluding that everything is pre-existent in nourishment (DK 59 A 46; Longrigg 1993, p. 65). Empedocles often becomes the target of Aristotle's criticism, mainly due to his false or incomplete accounts on matters which Aristotle eventually associated with the nutritive soul. Specifically, according to *de An.* II,4, Empedocles failed to refer to the soul as the formal agent of growth. Regarding plants, he claimed that growth takes place in two directions, downwards when they spread their roots in the ground because of the natural tendency of the earth contained in them; and upwards when they shoot in this direction due to the natural movement of the fire in them (*de An.* II,4, 415b29–416a2). Empedocles appears to have reflected on digestion as well. For him, we learn from Simplicius, food, after entering the mouth and being ground up by the teeth, is digested in the stomach by a process of 'putrefaction' (σῆψις); it is then carried to the liver, where it is turned into blood and is distributed throughout the entire body via the blood vessels (Simplicius, *in Aristotelis Physica commentaria* 371,33 = DK 31 B 61; Aetius, V,27,1 = DK 31 A 77; Longrigg 1993, p. 73). Another hot topic of debate was respiration. Empedocles' clepsydra analogy receives severe criticism in *de Juventute et Senectute, de Vita et Morte, de Respiratione* 13(7), 473a15–17 for lacking, in Aristotle's view, a clear reference to the purpose of respiration and the question of whether all animals perform that function or not. In fact, in his rather sweeping critique in *Juv.* 7(1), 470b6–13, Aristotle reproves all his predecessors for offering no or incomplete accounts on the subject (Althoff 1999, p. 78–85). Lastly, Empedocles was also interested in matters related to reproduction. Besides his ideas on sexual reproduction and embryological development, he seems to have attempted to establish a connection between nutrition and reproduction in his description of maternal milk as decomposed blood (*GA* IV,8, 777a7 = DK 31 B 68; Longrigg 1993, p. 74).

On the other hand, the various views propounded by medical authors of the 5th and 4th centuries BCE contributed significantly to the formulation of fundamental questions about nutrition, which they viewed as being inextricably linked to human health and well-being. According to *De Vetere Medicina* 3,4 (L. I,576) the medical art has long been rooted in, and closely associated with, dietetics, ever since human beings understood that, in order for them to benefit from their nourishment, they must consume foods that are in keeping with their nature; for consuming initially, like the rest of the animal kingdom, foods that were raw and uncompounded, they endured many, terrible sufferings because of their strong and brutish regimen. Two central themes to which both natural philosophers and doctors will repeatedly recur from now on are already found here: first, the importance assigned to cooking as a means of producing qualitative change in the food – as we have seen earlier, Empedocles spoke of some sort of putrefaction of food in digestion, a term which was gradually replaced by the term πέψις (DK 31 A 77); and second the use of metaphors of dominance to describe the way in which the forceful properties of food interact with the human body – a kind of struggle between two opposite forces trying to overcome each other; besides the depiction in Plato's *Timaeus* already referred to, cf. also *De Morbis* IV,2 (L. VII,544), Democritus DK 68 B 149, Galen *De fac. nat.* III,8 (K. II,173–174). This notion of dominance of one principle or force over the others and its association with matters of health and disease, appear in both the Pythagorean approach to health and disease in the language of opposites and harmony, and in the philosopher-physician Alcmaeon of Croton (5th c. BCE), who is said to have thought of health as the equilibrium of opposite forces in the body, and of disease as the result of the prevalence (μοναρχία) of one of them (DK 24 B 4).

As a consequence, the proportionate blending (σύμμετρος κρᾶσις) of opposite forces/qualities is now to be regarded as constituting health, internal balance and well-being in humans.

The elements or qualities contained in food were often associated with the material elements out of which the body is made up. In *De Morbis* IV,1 (L. VII,542), for example, we read that it is thanks to consumption of food and drink that the four types of fluid (ἰδέας ὑγροῦ) by which the human body is constituted, blood, bile, phlegm and water (ὕδρωψ), manage to maintain their presence in the body and eventually to continue to keep it alive. *De diaeta* I,3 (L. VI,472), to cite another example, gives us an important clue: fire (which is hot and dry) and water (which is cold and wet) are pointed out as the constituent elements not only of man, but also of all creatures (including e.g. plants, seeds) – fire being understood as the principle of movement and water as the principle of nutrition. If fire and water constantly interact and mingle with each other harmoniously, then movement and nutrition should also be thought of as mutually dependent and somewhat complementary processes in the living body, allowing it to continue to grow and maintain its healthy condition. In *De Carnibus* 6 we are told that the *pneuma* associated with inhalation nourishes the heat of the heart – an idea with which Aristotle explicitly disagrees in *Juv.* (12)6 – while in *De Carn.* 13 food is said to effect growth by irrigating the body, according to the like-to-like principle.

To take stock, the key issues that receive much attention from Hippocratic writers of the 5th and 4th centuries BCE and are subsequently addressed by Aristotle in his discussion of nutrition could be summarised as follows: In these medical texts a clear distinction is drawn between the body parts involved in the multifaceted process of assimilation of food and the other parts. In several cases, the ingested food is treated as opposing or 'attacking' the human body; so one is often advised to exercise great caution in consuming food, if the ability of the body, in particular of the abdominal area, to overcome it is not to be impaired. Digestion brings some sort of harmonisation of food with the body by weakening its forces; only once this has be done can the nutrients be distributed (usually via the veins) to the rest of the body and be eventually assimilated. Air is sometimes discussed in light of its ability to contribute significantly to the process of digestion. The importance of excretions is also recognised, as it is generally accepted that the ingested food will not be useful (i.e. nourishing) in its entirety, and that the body possesses inherent 'mechanisms' responsible for clearing it of the by-products of digestion.

3 Aristotelianism

If an inquiry into nutrition and the nutritive soul in Aristotle's biological and psychological works necessarily requires to broaden the scope of the investigation and to consider the way in which Aristotle's own natural philosophy reacts to and interacts with other philosophical and medical theories, this is even more so when it comes to look at the Late Antique, Medieval and Early Modern reception, explanation, rethinking, and further development of Aristotle's views. For these views often leave room enough for divergent and/or conflicting interpretations and raise a number of questions that in Aristotle's works remain unanswered and therefore engender and pave the way for forms and strategies of reception, assimilation, rethinking and/or criticism that are often characterised by a high degree of originality and heterogeneity.

For this very reason it would have been impossible to offer in this volume an exhaustive picture of how the centuries-long philosophical tradition that takes its defining inspiration and reference point from the work of Aristotle and goes under the name of Aristotelianism (re)thought nutrition and all the processes that pertain to it and (re)defined the nature, properties and functions of the nutritive soul. We therefore aimed for exemplariness rather than comprehensiveness and tried to collect in the second part of this volume contributions that shed light and some fresh insights into particularly meaningful, or controversial, or until now mostly neglected accounts of the nutritive soul offered by philosophers, theologians and doctors belonging to different times as well as cultural and linguistic milieus and being part of, or being tightly intertwined with, what we define as Aristotelian tradition.

What to our eyes these accounts have in common and make them meaningful could be summed up with the key-words 'contamination' and 'theoretical challenge'. Let us start with the first key-word: contamination. The history of Aristotelianism in its different declinations – Greek (Antique and Late Antique), Medieval (Arabic and Latin), Early Modern – is first of all a history of contaminations: between different philosophical traditions (e. g. between Plato's and Aristotle's philosophy in the Neoplatonic Greek commentators on Aristotle); between different disciplines (e. g., natural philosophy and the 'medical science' as shaped by Galen); between different approaches to the very act of thinking and inquiring truth (e. g., the 'philosophical' and the 'theological'). The reception of Aristotle's account of nutrition and the nutritive soul and all the further attempts to go beyond this account while (allegedly) staying faithful to a broadly conceived 'Aristotelian' theoretical framework offer a perfect case study to better appreciate the internal dynamics of these processes of contamination for reasons that should be clear to the reader of this introduction from what has been until now sketched. For, if Aristotle's account of the nutritive soul and, more generally, his notion of 'tripartite soul' had been developed by reacting, on the one hand, to the materialism of the Presocratic and medical Hippocratic accounts of nature, soul, body and bodily functions, and, on the other hand, to Plato's dualistic approach to the soul/body relationship and somewhat devaluating account of the appetitive soul, many Aristotelians had to face a somehow opposite challenge in dealing with questions concerning the nutritive soul. For, in rethinking Aristotle's views on nutrition and the nutritive soul, they mostly aimed (in a more or less programmatic way) to provide answers and offer accounts capable to bring these views to a higher level of coherence, perspicuity and theoretical cogency by 'contaminating' them with doctrinal elements stemming from other bodies of thought: from the Platonic and Neoplatonic theory of soul, from the Galenic theory of the 'natural faculties', from the Scholastic conception of the individual soul as substantial form.

This very process of contamination often results in a momentous theoretical challenge, and here we come to our second key-word. For reshaping Aristotle's account of the soul within a theoretical framework that integrates elements, for example, of the Platonic conception of the soul necessarily means to make an hylomorphic approach and a dualistic one interact and mingle with one another. This process of harmonisation is in some cases, and especially when it comes to accounts of the rational soul, produced by stressing some (actually or potentially) dualistic aspects of the Aristotelian theory of soul. But, when it comes to provide an account of the nutritive soul, which in Aristotle is in many respects the part of the soul in which the material and the formal aspects of the psycho-physiological processes are most tightly intermingled, this process of contamination and harmonisation turns into an actual challenge that requires theoretical solutions that in some cases prove to be highly original.

4 Synopsis of the Contributions

In his paper ("'Most Natural Among the Functions of Living Things': Puzzles about Reproduction as a Nutritive Function") **James G. Lennox** examines Aristotle's frequent claim that the nutritive and generative capacities of the soul are one and the same. This view, along with Aristotle's claim that to produce another like itself is the most natural of functions for a living thing, has been intensively debated since antiquity and has given rise to different interpretations. Lennox offers an explanation of how it is possible for the nutritive capacity, as a single capacity of the soul, to have two different functions, nutrition and reproduction, by paying special attention to their common goal, the continuation of being.

Besides reproduction, the nutritive soul, as efficient cause, is also responsible for growth and self-maintenance. Key questions related to this latter state, the state of being preserved as the sort of living thing one is, are seen in their proper light in **Mary Louise Gill**'s contribution ("Method and Nutritive Soul in Aristotle's *De Anima* II,4"). According to Aristotle's methodological plan in *de An.* II,4, one must first investigate the objects involved in the nutritive activity in order to be able to understand the activity itself. Understanding the activity would then enable the determination of the relevant psychic capacity. Hence, Gill devotes considerable space in her paper to examining the status of food in Aristotle's theory of nutrition and discussing the notions of blood and heat at work there, while drawing at the same time on important passages from Aristotle's other works, such as *PA* and *Metaphysica* Θ.

Talking about nutritive and generative 'materials' presupposes a reference to forms, since living matter cannot occur without form. Although it is admittedly difficult to dissociate form from matter in living things, in his study ("Nutrition and Hylomorphism in Aristotle") **Richard A. H. King** uses the example of nutrition to illustrate the 'distinctness' of a living thing's form and matter, or in other words 'the work of the soul physically', a concept which he deems necessary in order to understand the basis of the hylomorphism of *Metaph*. With *de Generatione et Corruptione* I,5 as a key-text for his discussion, in which Aristotle admits that (a) matter flows and (b) form grows, King explains why the current account of growth is a promissory note for an account of nutrition, and how the growing form serves at the same time as the agent of stability for the living body.

In the light of its reproductive capacity, the nutritive soul effects the production of both the male and female generative residues, semen and menstrual fluid respectively. In her essay ("The Female Contribution to Generation and Nutritive Soul in Aristotle's Embryology"), **Sophia M. Connell** decodes those factors that render the female's contribution a 'useful' residue of nutriment different from that of the male, and elucidates the significance of the former's contribution using the example of wind eggs. Connell finally solves an ontological problem related to the female's generative capacity: how is it possible for the female nutritive soul to be at the same time the generative soul, seeing that it cannot actually generate on its own?

According to one peculiar passage in *GA* II,6 (744b27–745a10), which is put at the centre of **Andrea Libero Carbone**'s investigation ("Why do not Animals Grow on Without End? Aristotle on Nutrition and Form"), nutrition and growth make use of two different 'kinds' of nutriment, one being the 'nutritive', the other being the 'growth-promoting'. What kind of food does the growth-promoting nutriment constitute, and why does it stop, from some point onwards, enabling growth in certain bodily parts? What role do bones play in delimiting the growth of the entire body? In providing answers to these important questions, Carbone gives us a clear picture of the way in which Aristotle understood and described one of the most common biological functions of living beings.

Staying on the issue of growth, **David Lefebvre** ("Looking for the Formative Power in Aristotle's Nutritive Soul") throws the spotlight on embryonic development. Lefebvre sets out to explore the issue whether in Aristotle's texts we can speak of a formative power in the nutritive soul, one that is responsible for the first constitution of the embryo. After investigating *de An.*, Lefebvre remarks that Aristotle makes no reference to such a power, precisely because he understands the formation of the embryo as a kind of growth. The same also holds for *GA*, in which, however, we are offered plenty of occasions, as Lefebvre stresses, to discuss issues such as matter at the beginning and at later stages of embryonic life, or the motions which initially 'constitute' the living being and promote growth at a later stage. Lefebvre concludes that even the evidence emerging from the study of *GA* eventually confirm the idea that in his embryologic account Aristotle remains faithful to the unity of the functions of the nutritive soul as presented in *de An.*

To what extent can Aristotle's views on digestion and nutrition be considered as original contributions, and what concepts did he adopt from the medical tradition? These are the central questions that motivate **Hynek Bartoš**' study ("Aristotle and his Medical Precursors on Digestion and Nutrition"), who discusses the relevant passages from the Hippocratic treatises *De Carn.* and *De diaeta* and highlights the significance of vital heat for the successful performance of the process of nutrition – a notion which Bartoš takes to be a Hippocratic relic in Aristotle's thought. Bartoš prepares the ground for the main body of his contribution by bringing forward the correspondences between the views presented in the above-mentioned Hippocratic texts regarding the status of the brain and Aristotle's relevant account in *PA* II,7.

Aristotle's concept of heat occupies also the most prominent place in **Giouli Korobili**'s contribution ("Aristotle on the Role of Heat in Plant Life"), which directs the spotlight on the much-neglected subject of the role heat plays in the life of plants. While there is scholarly consensus around the idea that for Aristotle all living things, in order to maintain their lives, need, among other factors, a principle of soul and natural heat, and that plants are ensouled beings endowed with nutritive soul, one crucial question still remains obscure: What does this heat actually do inside a plant, especially considering that plants present far less complexity of structure than animals and humans? Korobili attempts to give an answer to this question by offering an interpretation of the role of heat in the internal processes taking place throughout a plant's life cycle.

Robert Mayhew's contribution ("Reading and Sleep in pseudo-Aristotle, *Problemata* XVIII,7: On the Nutritive Soul's Influence on the Intellect, and *vice versa*") focuses on a key question that started very early to be debated in the Peripatetic milieu and that concerns the interaction of the nutritive part of the soul and the two other parts, especially the appetitive or perceptual part. In his essay Mayhew provides a commentary on pseudo-Aristotle, *Pr.* XVIII,7, which is especially concerned with the interaction between the nutritive part of the soul and the rational part and attempts to answer the question: "Why is it that some people, if they begin to read, sleep overtakes them even though they don't want to sleep, whereas others who want to sleep, are kept awake when they take up a book?" The complex explanations involve the effect of pneumatic movements and temperature on thought – which is somewhat surprising for Peripatetic texts, given Aristotle's account of sleep in his *De Somno et Vigilia*.

Issues concerning the relation between the nutritive soul and the other parts of the soul are also central to **Gweltaz Guyomarc'h**'s essay ("Dividing an Apple. Nutritive Soul and Soul Parts in Alexander of Aphrodisias"). At first sight Alexander does not seem to draw a distinction between parts of the soul and its powers or faculties. And yet, when approaching the nutritive soul in his *De anima*, Alexander claims the powers for growing and for reproducing are both linked or "yoked" (συνέζευκται) to the power for nourishing. The question is to understand how those capacities relate to each other: are they essentially one and the same? Is the difference between them only a conceptual one? And finally and more generally: if a soul is a kind of bundle of different powers, what makes the bond between them? Guyomarc'h argues that soul-powers are not just explanations of a fundamentally unique reality. The processes at stake (nutrition, growth, reproduction) are physically different, and the related soul-powers differ in essence, but also in their activities. Additionally, there is no mysterious bond, no additional 'yoke' behind a cluster of soul-powers that would bind them: a given soul is immediately identical with its powers and it is not a power of various powers. The main criterion by which one can account for the organisation and the unification of soul parts is the teleological criterion.

With **Tommaso Alpina**'s paper ("Is Nutrition a Sufficient Condition for Life? Avicenna's Position between Natural Philosophy and Medicine") we move into the field of Arabic Aristotelianism. Alpina deals with the nutritive soul by analysing the epistemological status of medicine as defined by Avicenna in his *Canon of Medicine* and the relation between medicine and natural philosophy. In providing the theoretical setting of the medical investigation in the first part of the first book of the *Qānūn*, Avicenna lists the things that the physician must accept on authority, because their existence has been already ascertained elsewhere (i.e. in natural philosophy). Nutrition, and the nutritive soul seem not to escape this paradigm: Avicenna provides a formal account of nutrition in the *Kitāb al-nafs* (Book of the Soul, i.e. the psychology of the *Kitāb al-Šifā'* [Book of the Cure]), and a mechanical account of it in the first book of the *Qānūn*. Alpina's paper raises the questions, whether it is really indisputable that the mechanical account of nutrition provided in medicine is subordinated to its formal account in natural philosophy and whether the treatment of the psychic faculties in the *Kitāb al-nafs* is the theoretical ground for the medical investigation devoted to them in the *Qānūn*.

A key-thinker of Latin Aristotelianism is object of investigation of **Martin Klein**, whose contribution focuses on "Digestive Problems: John Buridan on Human Nutrition". Medieval thinkers agreed that the human soul, being the substantial form of the body, is immaterial and yet the principle of fairly material operations. But how to make this plausible was particularly problematic in case of nutrition. For, how can food be substantially converted into the body as composite of matter and immaterial form? And how can an immaterial soul process such a material operation? These questions are particularly pressing for John Buridan, who identifies nutritive powers with the soul. In his paper Klein argues that Buridan conceives of nutrition as a merely material change, a view which is in line with his broader conception of substantial generation and the relation between a substantial form and its coming to existence in suitably disposed matter. Ultimately, the way in which Buridan accounts for nutrition turns out to be another example of a rising dualism between body and soul, pointing to developments some centuries later which will render substantial forms superfluous.

Christoph Sander's paper ("Magnetism and Nutrition. An Ancient Idea Fleshed out in Early Modern Natural Philosophy, Medicine and Alchemy") aims to trace the complicated history of two intertwined concepts, 'nutrition' and 'magnetism', which were closely related to each other in pre-modern times but appear to be unrelated from a modern perspective. Then, the concepts of 'specific attraction' and 'dispositional self-movement' were regarded as crucial to understanding the powers of a magnet and a living body. By uncovering the historical origin(s) of the relation between nutrition and magnetism, its rationale, its subsequent transformation and its dissolution, the historical concept of 'nutrition' will come into sharper view from the perspective of the history of ideas. At the same time, from the perspective of the philosophy of science, Sander's study presents a test case scenario for discussing the importance of analogies in the formation of scientific theory.

In her contribution ("From Food to Elements and Humors. Digestion in Late Renaissance Galenism" **Elisabeth Moreau** aims to explore the early modern reception of the Galenic theory of digestion in a major treatise on theoretical medicine: the *Physiologia* of the French physician Jean Fernel (c. 1497–1558). In his works, Fernel aimed to concile Galenic medicine and Aristotelian natural philosophy with the Platonic account of Marsilio Ficino in order to enhance the divine origin of life and the soul. Moreau's essay examines Fernel's explanation of digestion from both angles. First, she looks at his application of the Aristotelian theory of elements and mixture to digestion as a transformation of nutrimental matter. Second, she appraises the influence of Platonic philosophy on Fernel's interpretation of nutrition as a vital function directed by the soul, particularly its relation to occult qualities and the total substance. Special attention is also paid on food 'concoction' as a process of fermentation and coagulation. As argued by Moreau, Fernel explored these themes by synthesising the philosophy of Galen, Aristotle, and Avicenna and, just like other Renaissance Humanists did, by appraising medieval Latin-Arabic texts in light of ancient sources.

A very interesting perspective from which one can get new insights into the Medieval and Early Modern Aristotelian views on the nutritive soul is that investigated in **Bernd Roling**'s paper ("Standstill or Death. Early Modern Debates on the Hibernation of Animals"). Albertus Magnus in his commentary on the *Parva Naturalia* was maybe the first philosopher and naturalist to deal with the question of the hibernation of animals: How is it possible that nutrition of many creatures seems to be interrupted, but animals like bears or martens nevertheless continue to live and regain completely their vital energies in spring? Albert developed a model, with a kind of closed nutritive system in its centre, that became quite attractive for later natural philosophers. In Italy *physici* like Fortunio Liceti were debating Alberts ideas, later on especially the famous Danish polyhistor Ole Borch wrote a large treatise on the problem. Roling gives a survey of the debate, summarised by the encyclopaedical work of Karl von Bergen in 1752, taking the continuity of Aristotle and Galen in early modern medicine and zoology as starting point.

With **Andreas Blank**'s contribution ("Antonio Ponce de Santacruz on Nutrition and the Question of Emergence") we get into the field of Late Aristotelianism and get a look at how medical and philosophical traditions interact in a thinker like Ponce de Santacruz in dealing with questions concerning emergence and emergentism. Some scholars have argued that emergentism was clearly articulated in some ancient thinkers, including Aristotle, Galen and the Aristotelian commentators Alexander of Aphrodisias and John Philoponus. There is also a consensus that this view left some traces in medieval and Renaissance thought, often complicated by theories of celestial causation, only to fall into oblivion after the Pomponazzi affair up until the advent of the nineteenth-century British emergentists. The paper argues that this narrative can be challenged, and that emergentism remained a viable option in early seventeenth century. In particular, Blank argues that emergentist intuitions play a role in the discussion of nutrition in the natural philosophy of Antonio Ponce de Santacruz, royal physician to the Spanish king Philip IV.

We are deeply grateful to Dr. Serena Pirrotta, Prof. Ludger Jansen, Prof. Christoph Jedan and Prof. Christof Rapp for accepting this volume in the series 'Topics in Ancient Philosopy/Themen der Antike Philosophie'. Many contributions included in this volume were first presented at the conference *Nutrition and Nutritive Soul in Aristotle and Aristotelianism* held in Berlin at the Humboldt University on 22–24 March 2017. We would like to acknowledge the financial and institutional support provided on that occasion by the Research Training Group 'Philosophy, Science and the Sciences' of the Berlin Graduate School of Ancient Studies (BerGSAS) and the TOPOI Research Project 'Mapping the Vegetative Soul in Aristotle and Beyond'. A special thanks is due to Prof. Philip van der Eijk for encouraging us in pursuing this project and to Irma Handwerker for her invaluable help through all the stages of the organisation of the conference.

Last but not least we would like to acknowledge the very accurate and painstaking redactional work made by Dorothea Keller, the student assistant who helped us in preparing the final manuscript for the publication.

Giouli Korobili
Roberto Lo Presti

Bibliography

Althoff, Jochen (1999): "Aristoteles als Medizindoxograph". In: Philip van der Eijk (ed.), *Ancient Histories of Medicine. Essays in Medical Doxography and Historiography in Classical Antiquity*. Leiden, Boston: Brill, p. 57–94.
Claus, David B. (1981): *Toward the Soul. An Inquiry into the Meaning of ψυχή before Plato*. N. Haven, London: Yale University Press.
Johansen, Thomas K. (2012): *The Powers of Aristotle's Soul*. Oxford: Oxford University Press.
King, Richard. A. H. (2001): *Aristotle on Life and Death*. London: Bristol Classical Press.
Longrigg, James (1993): *Greek Rational Medicine: Philosophy and Medicine from Alcmaeon to the Alexandrians*. London, N. York: Routledge.
Perler, Dominik (ed.) (2015): *The Faculties: A History*. Oxford: Oxford University Press.

Aristotle

James G. Lennox
'Most Natural Among the Functions of Living Things'
Puzzles about Reproduction as a Nutritive Function

Abstract: Just before beginning his discussion of the nutritive soul, Aristotle asserts his Double Priority Principle (DPP): the objective correlate of each activity of the soul is prior to that activity, and each activity is prior to its corresponding capacity. This principle gives rise to a number of *general* puzzles (e.g., what sort of priority is being discussed, and is it the same priority in the two cases). But it also gives rise to a number of puzzles *specific* to the nutritive soul, resulting from Aristotle's claim that the nutritive and reproductive functions are both functions of the same capacity of the soul, and indeed, that nutrition and reproduction are one and the same capacity. In this paper, intended as propaedeutic to the other essays in this volume, I lay out the puzzles arising from tensions between the DDP and Aristotle's account of the nutritive soul, and in particular his claim that reproduction is a nutritive function.

Dying, as a natural consequence of the act of reproduction, is not an unusual phenomenon in the animal kingdom – it is widespread enough that there is a term for it: semelparity. For example, the female in many species of Octopus, while carefully protecting her developing brood in a protective lair, stops eating, essentially starving herself to death in the process. Aristotle was aware of this apparently self-sacrificial behavior of the female in caring for her fertilized eggs. As he reports in the *Historia Animalium*: "The females, having laid their eggs, brood over them, which results in the females becoming very weak; for they do not feed themselves during this period" (*HA* V,12, 544a13–15).[1]

What is normally thought of as nutrition, i.e. maintaining oneself by feeding, is, in such cases, given up in the interests of reproduction. And yet, as we shall see, Aristotle repeatedly insists that the nutritive and reproductive capacities are, in some sense, one and the same. In this paper I will first indicate just

---

[1] Aristotle was remarkably knowledgeable about octopus reproduction – he discovered hectocotylization, the use of one of the male's tentacles to transfer sperm into the female's body cavity. See *HA* IV,1, 524a4–11; cf. V,6, 541b8–12; V,12, 544a12–13. The latter two passages are worded as reports from fishermen.

https://doi.org/10.1515/9783110690552-003

how truly puzzling this claim is, and then go on to argue that Aristotle had both compelling biological and compelling metaphysical reasons for making this claim.

1 The Puzzles

I begin by worrying over two puzzles related to Aristotle's claim that the nutritive and the reproductive are one and the same capacity of the soul. The first puzzle is fairly obvious, and widely recognized. Immediately prior to the discussion of the nutritive soul, there is a methodological preamble, one which is said to apply to *all* of the capacities to be discussed.

> It is necessary for anyone who is going to conduct an inquiry into these things to grasp what each of them is, and then to investigate in the same way things closest to them as well as other features. And if one ought to say what each of these is, for example, what the intellective or perceptual or nutritive faculty is, then one should first say what reasoning is and what perceiving is, since actualities and actions are prior in account to potentialities. But if this is so, and their corresponding objects are prior to them, it would for the same reason be necessary to make determinations about, for instance, nourishment, and the objects of perception and reasoning. (*De Anima* II,4, 415a14–22; transl. Shields 2016)[2]

This principle of the priority of the object to the activity and of the activity to the capacity, which I will call the Double Priority Principle (DPP), creates an immediate conundrum as soon as Aristotle begins his discussion of the nutritive soul, in the very next sentence:

> So that one must speak first of nourishment and reproduction. […] This is both the first and the most common capacity of the soul, in virtue of which living belongs to all living things, a capacity whose functions are to reproduce and to make use of nutrients. (415a25–27)

One can approach the conundrum by considering this statement in light of the two different 'priorities' in the DPP:

---

[2] I here will simply flag some subtle variations in expression in this passage that I believe are significant and will discuss them later – note that in the first conditional, in the *protasis* all three major soul faculties are mentioned, but nutrition is omitted from the *apodosis*; and in the final sentence, about the correlates of these faculties, for the nutritive faculty we simply have τροφή, while for the other two we have αἰσθητόν and νοητόν. The former is ambiguous in a way that the latter two are not – they straightforwardly refer to the objects of perception and thought, while τροφή can refer either to food or to the nutritive capacity.

[i] Why is this a *single* capacity of the soul? Given that Aristotle is discussing two apparently quite different *functions*, reproducing and making use of nutrients, and given that functions are prior in account to capacities, should there not be *two* capacities corresponding to the two functions?

[ii] Given the second priority, of the corresponding 'objects' of the activities (τὰ ἀντικείμενα) to their activities, either the object of these two activities must in some way be the same, or there is not just one capacity of the soul here.

One might point to the case of perception as a way of suggesting that this is not all that puzzling. There are five different perceptible objects corresponding to five different modes of perceiving, but the perceptive soul is a single δύναμις of the soul. With perception, however, the correspondence between the object and the sense in each case is essentially the same: either directly or indirectly, the external 'sensible' acts on the sense organ, and all such information is conveyed to the seat of the common sense, the heart. There is no obvious common correlate for reproduction and nutrition, however; and the relevant organs involved in the two activities are about as different as organs can be. It is thus difficult to see how there can be a single corresponding object for these two functions. Moreover, while it at least seems plausible with thought and perception that these ἀντικείμενα are 'intentional objects', that model does not make sense in the case of nutrition and reproduction – the more so because it would have to be the *same* 'intentional object' (Johansen 2012, p. 93–102)! More puzzling still is the fact that Aristotle nowhere acknowledges that there *is* a puzzle: Immediately after stating the DPP, at 415a25–27, he says it is *therefore* necessary to first speak about nutrition and reproduction (περὶ τροφῆς καὶ γεννήσεως[3]); one sentence later he states that he is discussing a single most common capacity of the soul; and one sentence after that, he identifies using nourishment and reproducing as the *two* functions of this capacity. And we find this switching between singular capacity of soul and two quite different functions not only in *de An.* II,4, but in a related passage in *de Generatione Animalium*:

> And just as the things that come to be by craft come to be by means of instruments, but it is more true to say by means of their movement [note, sing.], and this is the activity [sing.] of

---

[3] It is very common for translators to ignore the distinction in the Greek between γέννησις/γεννήσθαι and γένεσις/γενέσθαι, a practice encouraged by the fact that (as here) there is often manuscript support for both. But the former is typically restricted to acts of biological reproduction, while the latter refers to any process of coming to be. When it is important to stress this, I will use restrictive terms such as begetting or reproduction for γέννησις and its cognates. Cf. Lefebvre, this volume.

the craft, and the craft is the shape of the things that come to be in another thing, in this way the power (δύναμις) of the nutritive soul [sing.], just as in the animals and plants themselves it later produces growth from nutrient using heat and cooling as instruments (for the motion of this [sing.] is in the plants and animals themselves, and each comes to be by a certain *logos*) so too that which comes to be by nature is constituted from the beginning. For the matter by which it grows and from which it is first constituted is the same, and so too the producing capacity [sing.] is the same. *Therefore if this is the nutritive soul it is also the reproductive soul*; and this is the nature of each thing, present within all plants and animals. But the other parts of the soul, while present in some animals are absent in others. (*GA* II,4, 740b25–741a3; emphasis added)

Aristotle does not see a puzzle here – if we (I) do, this may signal that he does not see the issue in the same way as we do. And there is a clue in this *GA* passages as to how Aristotle sees the relationship between nutrition and reproduction that we will come back to later: *both* the matter *and* the productive capacity (ἡ ποιοῦσα δύναμις) involved in growth and in coming to be are the same, and that fact appears somehow to license the identity thesis that we find puzzling.

2 A too Easy Solution

A common solution to this puzzle, presented most recently in Christopher Shield's Clarendon commentary to his translation of *de An.*, is to identify a common *goal* of the two activities. Pointing out that having one functional capacity playing two essentially distinct roles "seems to run afoul" of Aristotle's usual understanding of how capacities and their activities are related, Shields suggests that nutrition and generation are "twin aspects of the drive for self-preservation" (Shields 2016, p. 201 ad 415a22–b7).[4] There are, as we will see, a number of hints in this chapter that point the reader in this direction. However, seeing these two capacities as unified around the goal of self-preservation depends on resolving a third puzzle that arises in *de An.* II,4: If we are to think of these two functions as functions of a single capacity of the 'soul', why is it called the nutritive capacity, rather than the generative? After all, Aristotle seems to prioritize the *generative*

---

[4] So Ross (1961), p. 228 ad 415a22–b2: "what is less clear is his reason for treating nutrition and reproduction as activities of the same faculty (ll. 25–28). His reason is that both are forms of self-preservation. Nutrition is strictly so; reproduction is so in a way, since it is the production of a creature which is οἷον αὐτό (l. 28), 'like the producer'"; and Johansen (2012), p. 119: "since it is impossible for the same living being to remain one in number it participates in the divine and immortal by being the same kind, by remaining not itself but such as itself, using the key phrase from the *Symposium*." Notice that, for the most part, these passages simply restate Aristotle's position without explaining what is *prima facie* puzzling about it.

function over the using of nutrients. Note how the argument of the chapter proceeds:

> For the functions of this [first and most common capacity of the soul] are to generate and to make use of nourishment; *for* the most natural of functions for living things, as many as are complete and not deformed or spontaneously generated, is the production of another like itself, animal an animal, plant a plant [...] (415a25–29)

Note that, no sooner does Aristotle mention the *two* functions of ἡ θρεπτική ψυχή, nutrition and reproduction, than he goes straight to a discussion of the *reproductive* function; and the inferential γάρ suggests that doing so somehow follows from the fact that there are these two functions of the nutritive soul.

Moreover, late in the chapter, as he is wrapping up his account of this most common soul capacity, he makes this very suggestion: "Since it is just to name each thing after its end, and here the end is to beget another such as itself, the primary soul would be reproductive of another such as itself" (416b23–25 [Ross' 1961 text])[5]. Aristotle appears to be suggesting that the proper name for this soul should be the reproductive soul, not the nutritive soul! But the suggestion here is stated conditionally, and it is made in the context of characterizing the end as reproducing another such as itself, not as self-maintenance. And it is a suggestion Aristotle does not take up.

We find a similar prioritization of reproduction in an important passage in the *HA*, near the beginning of Aristotle's investigation into the different ways of life of animals. In fact, in this passage reproduction is said to be the *only* function that plants and sessile animals have – there is no mention of nutrition!

> And it is the same way with respect to the activities that constitute their way of life [as it is with their parts]. For as many plants as come to be from seed appear to have *no other function than to produce another like themselves;* and similarly, in certain animals [those without locomotion] too there is no other function to grasp apart from generation. Wherefore, while activities such as these are common to all, as soon as perception is added their ways of life differ in regard to mating (due to the [awareness of] pleasure) and with regard to birth and the rearing of young. (*HA* VII,1, 588b24–30)

---

5 For reasons having to do with his understanding of the argument, Torstrik (1862) moved ll. b20–23 so they appear after this passage rather than before. Ross accepted this emendation (Ross 1961, p. 231) but without changing the Bekker numbers. Shields also accepts it, but does change the line numbers (Shields 2016, p. 76 n. 20) so that this passage is at 416b18–19 in his translation.

This is a matter of prioritization; for later in this very passage nutrition *is* mentioned, along with reproduction as the central concerns of animals:

> So then, for these animals one part of life consists in activities related to giving birth, another part is concerned with nourishment; for the way of life and the efforts in all these animals is in fact concerned with these two things. And their food differs chiefly according to the matter out of which they are constituted. (589a2–7)

The topic of nutrition is taken up here because for such animals feeding their young is a critical aspect of successful reproduction. Otherwise, the first and most common function of living things, and the only one found in plants, is reproduction. The same message comes through clearly near the end of *GA* I, in discussing why male and female capacities are united in plants but separated in most animals.

> Nature fashions all this reasonably. *For plants have no other function or activity of their being* (τῆς μὲν γὰρ τῶν φυτῶν οὐσίας οὐθέν ἐστιν ἄλλο ἔργον οὐδὲ πρᾶξις οὐδεμία) except the generation (γένεσις) of seed, so that since this is done through coupling of male and female, nature has arranged them together by mingling them; this is why male and female are inseparable in plants. But the animal's function is not only to generate (τὸ γεννῆσαι ἔργον) (for that is common to all living things), but also to participate in some sort of cognition (γνῶσις) [...] (23, 731a25–32, emphasis added)[6]

In passages such as these, even the fact that plants also absorb nutrients from the soil in order to maintain themselves goes entirely without mention – it is explicitly said that their *only* function is to produce seed and to reproduce more plants just like themselves.

And yet: returning to *de An*. II,4, in an obvious reference to the DPP, after having offered a stirring teleological explanation for procreation, we read:

> Since the same capacity of the soul is nutritive and reproductive, it is necessary first to make a determination *about nourishment*; for it [this capacity of the soul] is demarcated from the other capacities by this function. (416a19–21, emphasis added)

---

[6] This passage highlights the way in which Aristotle uses the terms γένεσις and γέννησις – here the former term is used to characterize the production of seed, while the verb corresponding to the latter term is used of the biological function of making another like self. (See n. 3 above)

3 Most Natural of Functions

Enough with the puzzles – my aim for the remainder of this chapter is not so much to resolve them, as to provide a wider biological and metaphysical background that will make them less puzzling.

The first step is to take a careful look at the argument in *de An.* II,4 immediately following the statement of the DPP, which makes the case for reproduction (or formal replication, as I will call it) being the most natural of functions for complete living things; for it is here that we find the source of all three of the puzzles I've highlighted. I add numbers to the text to allow me to index my comments to specific points.

> For [i] most natural of the functions in living things, as many as are complete and neither deformed nor generated spontaneously, is the production of another like itself, animal animal, plant plant, in order that it may partake, as far as possible, in the always and the divine; for [ii] all [perfect living things] strive for this and [iii] do whatever they do in accordance with nature for the sake of this. (But [iv] that for the sake of which is double, the *of* which and the *for* which.) Now [v] since they are unable to partake of the always and the divine continuously, each one partakes in so far as possible, [vi] some more and some less, and [vii] it remains not itself but like itself, not one in number but one in form. (415a24–b7)

I begin with [i]: The production of another being like itself is, Aristotle claims, the most natural of functions for 'complete' or 'perfected' (τέλεια) organisms. Why 'most natural', and why reproduction, rather than nutrition or both nutrition and reproduction? This question is a central theme of chapter 6 of Thomas Johansen's recent study of *de An.* (Johansen 2012, p. 118–127). One reason why Aristotle might consider these functions *most* natural, Johansen argues, is that it is the paradigmatic example of the soul serving as the nature of living things. As I understand Johansen's argument for this claim, the nutritive soul is a paradigm nature in two different, though related, ways. First, it is the faculty of the soul where its role as inner principle of life is most clearly on display – the capacity to convert food into blood and to convert blood either into flesh and bone or into another organism is a life-sustaining power present in *all* organisms – which brings us to a second reason why Johansen thinks Aristotle designates it 'most natural'. It is the *universally possessed* faculty of the soul, and so not restricted to organisms that have perception and locomotion. But notice: both of these grounds would point to the use of nutrients, the capacity's *other* function, as most natural, not to reproduction.

A passage at the beginning of the *Politica* may give us further insight into why formal replication is a *most natural* function of soul:

The first necessity is for the union of those that cannot exist without one another, i. e. the union of female and male for the sake of generation[7] (and this [uniting] is not out of deliberation, but just as in the other animals and plants the longing to leave behind another such as oneself is natural) [...]. (I,1, 1252a26–34)

There are two points of interest in this passage. This striving to procreate is *not deliberative but natural*. One way, then, of unpacking the thought in our *de An.* passage that the act of reproducing is "most natural" is that it is a striving that is not 'deliberative' but (as we might say) instinctive. Aristotle has noticed, as biologists throughout history have, that a great deal of living activity is organized around the 'drive' to reproduce – and even in organisms that are able to deliberate, this desire seems to be 'non-deliberative' or natural. And Aristotle thinks he sees the value achieved in doing so – it is a way in which any living thing is capable of partaking in eternal being.

Then there is the opening claim in this *Pol.* passage: reproduction involves the union of male and female, described as those that cannot exist without one another. What does he mean by that? Perhaps a point that has been repeated many times in the history of biology: without the general phenomenon of the union of male and female for the sake of reproducing, the males and females that are uniting to reproduce would not exist, since they are each a product of such a union. In the final section of this paper I will turn to a passage in *GA* II that provides the metaphysical backing for just such a thought.

If, as I will conclude by suggesting, the function of formal replication is to allow the individual replicator to be eternally in the only way possible for a mortal being, it is a mode of self-maintenance with profound implications.

I begin by taking a closer look at the premises that support claim [i], that reproduction is most natural among the ἔργα of living things:

[a] Aristotle refers to *two* functions of the soul, but he identifies *only one* of them as most natural, and that is the function of *form replication*, the production of another being like the producer.

[b] This most natural of functions is performed by a perishable living being *in order to* partake of the everlasting and divine.[8]

The next clause, constituted of points [ii] and [iii] in the above translation, are introduced by a γάρ, which suggests they are offering a further explication of the 'most natural' claim: participation in the everlasting and divine is what all (perfect) living things strive for, and they do *whatever* they do *in accordance*

---

[7] Or begetting (γεννήσεως), if we accept Stobaeus.
[8] Johansen (2012), p. 110 rightly notes the almost identical formulation of this idea at *Symposium* 207c8–d5; cf. Shields (2015), p. 201, and Polansky (2007), p. 205.

with their natures (κατὰ φύσιν) for the sake of this. Points [v] – [vii][9] then make the case that, while perishable living things are unable to maintain themselves as numerically one being forever, they are able to maintain themselves forever *in a way*, by producing another being like themselves in form.[10] That is, we have an argument of the following form:

[1] Whatever complete/perfect living things do according to nature is done for the sake of partaking of the everlasting and divine.
[2] Perishable beings are unable to participate *continuously* in the everlasting and divine, i.e. by maintaining themselves one in number forever.
[3] They are able, to a greater or lesser extent, to participate in the everlasting and divine by reproducing, so that what remains is not numerically one being but formally one being.

What motivates this argument? Aristotle was both a biologist and a philosopher, and his work shows deep reflection on what unifies the many and varied activities on display in the living world. Careful observation of animals reveals that pretty much everything they do in accordance with their natures is done for the sake of self-preservation. Whether it is the use of their various senses to find food or to detect and avoid predators, or the use of their many and various modes of locomotion, typically activated and directed in response to information learned through their perceptual activity – everything they do is directed to preserving their lives. Of course, it is of the nature of life that these efforts may fail, but *continuity of being – staying alive – does seem to be the goal of these activities.*

It is easy to see activity devoted to reproduction as an exception. In cases such as the female octopus discussed earlier, the conflict between self-preservation and reproductive activity appears stark and direct.

Aristotle's argument in *de An.* II,4 is best seen as a profound attempt to deal with this problem: animals and plants are ensouled bodies. Individuals composed of the four earthly elements inevitably perish – the capacities needed to maintain themselves fail.[11] Organisms do what they do in the interests of self-

---

[9] Though it is important, in this paper I am not going to discuss the remark about the two ways in which 'that for the sake of which' is used. It will be discussed in the companion paper mentioned in n. 17.

[10] Cf. *GA* II,1, 731b24–732a1. As I will discuss, this passage supplies some of the metaphysical backing for this claim.

[11] It is important to remind ourselves that not all natural bodies or all hylomorphic compounds are perishable according to Aristotle; he repeatedly divides natural bodies into those that are eternal (i.e. heavenly bodies) and those that come to be and pass away (e.g. *de Partibus Animal-*

preservation, but by the very nature of organic being, they inevitably die. But, though numerically the same animal cannot live on forever, it is able to produce something one in form with it – indeed it exists only because it is a being with such a capacity. As Aristotle puts it: by means of reproducing, *each organism can continue to be in form, but not in number.* This allows every organism that is able to reproduce to *participate* in the everlasting and divine (and I read "divine" as a normative gloss on "everlasting"). Reproduction allows what makes the composite living thing actually to be what it is, its soul, to live on. And the soul is a *cause*, as Aristotle insists in the immediately following passage (often mistakenly treated as an "aside") *as the being* of ensouled bodies (ὡς ἡ οὐσία τῶν ἐμψύχων σωμάτων) – an expression immediately explicated in the following, metaphysically loaded, way:

> For in all cases the being (ἡ οὐσία) is the cause of being (τοῦ εἶναι), while in the case of living things to be is to live, and the soul is the cause and principle of life. (Τὸ γὰρ αἴτιον τοῦ εἶναι πᾶσιν ἡ οὐσία, τὸ δὲ ζῆν τοῖς ζῴοις τὸ εἶναί ἐστιν, αἰτία δὲ καὶ ἀρχὴ τούτου ἡ ψυχή, 415b12–14)

If I were to translate ἡ οὐσία as "the substance" as it so often is, the opening clause would sound less paradoxical than it does, but at the cost of being misleading. *De An.* II began by distinguishing three ways of being – as form, as matter, and as composite, and this passage first reminds us that soul is being in the sense of the form, and then that, as such, it is the source and cause of being – that is of living – for living things.

Everything done according to the nature of a living thing is done ἕνεκα partaking of the eternal and divine – and this is also what living things reach out or strive for. Ordinary nutritive activity is for the sake of this as well, but individual plants and animals can only maintain themselves by such activity for so long. It is reproduction that extends substantial being as form indefinitely into the future.

It is here, precisely, that much recent work aimed at understanding the relation of the nutritive and reproductive functions goes wrong. Aristotle is not saying or implying that the goal is "to reproduce the species" (Johansen 2012, p. 110); nor that "living beings seek immortality by their participation in the eternal species via generation" (Shields 2016, p. 201).

---

ium I,5, 644b22–26). The details of natural loss of vitality are a central theme of the latter half of *Parva Naturalia*, dealing with length and shortness of life, youth and old age, life, death and respiration.

The stated goal is *continued being* for the being that is reproducing – and in order to achieve that goal, certain living things are able to produce another living being alike them in form, and appear to do everything they do in accordance with their nature for the sake of that. They do not strive to "reproduce the species", nor is "participation in the eternal species" the means by which they seek immortality; rather they strive for participation in 'the everlasting and divine' by means of reproducing something like them in form. Form, i.e. soul, is the source and cause of being for living things – formal *replication* allows mortal individuals to continue *to be*, in a way.[12]

4 Being is Better than Non-Being

The metaphysical underpinnings of this argument in *de An.* II,4 are presented more clearly at the beginning of *GA* II. The argument, which runs from 731b24–732a12, rests fundamentally on Aristotle's Axiological Axiom: *Being is better than Non-being*.

In brief outline, the argument is this:
- Of things that are: Some are eternal, some can either be or not-be. (731b24–25)
- The beautiful/divine is naturally a cause of better in that which is capable.[13] (731b26–27)
- Perishable things are also capable of being, and participating in both the worse and the better, (i.e. they are not always becoming, as Plato claims). (731b27–28)

But how can they participate in the better? Well...
- Soul is better than body, and the ensouled is better than the soulless on account of soul. (731b28–29)
- [Reminder: Perishable beings are capable of being.]
- *Being is better than non-being, and [or i.e.?] living than non-living*. (731b30–31)

---

12 In Coates and Lennox (forthcoming) the thesis outlined here is provided with far more detailed support.

13 Capable of what, one might ask? The following sentence presents us with two obvious options: 'capable of being better or worse' or 'capable of being and not-being'. But given that ultimately, at 731b30–31, he states being is better than non-being, anything the existence of which is contingent is also capable of being better and worse by that very fact. So the ambiguity here may well have been intentional.

- [Another reminder: soul is the source and cause of living.]
- And on account of these causes there is a generation of animals. (731b31)

But how does the argument up to this point account for animal generation? What follows provides the needed argument, but it depends implicitly on the previously stated Axiological Axiom.

> For, since the nature of these cannot be eternal, that which comes into being is eternal in the way that is possible for it. Now it is not possible [for it to be eternal] in number (for the being of existing things is in the particular, and if [the particular] were such it *would* be eternal [in number]), but it is possible [to be eternal] in form. Wherefore there is always a kind – of human beings, animals and plants. (731b32–732a1)

Aristotle here denies that reproduction is aimed at becoming one with the eternal species. There is no eternal species, *unless* by that one intends a continuous series of replicating individuals. Two features of the conclusion of this passage point in this direction. First, the continued existence of these 'kinds' is a *consequence* of reproduction, *not* what it is for. The goal of reproduction is, as in *de An.* II,4, for the perishable individual to partake in eternal being in the only way possible for it. Second, note the plurals: a kind of human beings, of animals, and of plants. In a (much) earlier paper, I noted that the context here indicates that γένος (kind) should be understood in its etymologically primitive sense, given as its first meaning in *Metaphysica* Δ.28:

> [...] a continuous generation of things having the same form [...]; for example, we say 'for as long as the human γένος exists', which means as long as the generation of human beings continues. (1024a29–31)

When Aristotle concludes that reproduction insures that there is always a γένος of human beings, what it insures is a continuous generation of individual human beings.[14]

And now we must remember that our goal is to understand why Aristotle holds that reproduction and nutrition are two functions of one power or capacity of the soul, the nutritive capacity. And to do that we must return to *de An.* II,4, to see what Aristotle has to say about the *other* function of this capacity, nourishment.

> There is a difference, however, between being nourishment and being able to produce growth: *qua* a particular quantity, the ensouled being is capable of growth; *qua* a this

---

[14] See Lennox (1985), p. 70–71.

and a being (τόδε τι καὶ οὐσία), it is nourished. For what is ensouled preserves its being (σώζει γὰρ τὴν οὐσίαν) and *is*, so long as it is nourished; and it is productive of generation, not of that which is nourished, but rather something like what is nourished (οἷον τὸ τρεφόμενον), since its being already exists and nothing reproduces itself but rather preserves itself (γεννᾷ δ' οὐθὲν αὐτὸ ἑαυτό, ἀλλὰ σώζει, 416b12–18).[15]

Aristotle here insists on distinguishing the living thing *qua* capable of growth and *qua* capable of being nourished – the goal of nutrition, *qua* nutrition, is preservation of being. But keeping in mind the discussion of nutrition in *de Generatione et Corruptione* I,5, it is preservation of *formal, not material, being*, that he has in mind. Nutrition is constantly *replacing* matter while *preserving* form, just as continually adding logs to a fire preserves the being of the fire while replacing those logs that have been reduced to ashes (321b28–322a28; cf. Code 2004, p. 171–193, esp. 186–191).

Aristotle, however, immediately follows up this point by contrasting the living thing's preservation of its being by means of nutrition with reproduction – nothing reproduces itself, but preserves itself (γεννᾷ δ' οὐθὲν αὐτὸ ἑαυτό, ἀλλὰ σώζει, 416b17–18). Why would Aristotle want to insist on this point, given that he has been arguing throughout this discussion that nutrition and reproduction are functions of the same soul capacity? Part of the explanation, I believe, is the need to distinguish what he is saying from an apparently similar claim made by Plato at *Smp.* 207c–d. In that passage Plato has Diotima discussing reproduction as a way for mortal beings to share in immortality – but she then goes on to say:

> [...] although each individual animal is said to be the same so long as he continues to live, and therefore is called the same person from infancy until he becomes an old man; yet for all we call him the same, at no time are his parts the same, and he is always becoming a new man, while the old man is ceasing to exist, as you can see from his hair, flesh, bones, blood and all the rest of his body. *And not only with respect to the body, but also with respect to the soul.* (207d3–e2; emphasis added)

Aristotle is arguing that, to the contrary, the *form* of each of those uniform parts, and indeed of the ensouled body as a whole, is not in the constant flux described in Diotima's speech, but is preserved as the being it is. And his next step is to insist that *this preservation of formal being is also the point of reproduction*. He

---

[15] A full explication of this passage would require a discussion of the background theory of growth in *GC* I,5. Cf. King, this volume. Suffice to say that the capacity of the nutritive soul is simultaneously responsible for self-preservation, growth and generation of ensouled beings, but Aristotle's task in our passage is to distinguish these tasks.

begins that move by slightly modifying his description of the self-preserving nature of the nutritive function to stress that it is preserved "as the sort of thing it is" (ᾗ τοιοῦτον).

> ὥσθ' ἡ μὲν τοιαύτη τῆς ψυχῆς ἀρχὴ δύναμίς ἐστιν οἵα σώζειν τὸ ἔχον αὐτὴν ᾗ τοιοῦτον, ἡ δὲ τροφὴ παρασκευάζει ἐνεργεῖν· διὸ στερηθὲν τροφῆς οὐ δύναται εἶναι.

> Consequently, this principle of the soul is a capacity of the sort that preserves the thing which has it, as the sort of thing it is, while nutrition prepares it to be active. Hence, being deprived of nourishment it is unable to be. (416b17–20)

The being preserved in *self*-preservation is *such as* the being that is preserved in reproduction – and in both cases it is not the matter that is preserved, but the form. This leads Aristotle to raise the question that we found so puzzling earlier – why not call this the *generative* soul, since he so often seems to prioritize that function over nutrition.

> ἐπεὶ δὲ ἀπὸ τοῦ τέλους ἅπαντα προσαγορεύειν δίκαιον, τέλος δὲ τὸ γεννῆσαι οἷον αὐτό, εἴη ἂν ἡ πρώτη ψυχὴ γεννητικὴ οἷον αὐτό.

> Since it is just to name each thing after its end, and here the end is to reproduce another such as itself, it would be just to call this primary soul reproductive of something like itself. (416b22–25)

In the end, however, he does not adopt the policy that seems to be implied here, in all likelihood because of the priority of the actual – it is by means of the nutritive capacities of the parents that another being alike in form comes to be. On these grounds, the near identification of the two functions mentioned at the beginning of *de An.* II,4 is also a central aspect of the account of generation presented in *GA* II:

> For the matter by which it grows and from which it is first constituted is the same, so too the producing capacity is the same. *Therefore if this is the nutritive soul, it is also the reproductive soul* (εἰ οὖν αὕτη ἐστὶν ἡ θρεπτικὴ ψυχή, αὕτη ἐστὶ καὶ ἡ γεννῶσα) and this is the nature of each thing, present within all plants and animals. (4, 740b25–741a3, emphasis added)

This passage concludes a lengthy argument which begins with a paradox about blood and nutrients: if blood is the final nutrient for the parts of animals, and the heart is formed out of blood and has blood within it, but the heart is needed to convert nutrients into blood, where did that blood out of which the heart is formed come from? Aristotle resolves the paradox by arguing that the female contribution to generation is residual blood already present, and "[…] is potentially such as the animal is by nature, and the parts are present potentially, though none are in actuality" (740b19–21).

That residual blood provided by the female is the matter of the developing animal, while the male contributes the capacity of the nutritive soul (ἡ τῆς θρεπτικῆς ψυχῆς δύναμις, 740b30); and just as it later will use heating and cooling as instruments for producing growth, at the early stages of development it acts to 'constitute that which comes to be by nature'.

This way of understanding reproduction allows him, finally, to reject Plato's separation of form from the world of coming to be in *Metaph.* Z,8. He concludes a long argument against the futility of accounting for coming to be by reference to separate forms with these words.

> Therefore it is apparent that the cause consisting of the forms, as some are accustomed to discuss the forms, supposing some things besides the particulars, are useless in relation to both generations and beings; nor should they on this account be substances in themselves. Indeed, in some cases it is in fact apparent that the reproducer is such as that which is reproduced, not the same nor one in number with it, but one-in-form, as in natural things (for a human being begets a human being). (1033b27–32).

And the chapter summary underlines the point:

> So then, it is apparent that it is unnecessary to set up a form as paradigm (for forms might have been sought most of all in these cases; for these are most of all substances), but the parent is sufficient to produce and be the cause of the form in the matter. (1034a2–5)[16]

5 Conclusion

I began by identifying two puzzles that arise for Aristotle's identification of the capacity of nutritive soul with that of reproduction, given that he begins his study of the different soul capacities by positing his Double Priority Principle:
(1) Given that he refers to use of food and reproduction as two very different functions, why are there not be two corresponding capacities of the soul?
(2) Given that functions are to be identified by reference to their "objects" or correlates, either these two functions must have the same correlative or Aristotle is wrong to identify them with the same capacity.

---

16 A much fuller discussion of the metaphysics of reproduction can be found in Lennox (1985), p. 76–89.

As we explored his account of the nutritive soul in *de An.* II,4, a third, related puzzle came to the fore:

(3) Given the way in which the function of reproduction is prioritized in this discussion, if Aristotle insists on there being just one capacity of the soul here, why insist that it is the nutritive soul and not the reproductive or generative?

In response to the first two puzzles, I've outlined an argument to the effect that there is a single goal, to participate in the everlasting and divine, which is accomplished by the preservation of formal being – and that it is a natural goal for individual ensouled beings.[17] The very same capacity of the soul acting, by means of the instruments of heating and cooling, upon the very same material (fully concocted blood) preserves the formal being of fully actual living things via nutritive processes, and is productive of another being like it in form via reproductive processes.

There remains a tension in Aristotle's account that is highlighted by the third puzzle: the most common capacity of the soul is a nutritive capacity because generation depends on the efficient cause provided by the actual male parent and the fully prepared material of the female parent; but the goal of participation in the eternal and divine depends on the production of another being one in form with the parents – among the functions of living beings this is "most natural". When, near the end of *de An.* II,4, 416b23–25, Aristotle says "[s]ince it is just to name each thing after its end, and here the end is to beget another such as itself, the primary soul would be reproductive of another such as itself", I believe he is fully aware of this tension. Fair enough: the same tension is experienced by evolutionary and developmental biologists to this day.

Bibliography

Coates, Cameron, and Lennox, James G. (forthcoming): *Aristotle on the Unity of the Nutritive and Reproductive Functions.*

Code, Alan (2004): "On Generation and Corruption I. 5". In: de Haas, F.A.J., Mansfeld, J. (eds.), *Aristotle, De generatione et corruptione I. Proceedings of the Symposium Aristotelicum.* Oxford: Oxford University Press, p. 171–193.

Johansen, Thomas K. (2012): *The Powers of Aristotle's Soul.* Cambridge: Cambridge University Press.

---

[17] The full argument will be presented in a companion paper on which I am currently working with co-author Cameron Coates.

Lennox, James G. (1985): "Are Aristotelian Species Eternal?". In: Gotthelf, Allan (ed.): *Aristotle on Nature and Living Things*. Pittsburgh and Bristol: Mathesis Publications and Bristol Classical Press, p. 67–94. [Reprinted in: Lennox, James G. (2001): *Aristotle's Philosophy of Biology: Studies in the Origins of Life Science*. Cambridge: Cambridge University Press, p. 131–159.]

Polansky, Ronald (2007): *Aristotle's De Anima*. New York: Cambridge University Press.

Ross, William D. (ed.) (1961): *Aristotle's De Anima*. Oxford: Clarendon Press.

Shields, Christopher (transl.) (2016): *Aristotle: De Anima*. Clarendon Aristotle Series. Oxford: Clarendon Press.

Torstrik, Adolfus (ed.) (1862): *Aristotelis De anima libri tres*. Berlin: Weidmann Verlag.

Mary Louise Gill
Method and Nutritive Soul in Aristotle's *De Anima* II,4

Abstract: *De Anima* II,4 opens Aristotle's investigation of the various psychic faculties with the basic capacity shared by plants and animals alike: nutritive soul. He announces a three-step approach to be extended from nutritive soul to perceptive soul and intellect: to define a psychic faculty, first examine the functions it enables, and to understand the functions, examine the correlative objects – in the case of nutrition, food. This paper examines how the methodological prescriptions apply to nutritive soul and, focusing on the function of organic self-maintenance, argues that although the investigation of nutritive soul fits the stated program – food, self-maintenance, nutritive capacity – the order of investigation is the reverse of the order of explanation: the faculty is causally prior to the activity it enables, and the activity is prior to the correlative object, which is merely instrumental to that activity and to the psychic capacity. Moreover, the faculty is causally prior to its functions in three ways, as formal, final, and efficient cause. The paper does not speculate as to how the method extends to other psychic functions, but its analysis raises a question: Will Aristotle's treatment of the higher faculties fit the prescribed method, and if so how?

Aristotle's treatment of nutritive soul in *de Anima* II,4 invites careful scrutiny for two reasons. First, nutrition includes the most basic functions shared by all living creatures, plants and animals alike: reproduction, self-maintenance, growth and decay. Second, Aristotle treats nutritive soul before turning to higher faculties of select groups of creatures – especially perceptive soul distinguishing animals from plants, and reason restricted in the mortal sphere to human beings – and he insists at the start of II,4 that if we are to study the various psychic faculties, we must grasp what each of them is (τί ἐστιν), and then investigate what follows from that. He continues with a series of proposals about the right method for treating the three core faculties – nutritive, perceptive, and rational – encouraging the expectation that as the procedure applies to nutritive soul, so it will apply to perceptive and rational soul. A detailed study of nutritive soul should therefore reveal the vital status of nutrition as the chief manifestation of life itself and moreover pay dividends for correct procedure when tackling the higher psychic faculties.

This paper will examine how the methodological prescriptions apply to nutritive soul and argue that, although nutritive soul fits the stated program, it does so in an unexpected way. I shall not venture to speculate as to whether perception and reason fit the pattern more neatly, but my analysis should at least sound a cautionary note. Perhaps study of the perceptive and rational faculties will follow the expected pattern or in turn reveal similar or distinctive surprises.

1 Method

Aristotle announces at the start of *de An.* II,4 a three-step approach: to investigate any psychic faculty, one must determine what it is, and to determine what it is, one must examine the functions it enables, and to understand those activities, one must study the objects involved in that activity, and then, having started with the objects and continued with the functions, determine what the faculty is and the things following from that. Thus the proper order of the main investigation is objects, activity, capacity:

> The person intending to make an investigation about these things [the various psychic faculties] must grasp what each of them is (τί ἐστιν), then in light of that investigate what follows (περὶ τῶν ἐχομένων), and so on. If one must state what each of them is – e.g., what the intellectual capacity is (τί τὸ νοητικόν) or the perceptive capacity (τὸ αἰσθητικόν) or the nutritive capacity (τὸ θρεπτικόν) – prior to that one must state what thinking is (τί τὸ νοεῖν) and what perceiving is (τί τὸ αἰσθάνεσθαι), for the activities (αἱ ἐνέργειαι) and actions (αἱ πράξεις) are prior to the capacities (τῶν δυνάμεων) in account (κατὰ τὸν λόγον). And if so, and further one must have studied even prior to those the opposed objects (τὰ ἀντικείμενα), one would first need to make distinctions about those for the same reason, e.g., about food (τροφῆς), perceptible object (αἰσθητοῦ), and intelligible object (νοητοῦ). (415a14–22)

Aristotle gives a reason why one must investigate the function before the capacity: activities are prior to capacities in account (πρότεραι [...] κατὰ τὸν λόγον). *Metaphysica* Θ,8 examines various ways in which activity (ἐνέργεια) is prior to capacity (δύναμις), including priority in account (λόγῳ) (1049b12–17). The activity that realizes a capacity determines what the capacity is a capacity for – what the capacity equips its possessor to do or be. The activity or resulting state determines the content of the potency. So a capacity is defined with reference to the corresponding activity, for instance, sight with reference to seeing. Because the activity is prior to the potency in account, it is also prior to it in knowledge. We grasp what the capacity is (τί ἐστιν) by grasping the activity it enables.

Aristotle appears to extend his claim about priority in account to the correlative objects – food, perceptible object, and intelligible object – when he says:

> And if so, and further one must have studied even prior to those the opposed objects (τὰ ἀντικείμενα), one would first need to make distinctions about those for the same reason, e. g., about food (τροφῆς), perceptible object (αἰσθητοῦ), and intelligible object (νοητοῦ). (415a20–22)

Since he says that the correlative object is prior to the activity for the same reason, we expect the correlative object to be prior to the function in account. For example, thinking will be defined with reference to the intelligible object, and perceiving will be defined with reference to the perceptible object. To take one mode of perception, seeing will be defined with reference to its proper object, color. In the same way the activity or activities of nutrition will be defined with reference to food.

Our expectation that the proper object defines the function appears to be confirmed two sentences later:

> We must first talk about food and reproduction, since nutritive soul is the first and most common capacity of soul, the capacity in virtue of which life (τὸ ζῆν) belongs to all [ensouled] things. Its functions (ἔργα) are to reproduce and use food. (415a22–26)

Here Aristotle specifies a main function of nutritive soul as using food, bolstering the impression that the objects must be studied first for the same reason that one studies the function before the capacity: the function is defined with reference to its proper objects.

Yet evidence later in the chapter suggests, on the contrary, that food is posterior in account to the function it serves. After discussing food at some length, Aristotle states:

> Since nothing is fed that does not partake of life, the thing fed would be the ensouled body, insofar as it is ensouled, with the result that food (ἡ τροφή) too is relative to (πρός) an ensouled thing, and not accidentally (οὐ κατὰ συμβεβηκός). Being for food and for causing growth are different (ἔστι δ' ἕτερον τροφῇ καὶ αὐξητικῷ εἶναι); for insofar as the ensouled thing is a certain quantity (ποσόν τι), it [nutriment] causes growth (αὐξητικόν), but insofar as it [the ensouled thing] is a this (τόδε τι) and substance (οὐσία), it [nutriment] is food (τροφή). For it [nutriment] preserves the substance, and it [the ensouled thing] exists for as long as it is fed. (416b9–15)

Aristotle distinguishes the role of nutriment in two contexts, growth and self-maintenance. Although the same carrot can cause growth and preserve something as the substance it is, the being (εἶναι) for the carrot is different depending on the role it plays and the way the ensouled body is conceived – as a certain quantity (ποσόν τι) or as a substance (οὐσία) of a certain sort, a τόδε τι.[1] Nutri-

---

[1] Aristotle often claims that something one in number is more than one in being or account or

ment is posterior in account to the two nutritive functions and is even called by different names depending on which function it serves: αὐξητικόν ("productive of growth") or τροφή ("food"). He goes on to label nutriment in a third way in connection with reproduction, nutritive soul's most important function: γενέ-σεως ποιητικόν ("productive of generation") (416b15–16). What should we make of this conclusion in light of Aristotle's methodological prescriptions?[2]

Let us reconsider the passage on method and the initial portrayal of the nutritive functions as reproduction and using food (415a25–26). Notice that in the passage on method Aristotle hedges his claim by including the point about the objects in the antecedent of a conditional statement allowing us to understand the "if" as governing both the clause hypothesizing that the previous statement is true and a new supposition: "And if so (εἰ δ'οὕτως), and [if] further (δ' ἔτι) one must have studied even prior to those [i.e., capacity, activity] the opposed objects [...]" He then states the consequence using an optative as in the apodosis of a future less vivid construction: "one would need first to make distinctions (πρῶτον ἂν δέοι διορίσαι) about those for the same reason, e.g., about food, perceptible object, and intelligible object" (415a20–22).[3] The phrase "the same reason" could indicate that the correlative object is prior in account to the function, and therefore to the capacity. At the same time, the conditional statement may

---

form. *Physica* III,3, in discussing change from two perspectives, gives a helpful analogy (202a15–21, b5–16): the road from Athens to Thebes and the road from Thebes to Athens are one and the same in number (they are one physical road) but different in being or account, since the signposts occur in a different order depending on the direction of travel, and the path goes uphill in one direction, downhill in the other. For uses of these locutions in various texts, cf. *Ph.* I,7, 190a13–21 and *de An.* III,2, 425b26–426a1.

2 Some scholars, e.g. Shields (2016, p. 208–209), claim that Aristotle is guilty of a methodological circularity in his treatment of the objects of nutrition, since this passage ignores the advice of the procedural passage. My commentator in Toronto, Doug Campbell, pressed me to consider evidence elsewhere in Aristotle about relatives. *Categoriae* 7, 7b15–8a12, argues that some relatives seem to be simultaneous by nature (e.g., double and half), whereas others (e.g., perception and perceptible object) are not: the perceptible object is prior to perception, since destruction of the perceptible object removes the perception of it but not *vice versa*. Cf. *Metaph.* Δ,11, 1019a1–4, where Aristotle speaks of existential priority: A is prior to B, if A can exist without B but not B without A. To be sure, a carrot can exist without an animal to eat it. In Δ,11 Aristotle goes on to distinguish various sorts of priority in which the order of priority is reversed (1019a4–14). We shall pursue this reversal below.

3 Thanks to James Allen for calling to my attention additional linguistic evidence to strengthen the case.

leave room to reject the supposition or to allow for differences in the status of the correlative objects depending on the psychic capacity under investigation.[4]

I suggest that when in the next lines Aristotle identifies the nutritive functions as reproduction and using food, he is giving an initial sketch designed to orient the upcoming inquiry. The start of *de An.* II,2 recalls a procedural maxim he often repeats: in the order of inquiry, we start with what is more knowable to us and proceed to what is more knowable by nature, and then try to make what is more knowable by nature knowable to us. For instance, we begin with perceptible facts (the ὅτι) that living things take in nutriment, grow, persist, and reproduce, and then try to explain those facts by unearthing the cause (the διότι) (413a11–16), which is often hidden from view (cf. *Analytica Posteriora* I,2, 71b33–72a5; *Ph.* I,1, 184a10–b14; *Metaph.* Z,3, 1029b3–12). Such an approach would explain why Aristotle devotes considerable space to food in II,4, even though the nutritive object is ultimately posterior to the functions it serves. As II,4 will later show, food is an instrument used by the nutritive soul to feed the living body and is therefore posterior to the psychic faculty as well as its functions.

This paper will discuss the whole of *de An.* II,4, though not in the order of Aristotle's own presentation.[5] Instead, following his methodological plan, we start with nutritive objects, which are more accessible to us than the functions they serve, then discuss one of the three listed nutritive functions – self-maintenance – and finally address his leading question: what is the nutritive faculty – nutritive soul?

2 Food

In the second half of *de An.* II,4, while discussing nutritive soul as the efficient cause of the living body, Aristotle contrasts his views about food with those of

---

4 Wedin (1988, p. 13–14) emphasizes the importance of adhering to the prescriptions of the passage on method in II,4, and he mentions that the object must be conceived from the relevant perspective, citing an object of thought that is also an object of desire. The same is true about objects of perception, since one physical object (say an apple) can be touched, seen, tasted, smelled, and (in some situations) heard, and the function itself determines the properties of the object relevant to the sense modality. These observations suggest to me that the object is posterior to the function.

5 After the opening programmatic section and a section on the goal of reproduction with which I shall end, the bulk of the chapter discusses three ways the soul is the cause of the living body – as formal, final, and efficient cause. My paper is largely devoted to the long section on the soul as efficient cause, because that is where Aristotle discusses food and self-maintenance.

his predecessors, some of whom thought that opposite feeds opposite, others that like is fed by like, and concludes that both sides are partly right and partly wrong, depending on whether we are talking about first food or final food (416a19–b9). First food is raw and unconcocted and so unlike the living body it feeds, whereas final food has been worked up in the body and concocted and is therefore like the body fed. Since the discussion in *de An.* II,4 is highly abstract and the chapter sends us in closing to a proper treatment of food elsewhere (416b30–31), we turn to Aristotle's discussion of food in *de Partibus Animalium*.

2.1 Blood as Final Food

In *PA* Aristotle claims that blood is food in its final form in blooded animals (II,3, 650a34–35; cf. *GA* I,19, 726b1–3). Blood is fully concocted food, worked up in the heart and ready to replenish flesh and other parts of the living body.

PA treats blood as a uniform part of blooded creatures (II,2, 647b10–14), a stuff whose own parts are similar to one another, and describes it both as matter (ὕλη) of the whole body and as final food (ἡ ἐσχάτη τροφή) (II,4, 651a13–15). Blood counts as matter because it is the body potentially. Blood is potentially flesh, much as water is potentially a plant, or stones potentially a house (*PA* III,5, 668a13–27). Some scholars contend that blood is both the matter *for* a blooded creature and matter *of* it – that is, both the *preexisting* matter from which an animal and its bodily parts are generated, and the *constituent* matter that persists in the animal once it has been generated (Freeland 1987, p. 398–404, 406–407; Lewis 1994, p. 257–267; cf. Ebrey 2015, p. 62–68).

Aristotle evidently regards blood as at least the preexisting matter for blooded animals and their bodily parts.[6] In the first half of *Metaph.* Θ,7 he asks: When is something potentially a particular product? Not when the matter must still be transformed into matter of greater complexity (1048b37–1049a3). The matter is potentially the product when it has been sufficiently worked up that nothing needs to be added, subtracted, or changed about it before it is turned into the product (1049a8–12). A carrot fails to satisfy that condition, since it must still be chewed, transported to the stomach, heated and digested, and then transported to the heart to be further concocted into blood. Blood satisfies the Θ,7 condi-

---

[6] In reproduction the female of blooded animals contributes καταμήνια, a worked up residue of blood, as the matter from which an organism is generated. E.g. *de Generatione Animalium*, I,19 727b31–33; cf. *Metaph.* H,4, 1044a32–b3.

tion as food in its final form. Blood is the preexisting matter which, when further concocted, is transformed into flesh and other uniform parts of the living body.

Some evidence suggests that blood is also the constituent matter of blooded animals. Aristotle frequently speaks of the nonuniform parts, such as arms and legs, as ensouled and as persisting in name only (homonymously) when removed from the body.[7] In *GA* II,1, he extends the claim to flesh, a uniform part (734b24–27). Living flesh is the medium of touch. Although dead flesh has (for a time) the same dispositional properties as living flesh, since both are soft and squeezable, flesh must be properly hooked up to the soul to perform its function. *Mete.* IV,12 argues that all the bodily parts and materials, uniform as well as nonuniform, including even fire, are defined by their function in the living organism, though Aristotle says the function becomes less and less clear the simpler the matter (389b28–390b2).[8] The chapter tantalizingly mentions blood (αἷμα) at the end, along with flesh and semen, saying that we must determine what each of them is individually (καθ' ἕκαστον), and points ahead to the biological works (390b14–22). No mention of blood in the earlier passage when he says that material parts are defined by their function, but he mentions something that might seem to be more ultimate than blood when he declares that fire is among the things defined by its function (390a15–16).[9]

Other evidence in the biological works tells against the idea that blood is constituent matter of the living body. Aristotle says in *PA* that some things have proper (οἰκείαν) heat whereas others have foreign (ἀλλοτρίαν) heat; things with foreign heat have it as an accident (κατὰ συμβεβηκός), not in their own right (καθ' αὑτό). Blood has foreign heat; it is hot in the way that hot water is hot or hot iron is hot (II,2, 648b35–649a17). *PA* II,3 states that in one respect blood is in itself (καθ' αὑτό) hot, and in another respect not: heat is mentioned in the account of blood, in the way that whiteness is mentioned in the account of white man. White man is itself an accidental compound, however, not an intrinsically unified thing. Insofar as blood is hot because of an affection (κατὰ πάθος), it is not in itself hot (649b19–27). Heat is not part of blood's nature as blood.

The analogies illuminate Aristotle's conception of blood as matter and as food in its final form. Heat is a property of blood but one whose loss blood can survive. Whereas all ensouled parts of a living organism, nonuniform and uniform alike, are destroyed when the organism dies, blood can survive outside the body as the cold material it is in its own right, and it congeals (*PA* II,3,

---

7 E.g., *Meteorologica* IV,12, 389b29–390a2; *de An.* II,1, 412b18–22; *GA* I,19, 726b22–24; *Metaph.* Z,10, 1035b24–25.
8 I examine the teleology of *Mete.* IV,12 in Gill (2014).
9 Though see below: fire will take precedence over blood in Aristotle's account of nutrition.

649b28–35). The fact that heat is a mere affection of blood and not part of its nature tells against blood's status as the constituent matter of blooded animals.[10] Blood is present throughout the life of a blooded animal because animals must replenish themselves for as long as they live. It is the preexisting matter, not the constituent matter of blooded organisms.

2.2 Food and Heat

In the final part of *de An.* II,4, Aristotle distinguishes food as one of three factors involved in nutrition:

> Since there are three things – the thing fed (τὸ τρεφόμενον), that by which (ᾧ) it is fed, and the thing that feeds (τὸ τρέφον) – the thing that feeds is the first soul, and the thing fed is the body having that, and that by which (ᾧ) it is fed, the food. (416b20–23)

Food is an instrument by which (ᾧ) the soul feeds the body. A couple of lines later Aristotle distinguishes two distinct items that play an instrumental role:

> And that by which (ᾧ) it [the first soul] feeds is twofold, just like that by which it steers, both the hand and the rudder, and the one both moves and is moved, while the other is only moved (κινούμενον μόνον).[11] And it is necessary that all food can be concocted, and heat accomplishes the concoction. Hence every ensouled thing has heat. (416b25–29)

Here Aristotle mentions two instruments by which (ᾧ) the soul feeds the body – heat and food – similar to hand and rudder used by a steersman in managing a ship.

Early in his long discussion of nutritive soul as the efficient cause of the living body, Aristotle takes Empedocles to task for thinking that fire is simply re-

---

10 Lewis (1984), p. 262–264, gives a helpful analysis of heat as external to blood, but that evidence seems to me to tell against Lewis's thesis that blood counts as the constituent matter of an animal. Frey (2015), p. 376–377 and n. 4, argues that blood outside the body is homonymous blood, and therefore that blood in the body is alive. He concedes that Aristotle never explicitly says so.

11 Although κινοῦν μόνον is better attested by the MSS than κινούμενον μόνον, modern editors follow Philoponus (287,17–288,5) in preferring κινούμενον μόνον, because that version fits the analogy of the hand and the rudder. Simplicius (115,29–116,8) gets what I take to be the correct interpretation but regards it as weak because he is reading (κινοῦν μόνον). If Aristotle had written κινοῦν μόνον he would be referring to the nutritive soul, the first mover, which is unmoved. I agree with Philoponus (287,25–26) that the relevant heat is innate or vital heat (see n. 13 below). My interpretation of the passage agrees with Rodier (1900), p. 245–246, and Hicks (1907), p. 349.

sponsible for nutrition and growth. If that were so, he protests, organisms would fly apart, each element toward its own place, fire up and earth down (415b28 –416a9). On the contrary, says Aristotle, the soul is responsible, and heat is its helping cause. He mentions food in passing:

> The nature of fire seems to some people to be simply a cause of nutrition and growth, for it alone of bodies evidently is nourished and grows, and so both in plants and in animals one might suppose that this [fire] is the thing that does the work. Still, it is in a way (πώς) a helping-cause (συναίτιον), though not simply (ἁπλῶς) responsible (αἴτιον), but rather the soul [is responsible]. For the growth of fire is unlimited, as long as there is something that can be burned, whereas of all things constituted by nature there is a limit and formula (λόγος) of size and growth. These are [the responsibility] of soul, and not of fire, and of the definable form (λόγος) rather than the matter. (416a9–18)

Here fire is an efficient cause of growth but not the chief efficient cause. If fire were the primary cause, things would grow without limit as long as there is stuff – that is, food – to burn. Instead, soul is the first efficient cause of nourishment and growth. Soul limits and directs the activity of fire so that things grow to a size appropriate to their kind. Fire is a helping-cause (συναίτιον) and does the actual work, yet constrained by a psychic blueprint adjusted to the sort of organism fed (cf. GA II,6, 745a4–9).[12] If we juxtapose this passage and the one at the end of the chapter (416b25–29), we gain a full picture. Nutritive soul is the first efficient cause and employs two lower grade efficient causes as instruments to feed the organic body: heat and food.[13]

In *de Generatione et Corruptione* I,7, Aristotle distinguishes three grades of efficient causes: the last mover, which is moved in return by what it moves,

---

12 I speak of nutritive soul as a "psychic blueprint", following Mansion (1979), to capture the idea that the soul is responsible for changes that take place in the body without actually bringing them about. The soul is an unmoved mover. Think of the soul as a recipe listing the ingredients, indicating how and when to combine them, when and for how long and in what combinations to heat them up and cool them down, so as to achieve the called-for result, say a tree of a certain sort with an appropriate size. Thanks to Doug Campbell for pressing me to clarify this phrase.

13 I believe that Aristotle is speaking of vital heat, not ordinary fire, in both passages treating fire or heat as a tool for soul in *de An.* II,4 (and also when he speaks of the function of fire in *Mete.* IV,12, 390a15–16). Fire outside a living organism has its own nature and tends to move upward toward the periphery of the cosmos. That is not the function of fire within a living system as evidenced by Aristotle's joke in *de An.* II,4 at the expense of Empedocles. On vital/psychic heat in generation, see *GA* II,1, 732a16–20 and II,3, 736b33–737a7 (thanks to Tiberiu Popa for the references). On the loss of vital heat in organic decline and perishing, see *de Juventute et Senectute, de Vita et Morte, de Respiratione* 4, 469b6–20, and King (2001), p. 95–105. See also the detailed treatment of vital heat in Freudenthal (1995).

the agent that uses the last mover, and the first mover, an unmoved principle (ἀρχή) of motion (324a24–b4). For instance, when a doctor heals a patient, the last mover is the wine or medicine prescribed, something moved by the doctor administering it but also reciprocally moved by the thing it moves. There can, of course, be numerous intermediate movers between first and last mover. The first mover is the agent's knowledge of health guiding her work.[14] To judge from Aristotle's analogy in *de An.* II,4, the rudder is the last mover, the hand an intermediate mover, and the sailor's skill in navigation the first unmoved mover. As for the case he wants to illuminate with the analogy: food is the last mover, heat an intermediate mover, and nutritive soul the first unmoved mover. Not only is food posterior to the various functions it serves, food as last mover is also posterior to nutritive soul, the first mover it serves as an instrument.

Food as last mover is changed, indeed consumed, by what it feeds. Toward the end of his survey of the predecessors on food, before stating that both sides are partly right and partly wrong, Aristotle says the following about his own view:

> Further, food undergoes something by the thing fed, but not that [the thing fed] by the food, just as the carpenter does not undergo anything by the wood, whereas this [the wood] by that [the carpenter]; and the carpenter only shifts (μεταβάλλει) to activity (ἐνέργειαν) from idleness. (416a34–b3)

This passage with its announcement that the organism fed is not changed by the food consumed, any more than a carpenter is changed by the wood he carves, nicely orients our discussion toward our second main topic, the functions of nutritive soul. Why does Aristotle say that the body fed is not affected by the food it eats but merely shifts from idleness to activity?

3 Functions of Nutritive Soul

De An. II,4 discusses three main functions of nutritive soul, all of which depend on consuming food: reproduction, self-maintenance, and growth. Earlier we noted that food is relative to an ensouled thing and that its being differs depending on its role in nourishment proper (self-maintenance), growth, and reproduction. Reproduction and self-maintenance differ structurally from each another,

---

14 Several passages treat the form in the soul of the agent or art as a first mover, including *Ph.* II,3, 195b21–25; *Metaph.* Z,7, 1032b9–14, b21–23; and Λ,4, 1070b28–35.

because reproduction is a substantial *generation*, bringing something new into the world that was not there before, whereas self-maintenance is an *activity* that preserves the actor as the thing that it is.[15] We pause to consider this structural difference.[16]

3.1 Change and *Genesis*

Metaph. Θ presents two potentiality-actuality models, one concerned with change (κίνησις) and generation (γένεσις), the other concerned with activity (ἐνέργεια) and substantial-being (οὐσία)[17] – the topic Aristotle has been wrestling with since the start of *Metaph*. Z, where he examines being and especially substantial-being from the perspective of the categories. Aristotle turns in Θ,1 to δύναμις and ἐνέργεια, and he indicates that the model concerned with change, though it involves potentiality in its chief sense, is not the one most useful for his present project (1045b33–1046a2); then he announces at the start of Θ,6, when he turns to the second model, that he investigated δύναμις in connection with change because that discussion somehow bears on the present one (1048a25 –30). I take the present project to be the same as the one he embarked on at the start of Z: the investigation of being and especially substantial being.[18] Aristotle devotes the first five chapters of Θ to the first model, and he does so at least in part because the second model replicates all the features of the first with one key modification. That one modification results in a scheme quite different from the previous one. Consider the first model concerned with change and generation. Change for Aristotle involves an agent that brings about a change and a patient that undergoes it. When a thing is changed, it acquires a property previously lacked – e.g., a sick person regains health. The agent either has that property or, in artificial changes, has it in mind. Fire is hot and makes other things hot (*Ph.* VIII,5, 257b9–10); a sculptor has a sphere in mind and imposes that

---

15 In this volume see the paper by James G. Lennox on Aristotle's theory of reproduction.
16 I discuss this difference in more detail in Gill ([1990] 1994) and (2004).
17 I switch back and forth between translating οὐσία as "substance" and as "substantial-being" because the standard translation of the word as "substance" unfortunately obscures the connection with being, a connection that needs to be stressed when Aristotle contrasts becoming (γένεσις) with being (οὐσία). "Substantial-being" has its own disadvantages, because Aristotle speaks of both individual things and their form as οὐσία.
18 I owe this observation and my understanding of the strategy of *Metaph.* Θ to the groundbreaking work of Kosman (1984), developed further in his (2013), though we construe the two models differently. I develop further the interpretation here summarized in Gill (unpublished).

shape on a lump of bronze (*Metaph.* Z,7, 1032a32–b1, b21–23); and most important for nutritive soul, human generates human (e.g., *Ph.* III,2, 202a9–12; *Metaph.* Z,7, 1032a24–25; Λ,3, 1070a7–8). In the course of a change, the agent assimilates the patient to itself (*GC* I,7, 324a9–11) by transmitting to the patient the form the agent possesses. Aristotle, like many of his predecessors, adopts the so-called "transmission" theory of causation, also called the "synonymy" principle, that the agent of change has the property it transmits to the patient, and has it more eminently.[19]

In *Metaph.* Θ,1, Aristotle defines both active and passive principles (ἀρχαί) of change. The primary notion is an active δύναμις: "a principle of change in another thing or [in the thing itself] qua other" (1046a10–11). The corresponding passive δύναμις is "a principle in the passive subject itself of passive change by another thing or by [the thing itself] qua other" (1046a11–13; cf. *Metaph.* Δ,12, 1019a15–b15, b35–1020a6). Typically an agent, for example, a builder, acts on materials other than himself, such as bricks and stones, and generates a product – a house – distinct from the materials that go into its construction; or a doctor acts on someone sick and brings the patient into a different state, the state of health. Sometimes an agent acts on itself qua other, as when a doctor cures herself and, in virtue of her knowledge of health, brings herself from illness to health. Aristotle calls such cases accidental, on the grounds that the patient simply happens to be the same individual as the agent (*Ph.* II,1, 192b20–27). This sort of exception explains the qualification in Aristotle's definition of active and passive δυνάμεις: "a principle of change in another thing or *qua other*."

The first model features numerous ἐνέργειαι (or ἐντελέχειαι), actual things or states of things: the agent that brings about the change, the patient that undergoes it, the product that results from the change, the form of that product, and even the privative state from which the process begins. Moreover, in *Ph.* III,1–3 Aristotle defines change itself – the patient's transition from privative to positive state – as an incomplete ἐνέργεια of both the agent and the patient, though located in the patient (esp. *Ph.* III,1, 201a10–11; 2, 201b31–32; and 3,

---

[19] *APo.* I,2, 72a29–30. We find this idea in the Presocratics and Hippocratics and arguably in Plato. On the transmission theory of causation, see Lloyd (1976), Makin (1990), and Sedley (1998), p. 123–124, among others. Aristotle refers to the form of the agent as having the same form (ὁμοειδής) as the product at *Metaph.* Z,7, 1032a24–25; Z,9, 1034a21–24; and Θ,8, 1049b27–29. See the discussion of the synonymy principle in Burnyeat (2001, p. 33–38, esp. n. 59), where he argues that Aristotle should have called the efficient cause and its product συνώνυμος as he does at Λ,3, 1070a4–8. Aristotle more often calls them homonymous (ὁμώνυμος), and that strikes me as right, since the agent in artificial change does not share a form with the product but merely has the form of the product in mind.

202b5–29). That ἐνέργεια is incomplete, because its completion (the product) lies beyond the process and terminates it.

3.2 Activity and *Ousia*

Aristotle turns to the second potentiality-actuality model in *Metaph.* Θ,6, where he attempts to show what ἐνέργεια and δύναμις mean on the model more useful to the current project, and he gives several illustrations, including something awake to something asleep, and something seeing to something sighted with eyes shut (1048a30–b9). Later in the chapter he offers some criteria for distinguishing activities from changes. Whereas changes are incomplete because they aim at an end beyond themselves, activities are ends in themselves and complete as soon as they start and for as long as they last (1048b18–36).

Θ,8 opens with a statement about the various ways in which ἐνέργεια is prior to δύναμις and in the discussion pinpoints the vital difference between this model and the one concerned with change and *genesis*:

> Since we have distinguished in how many ways "prior" is defined, it is evident that ἐνέργεια is prior to δύναμις. I mean not only prior to the δύναμις that has been defined, which is said to be a principle of causing change (ἀρχὴ μεταβλητική) in another thing or qua other, but generally every principle of causing motion or rest (ἀρχῆς κινητικῆς ἢ στατικῆς). For nature (φύσις) too is in the same genus as δύναμις, since it is a principle of causing motion (ἀρχὴ κινητική), yet not in another thing, but in the thing itself qua itself.[20] ἐνέργεια is prior to every such δύναμις in both account (λόγῳ) and substantial-being (οὐσίᾳ); in time (χρόνῳ) it is prior in a way, and in a way not. (1049b4–12)

The penultimate statement in this passage reveals the crucial respect in which the second model differs from the first. On the second model the active principle of motion is not in another thing or the thing itself considered as other but in the thing itself *qua itself*. That single modification yields a scenario quite different from the one associated with change and generation. If something acts on itself qua itself, the motion is not like the doctor curing herself, acting on herself qua sick, and bringing herself into a state other than the one she was previously in. Instead the agent acts on itself qua itself – acts in virtue of certain active capacities and responds in virtue of corresponding passive capacities – and agent and patient engage in a joint activity that maintains the individual to which they belong as the individual it is. For instance, a French speaker in speaking French

---

[20] With Ross (1953) and Jaeger (1957), who both cite Bonitz, I delete 1049b8–9 γίγνεται [...] γάρ, following MS A^b and Ps.Alexander.

maintains and enhances her linguistic ability. Unlike a change, which takes place in the patient and results in a product that terminates the change, an activity takes place in the agent itself and preserves the agent in the state it is already in.

3.3 Self-Maintenance

Now consider a passage toward the end of *de An.* II,4, one line of which I quoted above to show that food is posterior to the function it serves. Aristotle has just distinguished nutriment as productive of growth from nutriment as food, and here he contrasts self-maintenance with reproduction, the third main function of nutritive soul:

> [B]ut insofar as it [the ensouled thing] is a this (τόδε τι) and substance (οὐσία), it [nutriment] is food. For it [the nutriment] preserves (σώζει) the substance, and it [the ensouled thing] exists as long as it is fed. It [nutriment] is also productive of generation, though not of the thing fed, but [of something] like the thing fed. That is because the substance of it [the thing fed] (αὐτοῦ ἡ οὐσία) already exists, and nothing generates itself, but preserves [itself]. Therefore such a principle (ἀρχή) of the soul is a potency (δύναμις) of a sort to preserve the thing having it (τὸ ἔχον αὐτήν) as such (ᾗ τοιοῦτον), and food prepares it [the ensouled thing] to be active (ἐνεργεῖν); hence, deprived of food, it [the ensouled thing] cannot exist. (416b13–20)

By consuming food an organism remains the active thing that it is, living its distinctive sort of life. The contrast with reproduction stands out: In reproduction the parent generates another like itself. Since nothing can generate itself, an organism, in acting on and maintaining itself, preserves itself as the very thing it is.

Though growth is a non-substantial change in the category of quantity, and so can be explained by Aristotle's first potentiality-actuality model as an increase in size, growth is also a stage of self-maintenance in the category of substance, and considered from that perspective, it is an activity dealt with by his second potentiality-actuality model. Once an organism acquires a heart or analogous organ, it already has its proper form, the soul principle. While reaching maturity the organism realizes its formal nature more and more as its instrumental bodily parts develop and its size approaches that of a mature specimen, as much as a doctor becomes a better doctor by practicing her art and learning from experience. As Aristotle told us in the passage on growth quoted above, fire is a helping-cause of growth, but nutritive soul is responsible, because the form and not the matter determines the proper limit of growth (416a9–18). Decay, which he omits in *de An.* II,4 but discusses elsewhere, is a later stage of self-maintenance when the organism gradually winds down and its size decreases. The organism's matter, not its form, is responsible for the organism's gradual loss of heat.[21]

4 What is Nutritive Soul?

We come finally to the controlling question of *de An.* II,4: What is (τί ἐστιν) nutritive soul? According to the opening passage on method, we understand the faculty by understanding its function, and we understand the function by understanding its correlative object. Aristotle declared that the function is prior to the faculty in account, and we saw that the claim echoes his statement about the priority of ἐνέργεια to δύναμις in account in *Metaph.* Θ,8. The actuality is prior to the δύναμις in account because the capacity is defined as the capacity it is with reference to the actuality it is a capacity for. At the same time, as we shall see, evidence in *de An.* II,4 casts doubt on the order of priority of function to capacity on causal grounds.

De An. II,4 states that the soul is the cause (αἰτία) and principle (ἀρχή) of the living body in three ways: as the source of motion (ὅθεν ἡ κίνησις), the final cause (οὗ ἕνεκα), and the substance (ἡ οὐσία) of the ensouled body (415b8 –12). Aristotle then discusses the soul as efficient, final, and formal cause in reverse order, and we have devoted the bulk of our discussion so far to the long final section on nutritive soul as efficient cause of the living body.

4.1 Nutritive Soul as Formal Cause

Aristotle's description of the soul as formal cause should surprise us in light of the earlier passage on method. Whereas there he claimed the priority of function to capacity in account, here he reverses the order of priority on grounds of formal causality:

> That soul is a cause as the οὐσία is clear, for the cause of being (τὸ γὰρ αἴτιον τοῦ εἶναι) for all things is the substance (οὐσία), and living (τὸ ζῆν) is the being (τὸ εἶναι) for living things (τοῖς ζῶσι), and the soul is the cause and principle of this. Further, the actuality (ἐντελέχεια) is a definable form (λόγος) of the thing that is in potentiality (τοῦ δυνάμει ὄντος). (415b12–15)

---

[21] See *Metaph.* Θ,8, 1050b6–28, and details about the loss of heat in *Juv.* 4, 469b6–20.

Aristotle makes two main points. First, the soul is the *cause of being*, and he identifies the being as the organism's function, living (τὸ ζῆν).²² This claim suggests that the criterion of priority paramount here is not the same as the one operative in the passage on method. Whereas there the activity is prior to the faculty *in account* (κατὰ τὸν λόγον), here the faculty is prior to the activity as its *formal cause:* the form makes the activity the sort of activity it is. To use Aristotle's distinction invoked earlier, the activity is more knowable to us, since it is the perceptible manifestation of the capacity, but the capacity is the cause of that manifestation and more knowable by nature.

Second, the soul as actuality (ἐντελέχεια) defines the thing in potentiality. The thing in potentiality is presumably the living body, since Aristotle opened the discussion of soul as cause by saying that it is cause of the living body (τοῦ ζῶντος σώματος) in three ways. This claim recalls his first of several definitions of soul in *de An.* II,1 as "the first actuality (ἐντελέχεια ἡ πρώτη) of a natural body that has life potentially (σώματος φυσικοῦ δυνάμει ζωὴν ἔχοντος)" (412a27–28) – a natural body he goes on to describe as instrumental (ὀργανικόν) (412a28–b6). The living body is among the things that follow soul and should be investigated once we understand what soul is. Remember the start of Aristotle's statement about method:

> The person intending to make an investigation about these things [the various psychic capacities] must grasp what each of them is (τί ἐστιν), then in light of that investigate what follows (περὶ τῶν ἐχομένων), and so on (περὶ τῶν ἄλλων). (415a14–16)

Adequate grasp of the nutritive soul will equip the inquirer to define the living body, its instrument.

4.2 Nutritive Soul as Final Cause

The soul is also the final cause of the living body:

> It is evident that the soul is a cause as that for the sake of which (οὗ ἕνεκα) as well. For just as mind makes for the sake of something, in the same way also nature (φύσις), and that is its end (τέλος). The soul is by nature such a thing for animals; for all natural bodies are instruments (ὄργανα) of the soul, and like the organs of animals, so also those of plants, because they are for the sake of the soul. The οὗ ἕνεκα is twofold: that [for the sake] *of* which and that *for* [the sake of] which (τό τε οὗ καὶ τὸ ᾧ). (415b15–21)

---

22 Aristotle famously speaks of substantial-form as the cause of being in *Metaph.* Z,17, H,2 and H,6.

De An. II,4 twice distinguishes two ways to be a final cause, and this is the second occasion. The first passage will be discussed in the final section of this paper. The interpretation of the distinction, as well as the two passages, is contested.[23] Aristotle nowhere says exactly what difference he intends, but his examples on different occasions suggest that he may have in mind more than one distinction.[24] His use of the genitive seems fairly secure: that for the sake *of* which (genitive) is the end (τέλος) or goal of a proceeding, whereas that *for* the sake of which (dative) is, as I understand it, something that profits in some way from that proceeding.[25] My description of the second notion is intentionally vague because the passages suggest that profiting is ambiguous between profiting from the use of tools in achieving some end, as an art or agent does, and benefiting from a proceeding by undergoing it, as a patient does.[26]

Aristotle first compares nature to art, as he often does in his works on natural philosophy. In artificial production an artisan has in mind the form of the product he aims to produce and replicates the form in suitable materials to yield that object. Similarly nature (φύσις) – the soul as δύναμις and first efficient cause of reproduction – replicates itself in suitable materials, and the form is its end (τέλος), identified in our passage as the soul. When Aristotle says that three causes – formal, final, efficient cause – coincide in an object, he usually also makes clear that the efficient cause of the proceeding is one in form, not one in number with the form it produces (e. g., *Ph.* II,7, 198a22–27). Organisms replicate according to their kinds, and the offspring acquires a form similar to that of its male parent. The soul of the male parent is the first efficient cause of reproduction, and the soul of the offspring is its τέλος.

---

[23] Recent treatments of the topic include Menn (2002), Johnson (2005, section 3.1), Rosen (2014), Johansen (2015), and Gelber (2018).

[24] In addition to the two instances in *de An.* II,4, see *Ph.* II,2, 194a27–b8; *Metaph.* Λ,7, 1072b1–4; *Ethica Eudemia* VIII,3, 1249b10–24; and *GA* II,6, 742a18–b6.

[25] Following the lead of Sedley (1991), p. 180 n. 3, I take the dative as a dative of interest, characterized by Smyth (1956), § 1474 as: "the person [or thing] *for whom* [or for which] something is done or in reference to whose case an action is viewed" (my additions in brackets). For a different view of the dative, as instrumental specifying tools used in some proceeding, see Gelber (2018).

[26] Traditionally scholars have taken the dative to specify the beneficiary. E.g., Hicks (1907), p. 340 says: "medicine (ἰατρική) has in view both τὸ οὗ, health, ὑγίεια, and τὸ ᾧ, the patient, ὁ ὑγιαίνων." Cf. Ross (1936), p. 509, on *Ph.* II,2, 194a35–36. Menn (2002) has argued that the dative specifies the user. Rosen (2014), p. 100, gives a helpful example to show that the user is distinct from the beneficiary: "I[f] a doctor is treating my friend and I fetch some bandages, I do this in order to benefit my friend, not the doctor; I do it in order to be useful to the doctor, not to benefit her."

Aristotle makes this claim and then explains: "The soul is by nature such a thing [i.e., a τέλος] for animals; for (γάρ) all natural bodies are instruments (ὄργανα) of the soul, and like the organs of animals, so also those of plants, because they are for the sake of the soul." Then follows the mention of two ways to be a final cause. Apparently the fact that the soul uses natural bodies as instruments explains its being an end (τέλος). Now to be sure, the male parent uses tools (especially σπέρμα) in replicating his form in the female material, but the soul Aristotle is talking about is the soul of the offspring – the soul resulting from the reproduction – not that of the male parent. The soul of the offspring becomes internalized in the developing embryo as soon as the heart or an analogous organ is formed, and from then on (and for as long as the heart continues to beat), the developing organism takes charge of its own further development and self-maintenance.[27] The final section of *de An.* II,4 argues that the soul is the inner efficient cause of an organism's life, and the section on the soul as formal cause identified the soul as the cause of being, of living a certain kind of life. Now Aristotle is arguing that the soul, by using the living body as its instrument, is for that reason an end.[28] I now want to show that the soul is also a τέλος in the first way, the goal for the sake of which the various life activities occur, activities it causes with the help of the living body as instrument. Let us look again at the last section of *de An.* II,4 and then again at *Metaph.* Θ,8.

4.3 Nutritive Soul as Mover and Goal of Self-Maintenance

We have already discussed much of Aristotle's lengthy treatment of nutritive soul as efficient cause, since that section deals with food, growth, and self-maintenance. Let me stress one main point from passages previously quoted from the last part of the chapter (416b20–23, b25–27). Three moving causes take part in feeding the body: food is the last mover, heat an intermediate mover – a helping-cause (συναίτιον) that does the actual work – and nutritive soul, the organism's own nature, is the first mover controlling the operation. This passage clearly shows that the soul is a final cause in the second way, as one that profits in its nutritive activity from the use of heat and food as tools. But does that activity also render the soul a final cause in the first way, as goal of that proceeding?

---

[27] See *GA* II,1–6 for details of Aristotle's theory of animal generation.
[28] Aristotle uses the word τέλος in reference to the user of tools as well as the goal of production in *Ph.* II,2, 194a27–36. Thanks to Jessica Gelber for correcting me on this point.

To understand how an organism's soul is both a mover and a final cause in the first way, we should look at another section of *Metaph.* Θ,8, in which Aristotle argues that ἐνέργεια is prior to δύναμις in substantial-being (οὐσία). This section (1050a4–b34) builds to ever more fundamental ways in which ἐνέργεια is prior to δύναμις in substantial-being. I focus on one passage that mentions life itself as a sample activity. Aristotle has just been talking about motions handled by the first potentiality-actuality model and now contrasts motions handled by the second model. Whereas ordinary changes take place in the patient and yield a product distinct from it (or at least a patient otherwise qualified than it previously was), motions explained by the second model have no separate product. Notice especially the penultimate sentence:

> But in those cases in which there is no other ἔργον [product] apart from the activity (ἐνέργειαν), the activity is present in them [the agents] (e.g., seeing in the one that sees, theorizing in the one that theorizes, and life (ἡ ζωή) in the soul, hence also happiness (εὐδαιμονία), since happiness is a certain kind of life. So it is evident that substance (ἡ οὐσία) and form (τὸ εἶδος) are ἐνέργεια. And according to this argument it is evident that ἐνέργεια is prior to δύναμις in substantial-being (τῇ οὐσίᾳ). (Θ,8, 1050a34–b4)

When the nutritive soul and nutritive body engage in the joint activity of nutrition, the activity takes place in the actor itself and, instead of changing it, sustains it as the actual thing that it is. The soul is at once an active δύναμις – a nature, first efficient cause of the activity – and the goal and actuality for the sake of which the activity occurs, the final cause as goal of that activity. The form as first actuality is sustained and enhanced by the various life activities it enables as δύναμις.[29]

5 Divine and Perishable Life

I end with two passages, one from early in *de An.* II,4 in which Aristotle distinguishes two ways to be a final cause, the other in the final section, both of which treat reproduction. After the opening section on method and before turning to the three ways that nutritive soul is a cause, Aristotle identifies the functions of nutritive soul as reproducing and using food and then explains:

---

[29] I agree with Menn (2002), p. 113 n. 44 and p. 121–128, that the soul is a final cause both as a user of tools (τὸ ᾧ) and as the goal (τὸ οὗ) of living, though he thinks that self-maintenance is an ongoing production, "*remaking* organs that are in continual decay" (italics in original) (p. 122). I have argued that self-motion is an activity, a preservation of the form that is already present in the living body.

> [F]or the most natural of the functions of living things [...] is to produce another like itself – an animal an animal, a plant a plant – in order to partake of the eternal and the divine as much as possible; for all [living] things desire (ὀρέγεται) that, and do as much as they do by nature for the sake of that. That for the sake of which (τὸ οὗ ἕνεκα) is twofold: that [for the sake] *of* which (τὸ μὲν οὗ) and that *for* [the sake of] which (τὸ δὲ ᾧ). Since, then, it is impossible to partake of the eternal and the divine by continuity, because no perishable thing can remain the same and one in number, insofar as each thing does partake of them, it does so in that way – one more, another less – and does not itself remain, but one like it does, not one in number, but one in form (εἴδει). (415a26–b7)

At the end of the chapter he again speaks of reproduction:

> Since it is right to call all things after an end (ἀπὸ τέλους), and to generate [another] like itself is an end (τέλος), the first soul would be able to generate [another] like itself. (416b23–25)

According to the first passage, the goal (τὸ μὲν οὗ) of reproduction, the end all living things pursue, is eternity and divinity, and the one that profits (τὸ δὲ ᾧ) as beneficiary from that yearning, carried out through reproduction, is not the individual itself, since it must perish, but its offspring, one like itself, and the descendants of its offspring, and thus, to the extent that mortal creatures can achieve immortality, they do so by preserving their form by perpetuating their species over time.[30]

Whereas reproduction preserves the form across generations and therefore benefits the species, self-maintenance preserves the form during the lifetime of the individual and so benefits the individual itself. Obviously, eating, growing, and reproducing do not suffice to maintain organisms of greater complexity than plants, but nutrition supports the whole range of activities that make an organism the kind of thing it is, and in our case that includes perceiving, walking upright, and thinking. Nutrition replenishes all the bodily parts that support an organism's way of life, and that complex life is the joint manifestation of the organism's active and passive δυνάμεις, activities that maintain the organism as the actual living thing that it is.[31]

---

[30] I follow Johansen (2015), p. 125–127, in regarding the species, rather than the individual, as the beneficiary. On the other side, with reasons for regarding the individual as the beneficiary, see Johnson (2005), p. 64–69, who also quotes several ancient commentators who favored that view.

[31] I presented versions of this paper at a conference on nutrition at the Humboldt University in Berlin in 2017, as part of a three-day Ph.D. seminar at the University of Oslo in 2018, and at a conference on Aristotle's Hylomorphism at the University of Toronto in 2019. I am particularly grateful for the probing commentary by Doug Campbell in Toronto, to my audience for helpful

Bibliography

Burnyeat, Myles (2001): *A Map of Metaphysics Zeta*. Pittsburgh: Mathesis.
Charlton, William (transl.) (2005): *Philoponus On Aristotle's On the Soul 2.1–6*. Ithaca: Cornell University Press.
Ebrey, David (2015): "Blood, Matter, and Necessity". In: David Ebrey (Ed.): *Theory and Practice in Aristotle's Natural Science*. Cambridge: Cambridge University Press, p. 61–76.
Freeland, Cynthia (1987): "Aristotle on Bodies, Matter, and Potentiality". In: Allan Gotthelf/James G. Lennox (Eds.): *Philosophical Issues in Aristotle's Biology*. Cambridge: Cambridge University Press, p. 392–407.
Freudenthal, Gad (2005): *Aristotle's Theory of Material Substance: Heat and Pneuma, Form and Soul*. Oxford: Clarendon Press, Oxford University Press.
Frey, Christopher (2015): "From Blood to Flesh: Homonymy, Unity, and Ways of Being in Aristotle". In: *Ancient Philosophy* 35, p. 375–394.
Gelber, Jessica (2018): "Two Ways of Being for an End". In: *Phronesis* 63, p. 64–86.
Gill, Mary-Louise ([1990] 1994): "Aristotle on Self-Motion". In: Mary-Louise Gill/James G. Lennox (Eds.): *Self-Motion from Aristotle to Newton*. Princeton: Princeton University Press, p. 15–34. First published in Lindsay Judson (Ed.): *Aristotle's Physics*. Oxford: Clarendon Press, Oxford University Press.
Gill, Mary-Louise (2004): "Aristotle's Distinction between Change and Activity". In: *Axiomathes* 14, p. 17–36.
Gill, Mary-Louise (2014): "The Limits of Teleology in Aristotle's *Meteorology* IV.12". In: *HOPOS* 4. No. 2, p. 335–350.
Gill, Mary-Louise (unpublished): "Aristotle's Hylomorphism".
Hayduck, Michael (ed.) (1882): *Simplicii in libros Aristotelis De anima commentaria*. Berlin: Reimer.
Hayduck, Michael (ed.) (1891): *Alexandri Aphrodisiensis In Aristotelis Metaphysica Commentaria*. Berlin: Reimer.
Hayduck, Michael (ed.) (1897): *Philoponi In Aristotelis De anima libros commentaria*. Berlin: Reimer.
Hicks, Robert D. (1907): *Aristotle De Anima*. Cambridge: Cambridge University Press.
Jaeger, Werner (ed.) (1957): *Aristotelis Metaphysica*. Oxford: Clarendon Press, Oxford University Press.
Johansen, Thomas K. (2015): "Blood, Matter, and Necessity". In: Ebrey, David (Ed.): *Theory and Practice in Aristotle's Natural Science*. Cambridge: Cambridge University Press, p. 119–136.
Johnson, Monte R. (2005): *Aristotle on Teleology*. Oxford: Clarendon Press, Oxford University Press.
King, Richard A. H. (2001): *Aristotle on Life and Death*. London: Duckworth.
Kosman, Aryeh (1984): "Substance, Being, and *Energeia*". In: *Oxford Studies in Ancient Philosophy* 2, p. 121–149.

---

questions and suggestions on the three occasions, and to Kathleen Cook and Jessica Gelber for valuable discussion.

Kosman, Aryeh (2013): *The Activity of Being: An Essay on Aristotle's Ontology.* Cambridge: Harvard University Press.

Lewis, Frank A. (1994): "Aristotle on the Relation between a Thing and its Matter". In: Theodore Scaltsas/David Charles/Mary-Louise Gill (Eds.): *Unity, Identity and Explanation in Aristotle's Metaphysics.* Oxford: Clarendon Press, Oxford University Press, p. 247–277.

Lloyd, Antony C. (1976): "The Principle that the Cause is Greater than its Effect". In: *Phronesis* 21, p. 146–156.

Makin, Stephen (1990): "An Ancient Principle about Causation". In: *Proceedings of the Aristotelian Society* 91, p. 135–152.

Mansion, Suzanne ([1971] 1979): "The Ontological Composition of Sensible Substances in Aristotle (*Metaphysics* VII 7–9)". In: Jonathan Barnes/Malcolm Schofield/Richard Sorabji (Eds.): *Articles on Aristotle: Metaphysics.* London: Duckworth, p. 80–87.

Menn, Stephen (2002): "Aristotle's Definition of Soul and the Programme of *De anima*". In: *Oxford Studies in Ancient Philosophy* 22, p. 83–139.

Rodier, Georges (1900): *Aristote Traité de l'âme.* 2 vols. Paris: Ernest Leroux.

Rosen, Jacob (2014): "Essence and End in Aristotle". In: *Oxford Studies in Ancient Philosophy* 46, p. 73–107.

Ross, David (ed./comm.) (1936): *Aristotle's Physics.* Oxford: Clarendon Press, Oxford University Press.

Ross, David (ed.) (1953): *Aristotle's Metaphysics.* 2 vols. Oxford: Clarendon Press, Oxford University Press.

Ross, David (ed.) (1956): *Aristotle's De Anima.* Oxford: Clarendon Press, Oxford University Press.

Sedley, David (1991): "Is Aristotle's Teleology Anthropocentric?" In: *Phronesis* 36, p. 179–196.

Sedley, David (1998): "Platonic Causes". In: *Phronesis* 43, p. 114–132.

Shields, Christopher (2016): *Aristotle De Anima.* Oxford: Clarendon Press, Oxford University Press.

Smyth, Herbert W. (1956): *Greek Grammar.* Cambridge: Harvard University Press.

Wedin, Michael (1988): *Mind and Imagination in Aristotle.* New Haven: Yale University Press.

R.A.H. King
Nutrition and Hylomorphism in Aristotle

M.F.B. *in memoriam*

Abstract: Nutrition provides Aristotle a way of distinguishing form and matter in living things. It also provides the way of binding form and matter. On the one hand, living activity takes part in continuously different matter, in any individual. On the other, nutrition, specified by any one indivisible kind of living thing, must take part in matter of the requisite kind. Between them these two facts explain the sense in which living is in a subject without being an attribute. For the soul is the cause of the living thing: it brings it about that there is a living body. Nutrition is the way that the soul achieves this. This explains how soul can then serve as the substance of living things in the *Metaphysics*.

1 Introduction

> For living organisms at least, the identity over a period of time is determined by the persistence of the same form in continuously changing matter [...] This, then, is Aristotle's theory of the continuity of the individual in whom the matter changes over a period of time. He himself makes the comparison with a stretch of water marked off by a measure – new water keeps on flowing through. The human or other animal form takes the place of the measure that might be set up to mark off a mile of river (*GC* I,5).

> But the way in which the analogy is only an analogy, which cannot be pressed is this: the matter of a substance e.g. of a living body, is in Aristotle's view in itself nothing but a potentiality; it is not e.g. actual flesh blood and bones except qua informed by the human or other animal form, or life or soul. (Anscombe 1961, p. 55–56)

Often, interpreters of Aristotle, like Elisabeth Anscombe, followed by John Ackrill (1972, ²1997, p. 170) and Bernard Williams (1986, and cf. Code 2004, p. 187) find it difficult to *distinguish* between form and matter in living things; it is also thought difficult to connect living form to living matter. Indeed, Aristotle suggests that there is no problem, when he says they are one, like a seal and the wax it is in (*de Anima* II,1, 412b6–9). For living matter, such as human flesh, cannot occur without human form. This appears to destroy the form-matter con-

---

Note: My thanks are due to Roberto Lo Presti and Giouli Korobili for organising a lively workshop. I am also grateful to audiences in Edinburgh and Tübingen for their probing of my views. The hope is, these views are now clearer for the criticism.

https://doi.org/10.1515/9783110690552-005

trast. This would be a disaster for Aristotelian physics, and hence for his metaphysics. His physics depends on this contrast not only for the account of change but also for the account of living things: the soul is the form, the body is matter. It is this last case that is crucial in his metaphysics. In this paper, I wish to show how living form is to be understood in *de An.* and other physical works, and suggest that this is the concept which allows Aristotle to use living things as the starting point of the enquiry into what there is, really, as such, in the *Metaphysica*. Thus, the *Metaph.* is not the place to go to understand what Aristotle is doing in *de An.*, except in the sense that it becomes clear there what he needs from physics. Conversely, without an understanding of *de An.*, *Metaph.* cannot be understood, in that our way into being as such proceeds through living things.

Aristotle's theory of nutrition (food: *trophê*, feeding: *trophê* or *trephesthai*) uses the theory that living things are composed of matter and form. Other theories of nutrition, which play a central role in early Greek ontology (Menn 2002, p. 2010), may not have this commitment, for example, Plato's in the *Timaeus*. In this paper, I argue that in fact Aristotle's concept of nutrition is presupposed at the moment when he argues that soul is the primary actuality of a natural instrumental body. For nutrition allows us both to *distinguish* between form and matter in living things, and to understand why soul – form or primary actuality – must be in matter: a) form grows and not matter, and this growth is the work of nutrition, the imposition of form on matter; b) while form does not exist without matter it does not (continue to) exist in the same matter, and so requires maintenance, replacement. While other forms of change, (coming to be, alteration, locomotion) occur in non-living things, growth, and hence nutrition, occurs uniquely in living things, and so can provide us with a physical, and not a logical account of the subjecthood of living body. And hence we have an argument for the status of the soul as form, and primary actuality of the natural instrumental body. This is then the actuality which, by preserving itself, is a fundamental feature of what is as such (*to on hêi on*), and so the entry into metaphysics ("primary philosophy") for us: living things are nourished along with us (*De Partibus Animalium* I,5, 644b29), hence we know much more about them that we do, for instance, about the stars.

In this paper, I begin by analysing a troubling argument at the start of *de An.* II,1, before going on to show how the analysis of growth in *de Generatione et Corruptione* I,5 precedes *de An.*, in the order of reading Aristotle's course on physics. I will then devote a little space to pointing to some passages in the *Metaph.* where the importance of nutrition for the projects of that work is visible.

2 The Origins of a Premise in *de An*. II,1

Let us start at the moment when Aristotle is working towards giving the most general account of soul in *de An.* II,1. This is a good place to start in that he here, if anywhere, *shows* that soul is form. Much of the apparatus he uses comes straight from the Organon (substance, the genera of being), but not all. Much hinges on the move from dialectic, the work of the Organon, to physics.[1] While *de An.* is a contribution to physics (I,1, 402a5–7), it also, of course, makes use of the topic neutral apparatus of the Organon. This move from dialectic to physics is made, in a general way, in the first lines of the chapter (412a2–10): we begin with the genera of being, one of which is substance, and end with distinctions between ways of being a substance, namely as form, matter and composite. For analysis of the composite into matter and form is the stuff of physics, and is never mentioned in the topic neutral Organon.

The importance of *de An.* II,1 is that soul is defined, in a loose sense, as the form or first actuality of a natural instrumental body (412b6–7). That is to say, we are asked to accept that soul is form not matter. The argument has, perhaps understandably, received a bad press. The problem is simply that one premise used appears to be that substance, *ousia*, is body. Yet the conclusion of the argument is that soul is form, which is another way of being *ousia* (412a7–9). It thus seems that we have here an argument which would support an attributive view of soul: soul is an attribute of a living thing. And that is very disappointing in a passage in which soul is meant to emerge as substance. Christopher Shields in his recent commentary on and translation of *de An.* (Shields 2016, p. vii) reckons there has been some 800 commentaries on *de An.*. It is noteworthy that this argument has resisted understanding.

Nutrition is the basis for Aristotle's theory of living things (cf. King 2001 and 2020): its mention is significant. But its significance here has not been realised.[2] I wish to show why mention is made in these lines of "nutrition through the

---

[1] For the nature and the importance of this distinction, see Burnyeat (2001), chapter 5: "The Organon as 'logical'".

[2] Cf. Polansky (2007), p. 151: "He uses the minimal requirement for life of mortal living bodies: they engage in nutritive life through their own power (cf. 411b27–30 and 413a30–32). He is introducing the necessary and sufficient conditions for ensouled life that he will only arrive at through argument in 413a24–b2." In other words, in Polansky's view, our passage is not the argument it appears to be. Shields (2016), p. 169–170 relies, rather like Hicks and Ross, on the three ways of being a substance (form, matter and composite), and adds a premise to exclude the soul being compound, while using a premise that the soul, unlike body, belongs to a subject. He ignores our premise 4 (see below) on nutrition entirely.

thing itself, and growth and decline" (412a14–15) by suggesting that in fact a premise is supplied in the argument for saying that soul is not a body: body is not (said) of something, rather it is "subject i.e. matter". This is clearly a crucial point in the argument for Aristotelian soul: the upshot, of course, is that soul is form (412a20), or more precisely, primary *entelecheia* of a natural body made of instruments (412b5–7). And this argument is not purely "logical", i.e. relying on forms of predication. It must be a physical argument, one using form and matter. Care is needed here. For, *in general*, physics requires form and matter. They are required for change in general, introduced at the most basic (learning) level in *Physica* I,7 and I,9. *Ph.* I is the introduction to Aristotelian physics, where we begin to learn to use form and matter in understanding change. What we want, however, is something more precise – body as matter, soul as form, indeed, as capacity and primary actuality of living things.[3]

I divide up the text of *de An.* II,1, 412a12–20 into a series of theses, numbered and lettered:[4]

(1) οὐσίαι δὲ μάλιστ' εἶναι δοκοῦσι τὰ σώματα, (1a) καὶ τούτων τὰ φυσικά·
(2) ταῦτα γὰρ τῶν ἄλλων ἀρχαί.
(3) τῶν δὲ φυσικῶν τὰ μὲν ἔχει ζωήν, (3a) τὰ δ' οὐκ ἔχει·
(4) ζωὴν δὲ λέγομεν τὴν δι' αὑτοῦ τροφήν τε καὶ αὔξησιν καὶ φθίσιν.
(5) ὥστε πᾶν σῶμα φυσικὸν μετέχον ζωῆς οὐσία ἂν εἴη, (5a) οὐσία δ' οὕτως ὡς συνθέτη.
(6) ἐπεὶ δ' ἐστὶ καὶ σῶμα καὶ τοιόνδε, (6a) ζωὴν γὰρ ἔχον, (6b) οὐκ ἂν εἴη σῶμα ἡ ψυχή·
(7) οὐ γάρ ἐστι τῶν καθ' ὑποκειμένου τὸ σῶμα, (7a) μᾶλλον δ' ὡς ὑποκείμενον καὶ ὕλη.
(8) ἀναγκαῖον ἄρα τὴν ψυχὴν οὐσίαν εἶναι ὡς εἶδος σώματος φυσικοῦ δυνάμει ζωὴν ἔχοντος.

I translate:
(1) Bodies are held most of all to be substances,
    (1a) and among bodies, the natural ones.
(2) **For** these are the starting points of the others.
(3) Some natural bodies possess life, (3a) others do not.

---

[3] Cf. Johansen (2012), p. 122–123, for an attempt to reveal the structure using simply *Ph.* and *de An.*, using a series of premises relating the definition of soul to hylomorphism, without anchoring the premises in the words of the text.

[4] Ross' 1961 text. Klaus Corcilius (2017) has returned to Aurelius' Förster's 1912 text of *de An.*, as being the last one to rely on a reliable examination of the mss. Our lines remain as they are in Ross.

(4) We call life nutrition through ⟨the living thing⟩ itself, and growth and diminution.
(5) **Thus**, every natural body which participates in life must be substance,
   (5a) substance in the sense of a composite.
(6) **Since** it is both body and of such and such a kind,
   (6a) for it possesses life, (6b) the soul may not be body,
(7) **for** body does not belong to things ⟨said⟩ of a subject, (7a) but rather as subject, i.e. matter.
(8) **Therefore**, it is necessary for the soul to be substance, in the sense of the form of a natural body possessing life potentially.

This is an argument. The relevant particles are printed in the translation in bold. My comments attach to the discrete theses:

(1) "Bodies are held most of all to be substances, (1a), and among bodies, the natural ones."[5] It is Aristotle's view of substance: the basis for the being of other things, a basis temporally, cognitively, and in being ("nature") that is at stake, hence not a view anyone can have expressed before.

(2) "**For** these are the starting points of the others." In what way are bodies starting points for all others? (1a) presents no problems: All things are made of earth, water, air and fire. Artefacts are made of bodies that have come about naturally. But why are bodies held to be substances (1)? Perhaps, the view we will meet later in this argument (7), namely that bodies are the subjects of other things, viz. of attributes.

(3) "Some natural bodies possess life, (3a) others do not." This premise deserves note, in that Aristotle thinks that there is a distinction between living things and other things, one that can, furthermore, be explained. (Note that the distinction made here between the animate and inanimate is not that reported from other thinkers in book I,2 403b25–27, namely by movement and perception). Life is not simply a brute fact about the world. Of course, the question remains just what kind of explanation may be possible, in Aristotle's view. That is the project of *de An.*

---

[5] We begin from a respectable opinion (*dokousi*): how do we move from opinion to truth? I ignore these questions about the status of apparently endoxically phrased premises, and how the transition to what is the case is achieved here. Note the move from *dokein*, via *legomen* in (4), to "must be" in (8). Such transitions in Aristotle deserve more attention than they have received. On the meaning of *endoxa*, see Reinhardt (2015).

(4) "We call life nutrition (*trophê*)⁶ through ⟨the living thing⟩ itself, and growth and perishing." Why does (4) occur here? On the face of it, simply to make a distinction among bodies, namely between those with and those without life, as in premise (3). Note that this distinction could be made without saying what life is. However, to *understand* this distinction, and so give the premise traction, we need to know what life is. We have here a preliminary definition, which will be confirmed, and made more precise, and explained by the following enquiry: we will learn that living can attach to other activities, without dislodging nutrition, growth and withering as the basic activity of living things. To see what this premise contributes, let us ask where this definition of life comes from. My suggestion is that this premise, at least: the understanding of growth and nutrition presupposed in the premise, comes from *GC*, and it occurs here because it gives us a physical grounding for hylomorphism in substance, i. e. in living things. This is the transition from dialectic to physics, which is made in the move from the greatest genera to nutrition. As has been pointed out by many readers, the science of living things is a part of, the culmination of, physics. *Physikê* is the Aristotelian name for the *epistêmê*, divided of course into different *pragmateiai*, studies. Form and matter have a specific application to living things, one that has proved elusive, as we saw at the outset. The point is that while matter is relative to form, always in Aristotle, there is a wide variety of relations between the two depending on what kind of forms, and hence matter, is at stake. This explains why form cannot be defined. By this I do not mean that certain forms cannot be defined. Clearly some can be, indeed, they are the best candidates for definition. What is at stake is whether form, quite generally in its relation to matter, can be defined. It cannot because this relation differs from case to case (cf. *Metaph.* Θ,6, 1048a35–b9). We learn this relation by considering cases. The definitory project depends on non-definitory procedures, namely the analogical grasp of form and matter, *dynamis* and *energeia*.

(5) "**Thus**, every natural body which participates in life must be substance" can be derived from (1) "Bodies are held most of all to be substances, (1a), and among bodies, the natural ones", since these bodies, living bodies, are a subclass of bodies.

In contrast, the proof for (5a) "substance in the sense of a composite" follows, in that it comes from (6) "**Since** it is both body and of such and such a kind, (6a) for it possesses life, (6b) the soul may not be body, (7) **for** body

---

6 *Trophê* here cannot mean *food*, it must be a verbal noun. The qualification "through itself", i.e. through the living thing itself is perhaps meant to exclude embryos when still dependent on the mother.

does not belong to things ⟨said⟩ of a subject, (7a) but rather as subject, i.e. matter."

(6b) is the conclusion we need further argument for, and this is provided by the next two premises, (7) and (7a). So, in turn, (7) and (7a) are the premises we need to find justification for. (7) may look like a "logical" premise, i.e. a topic neutral one, based on the way we talk about things: "body is not *said* of anything" is presumably the way we must expand the cryptical "body is not of anything". This last way of talking may be taken, however, for an *explanandum*, and not an *explanans*: On being told that body is subject and not attribute of anything, we want to know why, and in what sense. In *de An.* II,2, 414a19–27 we have a similar turn of phrase: "soul is of body, and not body". He then says that he has not yet considered the kind of body it is in. I take it this means: "not yet considered the kind of body it is in *precisely*", since he goes on to make the *general* remark that *entelecheia*, actuality, arises in a naturally suitable body. Where does Aristotle get the premise in this argument from, that body is a subject, and not one of the things (said) of a subject, but is instead rather like a subject i.e. matter? I think that the answer to this question lies in the way body, in this case living body, comes about, develops and is preserved. It cannot be a purely logical thesis, from the Organon, because those works do not deal in form and matter. The Organon cannot tell us *anything* about body.

This is a view that many commentators have eschewed. For example, Hicks and Ross would appear to follow the logical path. Hicks (1907, transl. p. 49) translates "And since in fact we have here body with a certain attribute, namely the possession of life, the body will not be the soul: For the body is not an attribute of a subject, it stands rather for a subject of attributes, that is, matter." And his analysis (Hicks 1907, p. 307 ad 412a11–b6) is as follows:

> The cogency of this reasoning depends upon two assumptions: (1) that οὐσία ἡ μάλιστα = σῶμα φυσικόν, (2) that σῶμα φυσικὸν ζωὴν ἔχον = ζῷον ἔμψυχον. The body of which it is said that it cannot be soul, is the animate body, which is the subject (ὑποκείμενον) of the attribute "life", which it is further assumed implies "soul". It could be wished that the last assumption had been definitely stated by Aristotle. There is yet one further assumption, viz. that soul is a substance (οὐσία).

Hicks' approach makes it an assumption of the passage that soul is *ousia*, not the demonstrandum. He continues (1907, p. 308 ad 412a17):

> If we know we have the two factors or components, if there is good reason to identify the one with matter, the other must be form, provided the analysis into two components was correct. Here it is into logical subject and essential predicate, together making "living body". To avoid mistakes, we must enquire if living is wholly distinct from body. For if

body implies life, we are not analysing properly, some part of the subject being itself a predicate in that case. But this, we are assured, is precluded: the body, the one component, is always subject and never predicate [...],

and, finally, Hicks quite rightly admits (1907, p. 309): "The inference from 'logical subject', never predicate to 'matter of a σύνολον' is a hazardous step."

This reading assumes that the argument in *de An.* II,1 is purely logical in character. So, in brief:[7] the body in question possesses life, so it is a body of a certain kind (*toionde*), to wit, by being alive. Hence what sets this body apart is being alive, and what makes it alive is soul, hence this soul belongs to the subject, or matter. Hence the soul is not matter. Since living things comprise matter and form, soul must be form.

This would be a purely *logical* reading, based not on Aristotelian physics, but on the way we talk. In fact of course, it is not purely logical, i.e. derived from the topic neutral dialectic of the Organon, since it mentions form and its correlate matter. So we must admit it has to be a physical reading of the premises. The problem lies in the application of the general premise: all things comprise matter and form, to this case, living things. So, I think there is still a *physical* gap to be filled on this reading. There is also the problem that soul then turns into an attribute of body, which is to be avoided in a passage arguing that the soul is substance. This point – in the readings of Ross and Hicks – is a great problem. For this passage is the passage arguing that soul is substance. However, as Michael Woods says.

> There are, of course, good reasons for insisting that the soul of an animal is not any sort of property, not for example, the property of being alive, or the instance of the property in the individual animal. Moreover, Aristotle is quite explicit in his rejection of the view that the soul is some sort of attribute in the *De Anima* and his rejection of the conception of the soul as a *harmonia*. (Woods 1994, p. 283)

---

7 Like Hicks, Ross appears happy with the view that soul is an attribute, although the message of the passage is that soul is substance. Ross (1961), p. 212–213, ad 412b16–20, says that it is absurd in our eyes that one should prove that soul is not body. But he summarises the proof which he sees here by saying: "In l. 17 he infers that soul is not a body, and the reason he gives ll.17–19 is that body is not an attribute, but a subject. The missing but easily supplied part of the proof is 'whereas soul (or besouledness) is, as we have seen, not a substance, but an attribute'. Put summarily, the argument in ll. 17–19 is 'soul cannot be body because it is that the possession of which distinguishes a living body from a lifeless one.'" Rather more recently, Williams also allows soul might be a property (1986, repr. 2006), p. 219 on 412a16–20: "The claim itself does not seem to introduce any particular item for the soul to be. It seems something more in the nature of a fact, or, possibly, a property."

The good reasons would, I think, include the fact that a mere attribute living might be taken from something without changing it essentially; and that is hard to swallow – quite apart from the Aristotelian arguments about *harmonia* (*de An.* I,4, 407a30–b34), "being fitted-together" that attribute which in some theories he says makes a body alive. The subject criterion gets us to body as substance. To get to form as substance we need another criterion, another distinguishing characteristic.

There are not two roles that body can play here, subject *and* matter, but there is a particular sense of being a subject. So *Hôs hypokeimenon kai hulên* must mean that body serves as matter. Matter is the way that body provides explanation here. This justification will also explain in what sense body is subject and form is "of" it: body is subject because the form – the process of nutrition, using food and the heat of the body, forms this body.

Notice what we are doing here: we are relying partly on what has gone before, but also on the more explicit (specific) account to come. This support comes both upwind and downwind from the thesis. Clearly, the more specific account cannot be *deduced* from the more general one.

In (8) "**Therefore**, it is necessary for the soul to be substance, in the sense of the form of a natural body possessing life potentially.", note, again, the extremely strong modality of the conclusion, in contrast to the doxastically qualified premises (1) "Bodies are held most of all to be substances, (1a), and among bodies, the natural ones" and (4) "We call life nutrition though oneself, and growth and perishing." I have translated the optatives in (5) and (6) ((5) "**Thus**, every natural body which participates in life must be substance, (5a) substance in the sense of a composite. (6) **Since** it is both body and of such and such a kind, (6a) for it possesses life, (6b) the soul may not be body.") to provide a suitable modality. Still, the conclusion we reach is by no means covered by the surface of the premises. Various problems force themselves on us. Why possess life "in potentiality" in (8)? More precisely, what justifies "in potentiality" here – where does it come from? They have the power to live. If we stick with actual living things, rather than seeds (as suggested by *Metaph.* Θ,7, 1049a15–6), then we must ask about their ability to live, which actually is provided by nutrition: "Each animal requires food for its existence" (*de Juventute et Senectute, de Vita et Morte, de Respiratione* 17(11), 476a16–17). Living things are able to live because they actually live; and this potentiality derives from nutrition.[8]

---

8 For an argument that nutrition is an activity, see King (2020).

To see the low regard Aristotle's argument here is held in, it is worth quoting at length from one recent commentary on *de An.*, for I take it that the verdict given is that Aristotle is fudging to hide his circular reasoning:

> This concise argumentation aims to eliminate both the composite body and its matter as candidates for soul. Since there are *living* bodies, if it is assumed that soul pertains to these and that not every body has life, soul cannot merely be body. Body may live or not live; consequently, something beyond body itself accounts for its life. Body is not what is predicated of a substratum to explain life, even living body is not so predicated, but body exists as substratum and matter of which life is predicated. Soul must be explaining why this body is such and such, that is, has life. Soul can hardly then be either body or living body. As just body, it would not explain why body has life, and as the composite living body, soul would also explain nothing. This would merely say vacuously that a living body is such because it is a living body (cf. *Metaph.* VII,17, 1041a10 – 28). For soul to provide an account of the life of the body – that is, it is posited to explain certain functions – it cannot just be the body or the composite living body. Why does Aristotle argue so tersely on such a crucial point? Perhaps the merest sketch suffices for his purposes since book 1 attacked the possibility that soul is a magnitude or body that could be in motion. If this has already been shown impossible, there is little need to belabor the point. But he is making the crucial assumptions: that life presupposes soul in the mortal being and that soul is a principle. The assumption that life connects with soul receives attention in the next chapter, and the assumption that soul is principle governs the whole treatment and may therein receive confirmation (see 402a6–7). Were Aristotle more explicit about the assumptions, much of the argumentation might seem question begging, and hence it behooves him to be brief and focused on the immediate issue. Body of any sort having been eliminated as candidate for soul, there remains as substance to serve as soul for the living natural body only its form. Soul is substance as form of a living natural body; this is genus and difference of the account. (Polansky 2007, p. 153 – 4 on 412a16 – 19, Polansky's emphasis)

If this is to be our verdict on this crucial argument, Aristotle's concept of soul appears indefensible. Let us try to do better than defend Aristotle by saying he hides bad reasoning in a cloud of concision.

(4) "We call life nutrition through oneself, and growth and perishing", immediately previously, cites a provisional definition of life[9] as growth, withering and nutrition. This suggests that he is moving from an account of nutrition, or growth and withering to an account of the body. Understanding how body is subject to soul, enables us to see how soul is substance. We should note how nutri-

---

[9] For the connection of this provisional definition with the life-cycle, and the full definitions of the phases of life in *Juv.* 24(18) see King (2010).

tion and two of its modes are mentioned together, as often.[10] This premise cannot come from the general philosophy of the Organon, since, notoriously, matter, and its correlate form, play no role in those works, as befitting works designed to be topic neutral. In *de An.*, nutrition is analysed into its factors as follows.[11] The primary soul nourishes, and it nourishes the body. This analysis seems to cause most problems. For how can soul be the agent of a process? I would argue that what this means is that the soul provides the limits to the nourishing, its structure, and so brings it about that some particular thing is nourished in the way that is appropriate to that thing. Saying that the soul is a moving cause is certainly not to say that it pushes or pulls, warms or cools: that is what the *organa* do, here: heat. Nourishment is with or by a) food (matter) b) heat (instrument) (416b21–32). The body here is what gets nourished.

Now, there are entirely general considerations for basing the need for form on accounting for change, as recounted by Aristotle in *Ph.* I,7 and I,9. Change runs between termini, which themselves are not subject to change. Note that these are general not in the sense that they are general axioms from which further conclusions can be deduced. They are general in the sense of being unspecific (cf. *Ph.* I,1, 184a21–25, and II,3, 194b16–23). But we need more, firstly because we need to understand why *living* bodies are "as subject, i.e. matter", not just any body. What makes a living body a living subject?

Let us then turn to *GC*, earlier in the order of reading or teaching (cf. *Meteorologica* I,1) than *de An.*[12] Elsewhere in *de An.*, Aristotle is less coy about referring back to *GC*, namely when he relies on the account there given of action and undergoing, in the account of perception in *de An.* II,5 (Burnyeat 2002,

---

10 *De An.* I,5, 411a30; II,4, 415b25–27; III,12, 434a22–26. *GC* I,5 discusses both growth and diminution, the distinction being simply that in the one stuff accedes, and the thing increases in quantity, in the other, stuff is lost, and the thing decreases in quantity. The basic process in both cases, present throughout life, is nutrition.

11 On the structure of nutrition, see the contributions by Mary Louise Gill and Andrea Carbone in this volume. See also King (2020) for a brief account of nutrition in *de An.* II,4, King (2001) for a full discussion.

12 See Burnyeat (2004) for discussion. The order is: *Ph., de Caelo, GC, Mete.* Then, after the account of *Mete.* I,1, 338a25–339a5, all we are told about the treatment of plants and animals is: "Once we have gone through these things, let us make theories in the way we have presupposed, if we are able to, about animals and plants, both generally, and separately." (διελθόντες δὲ περὶ τούτων, θεωρήσωμεν εἴ τι δυνάμεθα κατὰ τὸν ὑφηγημένον τρόπον ἀποδοῦναι περὶ ζῴων καὶ φυτῶν, καθόλου τε καὶ χωρίς, 339a5–8). Presumably, *de An.* is the general treatment of plants and animals (animals: I,1, 402a7, 10, plants: 5, 410b23), along with *Parva Naturalia* (cf. *de Sensu et Sensibilibus* 1, 436a11, on plants: *de Longitudine et Brevitate Vitae* 6, *Juv.* 5, 24(18)), followed by the specific accounts in *PA, GA, Historia Animalium*, and the lost *de Plantis*. Johansen (2012), p. 122, leaves out *GC* when collecting premises for his discussion of *de An.* II,4.

p. 37–40). In *de An.* II,1 there is no cross-reference. But given that *GC* comes before *de An.*, and that there is there an explicit account, as far as is necessary to distinguish growth and withering from other forms of change, i.e. coming to be, and alteration, it seems reasonable to expect that this is being used.[13]

3 *GC* on Growth: Matter Flows, Form Grows

Τὸ οὖν ὁτιοῦν μέρος αὐξάνεσθαι καὶ προσιόντος τινὸς κατὰ μὲν τὸ εἶδός ἐστιν ἐνδεχόμενον, κατὰ δὲ τὴν ὕλην οὐκ ἔστιν· δεῖ γὰρ νοῆσαι ὥσπερ εἴ τις μετροίη τῷ αὐτῷ μέτρῳ ὕδωρ· ἀεὶ γὰρ ἄλλο καὶ ἄλλο τὸ γινόμενον. Οὕτω δ' αὐξάνεται ἡ ὕλη τῆς σαρκός, καὶ οὐχ ὁτῳοῦν παντὶ προσγίνεται, ἀλλὰ τὸ μὲν ὑπεκρεῖ τὸ δὲ προσέρχεται, τοῦ δὲ σχήματος καὶ τοῦ εἴδους ὁτῳοῦν μορίῳ. (*GC* I,5, 321b22–28)

So, it is possible in relation to form that any part grows by the addition of stuff, but it is not possible relative to matter. For one must think of it as though one were to measure water using one and the same measure, for the thing coming about is always different. Thus, the matter of flesh grows, not by stuff acceding to every part, but by some leaving, and other stuff acceding, whereas it does accede to every part of the shape or form.

Here are two aspects of the *GC* I,5 account of growth (*auxêsis*) and diminution (*phthisis*):
a) matter flows;
b) form grows.

Quite how this works is a matter of some dispute, which we will turn to in a minute. Here before we delve into the text, we note how form and matter are distinguished. Since these distinct predicates – "grows", "flows" – each hold of two items in the same respect and at the same time, these two items must be distinct. This distinctness has to be explained by the process in which they are distinct, viz. growth. The image of the jug measuring continuously new measures of water will prove helpful. The measure is the form, the water the matter, an artifical illustration for the natural distinction. Anscombe, in the quotation we started from, understands the measure as a section of a river, past which always different water flows. This is an idiosyncratic understanding of measure here. The notion of a measure does not fit a length of river bank well; nor do we usually measure the water flowing in a river by a length of river bank. However, Anscombe's reading has the same results for our understanding of the separation between

---

[13] In perception, there needs to be an explicit reference, one could argue, since perception is merely *a kind of* action and undergoing: the account from *GC* needs adapting to fit the case.

form and matter: the water flows, and something persisting, the measure, whether riverbank or jug, is used to measure it.

The text continues:

> Ἐπὶ δὲ τῶν ἀνομοιομερῶν τοῦτο μᾶλλον δῆλον, οἷον χειρός, ὅτι ἀνάλογον ηὔξηται· ἡ γὰρ ὕλη ἑτέρα οὖσα δήλη μᾶλλον τοῦ εἴδους ἐνταῦθα ἢ ἐπὶ σαρκὸς καὶ τῶν ὁμοιομερῶν· διὸ καὶ τεθνεῶτος μᾶλλον ἂν δόξειεν εἶναι ἔτι σὰρξ καὶ ὀστοῦν ἢ χεὶρ καὶ βραχίων. Ὥστε ἔστι μὲν ὡς ὁτιοῦν τῆς σαρκὸς ηὔξηται, ἔστι δ' ὡς οὔ. Κατὰ μὲν γὰρ τὸ εἶδος ὁτῳοῦν προσελήλυθεν, κατὰ δὲ τὴν ὕλην οὔ. (321b28–34)

> This is clearer in the case of non-uniform parts such as the hand, because it grows in proportion. For the matter is more clearly different from the form there than it is in the case of flesh and uniform parts. Hence the flesh of a defunct human is held to be to a greater extent than his hand or arm. The result is that on the one hand any part of the flesh grows, and on the other, it does not. For it has acceded relative to the form, but not to the matter.

The point of turning to *GC* is to specify how the form-matter distinction applies to living bodies. While there may be several ways of using this distinction in the case of living things, I think there is one way which is fundamental, namely in growth and nutrition. On the surface, it looks as though matter grows: the flesh grows, thereby increasing the hand. Or at least: the body, that is matter and form, grows. But the passage ends by saying that something of the form has been added not of the matter. But note how the separation between form and matter is realised here: one measure, different loads of water. It is possible for any part to grow, when something is added, in relation to the form; it is not possible in relation to the matter. The matter does not grow, in that what grows is flesh, for example of a hand, and what makes a hand a hand is the form. The analogy with measuring water using a jug, ("a measure") suggests that the water does not remain: it is each time a different lot of water. A comparison need not fit in all points. One salient point is enough. We assume that something persists in growth: this distinguishes it from generation, coming to be *tout court*. So, something must persist. Michael Woods (1994, p. 287) insists that the one form must be able to occur in different kinds of matter, not merely in different parcels of the same kind of matter. The jug may hold water or wine, after all. But it is by no means clear that just any matter can fill just any form. Aristotle himself insists on the connection between *dunamis* and *entelecheia* (*de An.* II,2, 414a26–7): *entelecheia* is of such a nature that it comes to be in something which is potentially, i.e. the proper matter. When the *GC* passage on growth makes clear that the matter of flesh accedes and leaves continuously, it is surely the same kind of matter. The three conditions for growth are (321a19–24):

a) The growing thing, or a part of it, gets bigger.
b) Something has to be added to the growing thing.

c) The growing thing persists.

Although the chapter is officially about growth, i.e. change in quantity in keeping with the divisions between the genera of being, everything said here about growth and withering, bar change of size, should be said about nutrition too. This connection is made explicitly in the text (322a16–28), without giving the underlying explanation of nutrition. So, the account of growth is a promissory note for an account of nutrition. (Compare *GC* I,5, 322a22–8 with *de An.* II,4, 416b12–18.)

Living things are made of uniform and non-uniform parts. The latter, organs, grow by the growth of the former (321b16–22). The fundamental questions about growth, and hence nutrition concern the uniform parts. Uniform parts are both form and matter: what is the form? For example, flesh. What is the matter? Flesh, for example. So we have two aspects of the one thing, form and matter. Thus, form and matter are distinct. This claim is brought in here (321b19–22) with no argument: flesh and bone are "double" (*ditton*), in that both form and matter are called "flesh" or "bone". If this is true, then form and matter are being joined here, without argument, and without especial reference to nutrition. But this is only the first move in the explanation, as the text does say explicitly (321b17, *prôton*). On being told that flesh is both form and matter, we want to know more, and more is to come. We are then given an account of growth, and hence nutrition, in the uniform parts. It is through this that the organs grow, although the constituents are in flux, being continually replaced. The matter of nutrition is transformed into living stuff, with the requisite form. Thus, the form grows.

The point of this look back at *GC* I,5 was to show that form and matter are distinct in nutrition, and that body is the subject, that is, what *to auxêtikon*, the capacity to make grow, is in (322a10–16). This capacity is, of course, the soul. This is never said in *GC* I,5, since this text is about the basic forms of change in the Aristotelian cosmos. We then have two comparisons for the process of growth here. One is with mixture, on the basis that the product when water is added to wine is still wine, but more in bulk. The other is with a fire taking hold of fuel. This concerns the transformation, by heat, of nutriment into living stuff: there is a power in the flesh which makes flesh grow. Thus, the reason that form guides the process here is that it is the producer of the flesh. The second comparison with a fire is with the initial ignition of a fire: that is comparable not to growth but to coming to be. Aristotle does indeed think that nutrition is cooking or burning, that is why he thinks living things must be hot. Or else that it undergoes a kind of burning, thereby keeping the hot thing, the living thing, hot. So, the distinction is between a new fire, and an existing fire is important. Elsewhere he talks of the soul being kindled or set ablaze (*Juv.* 14(8), 474b13–14).

This capacity to make grow (*to auxêtikon*) is the form without the matter (322a28–33). These are the much-discussed final lines of *GC* I,5. They use the example of a pipe or tube[14] (*aulos*), as an organ which can grow by the accession of matter, which is potentially a tube. The power to grow, *to auxêtikon*, is the placeholder for the soul in *GC* I,5, which is otherwise not mentioned. This is clear in that the power to grow is identified with the form, which remains (322a33). This is one of the requirements of growth which we have already met. But it is also the crucial stable element in a world of change. It is stable, because it uses change, viz. the assimilation of nutriment to the living body to ensure its own existence.

4 Nutrition in the *Metaphysica*

My treatment here of many problems in Aristotelian *Metaph.* is summary, as needs be. Nutrition is never mentioned by name in the *Metaph.*,[15] let alone treated at length. So why think that it is germane?[16] The reason is simply that living things are those things which offer Aristotle the first steps towards the principle of being as such, and living things are distinguished from other bodies by their self-nourishment, growth and decline. Implicitly, under the guise of generation (*genesis*), nutrition is indeed a core concept. Not only is *genesis* a function of nutrition (*de An.* II,4, 415a26), it is the final end nutrition serves (416b25). Thus, the mention of the *genesis* of living things, as at Z,7, 1032a26–32 implies nutrition; there too, it is made clear that living things are substance (*ousia*) if anything is. Apparently, following the Platonic model,[17] *genesis* might be considered to have nothing to offer to *ousia*; that is, coming to be makes no contribution to what makes beings be. In this paper, we have seen that nutrition allows us to understand the problematic relation between form and matter, or, better, between *dynamis* and *entelecheia*, in the case of living things. Often, interpreters

---

14 See Kupreeva (2005), p. 133–135.
15 Incidentally, as what is necessary for life in Δ,5, 1015a20. The closely allied notion of *pepsis* "concoction" (see Lloyd 1996) is mentioned, in connection with unifying something, in contrast to it being a heap Α,8, 989a16; Z,16, 1040b8–9: "none of these things is a unity, but like a heap, before it has been concocted and one unity has come to be out of them." (οὐδὲν γὰρ αὐτῶν ἕν ἐστιν, ἀλλ' οἷον σωρός, πρὶν ἢ πεφθῇ καὶ γένηταί τι ἐξ αὐτῶν ἕν). Flesh, a product of nutrition, serves as an example of a unity, e. g. in Z,17, 1041b14, 21, 26.
16 See Kupreeva (2005), p. 104–105.
17 *Ti.* 27d–28b, but see *Philebus* 26d8–9.

find it difficult to *distinguish* between form and matter in living things, as we have seen already, since human form e.g. never exists without human matter (Z,11, 1036b26–32). And it is thought difficult to relate living form to living matter. Aristotle's theory of nutrition (*trephesthai*) presupposes the theory that living things are composed of matter and form and that it is the form that provides both growth and persistence. Other theories of nutrition may not have this commitment. Often, interpreters try to understand this relation in the *Metaph.* using purely logical concepts such as identity, inherence, particular and universal. But we need a physical basis for metaphysics, so we need to understand the work of the soul physically. As we have seen, this concept of nutrition is presupposed at the moment when Aristotle argues that soul is the primary actuality of a natural instrumental body in *de An.* II,1. For nutrition allows us to distinguish between form and matter in living things. And nutrition, growth and diminution is precisely the way that life is defined in this passage. The structure of growth and nutrition emerges earlier in the reading order in *GC* I,5, esp. 321b22–31, 322a28–34, as we have seen, and so can be presupposed in *de An.*.

While other forms of change (coming to be, alteration, locomotion) apply to non-living things, nutrition, and hence growth and diminution, apply uniquely to living things, and so can provide us with a physical, and not a logical account of the subjecthood of body, and hence for the status of the soul as form and primary actuality of the natural instrumental body. This allows us to slot soul into the role of primary being or substance in the metaphysics, insofar as we are dealing with those beings, such as ourselves, which are given to our inspection, while still fulfilling the criteria of being independent, and not subject to change, at least in one respect, their form or activity.

The point I think that must be made is that separation of form is to be understood in terms of things that exist separately (*de An.* II,1–2, 413a1–12). Care has to be taken, since the point of Aristotelian metaphysics, the study following physics, is that it uses physics, i.e. things that are matter and form. So, in one sense, form does not exist without matter. But, as we have seen in the case of nutrition, the form, e.g. a hand does not exist in the same matter, in that the matter is continuously exchanged in the process of being made into hand. So, what keeps matter and form separate is the process whereby form is imposed on matter, in other words nutrition. An extension of this process is generation, which in natural generation[18] moves from one form-in-matter to the same

---

[18] Cf. Buchheim (2001). Nature is the cause of these things coming together, being constituted, forming a unity namely by generation (cf. Θ,8, 1049b4–10) but *physis* is also a quite special structure in that form and matter, and both the starting point and the end point of generation are *physis*. Cf. *Ph.* II,1.

form in another matter: human generates human. But the basic process is the living thing's nourishing itself (*trephesthai*). This is underlined by Aristotle's attribution of generation to the nutritive soul (see Lennox, this volume).

The reason that living things do that is that in a certain respect, they are, to all appearances, permanent, furthermore, they ensure their own permanence. They thereby fulfil, in part, as it were, the demands set up in *Metaph*. E,1 for the subject of metaphysics (1026a10 – 13): they exist independently, unlike mathematical things, and they are not subject to change; even in growth, the form persists unchanged. Of course, they are not subject to change only in one respect. It is the form that does not change, as Aristotle concludes at the end of the intricate discussion of growth in *GC* I,5, as we have seen: the form remains. All this means is that growth does not replace one entity by another: growth does not make a human into a wolf. Nor does it make one human into another one; the coming to be of a human is generation, "coming to be", and not growth. The basic idea is that found expressed in *Metaph*. Z,17: the cause of the being of something is its *ousia* (1041b25 – 32). This cause of the living thing is the soul (*de An*. II,4, 415b8 – 14). Michael Woods (1994, p. 288 – 289), in his argument that the soul is a "this in this" (*Metaph*. Z,11, 1036b23), repeatedly appeals to the appropriate development of humans (his example of a living thing), appropriate, that is, to the kind of thing it is, and requiring the replacement of matter. In Aristotelian terms, which Woods, perhaps surprisingly, does not use, development is an achievement of nutrition, as we see in the definitions of the stages of the life-cycle in *Juv*. 24(18) (see King 2001 and 2009).

What does it mean to say that form is in matter, a this in this? This is an anti-Platonic slogan, of course: form, i.e. activity, takes place in matter,[19] or better, works on matter. This makes forms above and beyond the individuals superfluous, indeed detrimental, in that things that come to be would then not come to be a this such (*tode ti*), but only a suchlike (*toionde ti*). "A this such" is an individual falling under a sortal, which is thereby individual. "A suchlike" are things like one another, including copies of originals. A living thing exists in matter. If its form were separate from it, then it itself would only be a likeness of this form, such as the form, not the real thing but a copy, i.e. not itself a soul, but only like another form (Z,8, 1033b19 – 23). Instead of these useless forms, Aristotle points to the forms of natural things which reproduce themselves for "human generates human" (1033b26 – 32, cf. Buchheim 2001). All one needs, is something to produce the form in the matter (1034a4 – 6), and this individual, Kallias or Socrates, is the same in form, i.e. in its indivisible species, but different because of the

---

[19] On *logoi enhuloi*, see Buchheim (2009).

matter (1034a6–9). This of course raises the question we began from: how can one separate living form from matter, souls from bodies? As Aristotle says, it is hard to separate them notionally, since the form of human always appears in fleshes, bones and suchlike parts (Z,11, 1036b2–7). Part of the answer is that the matter flows, while the soul persists, in the individual; part of the answer is that in succeeding generations, through the changes of matter, forms also persist.

What living things actually do is live. This life is fundamentally nutritive activity. For that is what provides and maintains the organs, and their parts, involved in all other activity. Material living things require a *dunamis* to live, but they are able and obliged to provide this *dunamis* themselves, from outside themselves, by their activity. Their independence is based on them nourishing themselves, a form informing always new matter.

Bibliography

Ackrill, John (1997 [¹1972]): "Aristotle's Definitions of psyche". In: *Proceedings of the Aristotelian Society* 73, p. 119–133.
Anscombe, Gertrude E.M./Geach, Peter (1961): *Three Philosophers*. Oxford: Basil Blackwell.
Buchheim, Thomas (2009): "Was sind logoi enhyloi bei Aristoteles?" In: Föllinger, Sabine (ed.): *Was ist ‚Leben'? Aristoteles' Anschauungen zur Entstehung und Funktionsweise von Leben. Akten der Tagung vom 23.–26. August 2006 in Bamberg* [Philosophie der Antike 27]. Stuttgart: Steiner, p. 89–111.
Buchheim, Thomas (2001): "The functions of the concept of physis in Aristotle's Metaphysics". In: *Oxford Studies in Ancient Philosophy* 20, p. 201–234.
Burnyeat, Myles F. (2001): *A Map of Metaphysics Zeta*. Pittsburgh: Mathesis.
Burnyeat, Myles F. (2002): "De anima II 5". In: *Phronesis* 47 (1), p. 28–90.
Burnyeat, Myles F. (2004): "Introduction: Aristotle on the Foundations of Sublunary Physics". In: Haas, F.A.J. de; Mansfeld, J. (Ed.): *Aristotle, De generatione et corruptione I. Proceedings of the Symposium Aristotelicum*. Oxford: Oxford University Press, p. 7–24.
Code, Alan (2004): "On Generation and Corruption I. 5". In: Haas, F.A.J. de; Mansfeld, J. (Ed.): *Aristotle, De generatione et corruptione I. Proceedings of the Symposium Aristotelicum*. Oxford: Oxford University Press.
Corcilius, Klaus (transl./comm.) (2017): *Aristoteles. Über die Seele. De anima. Griechisch-Deutsch*. Übersetzt, mit einer Einleitung und Anmerkungen herausgegeben von Klaus Corcilius. Hamburg: Meiner.
Hicks, Robert D. (ed./transl./comm.) (1907): *Aristotle. De Anima*. Cambridge: Cambridge University Press.
Johansen, Thomas K. (2012): *The powers of Aristotle's soul*. Cambridge: Cambridge University Press.
King, Richard A.H. (2001): *Aristotle on life and death*. London: Duckworth.

King, Richard A.H. (2009): "The concept of life and the life-cycle in *De Iuventute*". In: Föllinger, Sabine (Ed.): *Was ist ‚Leben'? Aristoteles' Anschauungen zur Entstehung und Funktionsweise von Leben*. Stuttgart: Franz Steiner Verlag.

King, Richard A.H. (2020): "Nutrition". In: Connell, Sophia (Ed.): *Cambridge Companion to Aristotle's Biology*. Cambridge: Cambridge University Press.

Kupreeva, Inna (2005): "Aristotle on Growth. A Study of the Argument of 'On Generation and Corruption 15'". *Apeiron: A Journal for Ancient Philosophy and Science* 38.3, p. 103–159.

Lloyd, Geoffrey E.R. (1996): "The master cook." In: Lloyd, G.E.R.: *Aristotelian explorations*. Cambridge: Cambridge University Press.

Menn, Stephen (2002): "Aristotle's definition of the soul and the programme of de Anima." In: *Oxford Studies in Ancient Philosophy* 22, p. 83–139.

Menn, Stephen (2010): "On Socrates' First Objections to the Physicists (Phaedo 95e8–97b7)". In: *Oxford Studies in Ancient Philosophy* 28 p. 37–68.

Polansky, Ronald (2007): *Aristotle's De anima*. New York: Cambridge University Press.

Reinhardt, Tobias (2015): "On endoxa in Aristotle's Topics". In: *Rheinisches Museum* 158, p. 225–246.

Ross, William D. (ed.) (1961): *Aristotle's De Anima*. Oxford: Clarendon.

Williams, Bernard (1986, repr. 2006): "Hylomorphism". In: Burnyeat, M.F. (Ed.): *The sense of the past*. Princeton: Princeton University Press.

Woods, Michael (1994): "Human Being and Individual Soul". In: Scaltsas, T.; Charles, D.; Gill, M.-L. (Ed.): *Unity, identity and Explanation in Aristotle's Metaphysics*. Oxford: Oxford University Press, p. 279–290.

Sophia M. Connell
The Female Contribution to Generation and Nutritive Soul in Aristotle's Embryology

Abstract: In *de Generatione Animalium* (*GA*) Aristotle argues that both parents contribute to generation through differentiated products of the nutritive process, governed by nutritive soul. This appears to agree in general with the fact that the nutritive soul is the same thing as the generative soul, as set out in *de Anima*. This essay analyses the contribution of the female animal to generation as a nutritive residue and the result of her nutritive functioning. The female contribution to generation is made useful by its location and latent potentials: it ends up in the uterus ready to become all the parts of the new animal's body, once its own nutritive soul becomes actualised. After giving a comprehensive overview of the content of the female contribution as residue of nutrition, the last part of the essay articulates a challenge that this presents for Aristotle's account of nutritive soul. Since the female is unable to generate without the addition of the male generative residue, it would seem that her nutritive soul is defective, lacking the generative capacity that males possess. Articulating this problem requires a closer analysis of the connection between nutrition and generation in Aristotle philosophy. The essay finally concludes that because the female animal's soul attempts to perpetuate an animal the same in form into the next generation this is enough to render it generative as well as nutritive.

In Aristotle's theory of animal generation, the male acts as efficient cause of substantial change and the female as material cause (*de Generatione Animalium* I,20, 729a9–12, 22–34). The efficient cause brings form into matter, generating another animal the same in form as the male parent. The materials from the female animal are highly specialised, containing all the parts and the whole body of an animal the same in form as that female animal (*GA* II,4, 738b7–9, 740b19–21). In most animals,[1] these differentiated generative functions are car-

---

Note: I would like to thank David Lefebvre, Giouli Korobili and Pavel Gregorić for their comments and criticisms on earlier drafts of this paper.

[1] With the exception of insects. See *GA* I,22, 730b20, 25–26; II,4, 738b12–14.

ried out by the products of nutritive processes.² The nutritive soul in both parents effect the production of their differentiated seed, semen in the male and menstrual fluid in the female.

This essay will focus on the connections between nutritive soul and the female's spermatic contribution in Aristotle's embryology. The first section (1) sets out Aristotle's general theory of *sperma* as a residue of nutriment. The second section (2) explains how the female's contribution is a 'useful' residue of nutriment different from that of the male. Its usefulness is a combination of its (i) origins, (ii) place and (iii) eventual use. The third section (3) discusses the content of the female contribution. The fourth section (4) explains how the female contribution is connected to the nutritive soul of the embryo as it develops. The final section (5) details how the female condition might be thought to raise difficulties for the unity of the nutri-generative soul. The female retains a nutritive function while appearing to lack a generative one. Thus, Aristotle's thought that nutrition and generation are somehow the same needs to be carefully nuancing when it comes to explaining the nutritive soul of the female animal in all kinds. The conclusion (6) will bring together all of these analyses by elaborating further how the female's nutritive soul is generative.

1 Aristotle's Theory of the Female Contribution

The nature of the female contribution to generation is closely parallel to that of the male – both have the same origins in the body. In order to establish this account, Aristotle focuses on unknown theorists in *GA* I,17–18 who hold that *sperma* comes from "all the parts of the body". Balme (1972 [1992]) identifies six objections that Aristotle makes to this. In one prominent argument, Aristotle posits that if *sperma* were from the parts it would be a degenerative fluid (σύντηγμα). He presents his own theory as more viable – that *sperma* is a "residue" (περίττωμα).³ Initially, a residue is distinguished from nourishment (τροφήν) (724b25–26). Nourishment is then further divided into useful and useless: "every residue is (ἐστίν) either useful or useless nourishment" (725a4–5). *Sperma* is a part (μέρος) of this useful residue (725a11). It is the final (τὸ ἔσχατον) and most nourishing portion (725a14–28).

---

2 This accords with the strong association between generation and nutrition in both Aristotle and his contemporaries (Connell 2016, p. 127–141).
3 He also notes that *sperma* is a residue (περίττωμα) in *de Longitudine et Brevitate Vitae* 5, 466b9.

In order to make his point against the rivals, Aristotle must insist that, for them, *sperma* is a degenerative fluid (σύντηγμα or colliquesence). Aristotle uses the verb *suntêkô* to describe starvation as a process whereby an animal destroys itself like a fire burning itself out (*Long.* 5, 466b28–33). And so, this degenerative fluid is being broken down rather than being put together (*GA* I,18, 724b35–725a2).[4] Its presence, then, is a sign of disease.[5] It can be distinguished from residue because it does not have a natural place or receptacle in the body, instead flowing all over, causing trouble in seeking a way out (*GA* I,18, 725a34–35). This degenerative *suntêgma* shows that something unnatural is happening (725a28) and so cannot be the origin of the embryo since "nothing that is in accordance with nature (παρὰ φύσιν) ever comes from what is against nature (κατὰ φύσιν)" (*GA* I,18, 725a2–3). It would be odd for any theorist to consider *sperma* to be like that. Indeed, it may be that Aristotle's linguistic distinction unfairly masks the similarities between his theory and those of his rivals. So, for example, in the certain Hippocratic works, the *sperma* comes from the humours which are potentially constructive, containing powers of their own right.[6] Leaving aside whether Aristotle rightly characterised these opponents, let's now analyse more closely his own view that *sperma* is a sort of "residue".

Aristotle presents a new refined position on the connection of "residue" to nourishment in the *GA*.[7] More generally, he prefers to contrast the two. For exam-

---

[4] In instances of spontaneous generation, organised living being do come to be from waste material, such as dung (*Historia Animalium* V,19, 551a4, 552a16–18) and processes of putrefaction can result in living animals (*HA* V,1, 539a23; V,31, 556b26). Putrefaction (σηπομένης), however, would appear to be different from *suntêkô* as the former is external to living bodies. In the case of dung, this is a natural (useless) residue and so may have some positive qualities which aid in the production of simpler animals. In any case, these occurrences would not necessarily count as 'natural' according to Aristotle since they are due to chance (*GA* III,11).

[5] It may be unfair for Aristotle to assume that the process of 'melting' which medical writers associate with 'putrefaction' always produces a morbid secretion. For more on this point, see Coles (1995), p. 61.

[6] See e.g. *De genitura* III,1, L. VII,474–5. The Hippocratic works *De genit.*, *De natura pueri* and *De morbis IV* which are the most detailed account of the origins of *sperma* in the bodily parts/humour never use the term *suntêgma*. As Balme remarks: "The extant Hippocratic writings do not call seed a colliquation, but Aristotle considers their view tantamount to it" (Balme 1972 [1992], p. 146). The idea that Aristotle's view is not so far away from that of Presocratic "pangenesis" is a point originally made in a slightly different manner by Coles (1995), p. 59–61. See also Louguet on how Aristotle's argument also does not really work against Anaxagoras (Louguet 2015, p. 129–133).

[7] This may be because the *GA* focuses fertile on animals in the prime of life, when most of their residue is useful. Other biological works, in contrast, deal with animals at all stages of life and so spend more time considering useless residues. I thank Giouli Korobili for this point.

ple, in *de Partibus Animalium* bile is said to be a residue (περίττωμα) of the liver and that "residue is the opposite of nourishment" (*PA* IV,2, 677a27); it is contrasted with "healthy blood" which is sweet (as opposed to being bitter).[8] In the *GA*, in contrast, Aristotle allows for residues to be healthy and so they are not 'opposite' to nourishment in the sense of being pathological. Useful residues are (potentially) positive.

> All residue is either useless or useful nourishment. By 'useless' I mean that nothing natural can be further constructed out of it, but copious consumption of it harms greatly, and useful is the opposite. (*GA* I,18, 725a3–7)

Useless residues (usually termed just "residues" elsewhere) are the liquid and solid by-products of nutrition, and have a particular place to be stored before exiting the body.[9] These useless residues are a potential cause of disease (presumably disease of a different sort that that indicated by the presence of σύντηγμα). The useful residues, being contrary to the useless ones, must be those which can still contribute to the construction of natural parts and which can be consumed without any harm ensuing. The nutritive process begins, properly speaking, in the stomach, liver and spleen before a final concoction takes place in the heart; blood is the ultimate and most useful residue and *sperma* is a portion of this blood.[10] If we are led to wonder why, since it is useful, it is left over and not used to nourish, Aristotle offers an explanation as follows.[11] The useful final nourishment is meant to make up and maintain the adult's living body but an animal that is fully grown and reasonably well fed will have a little bit of this left over. This move ingeniously explains why animals that are too young, not well fed or in some other way infirm do not produce *sperma* or it is non-fertile

---

8 See also *GA* II,6: "residue is unconcocted stuff, and the most unconcocted thing in the body is earthy" (745b19–20).
9 Further descriptions of the anatomy associated with these waste products can be found at *GA* I,13, 719a29–720a11; *PA* III,8; *HA* I,2, 489a3–8; III,15; IV,2, e.g. 527a8, 529a14; V,5, 541a3–11.
10 See e.g. *GA* I,19, 726b2–3; for further references Boylan (1982), Althoff (1997) and Connell (2016), p. 141–151. The first entry of food often comes about through the mouth and the teeth play a role – but these organs do not begin any digestion or concoction only preparing the food to be cooked (*PA* III,1 and Lennox 2001, p. 243–246).
11 Lennox makes a similar point about Aristotle's characterisation of male and female contributions to generation as residues: "One might ask, why they are considered to be residues at all – why not suppose that the amount of blood produced was just the amount needed for nutrition and reproduction? Perhaps because, while semen and menstrual blood are occasionally used for reproduction, often they are not. They are residues because they only occasionally play a biological role" (Lennox 2001, p. 186–187).

(non-nourishing) (*GA* I,18, 725b19–25). It also explains why those who are too fat are infertile (726a3–7).¹²

The identification of *sperma* as a useful residue comes before Aristotle distinguishes male and female contributions.¹³ Although they are different (*GA* I,18, 724a36–724b8), the origin of the female *sperma* is the same as that of the male variety.¹⁴ The difference between them is slight – one of degree of concoction, the menstrual discharge being less concocted (*GA* I,20, 728a26–27; IV,1, 765b35–766a3). Recall that Aristotle defines *sperma* in general as "some part of the useful residue". This "most useful" residue "is the last from which come to be each of the parts [of the body]" (725a11–13). This description is actually most applicable to the female contribution that will become the body of the new living animal (*GA* II,5, 741b7; II,4, 738b26).¹⁵ However, thinking of the female contribution as a "useful residue" is complicated by two factors. First of all, it is often characterised in a negative way as waste and compared to pathological fluids. Secondly, Aristotle suggests that there are two grades of menstrual discharge. The next section (2) will consider the first problem. The second issue will be the focus of Section (3) on the content of the female contribution.

2 Is the Female Contribution to Generation a Useless or Useful Residue?

To make a case against his rivals, Aristotle presses hard on the idea that the new animal cannot be made up of something that is "against nature" (*GA* I,18, 725a2–3). Useless residues can also be unhealthy and contrary to nutrition and so it looks as if they could not become a natural embryo.¹⁶ Useful can be distinguished from useless and/or pathological residues by determining three things: (i) their origins; (ii) their location in the body; (iii) what they can be

---

12 Other than *sperma*, potentially useful resides are produced during the nutritive process on the way from raw food to internal ultimate nourishment (i.e. blood). These are then used to form non-essential body parts, such as extra fat, horns, hair etc. (See e.g. *HA* IX(VII),2). See n. 30.
13 The idea is that "everything comes to be from *sperma*, and that *sperma* comes to be from the parents (plural)." (*GA* I,17, 721b7–8; cf. I,18, 724a16–19). See also Lefebvre (2016), p. 44.
14 This is noted by many scholars, e.g. Coles (1995), Deslauriers (1998), Henry (2006), Mayhew (2004).
15 The male does not contribute any matter to generation (*GA* II,4, 738b20–26).
16 Lennox (2001) and Balme (1972 [1992]) touch on these issues in their commentaries. Claire Louguet (2017) is currently working on a detailed analysis of this section of the *GA*.

used for. Bona fide generative residues (i) originate in the digestive system as the final nourishment; (ii) are located in the generative organs (uterus, male genitals or female breasts, *GA* I,18, 725b3–4); (iii) can successfully generate a new animal. In making his case for *sperma* as useful rather than unnatural, Aristotle focuses on the male experience of ejaculation of seminal fluid. It must be useful since its removal results in exhaustion (*GA* I,18, 725a4–9, 725b18). But he seems aware that an opponent might claim that ejaculation actually results in a feeling of relief, similar to the removal of harmful fluids. He does not deny that this sometimes occurs, explaining that in those cases semen is mixed with diseased fluids, which it is a relief to get rid of (725b14–17, 726a14–16).[17] These superfluous fluids could even be *suntêgma* insofar as they do not have a place in the body but seek an exit where they can.[18] Later on Aristotle explains this phenomena.

> Likewise some animals are prolific and have abundance of *sperma* because they are able and some due to inability. The latter is due to much useless residue getting mixed in so that in some animals, disease occurs because the discharge (ἀποκάθαρσις) has no clear passage out. Some of these recover their health, others die. A diseased breakdown (συντήκονται) occurs as in the urine. (*GA* I,18, 726a10–15)[19]

The useless residues present along with the male *sperma* in these instances may be finding the correct location (ii) but do not meet criteria (i) or (iii).[20] Any *suntêgma* that is present fails on all three counts.

---

17 Although these emissions are said to be non-fertile, some mixing in of useless residues may be inevitable given the fact that they share the same passages out of the body (*GA* I,13, 720a9–10). See also *HA* IX(VII),1 "in all boys and girls who had residues in their bodies, when these are discharged together with the seed or the menses respectively, their bodies become healthier and more thriving with the departure of that which was impeding their health and nutrition" (581b29–33).
18 Aristotle seems to borrow some of these views from the medical tradition, for whom "[w]hether these emissions become morbid depended [...] on their subsequent free movement around the body and their occasional fixation" (Coles 1995, p. 62). See also the Hippocratic work *De locis in homine*.
19 At 726b25–30 Aristotle discusses something he refers to as "the spermatic colliquesence". This is usually taken to be an interpolation and to refer to gonorrhoea. See Balme (1972 [1992]), p. 146; Mich. Eph. *in libros de Generatione Animalium commentarium* 47,3, Galen *Definitiones medicae* 288, K. XIX,426.
20 Aristotle makes clear that useless residues normally occur along with *sperma* in males and females: "In those animals that emit semen, when the *sperma* of the male has entered, it sets the purest part of the residue, for most of the menses is useless (ἄχρηστον) fluid, just as the seminal discharge (γονῆς) of the male is mostly fluid" (*GA* II,4, 739a6–10).

In the male, pathological fluids exit in the ejaculate, in the female, there are different useless residues which gather in the uterus. Unlike the male, the female has a useless secretion that is not directly mixed in with her spermatic one but which leaves the body separately. This is discharge of white materials (*GA* II,4, 738a25–30). The so-called "whites" must have the same origins (i) as menstrual fluid and end up in the same place (ii) but being less concocted, cannot do what menstrual blood does and (iii) result in generation. Aristotle notes the following about both useful and useless female discharges:

> Both of these secretions of residue when of a moderate amount preserve the body, as it is purified of residues which are the cause of disease to the bodies. (*GA* II,4, 738a27–30)

The menses are put together with useless residues in the above passage because they become useless once they have not been employed in generation. In both sexes, once spermatic residues lose their effectiveness, they become waste products, which must exit the body or else they cause difficulties.[21]

Similar to the male experience of voiding his body of spermatic secretions, according to Aristotle, the loss of menstrual discharge results in fatigue and weakness (*GA* I,19, 727a3). There seems, however, to be another reason for the fatigue which is that *other* useful residues exit along with the menses. If these other residues were in a male system they would have gone to make up fit and healthy (and superior) parts and aspects of his body:

> Further [evidence]: females are not so full of veins and likewise are neater and smoother than males because the residues which go into these parts are discharged along with the menses. It is necessary to think as well that this is the cause of the lesser bulk and body size of females than males in the live bearing kinds. For only in these animals does the menstrual discharge flow externally, and most obviously in women. Women of all animals emit the most discharge. Because of this they are most obviously always pale and their blood vessels are less articulated, and they clearly have bodies that are falling short of males. (*GA* I,19, 727a16–26)[22]

The female systematically fails to add these extra useful residues to itself which makes it seem that this residue of nourishment is wasted. Also it seems accidental that they end up where they do. When the fine blood vessels that terminate in the uterus are overfull, which is due to the inability of the female, residues pass into the uterus (*GA* II,4, 738a10–15).

---

21 The health effects of the emission of semen and the evacuation of menstrual fluid is dealt with in much more detail in contemporary medical literature.

22 Aristotle posits that sexual dimorphism is most pronounced in the human kind (E.g. *HA* VIII(IX),1, 608b4–7; *GA* IV,8, 776b25–28). Here is a rationale for that supposed fact.

In some passages, Aristotle puts his theory of *sperma* as useful residues under some strain by implying that the female contribution is useless or even harmful. He says that males that have damaged generative organs and cannot produce seminal fluid, "suffer from looseness of the bowels caused by residue which cannot be concocted and converted into semen being secreted into the intestine" (*GA* I,20, 728a15–18). These residues are useless because the concoction has not taken place that would have made them into useful residues. The situation is then compared to the female state:

> Just as in the gut due to lack of concoction diarrhoea happens, likewise in the blood vessels haemorrhages and the flow of menstrual discharge occurs. For it is the same bloody flow, but that one is due to disease and this one natural. (*GA* I,20, 728a21–25)

Aristotle also likens menses to pathological haemorrhoids and nose bleeds (727a13–14). How could the natural and useful female residue be comparable to these three pathological ones? In fact, the female contribution can be distinguished from these. The residues in the bowels are obviously useless due to their origin (i), location (ii) and capacity (iii). Nose bleeds, haemorrhages and menstrual discharges all have the same origins (i) as the final form of nourishment.[23] They can also become parts of the body (iii) and thus have a use.[24] The only reason why these other fluids do not qualify as useful residues is their location (ii). Nose bleeds are in the nose and haemorrhages can occur anywhere in the body. And so it appears that in the case of the female *sperma* as useful residue depends on a principle of proper place to differentiate it from other similar liquids.

At times, Aristotle appears conflicted about the production of *sperma* in the female – is it an accident or purposive? It looks, on the one hand, like the striving of the female animal's nutritive soul to both maintain the being of the female animal herself and toward the generation of another animal the same in form, since it goes to the uterus to serve as the matter for the living body of the embryo. And yet he indicates that instead of nature aiming directly for that end, these residues accidentally end up in the uterus, where nature makes use of them.[25] However, there isn't anything in Aristotle's theory of male *sperma* to suggest that it is any more purposive. In the male body, the pure portion of the final residue is just

---

[23] This is presumably why if a woman has too many nose bleeds or haemorrhages there will be no residue left to contribute to the menstrual flow (E. g. *HA* IX(VII),11, 587b30–31).
[24] Although the blood of nose bleeds and haemorrhages could not have become a new animal, they could have maintained the parts of the body (or even built these up to be stronger).
[25] Leunissen's "secondary teleology". For further discussion of this point see Leunissen (2017), p. 143–145 and Connell (2016), chs. 8.6, 10.1.

there and thus used by nature for the sake of generation. For both male and female the location of their generative secretions is crucial to its influence.[26]

3 The Content of the Female Contribution to Generation

The female *sperma* is a useful residue because (i) it is the most refined portion of the final nourishment (blood), (ii) it resides in the uterus and (iii) it can become the body of the new animal. It is not clear at first what the content of the female contribution is. We know that the female *sperma* constitutes the material existence of the new animal, whereas the male fluid does not. Other key differences from the male *sperma* include the possibility of an initial differentiation into two parts, the more soul-like portion and the more nutritive one (as in wind eggs, see below) and that the female's spermatic contribution continues to nourish the offspring in *utero* in live-bearing animals (*GA* II,7, 745b29).[27]

In many animals the male sets the female contribution into parcels with coherent edges (*GA* I,20, 729a10; II,4, 739b21–24). In some cases, the female can make such parcels without male input. "In some animals, as in female birds, nature can generate up to a point: the female of these kinds do actually set a foetation, but what they set is incomplete, i.e. so-called wind-eggs" (*GA* I,21, 730a29–33). Wind eggs contain both white and yolk. For Aristotle the white portion of the egg is closer to form and "contains in itself the soul heat" (*GA* III,1, 752a2).[28] In contrast to certain unnamed opponents who argue that these eggs are the "relics of earlier impregnations" and contain the male input,[29] Aristotle insists it "is not due to the male and the female, the white being male and the

---

[26] "Each of the residues at the same time is in its proper place and comes to be a residue. Before that none [of them] unless by much force and against nature" (*GA* II,4, 739a3–4). The importance of the proper place is reinforced by Aristotle's insistence that male *sperma* cannot serve its purpose if it is deposited in the stomach rather than the uterus (*GA* III,5, 756b5–11). Although it is not explicitly defined as such, the uterus for Aristotle is a specialised organ of concoction (see e.g. *GA* I,12, 719a34). The female's uterus is thus analogous to the male sexual organ. For Aristotle, penile concoction during copulation only occurs in certain animals, see *GA* I,6, 717b24, 718a6–7. See also the Hippocratic work *De genit.* 1,2 for this idea.

[27] The female menses are not wasted during pregnancy in live-bearing animals but continue to be "useful" residues of nutrition. In those that lactate, the female residue continues to be useful after parturition.

[28] For a fuller discussion of the origin and principle of the new animal in bird eggs see *HA* VI,2.

[29] In fertilised eggs, "once the white and the yolk have been separated, they already possess the principle that comes from the male" (*GA* III,7, 757b13).

yolk female: both are from the female" (751b26–7). In addition, in describing the epigenetic development of the embryo, Aristotle distinguishes between "nourishing" (θρεπτικόν) and "growth-promoting" (αὐξητικόν) nourishment.

> Nature generates bones in the first construction, from the spermatic residue and as the animal grows, it gains growth from the natural nourishment, which supplies the supreme parts, but these are the mere leavings or residues of it. In all generation there is a first and a second nourishment, the nourishing and the growth promoting: the nourishing is that which maintains the being of the whole and the parts, the growth-promoting is that which contributes to making larger. (*GA* II,6, 744b27–36)

He notes that the secondary, growth-promoting nourishment comes "from the female or from outside" (745a4).

Dually differentiated wind eggs and the two sorts of nourishment used in embryonic development might make it seem that the female *sperma* comes in degrees. However, as we will see, neither phenomenon actually show that the female contribution has different parts; and there are many reasons to think that it does not. First of all, the two grades of nutriment that occur in development are not necessarily the same as the initial female *sperma*. Instead, these become differentiated later on in a process which involves the embryo's own nutritive soul. The good nourishment is used on the best parts and the less noble parts are constructed out of leftovers or residues. In *GA* II,6, after having detailed the eyes which he lists as "the final parts to be differentiated", Aristotle writes this:

> Each of the other parts come to be from nutriment, the noblest which share in the controlling principle from the most concocted and purest first nutriment, the necessary parts which are for the sake of those out of the inferiority and leftover residues. For like a good manager of a household, nature discards nothing out of which something useful can be made. In household management, the best of the nutriment produced is stretched to the free people, the worse and the residue of this to the household staff, and the worst aiding the nourishment of animals. Even as an external intellect makes these for growth, thus nature in generating things constitutes from the purest matter, flesh and the other sentient bodily parts; and from the residues, bones and sinews and hair (also nails and hooves and all such things). So these are the last to take shape, when residues of nature are generated. (*GA* II,6, 744b12–28)

This image suggests that in the gradual construction of the embryo, materials that are less and less good are used.[30] It would be surprising if Aristotle thought

---

30 This section of the *GA* is a more general illustration of "indirect" or "secondary" teleology, i.e. "the product of nature using leftovers to make some useful feature for the animal in ques-

that these less good materials came directly from the female. The fresh supplies of menstrual blood arriving in live-bearers via the umbilical cord (equivalent of the egg yolk) ought to be of the same uniform superior quality. One sure indication of this is Aristotle's belief in the possibility of superfetation where the menstrual discharge that comes after forms another viable embryo (*GA* IV,5, 773b8–25).

The uniformity of the female's spermatic contribution also makes best sense of it as the material cause. If it were already differentiated into different types of matter, then this would be too close the rival view that the parts are present in the *sperma* already. Instead, for Aristotle, the soul of the new animal brings about differentiation in the parts, exploiting the potentials present in the female contribution – which must be uniformly able to become *any* bodily part (*GA* II,4, 738b7–9), like pluripotent stem cells.

In adult animals, useful residues come in different degrees, since they can occur at different stages in the digestive process between external food and the ultimate internal nourishment (blood or its analogue). This is how the adult gains fat, hair, bodily bulk instead of spermatic residues, since that which would have ended up as a portion of the ultimate residue is diverted and used up early on in the digestive process.[31] An embryo, however, initially does not have any stomach, liver or spleen to bring about this gradual transformation. Even when these organs are present, they are not operative until after birth. The only digestive organ the embryo has is its heart, with which it completes the final concoction (*GA* II,6, 742b4, b35–7; II,5, 741b16–18). This heart then concocts the female's spermatic contribution into all the principle parts of the body, and when this digestive process is complete, only the scraps are left for certain less important parts.[32] It is only at a late stage in the construction of the embryo that the above left-overs become available (744b28); thus they are the embryo's and not the mother's residues.[33]

---

tion" (Leunissen 2010, p. 84). "Indirect" teleology is what Lennox calls this at *PA* IV,5, 679a1–30 and IV,3 (Lennox 2001, p. 291–292). For further discussion of this passage see the paper by Carbone in this volume.

31 Fat: *GA* I,18, 725b31–3; I,19, 727a33–37; bodily bulk: *GA* II,8, 748b20–24; IV,4, 771a28–30; hair: IV,5, 774a35–774b4.

32 This account of how the embryo is developed from blood must be part of what Aristotle points us to at *PA* II,3: "the way in which the parts derive their growth from blood, and the subject of nourishment generally, is more appropriately considered in the works on generation" (*PA* II,3, 650b8–10).

33 When Aristotle says that they come "from the female" (745a4), he must mean initially.

Next, the phenomenon of wind eggs also does not indicate that the female contribution is two-fold. This is mainly because wind eggs do not have the same potency as female menstrual blood. The wind egg is a failed animal which has been ruined by overheating. Male and female generative residues must strike the correct temperate balance in order to bring about generation: the female cools and the male heats (*GA* IV,2, 767a16–24).[34] Once the wind egg has been differentiated into two parts, it can no longer develop. The egg can only become fertile if copulation occurs *before* the white and yolk have separated (*GA* I,21, 730a5–8; III,7, 757b6–8). The female contribution, in contrast, retains a pluripotency and can still become the body of a new animal. The wind egg shows that if the female contributes too much heat, generation is not possible. However, it also provides the following insight about the female's nutritive soul: in this case, it is clearly striving to produce an animal the same in kind, differentiating out the portion of matter that is to become the animal's heart.

4 The Female Contribution and the Embryo's Nutritive Soul (*GA* II,5)

As we have seen, the female animal can overcook the initial portion of ultimate residue so as to separate out another portion. This means that the female, as Aristotle puts it, can "generate up to a point". But it cannot start the generation of an animal without male input. The male establishes the heart (or its analogue) which is then generative of the other parts. In the following passage, Aristotle phrases this as the wind egg possessing potential nutritive soul.

> If the female has the same soul and the matter is the residue of the female, why is the male required in addition? Why doesn't the female generate by itself from itself? The reason is that an animal differs from a plant through perception. It is impossible without the presence of sentient soul for there to a face or hand or flesh or any other part [of an animal] either actually or potentially, whether in some way or absolutely so. For that would be like a dead thing or a dead part. So if the male is the active agent of this sort of soul,

---

[34] "[T]hey [i.e. male and female] must stand in the right proportional relationship (συμμετρίας) to one another, since everything that is formed either by art or by nature exists in virtue of some due proportion. Now if the hot is too powerful it dries up fluid things; if it is very deficient it fails to make them set; what it must have in relation to the object which is being fashioned, is the mean proportional, and unless it has that, the case will be the same as what happens when you are cooking; if there is too much fire it burns up your meat, if there is too little it will not cook it – either way what you are trying to produce fails to reach completion. The same applies to the mixture of the male and the female."

where male and female are separated, the female is unable by itself to generate an animal from itself. For it was said that this is what it is to be male. However, spelling out this puzzle is reasonable, as it is clear in those wind eggs generated in birds, that the female is able to generate up to a point. Further, there is this puzzle: in what sense are these eggs said to live? For they are not so like fertilised eggs (for what comes to be from these are actually ensouled) but neither are they like wood or stone. For these eggs rot so that before that, they were in a certain manner living. It is clear that they have some sort of potential soul. But which sort? It must be the least, i.e. the nutritive sort. This is present in all animals and plants alike. Why are the parts and the animal not completed? Because the parts of the animal are not like those of the plant. Because of this, the male must share [the work]. For in these the male is separated. (*GA* II,5, 741a6–30)

The initial question arises because the female has the "same soul" as the male.[35] The nutritive soul is supposed to strive to make another animal like itself, so why wouldn't the female, which clearly has a nutritive soul, do this on its own? The answer at first seems to separate female and male into nutritive and sentient.[36] In fact, that reading is not necessary. The passage says that the male establishes "soul of this sort" (741a13–14 τὸ τῆς τοιαύτης ποιητικὸν ψυχῆς), i.e. an animal soul, with both nutritive and sentient faculties combined.[37] This is the sort of soul the female cannot generate on its own. As for defining male as the producer of sensory soul this hardly fits with previous definitions of male in the text, which all relate to agency in generation.[38] The male is that which is able to initiate the generation of a new animal of the same sort, i.e. with the same sort of soul, as the parents.

The passage, therefore, does not say that the female contribution to generation is only potential nutritive soul; it is referring to the wind egg phenomenon, where the female product is an overcooked failure. It does, however, gesture toward a close relationship between the female contribution and the nutritive soul of the embryo. After fertilisation, nutritive and sentient soul are present in the

---

35 They are the same in form (*GA* I,23, 730b34; *Metaphysica* I,9) and form in animals is their soul (*de An.* II,1, 412a20).
36 A certain interpretation of this passage has become ubiquitous, i.e. that the male contributes sentient soul (Henry 2016a). There is nothing inevitable about that conclusion. For scepticism see e.g. Connell (2016), p. 172–177, Carraro (2017), p. 285–286.
37 Sentient soul is ontologically inseparable from nutritive soul (*de An.* II,3, 415a1–2; cf. *PA* II,1, 647a25–27; *Somn. Vig.* 455b34–456a6).
38 *GA* I,2 says what it is to be male is to possess the principle of [substantial] change and generation (716a5–6; cf. IV,1, 765b14–15). *GA* II,4 defines the male as the "maker" (τὸ δημιουργοῦν) (738b21). We must also note that plants contain the male principle, and none of them are sentient, so maleness simply cannot be equivalent to the contribution of this sort of soul (I,23, 731a24–33; II,1, 732a12–14).

embryo but it will at first most strongly display growth, a function of the nutritive soul. Although the nutritive soul of an animal immediately begins to develop sense organs from a sensory core of the heart, it does not start off as sentient. The sense organs are at first dormant; and the embryo is mostly as-if asleep (*GA* V,1, 778b23), without any impressions that could form dreams (*de Insomniis* 3, 461a12–13). The most explicit articulation of the stage where nutritive soul is activated but sentient soul dormant come when Aristotle discusses the phenomenon of sleep. In the *GA* V puzzle concerning whether sleep or waking come first; Aristotle decided that sleep does, partly based on his embryology which specifies that embryos are in a sleep-like state.

> Because of the phenomena that as time goes on they become more and more awake, it is reasonable that the opposite, sleep, is the case in the beginning of their generation. Furthermore, the change from not-being to being comes to be through the intermediate and sleep seems to be naturally something like that, being like a borderline between living and not living, and the sleeper neither exists completely not does not exist. Life exists most of all in waking due to perception. If an animal has to have perception and it is first an animal when it first gets perception, then it must be thought that at the start of its constitution it is not asleep but in a sleep-like state, similar to what plants have. For it happens that at this time animals live the lifestyle (βίος) of a plant. (*GA* V,1, 778b25– 779a2)[39]

The animal embryo is taken to live a sort of plant-like life, devoid of sensation, pain or pleasure.[40] This similarity to plant life is further reinforced by the way in which the foetus feeds. The nutritive soul is first activated when the animal embryo when it begins to draw nourishment to itself (*GA* II,3, 736b10–11). This requires the sending out of a root and a shoot:

> Once the fetation is set, it acts almost like sown plant seeds. The principle also in those seeds is the first thing. And when this is distinguished, being potentially present in it earlier, a shoot and root are sent out from [this first actualised part]. From this nourishment is obtained. For the plant must grow. Likewise in the fetation, in the same way, all the parts are in it potentially, the principle has made the most progress. For this reason the heart is the first part to be separated in actuality. (*GA* II,4, 739b34–740a4)

Aristotle proceeds to explain that the embryo at this stage, because it is only potentially an animal, has to get its nourishment from elsewhere, and is again like a plant. Its initial lack of any stomach, liver, spleen or intestines which are not

---

[39] As also does the embryo inside the egg, *GA* III,2, 753b26–27.
[40] At *GA* II,3 Aristotle explains that "fetations" (τὰ κυήματα, the initial fertilised embryo of animals) are alive insofar as they "possess nutritive soul" (736a35–36).

among the first parts to be formed in the embryo, means that it cannot digest its own external food.[41] Furthermore, since it cannot locomote (740a25–7), it cannot get to its food; instead it has to send out a shoot to bring food to itself. The preparation of external food must still be undertaken by the mother's organs. The fact that the embryo's heart is operative first of all is where the comparison to the plant fails – for no plant has a heart or its analogue. The plant only draws in simple nourishment and grows in all directions (*de An.* II,2, 413a28–30). The heart is the centre of the animals' nutritive soul, and so maintains all the sentient parts of the body, as well as becoming the centre of sentient operations in due course. Plants never have this integrity.[42]

Although animals have a source of sensation from the very outset, the point of the continual comparison to plants is to emphasise that the embryo is not really properly alive yet, because in order to live, it must live the life of an animal, which is sentient.[43] From the female spermatic contribution, the next stage is the actualisation of the nutritive soul and the gradual development of the sentient body parts which can eventually actualise the passive potentials to sense (Connell 2016, p. 172–177). It is important, however, not to make too much of the nutritive/sensory division in embryological development. Although the fetation while it is developing the parts of its body, including the sense organs, will not sense, it is not nutritive only. Just as a child begins its life behaving like a non-human animal, but is not a beast, because it will become fully human (*Ethica Nicomachea* I,10, 1100a3), so also, the fetations of animals are not literally plants.

5 *Sperma* Production, Nutritive Soul and the Failure of the Female

Aristotle distinguishes the nutritive faculty from other soul functions in *de An.* and the *Parva Naturalia*. One convincing interpretation is that the nutritive soul is distinct for two reasons: (1) it can exist independently of the other parts and (2) there is an account of it that does not depend on reference to any other aspect of soul (Corcilius/Gregorić 2010; *de An.* I,5, II,2–4, III,9–10;

---

[41] Aristotle explains that plants use the earth as their external stomach, i.e. their food is precooked for them.
[42] The principle of soul in plants is diffuse which explains how cuttings can grow separately (see *de An.* I,5, 411b19–20).
[43] See the interesting analysis along these lines by Carraro (2017).

Somn. Vig. 1, 454a11–19). (1) is simply the point that some entities exist which only have this aspect or faculty of soul, i.e. plants (*de An.* II,2, 413a31–2; *GA* II,4, 1–3). For plants, nutritive soul is not a soul part but their entire soul (*de An.* I,5, 411b27–30; Corcilius/Gregorić 2010, p. 92). (2) depends on the idea that nutritive soul is what "maintains its possessor as such" (*GA* II,4; *de An.* II,4, 416b17–19; Corcilius/Gregorić 2010, p. 109). In this manner, the distinctness of nutritive soul from the other soul functions is secured. There is, however, a further question about its unity. The nutritive and generative soul are said to be one and the same (*de An.* II,4, 415a23, 416a19).[44] Thus, self-preservation is, in some sense, the same as the preservation or perpetuation of the kind. Scholars who have considered the nutri-generative soul sometimes attempt to explain how these two functions are really the same.[45] Thus, Corcilius and Gregorić write: "the object of the reproductive capacity is really the same substantial form that the nutritive capacity maintains for the individual living being by means of taking in and processing food. What the reproductive capacity does is perpetuate this form in another individual" (Corcilius/Gregorić 2010, p. 112–113). Here the focus is on the conceptual point and not the ontological one and female animals are not considered.

The fact that female animals do not generate on their own appears to show that there are certain (regularly occurring) animals that exists with nutritive but not generative functions.[46] On the above criteria, the two functions, far from being unified, would count as ontologically distinct. If it is true that female animals do not have a generative function, then it would seem that nutritive minus generative is ontological separable from nutritive plus generative. This ontological divide, then, destabilises the unity of this fundamental soul capacity.[47]

In order to mitigate this result, it is important to first get clear on the conceptual unity. This requires considering the structure of soul faculties. Aristotle explains that the different parts of soul have different objects, so, for example the sensory soul has sense objects that concern it.[48] The idea seems to be that there is one function if there is one type of object; perception and intellect have objects

---

[44] See also *de An.* II,3, 415a2–3; II,4, 415a27, 416b13–15, 26; *GA* II,1, 735a16–19.
[45] See especially Lennox in this Volume.
[46] This point also applies to those animals that are sterile, such as mules (*GA* II,8) and certain spontaneously generated kinds.
[47] On fundamental importance of nutritive soul see especially King (2001), p. 81, and Johansen (2012), p. 118.
[48] In fact, of course, it is more complicated than this since there are five varieties of sensory object and there is also the common sense and the capacity to store images (φαντάσματα). See Gregorić (2007).

that differ from the object of nutrition/generation – which is "food". For the nutritive soul, there are further complications concerning its object. First of all, food is two-fold, the external food that the animal consumes and the internal food that sustains its substantial being. The internal food, blood or its analogue, is what sustains the functioning body and its parts. This is made by the nutritive soul to be able to do maintain an animal of the particular sort that it is. Next, Aristotle specifies two objects of the nutri-generative faculty, "food" and "generation" (περὶ τροφῆς καὶ γεννήσεως, *de An.* II,4, 415a23). This may still yield conceptual unity as both are centred in the "substantial being" of the animal. This idea is elegantly elaborated in Thomas Johansen's book *The Powers of Aristotle's Soul*, who makes "the unity of nutritive soul" a problem to solve (Johansen 2012, p. 106–115). The "food" indicated in this definition is not external or raw food (which is unlike the animal) but the final nourishment which has become like it.[49] This means that "food" as the final nourishment, according to Johansen, is the same in form as the animal; "the object of nutrition is the form of the living being" (Johansen 2012, p. 109). This then explains how an animal the same in form can be generated; the form is in the residue of final nourishment which is "isomorphic with the living body whose form is the soul" (Johansen 2012, p. 102). But if this is the case, then the female *sperma* would seem not to fit. Although it is a pure portion of the final nourishment it cannot convey form to the new animal.[50] Does this constitute any reason to separate the nutritive and generative functions conceptually?[51] Let's consider more closely the connection between the nutritive soul, the final nutriment and form.

The idea of nutrition and generation as the same capacity requires thinking about the process of concoction (πέψις) (Johansen 2012, p. 106–115).[52] Through various complex stages of this process, the nutritive soul transforms the external food into internal nourishment, which is what each of the parts of the body require for its particular type of activity, often peculiar to the type of animal in question. In the *GA* Aristotle is pretty explicit about how specialised this finished nutriment is. Rejecting the idea that nutrition is brought about by a like-to-like action, he states: "instead it is [the fact] that the female residue is such as to be

---

[49] To the question of whether food is like or unlike the body, Aristotle answers that it is both. Before it is processed, it is unlike and after it is like (*de An.* II,4).
[50] The female, we are told, does not contribute form, e.g. *GA* I,2, 716a8; I,20, 729a11, 729a33; II,1, 732a10.
[51] Corcilius/Gregorić (2010) and Johansen (2012) who do not think that such a division is necessary but fail to discuss the complication produced when considering the female animal's nutritive soul.
[52] See also Lloyd (1996), ch. 4.

naturally the same as the animal, and the parts are in it potential, but none in actuality" (*GA* II,4, 740b19–21). The uniform and non-uniform parts develop simultaneously from this specialised material – and although Aristotle insists here that no parts exist "in actuality" (ἐνεργείᾳ) there is a way in which this final nourishment is an agent with a (latent) form. Food at this stage is like the animal, not in the way that like dead body parts are, since it is not possible for there to be any animal parts without soul in them (*GA* II,5, 741a10–12). As Aristotle explains in his *de Generatione et Corruptione*, food is not mere bulk but something that actively maintains the form of the animal (*GC* I,5, 322a20–3; Johansen 2012, p. 110). And by the form of a particular part like "flesh" Aristotle means living, sentient flesh (*GC* I,5, 322a10–13). Thus, the nourishment is similar to a living part. Furthermore, the process of concoction itself, something that happens only in living beings through their nutritive capacity, aims to produce the being of the kind, its form:

> [W]hen concoction has taken place we say that a thing has been perfected and has come to be itself [...] In some cases of concoction the end of the process is the nature of the thing – nature, that is, in the sense of form and the essence. (*Mete.* IV,2, 379b18–26; translation after Johansen)

This final nourishment appears to have agency insofar as it is the instrument of the nutritive soul. As Johansen notes, nutrition is a prime instance of the soul's self motion. The nutritive soul is not changed, but changes something external, i.e. food. It employs this food as an instrument to maintain its own being. As with any other instruments, this concocted food is both changed and changing. The final nourishment is capable of producing growth (*de An.* II,4, 416b13) and of producing generation (416b15). This description of the final nourishment certainly fits with the male's contribution to generation, which Aristotle likened to tools used to shape craft products (*GA* II,4, 740b26–30).[53] The equivalence between the action of the nutritive soul and the action of its instrument is clear when Aristotle remarks of the male contribution which is, of course, the final concoction of nourishment: "we specify either the semen or that from which the semen comes, since that which has the change in itself is no different from the one changed it" (*GA* II,1, 734b7–9). The problem is that the female contribution is also a further concocted portion of the residue of final nourishment. Her blood maintains the being of her body and is potentially all the living parts of another animal like herself (*GA* I,19, 726b16; III,9, 762b3–4). Since the male

---

[53] After all, "*sperma* is a residue of nourishment undergoing change" (*GA* II,3, 736b27–8; cf. Johansen 2012, p. 132–135).

and female of the kind have the same nutritive soul, why then can't the female perform this same magic and make another like itself? Aristotle is adamant that it cannot:

> Being a residue *sperma* is being changed by the same change as they by which the body grows through distribution of the final nourishment, when it enters the uterus it sets and changes the residue of the female by the same change which it itself happens to be changed by. For [the female matter] is a residue, and it has all the parts in it potentially, none in actuality. For it has those parts potentially which differentiate male from female. Just as from deformed parents sometimes deformed offspring are generated and something they are not, thus from the female sometimes a female is generated and sometimes not [a female] but a male. For the female is like a deformed male, and the menstrual fluid is *sperma*, but impure. For it does not have only one thing, the principle/start of soul. (GA II,3, 737a18–30)

The female is said to lack "the principle of soul". All animals, through the nutritive faculty, strive to be eternal in the way open to them, i.e. to generate another like in kind to themselves (*de An.* II,4, 415a22–415b7). Can it really be that the female lacks this drive to generate?[54] If the nutritive and generative functions are completely indistinguishable then the female would end up lacking the nutritive function which is nonsensical; female animals would not be alive.

One possible solution is to take the generative to be a sub-part of the nutritive soul. A soul part is conceptually separable if the account of it does not require reference to any other parts. When considering the generative function, it cannot be conceptually separated from the nutritive one, but is dependent on it.[55] The nutritive one, though, arguably does not entail the generative one. An animal can maintain its own being without producing another being like in kind. This is not, however, a conceptual distinction that threatens the unity of this soul part. Instead, it helps to clarify how they are "the same". They are not identical; rather the generative function is a sub-part of the nutritive, contained within it. On this reading, the female lacks this sub-part, which would help to explain why Aristotle describes her as "in a way maimed" (*GA* IV,2, 767a27–28; IV,6, 775a15–16). However, this option is unattractive in some ways. It does not explain how it is possible to prise apart the generative from the nutritive function given that both strive to maintain the form of the kind. Another possibility is to view the generative products of female animals as just that

---

[54] Generation is the 'most natural' function (*de An.* II,4, 415a27).
[55] As Corcilius/Gregorić note, "the account of the reproductive capacity does make reference, if only implicitly, to the nutritive capacity of the soul, so that we should not count it as a [conceptually separable] part of soul" (Corcilius/Gregorić 2010, p. 113).

– generative. And so, the nutritive soul of the female animal also strives to maintain the being of the kind into the next generation.

6 Conclusions

The object for the nutritive soul is not only food but also "generation" (*de An.* II,4, 415a23), meaning that nourishing is in essence the same thing as generating. Let's take as the example of a mature dolphin. In this case, what is nourished is this dolphin; nutritive soul aims to sustain a dolphin. What is generated is another dolphin; the same soul aims to sustain dolphinness (by generating another dolphin). A male animal, that is not defective, achieves the second related aim by providing the source of substantial generation. The female animal cannot do that. But this need not create an unbridgeable divide between the nutritive functions and aims and generative ones?[56] The female's nutritive soul does aim toward the generative goal – and generation couldn't happen if it did not. It must also have the generative variety of nutritive soul. The female achieves the generative aim differently from the male, by providing the materials that are ready to become all the parts of the body of an animal the same in form to her. As detailed in the previous section, internal nourishment is dynamic and aiming for the form of the kind. Just as the male's does, the female's generative residue is moving towards a new substance. This is very much apparent in the wind egg, produced by the female alone, which has already been undergoing a change toward the form. Because it rots, when the wind egg is not yet rotten, there is something striving toward that living state which is not present in the rotten egg – an in-between position.[57] And this must be what the female *sperma* is also like – given that it is in a ready state, a portion of the ultimate nourishment, poised to become all the parts of the functioning body. It is only by a conjunction of male and female that any new animal comes into existence and so, the crucial work undertaken by the female animal's nutritive soul ensures the existence of a new animal the same in kind as herself. The menses, then, are generative and constitute the female's attempt to generate another living being like itself.[58]

---

[56] Neither male nor female residue are "without soul" (*GA* II,3, 736a32–35).
[57] Like all useful residues of nutrition, it is both changed and changing. See n. 53 above.
[58] The female contribution also contains δυνάμεις which strive to generate a female animal resembling the female animal and her family (see *GA* IV,3, 767b33–768a9, 768a10–21).

Bibliography

Althoff, Jochen (1997): "Aristoteles' Vorstellung von der Ernährung der Lebewesen". In Wolfgang Kullmann/Sabine Föllinger (Eds.): *Aristotelische Biologie. Intentionen, Methoden, Ergebnisse*. Stuttgart: Franz Steiner, pp. 351–364.

Balme, David (1972 [1992]): *Aristotle De Partibus Animalium I and De Generatione Animalium I*. Oxford: Clarendon.

Balme, David (ed./transl.) (1991): *Aristotle. The History of Animals Books VII–X*. Cambridge, MA.: Harvard University Press.

Carraro, Nicola (2017): "Aristotle's Embryology and Ackrill's Problem". In: *Phronesis* 62. No. 3, pp. 274–304.

Coles, Andrew (1995): "Biomedical Models of Reproduction in the Fifth Century BC and Aristotle's *Generation of Animals*". In: *Phronesis* 60, pp. 48–88.

Connell, Sophia (2016): *Aristotle on Female Animals: A Study of the Generation of Animals*. Cambridge: Cambridge University Press.

Corcilius, Klaus/Gregorić, Pavel (2010): "Separability versus Difference: Parts and Capacities of the Soul in Aristotle". In: *Oxford Studies in Ancient Philosophy* 39, pp. 81–119.

Deslauriers, Marguerite (1998): "Sex and Essence in Aristotle's Metaphysics and Biology". In: Cynthia Freeland (Ed.): *Feminist Interpretations of Aristotle*. University Park, PA.: Pennsylvania State University Press, pp. 138–167.

Drossaart Lulofs, Hendrik J. (ed.) (1965): *Aristotle. De Generatione Animalium*. Oxford: Oxford University Press.

Gregorić, Pavel (2007): *Aristotle on Common Sense*. Oxford: Oxford University Press.

Henry, Devin (2006): "Understanding Aristotle's Reproductive Hylomorphism". In: *Apeiron* 39, p. 257–288.

Hett, Walter S. (ed./transl.) (1936): *Aristotle. On the Soul, Parva Naturalia, On Breath*. Cambridge, MA.: Harvard University Press.

Johansen, Thomas (2012): *Aristotle on the Powers of Soul*. Oxford: Oxford University Press.

King, Richard (2001): *Aristotle on Life and Death*. London: Duckworth.

Lennox, James G. (2001): *Aristotle on the Parts of Animals I–IV*. Oxford: Clarendon Press.

Leunissen, Mariska (2010): *Explanation and Teleology in Aristotle's Science of Nature*. Cambridge: Cambridge University Press.

Leunissen, Mariska (2017) *From Natural Character to Moral Virtue in Aristotle*. Oxford: Oxford University Press.

Louguet, Claire (2015): "Aristote et les théories pangénétiques du Ve siècle : enjeux métaphysiques d'un débat biologique". In: Victor Gysembergh/Andreas Schwab (Eds.): *Le Travail du Savoir / Wissensbewältigung – Philosophie, Sciences Exactes et Sciences Appliquées dans l'Antiquité*. Trier: WVT Wissenschaftlicher Verlag.

Louguet, Claire (2017) "La Generation des Animaux: Aristote et le Corpus Hippocratique". Colloque sur la vie chez Aristote, David Lefebvre, Université Clermont Auvergne, Clermont-Ferrand.

Mayhew, Robert (2004): *The Female in Aristotle's Biology: Reason or Rationalization*. Chicago: University of Chicago Press.

Peck, Arthur L. (ed./transl.) (1937): *Aristotle. Parts of Animals, Progression of Animals*. Cambridge, MA.: Harvard University Press.

Peck, Arthur L. (ed./transl.) (1942): *Aristotle. The Generation of Animals*. Cambridge, MA.: Harvard University Press.
Peck, Arthur L. (transl.) (1965): *Aristotle. Historia Animalium*. Cambridge, MA.: Harvard University Press.

Andrea Libero Carbone
Why do not Animals Grow on Without End?
Aristotle on Nutrition and Form

Abstract: In a dense passage of *GA* II,6, 745a4–10 Aristotle tackles the question of why, even though animals keep on nourishing themselves, they do not grow on without end. As childlike as this query may seem, the answer given in the passage is admittedly partial. Further, it requires Aristotle to take into account a rather complex network of topics, whose detailed study is, moreover, announced as forthcoming in his lost (or never written) writings on nutrition. These topics include, on the one hand, a fine-grained distinction between different parts and uses of nutriment and residues, and, on the other hand, an analysis of the relationships between growth and form, shape and size, essence and limit, outline and structure. In order to reconstruct the theoretical framework of what may have been Aristotle's fuller answer, then, we shall explore a number of passages of his psychological and biological works.

In a dense passage of *de Generatione Animalium* II,6 Aristotle tackles the question of why, even though animals keep on nourishing themselves, they do not grow on without end. The answer he gives there is admittedly partial and raises at least two perplexing difficulties. One pertains to the nature of the relationship established between the growth of bones and the increase of the animal's size. The other pertains to the criterium used for fixing the limits of size for animals. Further, it requires him to make reference to a rather complex set of topics, whose detailed study is announced as forthcoming in his lost (or never written) work on nutrition. These topics include the distinction between two kinds of nutriments, a fine-grained inquiry on how these nutriments and their residues are used by different parts at distinct stages of the animal's development and growth, and an analysis of the relationships between growth and form, shape and size, essence and limit, outline and structure. In order to reconstruct the theoretical framework of what may have been Aristotle's fuller answer, I shall explore a number of passages of his psychological and biological works where these topics are discussed.

In *GA* II,6 the question of why, even though animals keep on nourishing themselves, they do not grow indefinitely, is tackled as follows:

The bones, then, are made in the first conformation of the parts from the seminal residue. As the animal grows the bones also grow from the natural nourishment, being the same as that of the sovereign parts, but of this they only take up the superfluous residues. For everywhere the nutriment may be divided into two kinds, the first and the second; the former is nutritious (θρεπτικόν), being that which brings into being both the whole and the parts; the latter is concerned with growth (αὐξητικόν), being that which causes quantitative increase. But these must be distinguished more fully later on. The sinews are formed in the same way as the bones and out of the same materials, the seminal and nutritious residue. Nails, hair, hoofs, horns, beaks, the spurs of cocks, and any other similar parts, are on the contrary formed from the nutriment which is taken later and only concerned with growth, in other words that which is derived from the mother, or from the outer world after birth (ἐκ τῆς ἐπικτήτου τροφῆς καὶ τῆς αὐξητικῆς, ἥν τε παρὰ τοῦ θήλεος ἐπικτᾶται καὶ [τῆς] θύραθεν). For this reason the bones on the one hand only grow up to a certain point (καὶ διὰ τοῦτο τὰ μὲν ὀστᾶ μέχρι τινὸς λαμβάνει τὴν αὔξησιν) (for there is a limit of size in all animals (ἔστι γάρ τι πᾶσι τοῖς ζῴοις πέρας τοῦ μεγέθους), and therefore also of the growth of the bones) (διὸ καὶ τῆς τῶν ὀστῶν αὐξήσεως); if these had been always able to grow, all animals that have bone or its analogue would grow as long as they lived (εἰ γὰρ ταῦτ' εἶχεν αὔξησιν ἀεὶ καὶ τῶν ζῴων ὅσα ἔχει ὀστοῦν ἢ τὸ ἀνάλογον ηὔξανετ' ἂν ἕως ἔζη), for these set the limit of size to animals (τοῦ γὰρ μεγέθους ὅρος ἐστὶ ταῦτα τοῖς ζῴοις). What is the reason of their not always increasing in size must be stated later. (GA II,6, 744b27–745a10, transl. Platt)

This passage can be easily divided into two parts, both of which contain a closing reference to a further inquiry. The first part is focused on the distinction between two kinds of nutriments, while the second part deals with the development and growth of different parts of animals.

1 Two Kinds of Nutriments

Let us begin with examining the first part of the *GA* II,6 passage, where Aristotle distinguishes two different kinds of nutriments, the "nutritive" one (θρεπτικόν) and the "growth-promoting" one (αὐξητικόν), and stresses that the contribution they provide to the animal's constitution is different. While a further development of this topic is announced as forthcoming, such a distinction is not found elsewhere in the *Corpus*. It is useful to compare a passage of *de Anima* II,4 where Aristotle claims that

> [B]eing food and being capable of producing growth are different; for it is in so far as the ensouled thing is something having quantity that food is capable of producing growth, but it is in so far as it is a particular and a substance that it is food. For the ensouled thing maintains its substance and exists as long as it is fed; and it can bring about the generation not of that which is fed, but of something like it; for its substance is already in existence, and nothing generates itself, but rather maintains itself. Hence this first principle of the

soul is a potentiality as to maintain its possessor as such, while food prepares it for activity; for this reason, if deprived of food it cannot exist. (*de An.* II,4, 416b11–20; transl. Hamlyn)

No mention is made here of the two different kinds of nutriments. Rather, Aristotle stresses that food is substance and that it allows a living thing to maintain its substance and to reproduce into a new substance like itself. On the other hand, the reason why food is capable of producing growth is because the living thing has *quantity*, which means that it has a body whose extension can increase. More than that, when, elsewhere in the *Corpus*, they are alluded to together, nutrition and growth are pointed out as being activities of the same part of the soul.[1] For this reason, we should expect to find that Aristotle's point about the limit of animal's growth in *GA* II,6 rests importantly on the distinction between the two kinds of nutriments, as both topics are introduced in these terms there for the first and unique time.

In the immediately preceding passage, in *GA* II,6, 744b16–27, Aristotle introduces the metaphor of nature as a "good housekeeper" in order to illustrate how the earlier developing parts are formed out of the first nutriment.[2] A wise housekeeper gives the best food to the free men, the inferior food and the residues to the slaves, and the worst food to the domestic animals. Likewise, as Aristotle claims, the purest part of the first nutriment is used by nature for constructing flesh and the sense organs, while the residues are used for forming bones, sinews, hair, nails, hoofs and all similar parts.

We should highlight that the terms of the comparison don't match up perfectly, since while, on the one hand, the wise housekeeper distributes three different qualities of food, on the other hand, in his first outline, Aristotle sketches a simpler distinction between the purest food and the residues, so that no difference is drawn between the bones and the other earthy parts like sinews, nails, hoofs, horns and so on. A somewhat similar approach is found earlier in *GA* II,6, 743a11–26 where Aristotle first describes the formation of the hard and earthy parts: nails, horns, hoofs and other similar *external* uniform parts are formed by an excess of earthy stuff when the nutritive fluid *evaporates* (ἐξατμίζοντος τοῦ ὑγροῦ, 743a13), while bones and sinews, which are *internal* uniform earthy parts, are formed as the fluid *solidifies* (ξηραινομένης τῆς ὑγρότητος, 743a18) like earthenware pottery baked in an oven. Once again, no mention is

---

[1] See *de An.* III,12, 434a24–25; *de Juventute et Senectute, de Vita et Morte, de Respiratione* 4, 469a26–27; *GA* II,1, 735a16–21; II,4, 740b29–741a2; *Ethica Eudemia* II,3, 1219b38–39; *Ethica Nicomachea* I,6, 1097b34–1098a4.

[2] For a detailed analysis of this image see Leunissen (2010), p. 81–83, and Connell in this volume.

made here of a material qualitative distinction between two kinds of nutriments, while the contrast between the bones or the sinews and the other earthy parts is drawn in terms of different effects of the action of heath on the nutritive fluid.

In a sense, then, the further distinction between two kinds of nutriments appears to be the result of a progressive refinement towards a sounder explanation. Indeed, the statement according to which bones and nails or other like parts are made out of the same material, is vulnerable to the objection that such parts grow on indefinitely, while bones do not. Aristotle's move, hence, consists in pointing out that nails and other similar solid earthy parts are formed out of a different kind of nutritive matter, i.e. the growth-promoting one, which is "taken later" (ἐπικτήτου). It is provided to the embryo by the female parent, and, after birth, by the external food. Hence the parts which are formed out of the first nutriment, i.e. the fleshy parts made out of the purest matter and bones and sinews made out of the residues, grow towards a definite limit, while the other earthy uniform parts grow indefinitely, because they are made out of the residues of the growth-promoting nutriment. From our passage of de An. II,4, 416b11–20 we can also derive that, after birth, the better part of nutriment, i.e. external food, is used by animals toward the goal of maintaining themselves as substances, which means staying alive. Hence, all parts other than those made out of the residues of the growth-promoting nutriment stop growing at some point. But why? When do they stop and what makes them do so? What fixes this limit?[3]

2 Setting a Limit: the Bones' Growth

Let us turn to the second half of our passage of GA II,6, where these questions are partially addressed. At least at a first glance, Aristotle's argument looks like being somewhat circular. "For this reason", as he maintains, "the bones on the one hand only grow up to a certain point" (καὶ διὰ τοῦτο τὰ μὲν ὀστᾶ μέχρι τινὸς λαμβάνει τὴν αὔξησιν, 745a4–5). I take the reference of "for this reason" to be the fact that bones are formed out of the residue of the nutritive kind of nutriment, whose amount is limited. Evidence of this may be seen in the fact that the correlative δέ (745a11) to the τὰ μέν introduces a complementary explanation showing how hair and like parts keep growing on as long as the residue of

---

[3] This and other questions remained unanswered in Allan Gotthelf's discussion of our passage (Gotthelf 2012b, p. 113–114). Gotthelf has pointed out a number of important reasons why Aristotle's point should not be regarded as reductionist, but it seems to me that he doesn't really offer a positive alternative reading of this passage.

the external nutriment is available. Hence, what Aristotle provides here is an explanation in terms of material causation of the limits of the bones' growth and, consequently, of the fixation of the animal's size. This is explicitly stated in lines 745a8–9: "these set the limit of size to animals" (τοῦ γὰρ μεγέθους ὅρος ἐστὶ ταῦτα τοῖς ζῴοις). As he himself says in the preceding lines, though, "there is a limit of size in all animals" (ἔστι γάρ τι πᾶσι τοῖς ζῴοις πέρας τοῦ μεγέθους), "and therefore also of the growth of the bones" (διὸ καὶ τῆς τῶν ὀστῶν αὐξήσεως), and as he tells us in his concluding line, "what is the reason of their not always increasing in size must be stated later" (δι' ἣν μὲν οὖν αἰτίαν οὐκ ἀεὶ λαμβάνουσιν αὔξησιν λεκτέον ὕστερον). To sum up, then, the maximum size to which the bones can grow depends on the limit of the animal's size, which is in turn fixated by the bones. As this cannot possibly make sense, we should invoke the principle of charity and grant that Aristotle refers to a further explanation of the maximum size of both the bones and the whole animal. I will expand on this later.

For now, let us focus on why precisely the bones are said to set the limit of size to animals. As we have seen previously, in *GA* II,6, 743a18 bones are said to be formed like earthenware pottery baked in an oven. In *de Partibus Animalium* II,9, 654b29–32, though, Aristotle provides a slightly different metaphor, since he says that the fleshy parts are "moulded around" (περιπλάττουσι) the bones in the same way as artists model an animal out of clay "after having set up some hard body for support" (ὑφιστᾶσι τῶν στερεῶν τι σωμάτων). Now, of these two descriptions of the animal's development, the first one illustrates that the bones form out of the residue of the nutriment used for generating the fleshy parts, and that the bone's size is fixated by the general amount of available bodily matter. On the other hand, the second metaphor is consistent with the idea that the bones should be present as a necessary condition for the sake of the further development and growth of the fleshy parts, which cannot grow more than it is made possible by the support of the hard body of the bones. These descriptions can be consistently related to two different stages of the animal's development, namely respectively to the first formation of the embryo, and to the further growth of the body before and after birth. Remarkably, the image which is consistent with attributing a limiting role to the bones illustrates a teleological explanation of their function.

What does happen, then, when the bones stop growing? A passage of *Historia Animalium* III,18 may provide us with a clue to better understanding Aristotle' point:

> all animals get fatter when older than younger, and especially when, having attained their size in width and length, they grow also in depth. (Πιαίνεται δὲ πάντα πρεσβύτερα μᾶλλον

ἢ νεώτερα ὄντα, μάλιστα δ' ὅταν καὶ τὸ πλάτος καὶ τὸ μῆκος ἔχῃ τοῦ μεγέθους καὶ εἰς βάθος αὐξάνηται, *HA* III,18, 520b7–9, transl. mine)

We deal here with the "dark side" of nutrition, namely putting on belly fat. As it turns out, animals can actually keep growing after that youth has ended and the whole body has attained its size. Yet such increase is only in the depth-wise direction. This means that the size of the animal, whatever it is that fixes it, is related to the top/bottom and to the right/left axes of symmetry.[4]

We may now attempt an answer to the question why the role of fixing the size of the body is attributed to the bones. In fact, from the point of view of material-efficient causation, the natural heat, whose principle is in the region of the heart and of the lung, "determines growth from the middle along its proper movement" (ποιεῖ τὴν αὔξησιν ἀπὸ τοῦ μέσου κατὰ τὴν αὐτῆς φοράν, *PA* II,7, 653a31–32), which means along the top/bottom axis toward the top. Now, the backbone, which is the principle of all bones (*PA* II,9, 654a33), is defined as "what maintains the length and straightness of animals" (Ἡ γὰρ τὸ μῆκος καὶ τὴν ὀρθότητα συνέχουσα τῶν ζῴων ἡ ῥάχις ἐστίν, *PA* II,9, 654b13). And this is the reason why the material cause of the limit of an animal's size, whose growth is along the top/bottom axis, is fixed by the size of the backbone or of the analogous parts, because these principles of the bones or of the analogous supports maintain the length and straightness of the animal body along the top/bottom axis.

So much then for the explanation in terms of material necessity and for the role of the bones. What could be the further explanation of the maximum size of both the bones and the whole animal which is announced in our *GA* II,6 passage?

3 A Teleological Explanation of the Growth Limit

I can think of two possible consistently Aristotelian teleological explanations of why animals stop growing when they reach a certain size. The first one would be that animals grow up until they are able to reproduce themselves: this would imply that reproduction should be considered as that for the sake of which animals grow adult. Such a hypothesis would seem to be supported by the passages of *de An.* II,4, 415a22–415b7 and *GA* II,1, 731b24–732a1 where Aristotle claims that animals, which are not eternal as individuals, partake of eternity as

---

[4] On this topic see Carbone (2016).

kinds, thanks to reproduction. Yet Aristotle doesn't make explicit that reproduction is the aim and the finishing point of growth.[5] He only maintains that reproduction is the most natural function of completely developed living things (φυσικώτατον γὰρ τῶν ἔργων τοῖς ζῴοσιν, ὅσα τέλεια, *de An.* II,4, 415a26–27). As a matter of fact, however, a number of living things tend to reproduce themselves repeatedly over their lifetime and at different stages of growth. For this reason, the hypothesis of a teleological explanation of why animals stop growing based on reproduction would seem to be untenable.

A second possible teleological explanation could be that growth stops when the point is reached where there is the best balance between hot and cold for the exercise of the soul's functions. Such a hypothesis would seem to find support in the definition of youth as the "growth of the first cooling part" and of prime as the balance between the growth and decay of this growth (νεότης δ' ἐστὶν ἡ τοῦ πρώτου καταψυκτικοῦ μορίου αὔξησις, γῆρας δ' ἡ τούτου φθίσις, ἀκμὴ δὲ τὸ τούτων μέσον) in *Juv.* 24(18), 479a32–33. As Aristotle puts it, animals grow in size until they reach their prime, this last being actually defined as a goal (for the sake of the better), i.e. the stage when a point of equilibrium is attained between the growth-promoting force of heat and the balancing action of the cooling processes, thus providing the best conditions for performing the animal's activities and functions.[6] Hence this second hypothesis about a possible teleological explanation of the growth limit is a reasonable one. Yet, as we learn from Aristotle's treatment of growth in *de Generatione et Corruptione* and, more specifically, of biological growth in *GA* IV,3, "form grows proportionately and remains symmetrical" (ὥστ' ἀνάλογον αὔξειν καὶ διαμένειν ὁμοίαν τὴν μορφήν, 768b31–32). This means that balance is attained and preserved at each and every stage of growth, thus setting the limit of the relative size of parts and maintaining the body's symmetry. Balance is knowledgeably a matter of proportion, of relative quantity, and not of absolute size. As a consequence, this explanation still leaves somewhat open the question of how to define the limit of size increase of the whole body, for even if we assume that growth comes to an end when the cooling activity is enough to balance the growth promoting action of heat, one can still ask how the original amount of heat is defined.

---

5 See Rashed (2005), p. cvi, n. 1.
6 On this passage see King (2001), p. 134 f.

4 Size and Essence

Let us take up the question of the limit of growth from another point of view. As we have seen, in our *GA* II,6 passage Aristotle maintains that "there is a limit of size in all animals". Such a universal statement should incline us to think that what he announces at the end of the passage is a formal explanation.[7] In other words, what Aristotle seems to mean there, is that the proper size of an animal is inscribed in its essence. As I will try to show, such an explanation is primarily focused on the animal's bodily structure, as an individual and as a kind. It pertains specifically to Aristotle's biological *morphology*. As we know from *de An.* II,4,

> [W]hat is it that holds together the fire and the earth, given that they tend in opposite directions? For they will be torn apart, unless there is something to prevent them; but if there is, then this is the soul and the cause of growth and nourishment. [...] [I]n all things which are naturally constituted there is a limit and a proportion both for size and for growth; and these belong to soul, but not to fire, and to principle rather than to matter. (τῶν δὲ φύσει συνισταμένων πάντων ἔστι πέρας καὶ λόγος μεγέθους τε καὶ αὐξήσεως· ταῦτα δὲ ψυχῆς, ἀλλ' οὐ πυρός, καὶ λόγου μᾶλλον ἢ ὕλης, *de An.* II,4, 416a6–18; transl. Hamlyn)

This clearly means that the fixing of a size limit for animals pertains to the agency of soul as form. The main direction along which the defining agency of the soul is exerted corresponds to the opposite movements of earth and fire, namely to the direction of growth along the top/bottom axis. It seems to me likely that such explanation may be found summarized in a passage of Aristotle's extensive analysis of variability in animal offspring in *GA* IV, and more specifically in *GA* IV,4, where he focuses on the relationship between size and number of offspring:

> Just as for each of the completely developed animals there is a certain size, i.e. a limit of bigger and smaller (τι μέγεθος καὶ ἐπὶ τὸ μεῖζον καὶ ἐπὶ τὸ ἔλαττον), so that none could be generated either bigger or smaller, but it is within this interval of size (ἐν τῷ μεταξὺ διαστήματι τοῦ μεγέθους) that they have an excess or a deficiency compared to one another, and thus it is that one man is born bigger or smaller than another, or any other animal; so, also the seminal matter from which the embryo is generated is not unlimited towards both bigger and smaller (ἐξ ἧς γίγνεται ὕλης σπερματικῆς οὐκ ἔστιν ἀόριστος), so that an embryo cannot be generated from any amount whatever of it. (*GA* IV,4, 771b33–772a4; transl. mine)

---

[7] For a similar conclusion, yet based on a different reading of the text, see Gotthelf (2012b), p. 114–115.

What is highlighted here is that both the size to which an animal can grow, and the amount of matter from which an embryo is generated, must be definite and must lie within a specified interval.[8] Further, this passage clearly points out that the two facts are comparable, thus seemingly suggesting that their explanation is the same. Accordingly, as Aristotle points out in the same chapter, "the generative semen of a larger animal is necessarily very abundant, while that of smaller animals is few" (*GA* IV,4, 771a31–32), and "large animals have large embryos, proportionate to their size" (*GA* IV,4, 773b4–5). The explanation is that "both the potential which is acted upon and the heat which acts upon it are definite" (ὥρισται γὰρ ἡ δύναμις καὶ τοῦ πάσχοντος καὶ τῆς θερμότητος τῆς ποιούσης, *GA* IV,4, 772a28). On this basis we may conclude that even while the amount of available matter is a defining condition for the development of the embryo, both the minimal and maximal quantity of such matter and the interval within which the animal can grow are still defined by the animal's form.

But this still leaves us with a question. When is an animal too big (or too small, for that matter) to belong to a given kind? Let us focus on the examples of oversized and undersized animals given in *HA* VII(VIII),28, where Aristotle deals with differences among animals according to places:

> [I]n India, as Ctesias says (though he is not to be trusted) [...] the blooded animals and those that hide are all large. [...] [I]n the Red Sea the testaceans are all enormous. In Syria the sheep have tails a cubit broad, and the goats have ears a span and a palm long and in some the ears meet below towards the ground [...]. And in Egypt, whereas the other animals are larger than in Greece, such as the cattle and sheep, some are smaller, for example wolves and asses and hares and foxes and ravens and hawks, and some are about the same, for example crows and goats. The cause is said to be food, in that it is unstinted for some but scanty for others such as the wolves and the hawks, and so is the provision for carnivores since the small birds are scanty, and for hares and all that do not eat flesh because neither nuts nor fruits have a long season. In many places the climate too is a cause, for example in Illyria and Thrace and Epirus the donkeys are small, while in Scythia and the Celtic country they do not occur at all; for these animals winter badly. In Arabia the lizards are over a cubit long, and there are many mice bigger than the field mice, having forelegs a span long and hind legs as long as up to the first joint of the fingers. In Lybia the snakes develop a monstrous size, so it is said. (*HA* VII(VIII),28, 606a8–b10; transl. Balme)[9]

---

8 Cf. *Politica* VII,4, 1326a35–b2: "To the size of states there is a limit, as there is to other things, plants, animals, implements; for none of these retain their natural power when they are too large or too small, but they either wholly lose their nature, or are spoiled. For example, a ship which is only a span long will not be a ship at all, nor a ship a quarter of a mile long; yet there may be a ship of a certain size, either too large or too small, which will still be a ship, but bad for sailing" (transl. Jowett).

9 On how Aristotle deals with his sources in this passage see Li Causi (2003).

In general, in *GA* we find scattered evidences that the development of an embryo and the growth of an individual within the natural range proper to its kind can be influenced by environment, climate, food and other contingencies. Is it the case of the animals described here? Or do they belong to different kinds?

As Aristotle points out in his account of the determination of sex and of the resemblance to parents and to ancestors in *GA* IV,2–3, even though individuals tend to take after their parents and, first of all, to take after the male parent, the dynamics of the movements which are present in the seminal discharges is quite complex and leaves open the possibility of qualitative and quantitative variations due to a number of internal and external factors. Such variations can affect the size of one part of the body at different stages of its development and growth. For example, the overgrowth of one part of the body can occur in fully developed individuals as the result of a particular diet,

> as happens with athletes because they eat so much. For owing to the quantity of their food their nature is not able to master it in such a way that their form grows proportionately and remains symmetrical (ὥστ' ἀνάλογον αὔξειν καὶ διαμένειν ὁμοίαν τὴν μορφήν); therefore their limbs develop irregularly, sometimes indeed almost so much that no one of them resembles what it was before. (*GA* IV,3, 768b29–33; transl. Platt)

According to the theoretical framework that we have outlined, such metamorphoses in adults don't affect the body's structure constituted by the bones. Yet, interestingly, an oversized limb can also exceptionally be grown by an embryo. This happens when

> more matter gets constituted than is required by the nature of the part. As a result, it happens that one of the parts is larger than the others, as a finger or hand or foot or any of the other extremities or limbs. (*GA* IV,4, 772b15–18)

This is a structural variation, which means that also the bones are involved in the difference in size.[10] In both cases, however, the explanation is the same, namely that the generative movements relapse.[11] Such variations are clearly to be considered as anomalies, since the natural proportion and symmetry of the body are not preserved in the offspring.

---

[10] Cf. *Pol.* V,3, 1302b33–38: "[A]s a body is made up of many members, and every member ought to grow in proportion, that symmetry may be preserved; but loses its nature if the foot be four cubits long and the rest of the body two spans; and, should the abnormal increase be one of quality as well as of quantity, may even take the form of another animal" (trans. Jowett).
[11] See esp. *GA* IV,3, 768b16–769a6.

By contrast, oversized or undersized parts can of course be essential features of a kind, as it is the case for the birds' legs, necks and beaks (*PA* IV,12), the elephant's nose (*PA* II,16) or the arm of the octopus provided with only one row of suckers (*PA* IV,9). Here too, though, Aristotle seems to sharply distinguish what he considers an unproblematic size variability range from what is to be treated as a morphological limit-case. As is well known, according to Aristotle, within a given kind some parts of animals of different forms differ by the more and less, and all the previous examples illustrate how this type of quantitative differentiation can be associated with variations in size. However, while in a case like that of birds these differences are shared by a number of forms of the same kind and are also quite evenly distributed along the range, what stands out in cases like those of the elephant's nose, or of the single-rowed octopus arm, is that we deal with specific properties of one form of a kind. Hence, while size differences like those among birds can be plainly explained teleologically, more exceptional properties like the length of the elephant's nose require a more complex analysis of factors or cannot be given a teleological explanation, as it is the case for the tightness of the single-rowed octopus arm. As both examples turn out to be highly informative for our purposes, let us focus on them briefly, starting from the single-rowed octopus arm:

> Now while the other octopuses have two rows of suckers, one kind of octopus has a single row. This is because of the length and thinnes of their nature; for it is necessary that the narrow tentacle should have a single row of suckers. It is not, then, because it is best that they have this feature, but because it is necessary owing to the distinctive account of their substantial being. (*PA* IV,10, 685b12–15; transl. Lennox)

The impossibility of providing a teleological explanation of this difference in terms of a better structural fitness to function gives Aristotle the occasion for highlighting that size is *in itself* – i.e. morphologically and not necessarily also functionally – a definitional feature of animals.[12] The elephant's trunk is a clear example of an extraordinarily oversized part which goes far beyond the limits of recognizability:

> In the elephant, however, this part is most distinctive compared with the rest of animals – it is extraordinary in both size and potency. ('Ο δ' ἐλέφας ἰδιαίτατον ἔχει τοῦτο τὸ μόριον τῶν ἄλλων ζῴων· τό τε γὰρ μέγεθος καὶ τὴν δύναμιν ἔχει περιττή, *PA* II,16, 658b32–35; transl. Lennox)

---

12 On this passage see Gotthelf (2012a), p. 23–25; Pellegrin (2011), p. 54–55. On the analytical autonomy of Aristotle's morphology see Carbone (2016).

In *Pol.* V,9 Aristotle actually uses the example of the nose – which comes as no surprise, since it is one of his favourite examples because of the portrait of Socrates – in order to point out that when a part loses its proportions (τὴν μετριότητα) it ends up not being able to be identified as what it is (μηδὲ ῥῖνα ποιήσει φαίνεσθαι, *Pol.* V,9, 1309b23–31). That is why Aristotle emphasises the point that the elephant's trunk is actually a nose (*PA* II,16, 659a15), since despite its extraordinary size it is still able to perform its function. More than that, it is *because* of its extraordinary size that the elephant's trunk performs its function in the best possible way according to the elephant's nature. In other words, it is oversized for a purpose and for the sake of the better. Yet a further implication of considering certain parts as oversized, is that there must be a proportional size ratio between each part and the whole body, and this not only for each indivisible animal kind with relation to individuals, but also for wider kinds with relation to their sub-kinds. Once again, such a criterion is purely morphological and doesn't refer to a function.

5 Rigid Ranges

With this in mind, we should now be in a position to deal with the case of oversized or undersized animals. As we have seen, according to Aristotle, variations in body size at the individual level must remain within a certain range, which is proper to the kind. Exceptions are regarded as monstrosities, as it is the case with dwarfs, which are said to be "stunted in size" (πηροῦνται [...] τὸ μέγεθος ἐν τῇ κυήσει) during gestation (*GA* II,8, 749a5–6). Hence, even if individual variations within the allowed interval are possible, Aristotle's conception of size as a formal feature of animals implies that the size range for each animal kind must be *rigid*.

The best evidence of this is that such rigid size range is actually the keystone of Aristotle's explanation of multiparity in *GA* IV,4, 771a17–772b12. Since the amount of seminal discharge is proportional to the natural size of the animal, any excessive amount of it doesn't produce a bigger embryo, but gets divided up according to such proportion, while a lesser amount will be insufficient to produce an offspring. And in the same way, also in the case of uniparous animals, an excessive amount of seminal discharge doesn't produce a bigger embryo, but can exceptionally produce twins, which are also considered as an anomaly (*GA* IV,4, 772a4–8).

If though, as Aristotle maintains in *GA* IV,4, 771a17–b14, it is the largest live-bearing animals that produce one offspring, while the middle-sized ones produce few and the small-sized produce many, then this means that there is in fact a fixed size range not only of indivisible kinds, but also of wider kinds like the group which shares the common feature of being live-bearing. Evidences of this are abundant. In *PA* III,2, for example, elephants and camels are said to be oversized in relation to the other live-bearing animals, as such "excess in size" (μεγέθους ὑπερβολή) is their proper means of protection. But the generality level at which the size range is fixed can be much higher: for example, size is a determining condition with relation to the differences of locomotion among insects (*PA* IV,6), and the bloodless animals as a whole are said to be smaller than the blooded ones (*PA* III,4, 665a32).

If the size range of the kinds is to be considered as rigid, then we may conclude that the oversized and undersized animals described in *HA* VII(VIII),28 are not something like "populations" of individuals belonging to the same kind as their more familiar counterparts, which would have been transformed by different environmental and nutritional factors. Rather, they must be considered as different kinds, not unlike the kinds of pygmy humans and horses whose existence is mentioned in *HA* VII(VIII),12, 597a5.[13] Hence, according to Aristotle, smaller kinds of animals, for example, are fit to survive in an environment where less food is available, but such a correlation between nutrition and form is not to be interpreted in terms of evolutionary adaptation or even of epigenetic inheritance, which are not consistent with Aristotle's general conception of form and essence. The sense of such correlation is analytical, not historical.

According to this framework, though, it is still possible that there exists a kind of enormous serpents or of pygmy horses, and that in such cases the limits of size be fixed – in the qualified sense that I have described – by the fitness to environmental and nutritional factors. These factors are external. They provide the basis for a teleological explanation of the limit of size focused on the animal's ability to survive in his environment.

Yet, as Aristotle points out in *PA* IV,10, animals can be said to differ in a progressive increase not only according to their size, which is not only determined by the balance between earthy matter and natural heat, but also according to the orientation of their body with reference to the top/bottom axis, along which these elements move. Hence, for example, however large it may be, a quadruped cannot have a sufficient amount of natural heat to be upright, as this would result in losing its nature. This defining factor is related to the body plan inscribed

---

13 As Balme points out, "The pygmy humans and horses, whose existence Arist. accepts here, are regular 'kinds' (γένος, 597a8) and therefore are not connected with the 'pygmies' described at *GA* II 749a4, which are occasional deformities" (Balme 1991, p. 133 n. a).

in the animal's essence, and it is of paramount importance to the rigidity of the size range at higher levels of generality.

6 Conclusion

In *GA* II,6, 744b27–745a10, where the question of the growth limit for animals is explicitly addressed, Aristotle provides an admittedly partial explanation focused on material factors. It consists in showing that the parts which grow towards a definite limit are those made out of the nutritive kind of nutriment. Among these are the bones, which are made out of the residues of this same kind of nutriment, and whose limited growth sets a mechanical limit to the growth of the whole body along the top/bottom axis. Even if the topic of the growth limit is not explicitly addressed elsewhere, a sound Aristotelian teleological explanation of the growth limit might be based on a functional factor, in terms of what is "towards the better", namely by showing that growth stops at prime age, when a balance is attained between heat and cold in the body which provides the best conditions for the soul's activities. Nevertheless, as I have tried to show, there is solid evidence that according to Aristotle a definite size, or more precisely a rigid size range, is inscribed in the animal's essence and also pertains to wide groupings at different levels of generality. This claim provides the basis for a teleological explanation of the growth limit focused on qualitative, formal, and specifically morphological elements.

Bibliography

Balme, David M. (ed./transl.) (1991): *Aristotle. History of animals. Books VII–X*. Cambridge Mass.-London: Harvard University Press.

Carbone, Andrea L. (2016): "The axes of symmetry. Morphology in Aristotle's Biology". In: *Apeiron* 49, p. 1–31.

Gotthelf, Allan (2012a): *Teleology, First Principles, and Scientific Method in Aristotle's Biology*. Oxford: Oxford UP.

Gotthelf, Allan (2012b): "Teleology and Embryogenesis in Aristotle's Generation of Animals II.6", in *Teleology, First Principles, and Scientific Method in Aristotle's Biology*, Oxford University Press 2012, p. 90–117.

Hamlyn, David M. (transl./comm.) (1968): *Aristotle's* De anima; *Books II and III, with certain passages from Book I*. Oxford: The Clarendon Press.

Jowett, Benjamin (1885): *The Politics of Aristotle*, Oxford: The Clarendon Press.

King, Richard A.H. (2001): *Aristotle on Life and Death*, London: Duckworth.

Lennox, James G. (transl./comm.) (2001): *Aristotle. On the Parts of Animals I–IV*. Oxford: The Clarendon Press.

Leunissen, Mariska (2010): *Explanation and Teleology in Aristotle's Science of Nature*. Cambridge: Cambridge University Press.
Li Causi, Pietro (2003): *Sulle tracce del manticora. La zoologia dei confini del mondo in Grecia e a Roma*. Palermo: Palumbo.
Pellegrin, Pierre (ed./transl.) (2011): *Aristote*. Les Parties des animaux. Paris: Flammarion.
Platt, Arthur (ed.) (1912): "Aristotle, De generatione animalium". In: Smith, J.A./Ross, W.D. (eds.): *The works of Aristotle translated into English*. [Vol. V.] Oxford: The Clarendon Press.
Rashed, Marwan (ed.) (2005): *Aristote*. De la géneration et la corruption, Paris: Les Belles Lettres.

David Lefebvre
Looking for the Formative Power in Aristotle's Nutritive Soul

Abstract: Does Aristotle attribute to the nutritive soul a function specifically dedicated to the embryonic formation? The claim has sometimes been defended in the Aristotelian tradition or by modern scholars. This article intends to show that Aristotle never uses such formative power. The whole embryogenesis is the work of the nutritive soul inasmuch as it makes the living thing grow. Through a close reading of texts from *de Anima* II,4, *de Generatione Animalium* II,1, 4 and 6, we establish that (i) Aristotle distinguishes two stages in the generative process of the embryo: the generation (understood in a restrictive meaning) of the heart by an external principle (the male's seed) and the growth of the heart and of the whole embryo by the action of the internal principle located in the heart; (ii) Aristotle sometimes makes a more accurate distinction between the initial "constitution" of the organs and their growth. Even in this case, these two processes depend on the soul's "productive power", which is the same as the one by which the nutritive soul makes the living thing grow. There is no conflict on this point between the *de An.* and the *GA*. The growth of the adult animal and the whole formation of the embryo depend on a single function of the nutritive soul, the "use of the food". This result confirms the unity of the functions of the nutritive soul in Aristotle's psychology and embryology.

Aristotle's main account of the nutritive soul is to be found in *de Anima* II,4. However, for obvious reasons, *de Generatione Animalium* provides a number of new angles on this subject, mostly but not only, as we shall see, about the reproductive function of the nutritive soul. One could even think that Aristotle's detailed investigations into the process of embryonic formation lead him to partly reconsider the functions of the nutritive soul. I would like to show that this not the case and that Aristotle's account of the functions of the nutritive soul in the GA remains consistent with the definition given in *de An.* II,4. The case in point is what has been called from Galen, in the medical and philosophical tradition, the "formative power".[1] Aristotle himself never speaks of such power, but is it so

---
[1] Galen uses the expressions διαπλαστική or τεχνική δύναμις (*De facultatibus naturalibus* II,6, K. 15,12–13). Such power is located in the male seed, but see *De semine* II,5, K. 642,1–3. In *De fac. nat.*, the "formative power" has to be distinguished from the "alterative power" (*De fac. nat.* I,5,

clear that he doesn't need it? To make myself clear, I first provide three quotations borrowed from modern scholarship devoted to this topic. The first one is from Montgomery Furth's book *Substance, Form and Psyche. An Aristotelean Metaphysics*:

> Thus the "primary psyche", in which threptic and gennetic are conjoined, represents a twofold capacity of certain biological objects with regard to transtemporal persistence. The threptic, in the individual, is the basic faculty of *self-maintenance* across time, and works by metabolic exchange of materials with the environment, particularly the processing of nutrients or *trophê khrêsthai* [...]; this is explicitly *not* a generative process (for "nothing generates itself", 416b16–17 [...]), but a preservative one [...]. The gennetic, in the individual, is the faculty of *self-duplication* [...].[2]

The second quotation is from Richard King's *Aristotle on Life and Death*:

> Nutritive soul is identified first of all as that which produces growth in the existing individual; this is then *extended* to its function in the case of the initial constitution of the individual.[3]

The last one is from Sophia M. Connell's recent book *Aristotle on Female Animals, A Study of the* Generation of animals:

> Blood is the most finished of internal nourishment, and so *maintains* the being of the parts. The nutritive soul, which refines raw food into blood is then also able *to constitute* the initial body of an animal in instances of generation (*de An.* 416b14–16). In these cases, blood is that 'out of which the parts come to be' and that 'from which the parts directly come to be'. Although the parts 'come to be out of' blood, as explained in the previous section, the addition of material bulk is not what is primarily meant by this. (Connell 2016, p. 142)

K. 10,13–11,5), but sometimes Galen may attribute to the formative capacity a more specific function (see *De fac. nat.* II,3, K. 86,7–9). Stoics already use the same verb in this context. See Hierocles, *Elementa Ethica* 1a,9 where it is said that the seed (or more exactly the *pneuma* into which the seed is transformed) διαπλάττει τὸ ἔμβρυον; see also SVF II,462. See Gourinat (2008), p. 71; Ramelli/Konstan (2009), p. 37 n. 4–5. Inside the "reproductive faculty" of the "vegetative faculty", Avicenna has introduced a distinction between the "generative faculty" (responsible for the reproduction itself) and the "informative" or "formative" faculty (responsible for the formation of the embryo). See Alpina here p. 217–254, and an overview in Hirai (2011), p. 19–21.
2 Furth (1988), p. 160–161. Furth spells out "genetic" with double "n".
3 King (2001), p. 21. My emphasis. See also on the same page: "The nature of any living thing is its nutritive soul, and this is responsible both for forming the new living being, and for its increase in size when formed."

M. Furth, R. King and S. Connell stress two different aspects of the nutritive soul: Furth explains that the nutritive faculty, through its nutritive and "gennetic" capacity, has a "preservative" function and not a "generative" one. In his book, where he analyses extensively the nutritive faculty, Furth emphasises the idea of individual "self-maintenance" of the animal through the threptic function of this soul; he explains also that the "reproductive" capacity is "preservative" at the species level. However, Furth never deals with any function other than these two. R. King and S. Connell both add something more: the nutritive soul "constitutes" the "initial body" of the embryo. According to R. King, this power has to be seen as an "extension" and not as another function set apart from the two other functions of the nutritive soul; it could be defined as a modality of the nutritive soul, which is only activated at the beginning of the embryonic life of the animals. It might seem attractive to go one step further and to ascribe a constitutive power to the nutritive soul.[4] It must be emphasized of course that nowhere in *de An.* II,4 Aristotle ascribes such a power to the soul. In the *GA*, Aristotle seeks to explain how the nutritive soul operates as a principle in the formation of the offspring. But even there, Aristotle never talks clearly about the formative capacity of the soul. The power to constitute the initial body never has the same status as the two other functions of the nutritive soul, "to use the food" and reproduction (*de An.* II,4, 415a26). This essay will focus on this issue: is it possible to ascribe to the nutritive soul a formative capacity? How is this function related to the other capacities of the nutritive soul (growth, nutrition and reproduction)? Is this function to be linked with growth or rather with nutrition strictly speaking?

First of all, one has to make a distinction between two different and complementary views on this matter.[5] It is well known that the conceit of *dunamis*

---

[4] The trend to attribute to the nutritive soul a "generative" or a "formative power" is not rare. See for instance Code (2004), p. 193, on *GA* II,4 (discussed later).

[5] I am referring here very briefly to what is usually called in the Aristotelian scholarship hylomorphism and instrumentalism. There is no place to discuss this point here, nor the issue of the relationship between Aristotle's psychology and his biology. See Lloyd (1995); van der Eijk (2000), p. 61–62; (2005), p. 206–210; Menn (2002). My point here is only to suggest that the methodological difference between *de An.* and *GA* may explain why Aristotle doesn't use the same vocabulary to describe the capacities and the functions of the nutritive soul and of its instruments (whether they are the male seed, the *pneuma*, or the heat and the cold): in the *GA*, Aristotle tells us that there is in the male seed a *dunamis* that *produces* the embryo, something one can call a *productive power* (*GA* I,21, 729b5–6; II,4, 740b35), but it is unclear how the soul responsible for the nutrition and the reproduction can *produce* anything. In the *de An.* Aristotle never ascribes explicitly any formative or generating power to the soul. Here, I would like to examine to what extend the "use of food" can be the only principle responsible not only for the

(power, capacity) plays a key role in the *GA*.[6] According to Aristotle, the heat of the *pneuma* located inside the bubbles that makes the purest part of the male seed has the power to move a part of the menstrual fluid and then to start the embryogenesis (*GA* II,3, 736b35–737a1; Rashed (2018), p. 109–113 and Connell (2016), p. 146). So the male's contribution is not material, but it consists in the motions produced by the heat in the *katamênia* (*GA* I,19, 726b17–19; I,20, 729a34–b8; I,21, 730a14–15, etc.). There is no doubt that Aristotle ascribes the *dunamis* to move the matter to the male seed (or to the heat or to the *pneuma*). But the seed, the *pneuma* and the heat are all "instruments" of the soul.[7] Thus it is not to the soul itself, but to its instruments that Aristotle ascribes the power to initiate the motions in the material residue. One can make a distinction between the instrumental level, mostly used in the *GA*, where Aristotle focuses on the actions of the material instruments on the menstrual fluid and the psychological (or formal) level used in the *de An.* where Aristotle distinguishes the functions or capacities of the nutritive soul.[8] Now, it is well known also that Aristotle makes numerous analogies between the motions produced by the seed and the motions of the tools used by the carpenter or the craftsman in general.[9] So one has to make a distinction between two different views of the Aristotelian analysis on embryonic formation: when Aristotle deals with the powers of the nutritive soul and when he refers to the actions of the material instruments (seed, *pneuma*, heat); in this last case, these actions are described as a process of construction or shaping by means of analogies with the craftsman's activity. Thus, one has to distinguish the powers ascribed to the material instruments of the soul and the powers of the nutritive soul itself. The instruments of the nutritive soul have the powers to shape the embryo by setting the menstrual fluid in motion, whereas one doesn't find any formative power in the nutritive soul itself.

---

growth, but first of all for the *production* of the embryo. In this case, it would become useless to ascribe to the nutritive soul a special function dedicated to the formation of the embryo.

6 See Lo Presti (2014); Lefebvre (2018a).

7 Aristotle qualifies as instruments of the nutritive soul (i) the sexual organs (*GA* I,2, 716a31–35; 16, 721a26); but also (ii) the male seed: nature uses the seed as an instrument that possesses "motion in actuality" (*GA* I,22, 730b21), like the tools of the carpenter transmit the motions of the art; and (iii) the *pneuma* (*GA* V,8, 789b8–9) i.e. (one may conclude) the heat of the *pneuma* (II,6, 742a14–16). Lloyd (1995, p. 150) mentions also "mouth, teeth, lips, tongue, stomach, liver, omentum, mesentery, and the whole digestive tract." – We follow Menn (2002), p. 113, when he says that "much of the programme of the *De anima* and the *Parts* and *Generation of Animals* is to work out the picture of the body as ὄργανον of the soul and use it to explain particular vital activities."

8 *De An.* tolerates exceptions as for instance III,10, 433b19–21.

9 *GA* I,18, 723b29–30; 22, 730b4–32; II,4, 740b25–34; 6, 743a25–26.

However, it appears that the analogies Aristotle makes between the art using tools to shape a matter and the nutritive soul using the vital heat to move the menstrual fluid aim to interpret the nutritive soul as power to build, to generate or to constitute something. But the nutritive soul doesn't have such a generative power. As we shall see, the point is that Aristotle doesn't make any real distinction between the power to make the living thing grow and the power to constitute the initial body of the embryo. Growth appears as the same thing as the constitution of the embryo or as the continuation of its constitution. I will first discuss these issues within the *de An.* context before considering them in the perspective of the *GA* (II,1, 4 and 6).

1 *De An.* II,4 and the Formative Power of the Soul

Let's begin examining the point from *de An.* II,4: does Aristotle deal with any formative capacity of the soul in this chapter? If he doesn't explicitly speak about it, does he leave the possibility open to find a place for this capacity? At first sight, not only Aristotle doesn't speak about that capacity, but he seems also to make it difficult to find a place for such a capacity.

In *de An.* II,4, Aristotle ascribes only two functions to the nutritive soul: "reproduction and the use of the food" (γεννῆσαι καὶ τροφῇ χρῆσθαι, 415a26). Everywhere in this chapter, Aristotle uses γεννῆσαι, which means "to beget" (first of all for the male) and "to bear" (for the female); the adjective γεννητικός, also used by Aristotle in this chapter, has the same meanings. This function clearly designates the capacity "to produce another thing like oneself" (415a28), plant or animal. Aristotle lays stress on the fact that the living being produces *another* being, "not one in number but one in species" (415b6–7). The same point is made particularly clear latter on, in the same chapter, just after Aristotle has made a distinction between τροφή and αὐξητικόν. The nutrient (τροφή) is the unique "correlative object" of the two functions, nutrition (τροφή) and reproduction.[10] There is τροφή strictly speaking only if the living

---

10 See II,4, 416b20–23. "Correlative object" translates ἀντικείμενον (415a20). The notion of τροφή is subject to further elaboration in the course of II,4: (i) depending on whether the τροφή is digested or not (416b3–7); (ii) depending on whether τὸ τρεφόμενον is a defined quantity or a substance; Aristotle confines τροφή to this latter case (416b11–13); (iii) depending on whether the τροφή, i.e. "that with which" (ᾧ τρέφεται) the living body is nourished is "moving

being is a particular and determinate being. In this case, the τροφή "maintains" the substance itself; but τροφή is also what "produces generation not of what is fed, but of something like it" (416b15–16). The reason why there is only one "correlative object" for the two different activities (nutrition and reproduction) is that there is one single thing (the nutriment at different degrees of concoction) used for both functions: (i) the nutrition of the living thing and its growth and (ii) the production of the generative residue (the male seed or the menstrual fluid). Both functions imply the existence of the *ousia* (416b13–16), i.e. the living body itself. Aristotle emphasizes this point: "nothing begets itself" (416b16–17). Aristotle makes the same claim in *de Motu Animalium* 5: nothing is cause for its own generation, because it would suppose that "the mover must pre-exist the moved, and the begetter the begotten. But nothing is prior to itself" (*MA* 5, 700a35–b3, transl. Nussbaum 1978). Thus the nutritive principle has two capacities: to preserve (σώζειν) a being that is already living (and not to generate it) and to produce the residue (the male seed or the menstrual residue) for the generation of *another* living being. The distinction between the three elements of the nutrition made in *de An.* II,4, 416b20–23 goes in the same direction. Aristotle distinguishes "what is fed" (τὸ τρεφόμενον), "that with which it is fed" (ᾧ τρέφεται) and "that which feeds" (τὸ τρέφον). "That which feeds" is the "primary soul", namely the nutritive soul; "what is fed" is the body that possesses this soul, and "that with which it is fed" is the food or, more exactly, the nutrient already concocted. So, what is needed for the nutrition is clearly a "*body*" ensouled (see 416b22 where Aristotle says: τὸ ἔχον ταύτην σῶμα; and 416b18–19). Hence, what appears at first sight in *de An.* II,4 is that Aristotle doesn't leave any room for the formative function of the living being itself we are looking for. Nutrition and begetting imply the pre-existence of the living being, its "anteriority". The "ensouled body" must exist before nutrition; the nutritive soul doesn't "constitute" it.

Aristotle' silence on a formative function of the nutritive soul is confirmed by Alexander of Aphrodisias' own treatise *On the Soul*. Indeed, Alexander follows closely the order of Aristotle's analysis (Caston 2012, p. 2), and he completes his commentary on *de An.* II,4 without a single word on a formative function of the soul. Nevertheless, at the end of the part devoted by Alexander to the nutritive soul, we find two sections that relate more loosely to Aristotle's text. The second one deals with the fact that organic parts detached from a plant continue to develop themselves (37,4–38,11), whereas animal parts die after they have

---

and moved" (the heat makes the concoction) or only "moved" (the food) (416b26–28; reading ᾧ τρέφεται in 417b26 with Corcilius 2017, *ad loc.*).

been cut off; the first section (36,19–37,3) is a short embryological excursus on the power of the nutritive soul located in the male seed. Alexander deals briefly with a topic that is actually more akin to the *GA:* he explains that the nutritive soul in the male semen is also the cause of the "constitution" of the embryo in the matter. Alexander uses the same words as Aristotle in order to express the first "constitution" of the embryo (σύστασις, συνιστάναι):[11] "The soul and power for nourishing [oneself] is responsible for the initial formation of the animal's body as well as for its being, increase, and growth."[12]

This comment is not brought about by anything in Aristotle's text. Alexander claims that the nutritive soul is not only responsible for the growth of the embryo, but first of all for its *being* and for its first organic organisation. This comment introduces something quite new that we don't find in Aristotle's own analysis on the powers of the nutritive soul in *de An.* II,4. As noted above, the activity of the nutritive soul, as described in *de An.* II,4, clearly implies that the living being already exists. The point is to know why Alexander provides this additional commentary. It is consistent with Alexander's purpose in his treatise. The exegete wants to give a systematic account of the Aristotelian conception of the soul, which leads him to bring together materials from both treatises, *de An.* and *GA* (Caston 2012, p. 3). Here, he wants to explain what role the nutritive soul plays in the embryogenesis, something that we don't find explicitly in the *de An.*, so that he goes beyond what Aristotle says in *de An.* II,4 and he uses knowledge that comes from the *GA*. However, Alexander's comments are not as precise as we might wish. He doesn't ascribe to the nutritive soul any particular "constitutive" function. He just says that this soul is responsible for the being, the organisation and the *growth* of the embryo so that we must conclude that the being and the organisation of the embryo come from the threptic function of this soul (the use of the τροφή). But Alexander doesn't make any clear distinction between the generation of a new living being, the first organisation of the embryo and its growth. So, a conclusion we can draw from Alexander's texts is that *de An.* is maybe not the right place to look for a formative function of the nutritive soul, since Aristotle's concern in the *de An.* is not the *genesis* and the formation of the animals. So, we'll have to turn to the *GA*.

---

11 This process never receives properly a definition in the *GA*. The verb designates the setting of the female secretion due to the heat of the male seed: what is homogeneous comes together, separates from the fluid, makes a unity and receives a shape. See *GA* I,20, 729a10–14; II,1, 733b20–21; II,4, 739b20–28.

12 Alexander of Aphrodisias, *De Anima* (36,19–21 Bruns): ἔστι δ' ἡ θρεπτικὴ ψυχή τε καὶ δύναμις αἰτία καὶ τῆς συστάσεως τὴν ἀρχὴν τῷ τοῦ ζῴου σώματι, ὥσπερ οὖν καὶ τοῦ εἶναί τε καὶ τῆς ἐπιδόσεώς τε καὶ αὐξήσεως. I quote Caston's translation in Caston (2012).

However, a more careful reading of *de An.* II,4 enables us to revise this first view. First of all, as already said, in II,4 (with a single exception at 416b15), Aristotle always uses γεννῆσαι, γέννησις or γεννητικός, all words that designate the generation of *another* being in the strictest sense, i.e. reproduction. This is not "generation" in the sense of "becoming something" (γένεσις). Aristotle doesn't use words of the family of γέννησις (with the double Greek letter, nu) to designate the formation of the homogeneous parts (the semen, the menstrual residue, etc.) or the embryonic development. In these cases, he uses γίγνεσθαι. So, we have to make a distinction between two kinds of generation: γέννησις, reproduction or begetting, which is a function of the nutritive soul of both parents, and γένεσις (or σύστασις), the formation or the constitution of the body of the living being itself. The point is to know how the σύστασις of the embryo depends on the nutritive soul. Is there any formative capacity responsible for it? With a closer look at *de An.* II,4, it appears that this capacity is not missing. Two options seem available.

(1) In *de An.* II,4, when he is criticizing Empedocles' mechanical and materialist explanation of the growth by the movement downwards of the earth or upside of the fire (415b28–416a2), Aristotle talks of the formation of the parts of the living being. In this context, he uses the word αὔξησις (growth) to designate precisely the development of the roots downwards, in the earth, and the development upwards of the branches, roots and branches being the organs of the plant. As Aristotle explains against other naturalistic theories of nutrition, the cause of the growth of the living thing is not only fire, but also the soul (416a13–18). Thus, contrary to what we said, one may claim that, in this chapter, Aristotle actually deals with the capacity that *constitutes* the primary body. One could reply that Aristotle is talking here about a specific kind of change, the increase of bulk of the different parts of the plant. This is not completely true. In other places in the *de An.*, Aristotle uses growth (αὔξησις) as the typical feature of the nutritive life. When he gives his definition of life in *de An.* II,1 (412a14–15), life is defined (without reference to plants) as "self-nutrition, growth and decay" (τὴν δι' αὑτοῦ τροφήν τε καὶ αὔξησιν καὶ φθίσιν).[13] From an external view, the becoming of a living being doesn't seem to be anything more than growth, and life for plants and most animals is growth and decay.

Thus, what Aristotle means by "growth" in these texts is not only a quantitative increase, but a double process that is in the same time σύστασις and αὔξησις. *De An.* III,12, 434a22–26 introduces a distinction relevant for our discussion, that we'll find latter more elaborated in the *GA*: this is the distinction be-

---

13 See also *de An.* II,2, 413a27; III,9, 432b8–11 and 12, 434a22–26.

tween γενόμενον (to be born, to exist) and αὔξησιν ἔχειν (having growth). Aristotle sets there a major principle of his embryology: "it is necessary that anything that was born (once it was born) grows" (434a24–25). In this context, the growth designates precisely the differentiation and the development of the different parts of the living being. It is worth noting that Aristotle names *growth* the "constitution" of the living being by itself. Indeed, one might be surprised to learn that Aristotle conceives the formation of the living being as a kind of *growth*. Of course, growth is not to be seen only as a quantitative change, as the difference Aristotle makes between τροφή (nutrition) and αὐξητικόν (what makes something grow) shows it. The difference is not between two different things, but it consists in two different ways to describe the function of the same thing: whether one takes the living being, which is fed, as a certain quantity or as a substantial and individual being. Aristotle explores this difference in two different texts. In *de An.* II,4 (416b11–20), the difference is about two definitions of what is fed: as ποσόν τι or as a τόδε τι καὶ οὐσία. As a certain thing having a quantity, the food is something that produces growth and it produces then a change in the category of quantity; as a particular being and a substance, there is nutrition; in this case, the nutritive soul "maintains" the substance itself (σώζει γὰρ τὴν οὐσίαν, 416b14). The end of *GC* I,5 reveals partly corrupted unfortunately.[14] But we can at least grasp the meaning of this difference (322a20–28). What makes the difference between τροφή and αὔξησις is the following: αὔξησις implies that what is fed is potentially both a certain quantity and a certain quantity of flesh (e. g.); on the contrary, τροφή implies only that what is fed is potentially flesh (Code 2004, p. 191). In other words, τροφή is only a conservative and a qualitative process, whereas αὔξησις is altogether a qualitative and a quantitative process that implies an increase in bulk. That's the reason why αὔξησις is chronologically first in the life of a living being; it stops after a certain period of time (different according to the kinds of living beings); on the contrary, τροφή still proceeds after the end of the αὔξησις, during the diminution of the size of the living being, until the end of its life. These three activities belong to the nutritive soul: τροφή, αὔξησις and reproduction. From a chronological point of view, growth comes first; τροφή and generation come after. At a certain stage (or age) of the animal's development, the nutritive soul stops using the τροφή to make the body and uses it to maintain it and to produce the generative residue.[15] So, the growth of the living thing, as a whole

---

14 On *GC* I,5, 322a28–33 see Rashed (2005), p. cx–cxiv and p. 128 n. 2.
15 *GA* I,18, 722b22–23. This is an ideal case, since the residue of the τροφή may become seed or fat (*GA* I,18, 725b25–726a6; II,7, 746b25–29).

and in each of its parts, ceases at a determined point; the growth of living being is not an infinite process, as the growth of the fire;[16] it gives to the living being and to all its parts its natural shape and form; its gives to each of its parts the size that enables them to realize their function and to work on their own, so far the parts cannot achieve their function if they do not have a determined size. Thus, growth appears as a teleological process controlled by the nutritive soul (de An. II,4, 416a18). However, even in these conditions, even if the growth provides their natural and final proportions to the parts of the living being, is it possible to say that the nutritive soul "constitutes" the parts of the living being *only* by making them grow?

(2) Another option could be available. If τροφή maintains a substance that already exists, this "conservative" function might be understood as formative. One may look at the conservation as formative, so far, even at the very beginning of the embryonic life, it exists a living thing that has to be conserved and maintained: the heart. As soon as the heart has been generated, the whole *ousia* exists potentially. So, one could think that the nutritive principle, as soon as it exists in the heart of the embryo, preserves its own existence by constituting the embryo. It doesn't seem the right track however, simply because the τροφή implies that the living being has reached its natural and final size, which is not the case here. The constitution may not be reducible to growth, but it is not the same thing as the conservative process, which the τροφή is responsible for. – Aristotle's discussions in the *GA* will provide a better understanding of the role growth plays in the constitution of the embryo.

2 Growth and the Productive Power in the *GA*

The way the beginning of the embryonic formation is described in *GA* II,1 and 4 confirms the function of the growth according to Aristotle. These two chapters of the *GA* are very closely connected; for what is explained for all kinds of animals in II,1 is applied in the case of human being in II,4.

2.1 Generation and Growth in *GA* II,1

In *GA* II,1, Aristotle's purpose is to define the principle of generation. In 735a12–26, this task has been completed. Aristotle has given the solution that

---

[16] *De An.* II,4, 416a13–18; *GA* II,6, 745b4–9 and IV,4, 771b33–772a2.

explains how a principle, which is outside, can be principle of the embryonic generation inside the female's body. Generation is caused by an external principle, external to the womb and more precisely to the menstrual residue: the moving cause is ultimately the nutritive soul of the father outside. The model used to explain how the motion provided by the father can be transmitted to the semen and, ultimately, to the menstrual fluid is an automaton. The following text bears precisely on the transition from the external change caused by the father (homonymous to the living being generated) to the internal change:

> Now this generative process is not caused by any of its parts, but by that which proximately moved it from outside. For nothing generates itself. But once it has been produced, it proceeds to increase itself (οὐθὲν γὰρ αὐτὸ ἑαυτὸ γεννᾷ· ὅταν δὲ γένηται αὔξει ἤδη αὐτὸ ἑαυτό). Therefore, some first thing is produced, not everything at once. And the first thing to be produced must be that which contains the source of increase; for all alike, whether plant or animal, possess this, the nutritive (τὸ θρεπτικόν). (And this is what is generative (τὸ γεννητικόν) of another like oneself; for that is the function of every naturally perfected thing, both animal and plant). It must be so, because once a body has been produced it must be increased (ἀνάγκη δὲ διὰ τόδε ὅτι ὅταν τι γένηται αὐξάνεσθαι ἀνάγκη). Therefore, although it was generated by that which is synonymous (a man by a man), it is increased by means of itself (αὔξεται δὲ δι' ἑαυτοῦ). It itself must be something, therefore, if it causes increase (αὐτὸ ἄρα τι ὂν αὔξει). Now if it is one particular thing, and this is first, it must be the first to be produced. Consequently if the heart is the first to be produced in certain animals (and the part analogous to it in those that do not have a heart), the source must be from the heart in those that have one, and from the analogous part in the others. (GA II,1, 735a12–26, transl. Balme 1992)

Four important claims are established here: (1) Aristotle has asked how each part of the embryo is formed (734b19). He first replies that the principle of the embryonic development is not whatever part of the embryo (so far "nothing generates itself"), but an "external first mover" (735a13), homonymous to the embryo. So, the first embryonic part is produced by the external mover. (2) Once this first thing has been produced, the embryo increases by itself. Aristotle repeats this principle: "once it has been produced, it proceeds to increase itself" (735a13–14) or "once a body has been produced, it must be increased" (735a19–20).[17] (3) Since all the *other* embryonic parts are not produced simultaneously, but one after the other, the first part produced by the external mover must possess the principle of the growth of *all the other* parts. This part is the nutritive part. Aristotle holds here a logical proof of the priority of the generation of the heart as a principle of growth.[18] (4) As one can see that heart is the first

---

17 We find the same principle in GA II,4, 740a20–21.
18 See GA II,4, 739b33–740a5, 740b3, etc.

part produced in the blooded animals, the principle of growth comes from the heart (or from whatever analogous part for the animals without blood).[19]

Once again, it is noteworthy that Aristotle calls the first internal change of the embryo *growth*: immediately after the *genesis* of the heart begins the *growth* of the other parts of the living being, even if no part (except for the heart) exists yet. Given the way Aristotle defines growth elsewhere (*de An.* II,4 and *de Generatione et Corruptione* I,5), the explanation of the embryogenesis as *growth* may seem oversimplified. Growth requires that something already exists (flesh, bones, etc.). Something else goes against the idea that growth is enough to "constitute" the embryonic body. Aristotle argues at length in *GA* I,18 against preformationism. Preformationism requires only quantitative change in order to constitute the embryo (*GA* I,18, 723a9–24). So the parts have to pre-exist, as growth cannot by itself generate the parts. There is only one part for which growth is enough, the heart. Aristotle has said that the principle of change was the first part generated and this is also the first part to grow. But for the other parts of the body, another kind of change seems to be required to explain the constitution of the body.

But this is not Aristotle's conception of growth here. He argues for a two-step conception of the changes involved in the generation of animals: (1) first a change whose cause is an homonymous external moving cause; it produces only one thing, something that possesses the nutritive principle, the heart; this is the only *generation* strictly speaking; (2) then, an *internal* change caused by an internal principle, the heart itself, clearly expressed by the clause αὔξεται δὲ δι' ἑαυτοῦ (735a21, 735a14).[20] Aristotle mentions growth as the single internal change caused by the heart. We can compare on this point *de An.* II,4 and *GA* II,1 where the same clause appears. In *de An.* II,4, Aristotle says: "nothing generates

---

19 In the middle of the text (735a17–19), Balme (1992) put in brackets a clause that is not directly related with the point Aristotle is making here. However, this clause is far from being irrelevant. The fact that the nutritive part is also generative is an additional proof that it is the first part; if generation is the first and most common function to all living beings and if the nutritive part is also reproductive, then the nutritive part has all the more reasons to be the first part produced in the living being. In this way, Aristotle joins closely together the nutritive and reproductive function of the first soul. We'll see the same move in *GA* II,4 quoted below.

20 On these two stages, see Preus (1975), p. 112, Code (1987), p. 54–58 and Johansen (2012), p. 140–143. This distinction avoids the consequence that Aristotle believes that "individual natural things have in themselves the principle of their own genesis" (Kelsey 2003, p. 71) or that animal generation is not natural, since nature is an inner principle and "nothing generates itself". See Lefebvre (2018b), p. 82–86 on the particularity of sexual generation as a natural change. In *GA* I,18, 724b8, Aristotle clearly emphasises this particularity by saying that "generation from contraries" (sexual generation) "is *also* natural generation."

itself, but rather maintains itself" (416b16–17, transl. Hamlyn 1993). In *GA* II,1, growth takes the place of self-maintenance. So, growth has not only to be regarded here as an increase of bulk, but also as the whole process of constitution of the embryo (what we call morphogenesis and organogenesis); growth is responsible for the constitution (σύστασις) of all the vital parts (homogeneous and heterogeneous) of the embryo, and for the increase of these parts.

So, *GA* II,1 gives the following picture of the role of the nutritive soul. Two stages with three (numerically) different nutritive souls are implied in generation. (1) First of all, the two nutritive souls of both parents, male and female (in the case of sexual reproduction). According to their "reproductive" function, the aim of the nutritive soul of each parent is to generate another one like it; that's why they produce a generative residue. The heat contained in the *pneuma* of the male seed is the moving cause of the generation of the heart. This generation (whose first moving principle is external) is the one and only generation strictly speaking. This part of the generative process is clearly enough described by Aristotle: the nutritive soul of both parents involved in this process is the nutritive soul in its reproductive function (γεννητική, *de An.* II,4, 416a19); the kind of change is clearly designated by Aristotle as a generation. The short treatise on *Life and Death* gives us a confirmation on this point. Generation is defined as πρώτη μέθεξις ἐν τῷ θερμῷ τῆς θρεπτικῆς ψυχῆς, "the first participation by the nutritive soul in heat".[21] (2) So far "what is generated grows", the first part to be generated has to contain the principle of growth; but the principle of growth is the nutritive soul of the embryo. In *GA* II,1, the nutritive soul is responsible for the growth of the embryonic body. Since, according to *de An.* II,4, the nutritive soul has two main functions ("reproduction and the use of food"), we must conclude that the formation of the embryo (what Aristotle has called here its "growth") depends on the nutritive soul that is in the embryonic heart, as far as it uses the food to insure the growth of the body. Aristotle explains in *GA* how the heart finds its nourishment first in the menstrual residue left after its own conception and afterwards in the uterus where the heart sends

---

21 *De Juventute et Senectute, de Vita et Morte, de Respiratione* 24(18), 479a29–30: Γένεσις μὲν οὖν ἐστιν ἡ πρώτη μέθεξις ἐν τῷ θερμῷ τῆς θρεπτικῆς ψυχῆς, ζωὴ δ' ἡ μονὴ ταύτης. The point is to know whether generation is defined as (a) the "first participation" of the nutritive soul in the heat or as (b) "the first participation ⟨of the living being – implied⟩ to the nutritive soul in the heat"; or (c) if τῆς θρεπτικῆς ψυχῆς qualifies the heat, as the first participation ⟨of the living being⟩ in the heat of the nutritive soul. Cf. King (2011), p. 134–135; Freudenthal (1995), p. 115 translates "mediated by warm substance".

off the umbilicus.[22] Thus, in *GA* II,1 Aristotle's purpose is to interpret the whole embryogenesis, including the differentiation of the organs, as a growth process caused by the nutritive soul in the heart. On this point, *GA* II,4 provides new insights and some further complications.

2.2 The Productive Capacity of the Nutritive Soul in *GA* II,4

Let's turn now to *GA* II,4, 740b24–741a3. Aristotle's concern is to give an account of the embryonic formation in the case of human being:

> Ὕλην μὲν οὖν παρέχει τὸ θῆλυ, τὴν δ' ἀρχὴν τῆς κινήσεως τὸ ἄρρεν. ὥσπερ δὲ τὰ ὑπὸ τῆς τέχνης γιγνόμενα γίγνεται διὰ τῶν ὀργάνων – ἔστι δ' ἀληθέστερον εἰπεῖν διὰ τῆς κινήσεως αὐτῶν· αὕτη δ' ἐστὶν ἡ ἐνέργεια τῆς τέχνης, ἡ δὲ τέχνη μορφὴ τῶν γιγνομένων ἐν ἄλλῳ – οὕτως ἡ τῆς θρεπτικῆς ψυχῆς δύναμις, ὥσπερ καὶ ἐν αὐτοῖς τοῖς ζῴοις καὶ τοῖς φυτοῖς ὕστερον ἐκ τῆς τροφῆς ποιεῖ τὴν αὔξησιν, χρωμένη οἷον ὀργάνοις θερμότητι καὶ ψυχρότητι (ἐν γὰρ τούτοις ἡ κίνησις ἐκείνης, καὶ λόγῳ τινὶ ἕκαστον γίγνεται), οὕτω καὶ ἐξ ἀρχῆς συνίστησι τὸ φύσει γιγνόμενον. ἡ γὰρ αὐτή ἐστιν ὕλη ᾗ αὐξάνεται καὶ ἐξ ἧς συνίσταται τὸ πρῶτον, ὥστε καὶ ἡ ποιοῦσα δύναμις ταὐτὸ τῷ ἐξ ἀρχῆς· μείζων δὲ αὕτη ἐστί. εἰ οὖν αὕτη ἐστὶν ἡ θρεπτικὴ ψυχή, αὕτη ἐστὶ καὶ ἡ γεννῶσα· καὶ τοῦτ' ἔστιν ἡ φύσις ἡ ἑκάστου ἐνυπάρχουσα καὶ ἐν φυτοῖς καὶ ἐν ζῴοις πᾶσιν […].

> The female, then, provides the matter, the male the source of the movement. And just as things which are produced by the agency of an art are produced by means of tools—it's truer in fact to say 'by the movement of the tools' (this (movement) being the actuality of the art, and the art being the shape of that which comes to be in another)—in this way the nutritive soul capacity, just as, in the animals and plants themselves later on, it produces growth from the food using as tools heat and cold (for the soul capacity's movement is in these, and each thing is produced according to a certain definition), in this way also at the very beginning the thing produced by a nature is constituted. For it is the same matter by which it grows and from which it is first constituted, so that the productive capacity is also the same as the initial one (but greater than it). If then this is the nutritive soul, it is also that which generates, and this is the nature of each ⟨organism⟩ and present in all plants and animals […].[23]

---

[22] *GA* II,4, 739b33–740a1, 740a17–23, 740b2–8; II,7, 745b22–29. On the stages of the "gradual fœtal development", see Connell (2016), p. 146–149.

[23] The Greek text comes from Drossaart Lulofs' edition (1965). We quote the translation of Gotthelf (2012, p. 98) with a small change: just after "productive capacity", Gotthelf adds between angle brackets "for growth" ("so that the productive capacity ⟨for growth⟩ is also the same as the initial one.") We disagree with this addition: this capacity is not only for growth but also for the constitution of the embryo.

This passage, of central importance to our current issue, has been the subject of a great deal of discussion.[24] The grammatical construction of the main sentence is a bit "convoluted", as it has been noted rightly by Gotthelf (2012, p. 38). Aristotle has joined together two parallels in one long clause. As we already noticed, the first analogy is common in the treatise: Aristotle makes an analogy between the activity of an art producing an artefact and the way the nutritive soul uses its instruments to produce growth. The point is precisely that the nutritive soul produces growth at two different times; this difference introduces the second analogy. Both analogies are significant, but the second analogy is especially important for us. There Aristotle makes an analogy between the way the nutritive soul makes an adult living being grow (a plant or an animal) and the way, at the very beginning of the embryonic life, the same nutritive soul produces (συνίστησι, 740b33) this living being. Aristotle underlines clearly the chronological difference by the ὕστερον and ἐξ ἀρχῆς (740b30, 33). This second analogy is supported by an argument: so far the matter that makes the adult living being grow (ᾗ αὐξάνεται) and the matter from which the embryo is constituted (ἐξ ἧς συνίσταται) are the same, the "productive capacity" (ἡ ποιοῦσα δύναμις) must also be the same in the two cases, and this "productive capacity" seems to be "the capacity of the nutritive soul" (740b29). The first analogy is between two different things: the movements of the tools used by an art making an artefact from the matter are compared to the movements produced by the two instruments (the hot and the cold) used by the nutritive soul making an embryo from the food. Aristotle emphasises the similarity between the two processes: art is a form in the soul of the craftsman outside the thing produced; its activity is to be found in the motions of the tools; the nutritive soul is itself a form operating from inside through its own instruments (the *pneuma*, the hot and the cold).[25] The second analogy is between two different but consecutive activities of the same capacity, the nutritive soul.

Before going into more detail about some difficulties, let's outline the overall meaning of this text. The broader context of the passage is the following: Aristotle is criticising the claim that the differentiation of the parts of the animal (the διάκρισις, 740b13) can be explained through the general principle "like makes its way to like". Against this thesis, Aristotle's first reply is to remind us that all the parts of the animal are potentially and not yet in actuality in the female's residue. Thus, what is needed is the moving cause to actualise these parts. Aristotle mentions first the principle of motion provided by the male seed and right after

---

24 See King (2001), p. 21; Pellegrin (2018), p. 83.
25 See above n. 7.

this introduces the comparisons we read in the text quoted above. Its main idea is that the productive capacity (ἡ ποιοῦσα δύναμις, 740b35) that makes the animal grow is the same at the beginning of the embryonic formation, namely the nutritive soul. The point is to know what this capacity makes; as it is said, the "capacity of the nutritive soul" makes the animal grow through the food. This doesn't mean that the constitution is the same thing as growth. It only signifies that, by the same matter and by the same motions, the same capacity of the nutritive soul first constitutes "what is produced" and then makes it grow. It means that there are not two different capacities or two different souls but that the same capacity of the same nutritive soul is responsible first for the constitution of the embryo and then for its growth. We can isolate three different claims here: (1) the same matter is used by the "productive capacity" at these two different stages of the life (at the very beginning and after); (2) this capacity operates in the same way at the beginning and after; (3) the capacity that is responsible for the embryonic "constitution" and for the growth of the living being are the same.

(1) The first claim (there exists a single matter at the beginning of the embryonic life and later on) doesn't seem controversial. It is used as a well-known principle in *GA* IV,8, 777a4–8:

> It is clear that milk is possessed of the same nature as the secretion out of which each animal is formed (this has in fact been stated already): the material which supplies nourishment and the material out of which Nature forms and fashions the animal are one and the same. And this material, in the case of blooded animals, is the bloodlike liquid, since milk is concocted, not decomposed, blood. (transl. Peck 1942)[26]

By "matter", Aristotle is not referring here to the food before its concoction, but to the food already elaborated and concocted, namely the blood (or its analogue). The food can have various aspects: it can be the part of the menstrual residue not yet used in the constitution of the embryo and that can be used as first food (*GA* II,4, 740b2–8); it can be also any kind of food "from outside", as Aristotle says, for instance milk (*GA* II,4, 739b25–26) which is "concocted blood". What this first claim means is that blood is the only matter used for the growth and for the embryonic constitution itself. Aristotle doesn't say anything different when he claims in *de An.* II,4 that there is only one "correlative object" for the different functions of the nutritive soul.[27] Here is the unity of the different functions of the nutritive soul in *de An.* II,4: the "use of the food" is the only real activity of the nutritive soul; its goal only changes: to con-

---

26 See also *Juv.* 3, 469a1–2; *PA* II,4, 651a13.
27 See above n. 10 and Johansen (2012), p. 107.

stitute the embryo, to make the body grow, to maintain the body alive or to produce the residue for the reproduction. However, it may seem that the nourishment of the living being and the constitution of its body are two different activities.

(2) According to the second claim, on the contrary, these two activities are the same. The "productive capacity" produces growth "afterwards" through the same movements it has "constituted" the living being at the beginning. These are the motions of the "instruments" (the *pneuma*, the hot and the cold) controlled by the nutritive soul (as the motions of the craftsman's tools are controlled by the art in its soul). The λόγῳ τινί (740b32–33) refers to the nutritive soul as formal cause. As the art, the nutritive soul is not only a moving cause (a productive capacity), but also a formal cause that controls the constitution and the growth of the parts. So, one must understand that the nutritive soul has a "productive capacity"; this productive capacity consists in the motions produced by the *pneuma* in the matter; by the same motions, the nutritive soul constitutes the animal (or the plant) and makes it grow later.[28]

(3) The third claim is the most difficult one. Aristotle's main purpose in this passage is to ascribe to the nutritive soul the function of being able to "constitute" (συνίσταται, 740b35) the embryo. His strategy is to show that, since, at the beginning and afterwards, the matter and the operations of the tools are the same, the "productive capacity" is also the same.[29] However, it is obvious that this capacity doesn't make the same thing when it "constitutes" the living being at the beginning and when it makes it grow afterwards. We understand now the link between the two analogies: the purpose of the first analogy between the production of an artefact by an art and the way the nutritive soul makes something grow is precisely to understand the grow-promoting work of the nutritive soul as generative-work.[30] From this point of view, we can notice a change between GA II,1 and GA II,4: in GA II,1, the nutritive soul is responsible for the growth of the first part to be generated, the heart, and Aristotle doesn't make mention of anything else. In GA II,4, he introduces a new step – the "con-

---

28 One can compare GA II,4, 740b35 and GA I,21, 729b6–8.
29 It is worth noting that Aristotle makes the same claim in *Ethica Nicomachea* I,13, 1102a32–b2: this is the same growth-promoting potentiality that belongs to the embryos and to the fully developed animals. But there are two differences with the text from GA II,4: in EN, he doesn't make any mention of a "productive capacity", and he supports this claim on the fact that it is more "reasonable" (εὐλογώτερον, 1102b2) to say that this is the same nutritive potentiality that belongs to the embryo and to the adult than "another other".
30 See Menn (2002), p. 122: "Aristotle wants to bring out the art-like aspect of nutrition by reconceiving it on the model of generation."

stitution," what we could call the σύστασις or the διακόσμησις of the embryo (*GA* II,4, 740a8). Growth and constitution are explicitly distinguished. So, the range of capacities of the nutritive soul is wider than we thought. That's the reason why Aristotle says that this soul has a "productive capacity" (ἡ ποιοῦσα δύναμις). The Greek ποιοῦσα designates not only the growth, but also the constitution – the whole process of the embryogenesis. Thus, the embryogenesis appears as the succession of four different overlapping processes: (a) the "generation" of the heart (generation being used here in a restrictive sense); (b) the growth of the heart; the heart is the first part to be generated by an external moving cause, but it grows "by itself", by an internal moving cause, the nutritive soul of the new living body; (c) then, the "constitution" of the different parts of the body, uniform and non-uniform, and (d) the growth of these parts *in utero*.

Before going further, we need to discuss two textual issues. The text from *GA* II,4 contains two controversial clauses. What does Aristotle mean by μείζων δὲ αὕτη ἐστίν (740b36) and αὕτη ἐστὶ καὶ ἡ γεννῶσα (740b37)? Concerning the first issue, the point is to understand what exactly is μείζων.[31] Given that Aristotle was saying just before that the matter and the capacity (δύναμις) are the same from the beginning, there are two possibilities: μείζων may be the matter itself or the capacity. In both cases, Aristotle would be making a quantitative difference: the word μείζων is quite vague; it means a difference of degree, a difference of size or of strength; it can signify that matter is bigger or that the capacity is greater or stronger. In both cases, this precision is a way to specify the claim that the matter or the capacities are the same; they are the same but one is more "important" or "bigger" than the other. That's the reason why it is important to understand exactly what this means.

One could think that the matter (namely the blood) is "more important" (in quantity or in strength). In this case, Aristotle says that the embryo is generated in a part of the menstrual residue especially pure and hot. But, from a grammatical point of view, since δύναμις is the nearest word in the clause, it is easier to understand that what is μείζων is the capacity (the ποιοῦσα δύναμις). In this case again, there are two possibilities. (i) Since Aristotle is talking about growth, it seems natural that the clause means that, even though the capacity is the same (because the matter is the same), this capacity is full-grown or greater *after* the growth, precisely because the living being has grown. It implies that,

---

[31] Editors have thought that τῷ ἐξ ἀρχῆς with the dative is a bit harsh. Louis (1961) has preferred the reading of the ms. S (τὸ ἐξ ἀρχῆς): "la puissance active est aussi la même dès le principe." For the same reason, Wimmer and Bitterauf have made other corrections, that led Wimmer followed by Peck (1942) to bracket τῷ ἐξ ἀρχῆς· μείζων δὲ αὕτη ἐστίν (740b36). But nothing supports this move. With Drossaart Lulofs, we keep the dative and the sentence that follows.

as the living thing is becoming greater, the capacity of its nutritive soul also becomes greater. By greater, we may understand that the capacity is more efficient after the growth, because it needs to transform or elaborate food in blood, whereas, at the beginning, the same capacity has only to use a food already concocted and elaborate, namely the blood itself (Pellegrin 2018, p. 85–86). (ii) On the contrary, one may believe that what is "greater" is the capacity at the beginning of the life of the embryo, because it is much more difficult to "constitute" the different parts of the living thing than to preserve them.[32] In GA II,6 (743a17–26), Aristotle insists on the fact that the "constitution" of the parts, particularly the bones, implies a high degree of internal heat, like in an oven, to make possible their formation. Against this hypothesis, we may remember that, at the beginning, the embryo lives like a plant; Aristotle repeatedly uses this analogy;[33] and Aristotle has explained in PA II,3 (650a20–27) that plants, "by means of their roots", find in the hearth their nourishment already elaborate and worked-up (the hearth is analogous altogether to their stomach and to their mouth). It is difficult to choose between these two readings. To some extent, it doesn't matter, because for Aristotle the point is just to say that there isn't a big difference between the productive capacity at the beginning and after: one is just more powerful than the other. However, it makes better sense to understand Aristotle to be saying that the productive capacity is μείζων, i.e. more powerful, once the living being has been constituted. Once the animal has been constituted, the productive capacity of its nutritive soul is more powerful because the animal has also more vital heat inside (for instance, once they are adult, the viviparous have their lungs fully developed and then are hotter than at the beginning of their life: GA II,1, 732b31–33).

The second difficulty is as follows: when Aristotle says that the nutritive capacity is also "generative" (γεννῶσα, II,4, 740b36), it can mean two different things. Firstly, one can understand that this capacity has the power to "generate" the parts of the bodies; this is precisely what Aristotle has said just before: the nutritive soul is able also to "constitute" the parts of the living being. This reading is adopted (wrongly to my mind) by Devin Henry for instance. In his comments on this text, Henry explains that the heart contains a "generative principle". I quote: "Aristotle eventually identifies the embryo's generative nature with its soul, which is said to be 'the active power' that forms the parts of the body in

---

[32] This reading is adopted by Menn (2002), p. 122 n. 52, and Pellegrin (2018), p. 83.
[33] GA II,4, 739b33–740a3; 7, 745b22–29; V,1, 779a1–2.

the beginning" (Henry 2009, p. 374).³⁴ But Aristotle cannot be saying that if the productive capacity is the nutritive soul, it is *also*³⁵ the faculty that generates the embryo (so far this is what means that the nutritive soul is the productive capacity). For this reason, I think we must understand Aristotle to be saying here what we already know: the nutritive soul has also a *reproductive capacity of another being* (and not a generative capacity of the embryonic body). The use of γεννῶσα (slightly unusual instead of γεννητική) should not make us hesitate. It always means to beget or to bring forth, to bear (not to generate an embryo, as Henry thinks). Aristotle is saying the same thing as in *GA* II,1 (735a17–19);³⁶ the two texts are parallel: the nutritive soul not only has a "productive capacity" but also a reproductive one, which aims to generate another being like itself. Aristotle's purpose here is precisely to join together the functions of the nutritive soul he has identified: the growth, the constitution of the embryonic body (which both form here "the productive capacity") and reproduction. All this constitutes the "nature" that belongs to all the living beings, whereas only some living beings have the other parts of the soul (II,4, 740b37–741a3).

We can now return to our excerpt from *GA* II,4. With regard to our original question, this passage makes two important claims. (1) There is one and only "productive capacity" that is responsible for the first constitution of the embryo and for its growth. This capacity is the nutritive soul itself using its instruments (the hot and the cold) to move the menstrual fluid. We don't have to believe that there is in the nutritive soul some special power to generate the embryo: the nutritive soul is a "productive capacity" that first constitutes the embryo and then makes it grow; but the constitution and the growth are made by the same motions produced by the same instruments operating on the same matter. (2) In embryogenesis, one must make a distinction between two stages: the "constitution" and the growth. This claim is not fully elaborated in the text. The constitution is the differentiation of the parts of the living being, i.e. the actualisation in the menstrual fluid of the embryonic parts (firstly the heart) through the motions produced by the hot and the cold. But the difference between this "constitution" and growth is obviously not the most important thing for Aristotle. The same text from *GA* II,4 gives the reason why: the motions that make the living thing grow are the same as the constitutive motions. What that means also is that the male seed, as it is a residue of the blood of the father, is the bearer of the motions re-

---

34 See also Pellegrin (2018), p. 83: "What is here called the generative power of the nutritive soul is the organogenetic faculty that the living being possesses in itself once the principle of the embryo has been formed."
35 Cf. the καί at 740b37.
36 See text quoted above n. 19.

sponsible for the growth of its father. So, to some extent, the formation or the constitution of the embryo in the menstrual fluid is the continuation of the growth of the father (in another matter but through the same movements). As Aristotle says it explicitly in *GA* II,4, the constitutive motions are the same as the motions responsible for the growth, not only the growth of the embryo, but also the growth of its own father.[37] A passage from *GA* II,3 (737a18–24) makes that point clear:

> As semen is a residue, and as it is endowed with the same movement as that in virtue of which the body grows through the distribution of the ultimate nourishment, when the semen has entered the uterus, it "sets" the residue produced by the female and imparts to it the same movement with which it is itself endowed. The female's contribution, of course, is a residue too, just as the male's is, and contains all the parts of the body potentially, though none in actuality […]. (transl. Peck 1942)

A last text from *GA* II,6 sheds some light on this point.

2.3 Constitution, Growth and Nutrition in *GA* II,6

One of the main results of *GA* II,4 is to make a place for a special stage before the growth: the "constitution" of the parts of the living thing. It is made by the same capacity of the nutritive soul, by the same motions of its material instruments, through the same matter; but Aristotle differentiates between the constitution of the embryo and its growth. This account is slightly revised by Aristotle in *GA* II,6. There, Aristotle deals with the formation of the parts that come to be actualized after the heart:

> The bones, then, are formed during the first stage of construction (ἐν τῇ πρώτῃ συστάσει) out of the seminal residue, and as the animal grows, they grow too. Their growth is derived from the natural nourishment, which is the same as that which supplies the supreme parts; only they get merely the leavings and the residues of it. In every instance, of course, there is nourishment of two grades present: (1) nutritive (θρεπτικόν), that is to say, which provides both the whole and the parts with being; (2) growth-promoting (αὐξητικόν), that is to say, which causes increase of bulk. These will have to be more particularly distinguished later on. The sinews are constructed in the same way as the bones, and out of the same materi-

---

[37] This picture is obviously complicated by the fact that, as Aristotle shows in *GA* IV,2–4, the male's seed, for a variety of reasons, cannot "impart" to the female contribution *exactly* "the same movement with which it is itself endowed". But in *GA* II,3, Aristotle still uses κίνησις in the singular and doesn't take into account the diversity of movements carried by the male's seed and existing, to some extent, in the female's contribution itself.

als, viz., the seminal or nutritive residue. As for nails, hair, hoofs, horns, bills, cocks' spurs and any other such part, these are formed out of the supplementary or "growth-promoting" nourishment (ἐκ τῆς ἐπικτήτου τροφῆς καὶ τῆς αὐξητικῆς), this additional nourishment being obtained from the female, and from outside. (*GA* II,6, 744b27–745a4, transl. Peck 1942)[38]

Just before this text (744b11–27), Aristotle has made a distinction between two kinds of parts of the animals: the first one (the "most noble") are made from the most elaborate and pure parts of the food; the body parts that are only necessary for these parts are made from the residue of what has been used for the noble parts. In the case of animals, the noble parts are the "flesh and the other sense organs", which are made from the purest part of the menstrual blood, whereas the bones and all the parts necessary for the sense organs are produced from the residue. Bones are made from the residue of the "first food". Aristotle introduces then a difference between two kinds of food: the αὐξητικόν and the θρεπτικόν. The bones and the sinews are made from the residue of the nutritive food; on the contrary, the other parts (as nails, hair, hoofs, etc.) are made from the residue of the nourishment that is used for growth. In *GA* II,1 Aristotle has linked the constitutive function of the nutritive soul to the growth-promoting power of the soul. Here Aristotle makes a new distinction between two kinds of food: the nutritive "provides both the whole and the parts with being", whereas the "growth promoting causes increase of bulk". Aristotle seems to make the θρεπτικόν responsible for the constitution of the parts of the embryo. The θρεπτικόν doesn't preserve the living thing in its being, as in *de An.* II,4, but provides to the offspring the *being* of the whole and of the parts. This text introduces a distinction between two kinds of food according to the parts (the most noble and the necessary parts); it doesn't in itself undermine the distinction made in *GA* II,4 between the two stages of the embryonic generation: the constitution and the growth. It makes clear that the growth of the embryo requires the collaboration of two kinds of food, the αὐξητικόν and the θρεπτικόν.

To conclude, it appears now that Aristotle nowhere isolates any special function of the nutritive soul dedicated to the "generation" or the "formation" of the embryo, either in *de An.* II,4 or in the *GA*. From this point of view, the three texts quoted from the *GA* reveal that Aristotle is consistent with the definition of the nutritive soul given in *de An.* II,4. Even if, in the *GA*, Aristotle introduces different stages in embryogenesis, they all depend on the same two functions of the nutritive soul distinguished in *de An.* II,4. This means that, according to Aristotle, the two functions distinguished in the nutritive soul in *de An.* II,4 (the reproduc-

---

[38] On this text, see the explanations of Leunissen and Gotthelf (2010), p. 342–345.

tion and the "use of the food") are sufficient for explaining the four following processes: (a) the separation of the heart in the menstrual fluid, the heart being the first part generated by the motion produced by the heat of the male seed; (b) the separation or differentiation of the different parts, this διάκρισις being itself produced by the *pneuma* used by the nutritive soul in the heart; (c) the growth of these parts in the womb and (d) the growth of the living being once born, both processes caused by the use of the food made by the nutritive soul of the new living being. The fact that Aristotle uses analogies between the nutritive soul or its material instruments and the craftsman's work doesn't imply that the nutritive soul has a particular capacity to produce the embryo. What appears to us as the generation of new living being is conceived by Aristotle as a kinetic process, the communication of the motions (by which the nutritive soul of the father makes him grow and ensures his own conservation) to the nutritive soul of the embryo through the seed of the father and the heart of the embryo.[39]

Bibliography

Balme, David (transl.) (1992, 1972): *Aristotle, De Partibus Animalium I and De Generatione Animalium (with passages from II. 1–3)*. Oxford: Clarendon Press.

Caston, Victor (transl./comm.) (2012): *Alexander of Aphrodisias, On the Soul, Part 1, Translated with an Introduction and Commentary*. London, New York: Bloomsbury.

Code, Alan (1987): "Soul as Efficient Cause in Aristotle's Embryology". In: *Philosophical Topics* 15, p. 51–59.

Code, Alan (2004): "*On Generation and Corruption* I. 5". In: de Haas, Frans/Mansfeld, Jaap (eds.): *Aristotle*, On Generation and Corruption, Book I, Symposium Aristotelicum. Oxford: Clarendon Press, p. 171–193.

Connell, Sophia M. (2016): *Aristotle on Female Animals. A study of the* Generation of animals. Cambridge: Cambridge University Press.

Corcilius, Klaus (transl./comm.) (2017): *Aristoteles, Über die Seele, De anima, Griechisch-Deutsch, Übersetzt, mit einer Einleitung und Anmerkungen herausgegeben*. Hamburg: Meiner.

Drossaart Lulofs, Hendrik J. (ed.) (1965): *Aristotelis* De Generatione Animalium *recognovit brevique adnotatione critica instruxit*. Oxford: Clarendon Press.

Freudenthal, Gad (1995): *Aristotle's Theory of Material Substance. Heat and Pneuma, Form and Soul*. Oxford: Oxford University Press.

Furth, Montgomery (1988): *Substance, Form and Psyche: An Aristotelean Metaphysics*. Cambridge: Cambridge University Press.

---

[39] I would like to thank Sophia Connell and Sean Kelsey for their careful reading and valuable comments on a first draft of this article.

Gotthelf, Allan (2012): "Teleology and Embryogenesis in Aristotle's *Generation of Animals* II.6". In: *Teleology, First Principles, and Scientific Method in Aristotle's Biology*. Oxford: Oxford University Press, p. 90–116.

Gourinat, Jean-Baptiste (2008): "L'embryon végétatif et la formation de l'âme selon les Stoïciens". In: Brisson, Luc/Congourdeau, Marie-Hélène/Solère, Jean-Luc (eds.) : *L'Embryon, Formation et animation*. Paris: Vrin, p. 59–77.

Johansen, Thomas Kjeller (2012): *The Powers of Aristotle's Soul*. Oxford: Oxford University Press.

Hamlyn, David W. (transl.) (1993): *Aristotle, De Anima, Books II and III*. Oxford: Clarendon Press.

Henry, Devin (2009): "Generation of Animals". In: Anagnostopoulos, Georgios (ed.): *Blackwell Companion to Philosophy, A Companion to Aristotle*. Cambridge: Cambridge University Press, p. 368–383.

Hirai, Hiroi (2011): *Medical Humanism and Natural Philosophy. Renaissance Debates on Matter, Life and the Soul*. Leiden/Boston: Brill.

Kelsey, Sean (2003): "Aristotle's Definition of Nature". In: *Oxford Studies in Ancient Philosophy* 25, p. 59–87.

King, Richard (2001): *Aristotle on Life and Death*. London: Duckworth.

Lefebvre, David (2018a): *Dynamis, Sens et genèse de la notion aristotélicienne de puissance*, Paris: Vrin.

Lefebvre, David (2018b): "Aristotle's *Generation of Animals* on the separation of the sexes". In: Sfendoni-Mentzou, Demetra (ed.): *Aristotle – Contemporary Perspectives on his Thought*. On the 2400th Anniversary of Aristotle's Birth. Berlin/New York: De Gruyter, p. 75–93.

Leunissen, Mariska/Gotthelf, Allan (2010): "What's Teleology go to do with it? A Reinterpretation of Aristotle's *Generation of Animals* V". In: *Phronesis* 55, p. 325–356.

Lloyd, Geoffrey E.R. (1995): "Aspects of the Relationship Between Aristotle's Psychology and His Zoology". In: Nussbaum, Martha Craven/Oksenberg Rorty, Amélie (eds.): *Essays on Aristotle's De Anima*. Oxford: Oxford University Press, p. 147–168.

Lo Presti, Roberto (2014): "Informing Matter and Enmattered Forms: Aristotle and Galen on the 'Power' of the Seed". In: *British Journal for the History of Philosophy* 22, p. 929–950.

Louis, Pierre (ed./transl.) (1961): Aristote, *La Génération des animaux*. Paris: Vrin.

Menn, Stephen (2002): "Aristotle Definition of the Soul and the Programme of the *De anima*". In: *Oxford Studies in Ancient Philosophy* 23, p. 83–139.

Nussbaum, Martha Craven (transl./comm.) (1978): *Aristotle's De Motu Animalium, Text with Translation, Commentary and Interpretive Essays*. Princeton: Princeton University Press.

Peck, Arthur L. (transl.) (1942): *Aristotle, Generation of Animals*. Cambridge, Mass., London, England: Loeb Classical Library.

Pellegrin, Pierre (2018): "What is Aristotle's *Generation of Animals* about?". In: Falcon, Andrea/Lefebvre, David (eds.): *Aristotle's Generation of Animals, A Critical Guide*. Cambridge: Cambridge University Press, p. 77–88.

Preus, Anthony (1975): *Science and Philosophy in Aristotle's Biological Works*. New York: Olms.

Ramelli, Ilaria/Konstan, David (2009): Hierocles The Stoic: *Elements of Ethics, Fragments, and Excerpts*. Atlanta: Society of Biblical Literature.

Rashed, Marwan (ed./transl.) (2005): *Aristote, De la génération et la corruption*. Paris : Les Belles Lettres.

Rashed, Marwan (2018): "A Latent difficulty in Aristotle's Theory of Semen. The Homogeneous Nature of Semen and the Role of the Frothy Bubble". In: Falcon, Andrea/Lefebvre, David: *Aristotle's* Generation of Animals, *A Critical Guide*. Cambridge: Cambridge University Press, p. 108–129.

van der Eijk, Philip (2000): "Aristotle's Psycho-Physiological Account of the Soul-Body Relationship". In: John P. Wright/Paul Potter (Eds.): *Psyche and Soma, Physicians and Metaphysicians on the Mind-Body Problem to Antiquity to Enlightenment*. Oxford: Oxford University Press, p. 57–77.

van der Eijk, Philip (2005): "The Matter of Mind: Aristotle on the Biology of 'Psychic' Processes and the Bodily Aspects of Thinking". In: Id. (ed.), *Medicine and Philosophy in Classical Antiquity, Doctors and Philosophers on Nature, Soul, Health and Disease*. Cambridge: Cambridge University Press, p. 206–237.

Hynek Bartoš
Aristotle and his Medical Precursors on Digestion and Nutrition

Abstract: In this essay I focus on the Hippocratic *De carnibus* and *De diaeta* and make the case that both texts, each in its peculiar way, represent a specific medical tradition upon which Aristotle draws in his accounts of nutrition and digestion and against which he introduces his own refined version of the concept of innate heat/fire.

Aristotle's rare discussions of digestion and nutrition are laconic and often conclude with a reference to other studies, an exhortation to further research, or a remark that the presented account is only preliminary.[1] Nonetheless, in all these accounts Aristotle draws on the assumption that life and the possession of soul depend upon some degree of heat, "for digestion, by which animals assimilate their food, cannot take place apart from the soul and heat; for all food is rendered digestible by fire" (*de Juventute et Senectute, de Vita et Morte, de Respiratione* 14(8), 474a25–29). Moreover, he believes that the vital fire (or heat), by means of which all food is concocted,[2] is itself fed on moisture or water (*de Partibus Animalium* II,3, 650a3–4, *Meteorologica* II,2, 355a5, *de Anima* II,4, 416a25–27) and that it burns and draws into itself the nutritive moisture from the body as long as the animal lives (cf. *Mete.* IV,1, 379a23–26).[3] Accordingly, he takes for granted that all animals are "*naturally* moist and warm" (*de Longitudine et Brevitate Vitae* 5, 466a18–19)[4] and that the warm moisture causes their "growth as well as life" (*Long.* 5, 466b22–24).

There is no doubt that Aristotle has found in the concept of vital heat/fire an instrument "through which he can explain the myriad of functions in the living organism, from its very generation to its passing away" (Mendelssohn 1964, p. 13). "But to see Aristotle ascribing such an important role to the notion of

---

1 Cf. *de An.* II,4, 416b30–32; *PA* II,3, 650a30–31 and 650b8–11; III,14, 674a19–22; IV,4, 678a17–20; *de Generatione Animalium* II,6, 744b28–37.
2 For various meaning of "concoction" in Aristotle, see Peck (1942), p. lxii–lxvii, Wilson (2020) and Popa (2020).
3 The same holds for plants as well, since they also have nutritive soul and "are nourished and continue to give, as long as they are able to absorb food" (*de An.* II,1, 412b29–30, transl. Hett).
4 Cf. *Historia Animalium* I,4, 489a20–21, transl. Peck (modified): "Now every animal contains moist, and if it is deprived of this either in the natural course or forcibly, it perishes."

vital heat is surprising", points out Freudenthal (1995, p. 70) and raises the following questions of particular significance: "What are Aristotle's grounds for his strong claims on behalf of vital heat? And why are these claims not stated explicitly and made the object of a full-fledged theory?"

Freudenthal, who focuses on the natural upward motion of the vital heat and its *macrocosmic* dimension that establishes Aristotle's *scala naturae*, identifies a tradition of theological cosmology attested in the Hippocratic *De carnibus*, the "Pythagorean Notebooks", and in the fragments of Diogenes of Apollonia and of Aristotle's *de Philosophia*. All these accounts "gravitate about one and the same doctrine" which is monistic, claims Freudenthal, and "evidently a Heraclitean legacy". "Of this heat-centred, theologically and cosmologically founded physiology we find definite traces in Aristotle's biological treatises [...] the vital heat of Aristotle's biological treatises is a de-theologized version of the earlier divine *thermon*", concludes Freudenthal (1995, p. 98).

If we focus on the *microcosmic* dimension, i.e. on the vital heat in animal bodies and its role in their growth, digestion and nutrition, an alternative explanation comes into consideration. The Roman encyclopaedist and historian of medicine Celsus ascribes the idea that food is concocted by the heat to Hippocrates,[5] and it is therefore reasonable to consider the possibility that Aristotle adopted this specific view from the medical tradition. As far as I know, this possibility has been only briefly suggested by W. Ogle but never developed in detail.[6] Accordingly, in this essay I aim to elaborate this hypothesis and to make the case that Aristotle's conception of innate heat was considerably inspired by a specific medical tradition attested in the so-called "Hippocratic" writings.

Two Hippocratic texts, namely *De carn.* and *De diaeta*, have been suggested as the most relevant sources for Aristotle's concept of fire/heat by W. Kullmann (2007, p. 303), T. Popa (2014) and other contemporary scholars.[7] Most recently, G.

---

[5] Celsus (*De Medicina* I, prooem. 20, transl. Spencer): "[...] some following Erasistratus hold that in the belly the food is ground up; others, following Plistonicus, a pupil of Praxagoras, that it putrifies; others believe with Hippocrates, that the food is cooked up by heat (*per calorem cibos concoqui*)."

[6] W. Ogle (1882, p. 159) in his commentary to Aristotle's *PA* II,3 writes: "the opinion that digestion is due to heat appears to have originated with Hippocrates, and was adopted by Aristotle and also by Galen."

[7] The relevance of *De carn.* to Aristotle has been discussed in detail by Freudenthal (1995), p. 74–105, and Oser-Grote (2004), p. 27–33. As for the suggested correspondences between *De diaeta* and Aristotle's zoological accounts, see van der Eijk (1995 and 2007); Bartoš (2015), p. 241–286. See also Ch. Lefèvre (1972), p. 182–214, who developed a hypothesis that Aristotle in his early years adopted the concept of vital heat/fire from the Hippocratic *De diaeta*. In contrast to Lefèvre, my aim is to argue that Aristotle's inspiration drawn from medical tradition

Betegh (2020, p. 53) concludes that in comparison with all other extant pre-Aristotelian evidence, "it is precisely in these Hippocratic texts that we find the motive force of heat and fire stated most explicit, and used as an explanatory principle in the most systematic way". Following these suggestions, in this essay I focus on the evidence of the Hippocratic *De carn.* and *De diaeta* and make the case that both texts, each in its peculiar way, are highly relevant to Aristotle for they conveniently represent the tradition (otherwise poorly documented in pre-Aristotelian evidence) upon which Aristotle draws and from which he adopts some of the most essential ideas of his own accounts of nutrition and digestion.[8] I suggest that reading Aristotle's accounts against these texts can shed new light on Aristotle's rather selective discussions of the topic. A consensus with respected medical authorities, on the one hand, gives Aristotle sufficient grounds for making strong claims without the need of their justification; a disagreement with the same authorities on particular details of the theory, on the other hand, can explain why Aristotle in his brief and dense accounts of the topic meticulously discusses seemingly minor and far-fetched points.

Before focusing on passages directly related to digestion and nutrition, in the first part I shortly discuss Aristotle's account of brain in *PA* II,7 which reveals a number correspondences with the two Hippocratic texts and illustrates that they both (especially when in combination) provide a remarkably complex picture of the pre-Aristotelian tradition.[9] In the second part I focus in more detail on *De carn.* and its main ideas related to digestion and the digestive system,

---

should not be restricted to one single medical source (i.e. *De diaeta*) and to Aristotle's early years only.

[8] Given the fact that I have already discussed in detail the relevant passages from *De diaeta* elsewhere (most recently and systematically in Bartoš 2015), in this paper I only briefly summarize my understanding of the text (with references to my previous publications) and devote special attention to *De carn.*, one of the most valuable (but unfortunately often neglected) non-fragmentary evidence of pre-Platonic philosophy of nature.

[9] Both Hippocratic authors admit to draw on preceding tradition (*De carn.* 1; *De diaeta* I,1), in both texts there are evident traces of pre-Socratic philosophy. The influence of Heraclitus is especially significant in *De diaeta*, the account of which can be interpreted as "a rehabilitation of the doctrine of Heraclitus" (Peck 1928, Appendix, p. xii–xiv; cf. Bartoš 2015, p. 113–129). Aristotle explicitly mentions Heraclitus and Hippasus of Metapontum in *Metaphysica* I,3, 984a8–9, among earlier proponents of the idea, and also Democritus in *de An.* I,2, 404a1–2 and 405a7–9. In *Metaph.* XIII,4, 1078b19–21, Aristotle praises Democritus as the only natural philosopher who had some knowledge of the essence and "in a way defined the hot and cold" (ὡρίσατό πως τὸ θερμὸν καὶ τὸ ψυχρόν). It is also noteworthy that according to Diogenes Laertius (IX,47) Aristotle himself wrote a treatise entitled "Causes Concerned with Fire and Things in Fire" (Αἰτίαι περὶ πυρὸς καὶ τῶν ἐν πυρί).

and in the third part I return to Aristotle and discuss his accounts of growth and digestion in the light of the Hippocratic evidence.

1 Brain and the Balance of Heat

Brain was a rather puzzling organ for Aristotle[10] as well as for his predecessors[11] and its function has been a matter of scientific disputes. Aristotle opens his account of the organ in *PA* II,7 with a discussion of its material nature, claims that it is "the coldest of the parts within the body" (*PA* 652a27–28), and supports it with the observation that "the coldness of the brain is apparent to the touch; and furthermore, it is the most bloodless of all the moist parts in the body (indeed it has no blood at all in it), and the driest" (*PA* 652a33–36). When he turns to the question of its function, he briefly refuses the possibility that brain contributes to perception, and then introduces his own teleological explanation: "it is present in order to preserve the animal organism as a whole" (*PA* II,7, 652b7). To specify in what sense brain preserves the organism "as a whole", Aristotle shortly explains, with a reference to previous tradition, the concept of vital heat and the idea of thermic balance in the body:

> For while some crudely posit fire or some such potential to be the animal's soul, it is perhaps better to say that soul is constituted in some such body (ἐν τοιούτῳ τινὶ σώματι). This is because among bodies the hot is the one most able to assist with the functions of the soul; for nourishing (τρέφειν) and producing change (κινεῖν) are functions of soul (ψυχῆς ἔργον ἐστί), and these things come about most of all through this potential. Saying fire is the soul, then, is like saying the saw or auger is the carpenter or carpentry because the function is accomplished when they are near each other. (*PA* II,7, 652b7–15, transl. Lennox – modified)

In this short passage one can identify at least three remarkable traces of ideas advocated in the Hippocratic *De diaeta*. First, the very same two capacities in which fire assists the soul on Aristotle's account, namely nourishing (τρέφειν) and producing change (κινεῖν), are identified in the Hippocratic text as two vital aspects of human nature, represented by water and fire,[12] which need to be kept in balance by means of food, drinks, exercises and other dietetic means (*De diaeta* I,2). Secondly, when the Hippocratic author attempts to relate

---

10 Cf. Lennox (2020).
11 Cf. van der Eijk (2005), p. 119–135; Bartoš (2018).
12 Cf. *De diaeta* I,3: "fire can move all things always (πάντα διὰ παντὸς κινῆσαι), while water can nourish all things always (πάντα διὰ παντὸς θρέψαι)."

the "the strongest and hottest fire" to soul in chap. 10, he uses the same phrase (ἐν τούτῳ) as Aristotle recommends in our passage as an alternative to the identification of animal's soul with fire.[13] And finally, the striking and seemingly farfetched analogy between the works of fire in the body and the craft activities of sawing and auger-boring can be understood as a critical remark on the use of the same analogies in the Hippocratic *De diaeta* (in chap. 6, 7 and 16),[14] to my knowledge the only pre-Aristotelian text drawing the same two analogies to illustrate that the power of the fire is permanently counterbalanced by its opposite (i.e. the water which is essentially cold). Accordingly, Aristotle seems to correct the particular Hippocratic analogies by suggesting a distinction between the carpenter, which is the soul on his own account, and his tools, which represent the fire/heat and the water/cold in the Hippocratic analogy as well as in Aristotle.[15]

Having thus clarified his own conception of vital heat in continuation with preceding tradition, Aristotle concludes his argument concerning the function of the brain as follows:

> That animals must partake of heat is clear from these considerations; and since everything requires an opposing counterweight in order that it achieve the moderate state and the mean (for the mean possesses substantial being and the defining account, while each of the extremes separately does not) – because of this nature has devised the brain in relation to the heart's location and heat. And it is for the sake of this that this part, with the combined nature of earth and water, is present in animals. For this reason too all the blooded animals have a brain, while virtually none of the other animals does, excepting those that have brain by analogy, such as the octopus;[16] for they all have little heat on the account of being bloodless. (*PA* II,7, 652b15–26, transl. Lennox – modified)

---

13 Cf. *De diaeta* I,10: "The hottest and strongest fire, which controls all things, ordering all things according to nature, imperceptible to sight or touch, wherein (ἐν τούτῳ) are soul, mind, thought, motion, growth, decrease, mutation, sleep, waking. This governs all things always, both here and there, and is never at rest." For more details on this passage, see Bartoš (2015), p. 257–260.
14 Cf. Bartoš (2012). If Aristotle aimed only at illustrating the difference between the fire as an instrument and the soul as a craftsmen, one would expect him to speak about a fire in a fireplace or oven, operated by soul, the master-cook who always knows his goals and how to achieve them, and who tends the fire in the oven, turns the heat up or down according to the situation and desired product, etc. Cf. Aristotle, *GA* II,6, 743a18–21, transl. Peck (modified): "The sinews and bones are formed, as the moisture solidifies, by the agency of the internal heat; hence bones (like earthware) cannot be dissolved by fire; they have been baked as it were in an oven by the heat present at their formation."
15 Cf. Bartoš (2015), p. 257–259.
16 On the other hand, Aristotle repeatedly ascribes brains to all cephalopods in *HA* (I,16, 494b25–29; IV,1, 524b2–4, and 524b32–34). For a discussion of this discrepancy, see Lennox (1996) and (2020), p. 238–242.

Aristotle's argument draws on four assumptions: (i) "animals must of necessity have in them a certain amount of heat"; (ii) "everything requires an opposing counterweight in order that it achieve the moderate state and the mean", accordingly also the heat in the body needs some counterweight; (iii) most of the heat is located in the heart; (iv) the brain is the coldest part in the body and therefore – at least in this particular argument[17] – the most convenient candidate to be identified as the organ counterbalancing the heat of the heart.

The first assumption is shared by both Hippocratic authors under consideration as well as several other medical authors of the time.[18] The second one, i.e. that all the elemental constituents of the body need to be kept in balance, that each constituent is necessarily counterbalanced by its opposite, is attested in a limited number of Hippocratic texts, especially those discussing dietetic principles (such as *De natura hominis*, *De morbo sacro*, *De aere aquis locis*, and *Aphorismi*).[19] The author of *De diaeta*, the most elaborate ancient account of dietetics, combines the second assumption with the first one right at the outset of his elemental account of human (and animal) nature (*De diaeta* I,3).[20]

As for the remaining two assumptions, namely that the heat is localized in the heart and the cold in the brain, *De diaeta* is of no help due to its author's vagueness concerning internal anatomy and the fact that he entirely omits from his account heart (as well as lungs and many other vital organs), and brain is mentioned only once, among edible parts of animal bodies with remarkable nourishing qualities (brain is classified as a "bloodless" and particularly moist organ with the strongest nourishing effect on the human body, *De diaeta* II,49, Joly-Byl 172,7–10). The Hippocratic *De carn.*, on the other hand, expresses both assumptions sufficiently clearly: brain is "the metropolis of the cold" (*De carn.* 4), "the heart and the hollow vessel (ἡ κοίλη φλέψ) have the most heat" (*De carn.* 5). In other Hippocratic texts we find only little support for these

---

[17] It should be noted that in *PA* III,6 Aristotle on the same assumption (i.e. that "it is necessary to have some means for cooling the heat" in the body) draws a conclusion that animals that breath have lungs for the sake of cooling (668b33–669a1).

[18] E.g. *De nat. hom.* 12; *Aphor.* I,14 and 15; *De liquidorum usu* 2; *De arte* 12, *De morbis* I,11. As for Aristotle's claim that all animals are "*naturally* moist and warm, and that life too is of this nature, whereas old age is cold and dry" (*Long.* 5, 466a18–20), see the Hippocratic *De nat. hom.* (ch. 12) and *De diaeta* (I,33).

[19] Few lines after our passage Aristotle for this conception employs the expression "proportioned blend, *symmetros krasis*" (*PA* II,7, 652b33–35), which indicates that he uses here technical language of dietetic debates (cf. Bartoš, forthcoming). E.g. in the Hippocratic *Aphorisms* we read that only those women can conceive who have "a proportional blend (*krasin symmetron*)" of hot, wet, dry and moist elements in their body (*Aphor.* V,62, L. IV,554–556).

[20] Cf. Bartoš (2015), p. 72–82.

two assumptions[21] and, as far as I know, none attests both assumption in combination.[22] Correspondingly, the Hippocratic author endorses the *hoti* part of Aristotle's argument, i.e. that the brain is indeed the coldest part of the body and the heart the hottest one, and thus paves the way for the idea that these organs create in the body a kind of thermic counterbalance,[23] a remarkable speculation of crucial importance in Aristotle's teleological accounts of the two organs.

In summary, if read against the evidence of the two Hippocratic texts, Aristotle's brief account of brain can be understood as follows: he starts with a discussion of the relevant facts and, on the basis of empirical observations as well as theoretical considerations, refutes rival theories (i.e. that brain contributes to perception).[24] Then he identifies the most appropriate first principles suggested by his predecessors (i.e. the concepts of vital heat and thermic balance) and introduces corrections and adjustments to the specific conceptions and formulations of the most advanced accounts available at the time (such as the Hippocratic *De diaeta* and *De carn.*). Finally, he suggests an explanation in which he adopts from this tradition the most relevant conceptions and alleged facts (including the assumption that the heart is the seat of the heat and the brain the

---

[21] In contrast to the belief that brain is the coldest part of the body (shared by Aristotle and the Hippocratic author of *De carn.*), it is remarkable that according to *De diaeta* the brain belongs to the moistest (rather than coldest) parts of animal bodies (*De diaeta* II,49), and according to *Aphor.* V,18 cold is actually harmful to the brain. In other words, Aristotle's belief is neither an indisputable piece of empirical experience nor a generally accepted opinion.

[22] Apart from *De carn.*, the closest parallel to Aristotle's account is in the Hippocratic *De liquid. usu* 2,1, were we read that brain and other cold parts on the periphery of the body are located "far away from the body's own heat (τοῦ οἰκείου θάλπεος)", a claim which seems to imply that this heat, or its source, is located somewhere around the centre of the body.

[23] Cf. Aristotle, *GA* II,6, 743b26–32; *PA* III,11, 673b9–11 and IV,10, 686a5–10.

[24] Hippocratic authors occasionally mention the connection between the brain and individual sense organs (*De locis in hom.* 2; *De diaeta salubri* 8), some presuppose that motor functions and speech are correlated to the brain (*Coacae praenotiones* 489), but only few submit a complex theory about the central role of brain in perception and cognition. Such a theory has been ascribed already to Alcmaeon of Croton, Plato's Socrates mentions brain as one of the candidates to be the part of body responsible for thinking (*Phaedo* 96b2–8), and the most elaborated formulation of the theory is attested in the Hippocratic *De morbo sacro* (16–17). The author of *De carn.* explicitly denies that it would be the brain which echoes, for the brain, as he argues, is moist and moist things do not resonate: "There are many proofs that what is driest echoes best; and when a thing echoes best, we hear best" (*De carn.* 15, Potter 152,23–25). Even though the author is well aware of the connection between the eyes and the brain (*De carn.* 17), he does not assign to the brain any specific role in the process of vision. Nevertheless, there is a perceptive role which the author ascribes to the brain: "The brain, being itself moist, perceives the smell of dry things, by drawing the odour along with air through the bronchial tubes which are dry" (*De carn.* 16, Potter 154,6–8).

seat of the cold), and – after necessary adjustments – recombines them in teleologically structured arguments. As I shall argue, in his accounts of digestion and nutrition Aristotle adopts a similar strategy.

2 Nourishment and Digestion in the Hippocratic *De carnibus*

The author of the Hippocratic *De carn.* provides one continuous account based on a single hypothesis, accompanied by a limited set of explanatory principles which suffice him to explain a broad range of natural processes, starting with cosmogony, animal and human embryology and concluding with human diseases. As P. Potter (1995, p. 129) puts it, "the theory put forth in the first part of the treatise and developed with such energy and care, marks its author as one of the most extraordinary scientific minds of the Corpus; the reasoning here is comparable in power and originality to anything we find in *Generation-Nature of the Child, Diseases IV, Regimen,* or the *Sacred Disease.*"

The author introduces his enterprise in the first chapter as a medical account aiming at explanations for various natural phenomena based on "one common principle":

> In this treatise I shall employ assumptions that are generally held – both those of my predecessors and my own – since it is necessary to establish a common starting point (*koinê archê*), if I wish to compose this treatise about medicine. About what is in the heavens I have no need to speak, except insofar as is necessary in order to explain how man and the other animals are formed and come into being, what the soul (*psychê*) is, what health and sickness are, what in man is evil and what good, and where his death comes from. (*De carn.* 1, transl. Potter 132,1–10)

Out of the topics enumerated in this passage, only the first one (i.e. "how man and the other animals are formed and come into being") is really covered in the treatise. It is noteworthy that soul (*psychê*) is also mentioned among the phenomena explainable by means of the common principle, i.e. the heat, although the expression *psychê* is never mentioned again in the treatise.

Heat as the main cosmic agent and the main explanatory principle is introduced in the second chapter (*De carn.* 2, Potter 132,12–14): "I believe that what we call heat is in fact immortal, that it perceives all things, and sees, hears and knows all that is and all that will be." The immediate connection of the heat with the sense perception ("perceives all things, and sees, hears and knows [...]") plays no significant role in the accounts of hearing, smell, seeing and speech in chap. 15–18. The heat's role consists in forming and arranging the particular bodily parts in such a way that their structure mechanically brings perception to a man. Apart from chap. 2 and the first part of chap. 3, the author does not say much about macrocosmic processes and is fully devoted to the explanation of the generation and growth of the human body. Heat is the most prominent element and moving force of the natural processes involved in embryonic development, in which cold also plays a significant role (for instance the formation of liver is explained in chap. 8 as a result of the process in which "cold gained mastery over the heat").[25] All organs and structures of the human body are described as a result of the effect of the heat (and the cold) on various materials. By means of heat and cold, together with dry and wet, the author defines other corresponding qualities, most significant of these are the "fatty" (*liparos*), which is hot and dry, and the "sticky" or "gluey" (*kollôdês*), which is cold and wet (chap. 3, 4, 5, 7, 8, 9, 10, 12, 13, 14, 17). In chap. 3 the author summarizes the "fundamental processes" (Craik 2015, p. 45) involved in the formation of the various bodily parts and structures as follows:

> These [processes] occur according to the following [principles]: Cold condenses, heat melts and, over a long period of time, also dries. Where any fat is present with other materials, heat burns it up and dries it very quickly; where, on the other hand, gluey material is present with cold but without fat, what is gluey refuses to be burnt up, but, on being heated for a time, condenses (i.e. into a membrane).[26]

The most repeated pattern used in the author's embryology is that the process of heating and drying results in the formation of a new structure limited by solidi-

---

[25] It is noteworthy that the Hippocratic author, like Aristotle, explains the formation of the flesh as a result of the agency of cooling: "About muscles the explanation is the same: cold brought what was moist to a stand-still, congealed it, and turned it into muscle (τὸ μὲν ψυχρὸν ἔστησε καὶ ξυνέπηξε καὶ ἐποίησε σάρκα), while what was gluey formed canals; in these canals there is moisture just as in the large vessels" (*De carn*. 9, Potter 144,25–146,4). Cf. Aristotle, *GA* II,6, 743a8–11, transl. Peck 1942: "As the nourishment oozes through the blood-vessels and the passages in the several parts (just as water does when it stands in unbaked earthenware), flesh (σάρκες), or its counterpart, is formed: it is the cold which 'sets' the flesh (ὑπὸ τοῦ ψυχροῦ συνιστάμεναι), and that is why fire dissolves it (διὸ καὶ λύονται ὑπὸ πυρός)."

[26] *De carn*. 3, Potter 136,21–27: περὶ μὲν τουτέων οὕτως· τὸ μὲν ψυχρὸν πήγνυσιν· τὸ δὲ θερμὸν διαχέει, ἐν δὲ τῷ πολλῷ καὶ ξηραίνει χρόνῳ· ὅκου δὲ ἂν τοῦ λιπαροῦ ξυνῇ τι τουτέοισι, θᾶσσον ἐκκαίει καὶ ξηραίνει· ὅκου δὲ ἂν τὸ κολλῶδες ξυνίῃ τῷ ψυχρῷ ἄνευ τοῦ λιπαροῦ, οὐκ ἐθέλει ἐκκαίεσθαι, ἀλλὰ τῷ χρόνῳ θερμαινόμενον πήγνυται. Potter's translation of the opening clause ("these processes occur according to the following principles") uses the expressions "processes" and "principles" which are not in the Greek text. Although a perfectly possible translation, it may cause a confusion concerning the author's aim to find one common principle (*koinê archê*) of all processes under discussion.

fied envelope. This envelope, a new interface in the body, is called a "coat" (*chitôn*), "membrane" (*mêninx*), "skin" (*derma*), or "caul" (*hymên*). Similar accounts, in which the inner bodily structures like bones, tendons, and vessels are formed, as it were, "in a sort of gigantic kitchen, where fire burned and roasted and liquefied the earth" (Jouanna 1999, p. 279), can be found also in the Hippocratic *De diaeta* and in Aristotle's *GA*.[27]

It should be noted that the summary of the fundamental processes in chap. 3 defines the capacities of the heat and the cold to transform various kinds of material, but says nothing either about the growth of the bodily parts in terms of feeding, or about the specific relation between the two elements. Digestion as a topic for discussion is introduced for the first time in chap. 6, in which the author explains the constant movement (i.e. pulsation) of the heart and announces a crucial assumption of his account, namely that the heat feeds on the cold:

> The origin of this movement can also be understood from another example: if you light a fire inside a room when no wind is blowing in, the flame moves, sometimes more, sometimes less, although no wind is stirring that we are able to perceive. And we know that the nourishment (*trophê*) of the hot is cold. For example the fetus in the belly continually sucks with its lips from the uterus of the mother and draws nourishment (*trophê*) and breath (*pneuma*) to its heart inside, for this breath is hottest in the fetus just at the time that the mother is inspiring. To this [i.e. the heart of the fetus] and to the rest of the body, too, heat gives movement, and to all other things. (*De carn*. 6, transl. Potter 142,7–13)

The example with the flame in the lamp illustrates that the heat not only feeds on the cold but also that it actively seeks for its nourishment and draws it from its environment. The author assumes that the same happens with the heat in the heart of an embryo, and his remark that "this breath is hottest in the fetus just at the time that the mother is inspiring" evidently suggests a direct connection between the heat in the heart and the process of respiration. Nonetheless, he unfortunately does not develop this idea any further.[28] Instead, he claims that the

---

[27] E.g. *De diaeta* I,9 (transl. Jones 1923): "[...] owing to the movement and the fire it [i.e. the fetus] dries and solidifies; as it solidifies it hardens all round, and the fire being imprisoned can no longer draw to itself its nourishment in sufficient quantity, while it does not expel the breath owing to the hardness of its envelope. So it consumes the available moisture inside. Now the parts in the compacted, dry mass that are solid in substance are not consumed by the fire for its nourishment, but they prove powerful, and as the moisture fails they become compact, and are called bones and sinews." As T. Popa (2014, p. 1–20, 8, n. 17) suggests, the Hippocratic *De diaeta* and *De carn*. "anticipate somewhat similar material accounts in Aristotle's *Meteorology* IV and *Generation of Animals* II".

[28] It is especially remarkable that the author speaks on the same footing about "nourishment and breath", two distinguishable conditions of life, both of which he aims to explain. A number

heat draws into itself not only "breath" but also "nourishment", which is an addition of crucial importance for the remarkably complex discussion of the formation of teeth in chap. 12 and 13.

The author explains how teeth come into being (i.e. they are "dried and burnt up" by the heat out of "gluey and fatty" material), why they are harder than other bones ("because they contain no cold"), and where do they get nourishment from (the primary teeth "grow from the regimen in the uterus, and after birth from milk, in the child at the breast, when these have fallen out, the permanent teeth take their growth from foods and drinks", *De carn.* 12, Potter 148,8–22). Then he turns to a specific question concerning the relation between the growth of the secondary teeth and other parts of the body, namely bones, and explains what actually happens with the ingested "foods and drinks" before they become nourishment for the growth of the teeth:

> Teeth are formed later than the other bones for the following reasons. In the jaws there are vessels which provide only them among bones with nourishment from the lower cavity (*koiliê*). Now bones increase, as do all other parts, by adding what is similar to their own particular quality. For when food and drink collect in the stomach and intestines, and are heated, the vessels arising there draw off the finest and moistest part, leaving the thickest part behind, which turns to faeces in the lower intestines. As the foods are heated, then, these vessels draw off the finest part from the stomach and the upper intestines – the part above the jejunum – and as the foods pass through the jejunum into the lower intestines, they solidify and become faeces. When the nourishment arrives, it gives up the particular quality corresponding to each part, for it is by being watered by this nourishment that every part increases, the hot, the cold, the gluey, the fat, the sweet, the bitter, the bones, and all the other parts that are in a person.

---

of other Hippocratic authors also take for granted that humans and other (respiring) animals need not only food and drink in order to survive but also (and on some accounts even more essentially, cf. *De flatibus* 3) air, but only few recognize the relevance of respiration to digestion, and no one, as far as I know, succeeds in explaining both. The author of *De diaeta*, who defines water as the only element nourishing the innate fire (I,4), also recognizes the significance of *pneuma* for metabolic processes and in agreement with *De carn.* presupposes that the embryo "draws to itself its nourishment from the food (τὴν τροφήν) and breath (καὶ πνεύματος) that enter the woman" (*De diaeta* I,9, see also I,13), although he does not sufficiently clarify its relation to the fire and does not provide any explanation of breathing. The author of *De natura pueri* assumes that vital heat is nourished by cold breath (*De nat. pueri* 1 = 12 L. VII,486; and 11 = 22 L. VII,514). In the Hippocratic *De alimento* we read that *pneuma* feeds the lungs, or perhaps something in the lungs (29). The author of the Hippocratic *De morbis* IV,47 (Loeb 138–140, transl. Potter 1995) mentions the cooling role of pneuma in his description of the moment death: "And the body, not being able on account of its weakness – since all its moisture has become morbid – to draw breath (*pnoê*) in order to cool (*hôs diapsychêtai*) what is in the cavity, will expire all the living part (*to zôtikon*) of its moisture, and so the person dies."

> Here is why teeth are formed at a later time. I mentioned above that the jaws alone of the bones have vessels inside themselves, and this is why more nourishment is drawn to them than to other bones. Since they have more nourishment coming in a more massive afflux, they continue to increase by adding what is similar to their own particular quality, until a person has reached adulthood. A person reaches adulthood when he has acquired his definitive form, and this generally occurs between seven and fourteen years of age. In that time all the teeth, including the largest, are formed, once those that came into being from the nourishment in the uterus have fallen out. He continues to growth into the third seven-year period, in which he becomes a young man, and even until the fourth and fifth seven-year periods. In the fourth seven-year period two more teeth are formed in many persons, and these are called the wisdom teeth. (*De carn.* 13, transl. Potter 148,23 – 152,6)

In his answer to the question why teeth are formed later than the other bones the author introduces a general assumption according to which all bodily parts grow by "adding what is similar to their own particular quality", which some commentators of the passage regard as inconsistent with his previous claims, namely with the assertion in chap. 6 that the heat is fed on the cold (Willerding 1914, p. 63; Deichgräber 1973, p. 48), or with the "chemical principles" introduced in chap. 3, as Potter (1995, p. 130) suggests. Both objections, I suspect, confuse two different aspects of the topic and two levels of the author's description, namely (a) the level of the elemental qualities, on which the heat feeds on cold, and (b) the level of the products made by the heat and cold, namely the bodily parts (especially the "homoiomerous" parts in Aristotle's terminology),[29] including the nutriment (i.e. the digested foods and drinks) which is transported via the blood vessels from the lover cavity to all other bodily parts. The fact that each of these levels of description requires a different kind of explanation, is not an inconsistency but rather a remarkable scientific achievement which plays a significant role also in Aristotle's account, as we shall see shortly. On my reading, there are three distinct phases of the digestive process in the Hippocratic account:

(1) In the first phase "food and drink collect in the stomach and intestines, and are heated". The food is concocted in the stomach and the upper intestines and processed into an intermediate product, which will be further separated and processed in the following stages.
(2) In the second phase, the crucial role is ascribed to the blood vessels attached to the stomach and the upper intestines which draw into themselves the "finest and moistest part" of the intermediate product ("of the heated food", θερμαινομένων τῶν σιτίων), while the rest moves through the jeju-

---

[29] Cf. Aristotle, *Mete.* IV,12, 390b3 – 14.

num into the lower intestines, it solidifies and becomes "faeces".[30] Thus the "foods and drink" mixed together and processed in the stomach and the upper intestines are divided into "faeces" on the one hand, and "nourishment" (*trophê*) on the other. This "nourishment" is distributed around the body via the blood vessels.

(3) The digestive process is completed when the "nourishment" reaches the terminal recipients (i.e. the parts which are nourished, for instance the growing teeth) and supplies each part with the corresponding quality (ἰδέα/εἰδέα) necessary for its growth.

In the Hippocratic account, only the final nourishment (in phase 3) feeds the particular parts of the body according to its character, i.e. like to like ("the particular quality corresponding to each part", "what is similar to their own particular quality"). Accordingly, two different kinds of "nourishment" must be distinguished in the Hippocratic account,[31] two different types of growth and two different patterns of growth: the cold is defined as (A) the "nourishment" of the heat, the heat feeds on its contrary: cold. There are two sources of the nutritive cold upon which the innate heat feeds: the inhaled air (which travels, according to the Hippocratic author, up to the heart, in which the greatest part of the heat in the body resides), and the food and drink collected in the stomach. As for the latter source, concoction in the stomach (and elsewhere) is a kind of by-product of the action of the heat upon the foods and drinks, and (B) the "nourishment" of the body is a kind of waste or leftover from the heat's metabolism.[32] As the cold quality contained in the foods and drinks is progressively consumed by the heat in the stomach, the foods and drinks disintegrate and melt. Given the fact that nutritive part of the concocted food is transported via

---

[30] Cf. Aristotle, *Mete.* IV,3, 381b12–14, transl. H. D. P. Lee: "For digestion takes place in the upper belly (*koilia*) and the excrement decays in the lower. The reason for this we have explained elsewhere."

[31] The author of *De carn.*, I suggest, holds a similar position as attested in *Alim.* 29: "The lungs draw a nourishment which is the opposite of that of the body, all other parts draw the same." Both authors consciously employ both patterns of change ("like to like" and "contrary is fed by contrary") in their account of nutrition and neither feels obliged to restrict to one of them only.

[32] Cf. Aristotle, *PA* II,2, 647b21–28, in which he differentiates between (i) the uniform/homoiomerous parts which act as material to the instrumental/anhomoiomerous parts (e.g. flesh, sinews, bones), (ii) the "fluid" parts that act as nutriment for the first kind of parts (e.g. blood), and (iii) the residues produced from the second kind of parts, for instance the solid as well as fluid excrements (or secretions like pus and rheum, cf. *Mete.* IV,2, 379b30–32 and IV,3, 380a21–22). See also Aristotle's account of plants in *PA* II,10, 655b35–37, according to which plants yield fruits and seeds instead of residues.

the blood vessels, and that the author uses the image of irrigation ("being watered (ἀρδόμενα) by this nourishment"),[33] one can even assume that this final nourishment is the blood. Although the author mentions blood in no more than two passages, these are highly significant. First, to support his claim that bodily moistures congeal (unless they include something gluey or fatty) as soon as the cold in them masters the heat, the author refers to the following evidence: "This is my proof (τεκμήριον). When you slaughter a sacrificial animal, as long as its blood remains warm (θερμόν) it is liquid (ὑγρόν), but when it becomes cold, it congeals (ἐπάγη)".[34] And second, a similar experience with cutting, this time into human bodies, is mentioned in chap. 9 in which he speaks of a "demonstration":

> There is heat in the whole body, in fact it constitutes the greatest part in the body; in the moisture of the body, again, there is much cold; in fact so much cold that it could congeal this moisture (πῆξαι τὸ ὑγρόν), if it were not overwhelmed and so liquified by the whole body's heat (διακέχυται ὑπὸ τοῦ θερμοῦ).
>
> Demonstration (ἀπόδειξις) that the moist part in the body is kept hot. If someone should make a cut in a person's body anywhere he wishes, hot blood will flow out and, as long as it is hot, it will be liquid. But when this blood is cooled by the internal cold and the external cold, a skin or membrane forms; if you remove this and let a little time pass, another skin will be seen to form; and if you keep on removing this, still another skin will form, because of the effect of the cold. (*De carn.* 9, transl. Potter 146,5–20 – modified)

Accordingly, the author takes blood as paradigmatic bodily moisture in which there is much of the nutritive coldness, and due to this coldness blood on its own tends to congeal. Nevertheless, as long as it is in the living body, the innate heat counterbalances the cold in the blood and thus keeps it fluid, but when it leaves the body, the innate coldness in the blood, supported by the coldness coming from the environment, overpower the heat and blood congeals.[35]

In summary, in every part of the body there is some vital heat: there is some in the peripheral limbs and organs, which draws into itself as its nourishment the cold moisture from the veins, and in the veins and other blood vessels there is even more of the heat. Most of the heat is in the heart and aorta, which explains why these blood vessels draw into themselves the nutritive moisture from its sources, i.e. from the stomach and upper intestines. As the nutritive moisture travels through the body, at each level of the process, the heat present

---

[33] Cf. Plato, *Timaeus* 80d–e.
[34] *De carn.* 8, Potter 144,10–12.
[35] Cf. Aristotle, *PA* II,4: "blood is solidified by cold" (651a7–8); "and while it is in the body the blood is moist/fluid (ὑγρόν ἐστι) on account of the heat that is present in animals" (651a10–12).

in the particular organ partly consumes the cold in the moisture and thus further concocts it, while the leftover travels further via the connecting blood vessels towards more peripheral parts of the body, being sucked by the heat in these peripheral parts. Thus in all parts of the body the always hungry heat draws into itself the available nourishment, which is heated up and further decomposed as the heat partly consumes the cold quality. This mechanism can explain, I believe, the whole process of digestion, up to the point where the most concocted nourishment reaches the bodily parts at the periphery of the nutritive system, for instance in particular bones, muscles, hair or teeth. These terminal parts are the final recipients of the "nutrition" in the second sense, i.e. as the nutrition upon which the body feeds and out of which it grows according to the "like to like" principle (e.g. hot parts are fed by hot nourishment, cold or gluey parts are fed by cold or gluey nourishment, etc.). In this sense, one can say that blood is the terminal form of nourishment.

3 Aristotle on Nourishment and Digestion

The notion of blood as the final form of nourishment is probably the most elaborated topic in Aristotle's accounts of nutrition and digestion. Assuming that "nourishment is matter, and blood is the last stage of nourishment", blood can be understood as "the matter of the entire body" (PA II,4, 651a14–15).[36] Accordingly, blood as the first material cause of all other constituents of animal bodies is the first uniform part discussed in PA.[37]

In PA II,3 Aristotle first specifies the natural qualities of the blood: it is hot[38] as well as moist,[39] and then he proceeds to the following argument concerning the digestive system as a whole:

> Since everything which grows must take in nourishment, and nourishment is in every case from moist and dry, and the concoction and transformation of these things comes about

---

[36] In a similar vein the nature (*physis*) of menstrual blood can be classed as the "prime matter" (*protê hylê*) of the bodies of blooded animals (GA I,20, 729a32–33).
[37] Cf. HA III,2, 511b11–12, transl. Peck (1965) – modified: "Now, since blood and blood-vessels, as natural substances, give the impression of being fundamental (ἀρχῇ ἔοικεν), we must discuss them first [...]."
[38] PA II,3, 649b26–29: "heat is included in the *logos* of blood [...] blood is essentially hot."
[39] Some uniform parts – such as blood – must act as nutriment for the others, "for everything that grows gets the material for its growth from what is moist" (PA II,2, 647b26–27). Cf. PA II,4, 651a11–13: "blood is moist on account of the heat which is there"; GA I,18, 726a19–20: "the nourishment of all animals tends to be moist".

> through the potency of the hot, owing to this cause if no other all the animals and plants must have a natural origin of heat – and this, like the preparation of the nutrients, is shared by numerous parts. For it is evident that the first nutritive service in animals is performed by the mouth and, for those animals whose nutrients must be cut up, by the parts within it. But this is in no sense a cause of concoction, but rather of good concoction; for their division into small pieces facilitates the preparation of the nutrients by the heat. The work of the upper and lower gut forthwith concocts the food with the aid of the natural heat. (PA II,3, 650a2–14, transl. Lennox)

Aristotle's account is formulated as a chain of hypothetical necessity: (i) growth presupposes nutrition, (ii) nutrition is produced by means of "the potency of the hot", and (iii) therefore every growing organism must necessarily have some source of the heat, which Aristotle, in accordance with the author of *De carn.*, locates around the heart. And since he takes for granted that the heat/hot is shared in the body by numerous parts, also digestion must be shared by numerous parts.

Aristotle gives prominent attention to the question whether mouth, by means of which animals take in food, is such a part.[40] Although it is indeed the very first bodily part which works on food (αἱ ἐργασίαι τῆς τροφῆς), and despite the fact that there is some heat in the mouth as it is in all bodily parts, mouth does not participate in digestion as such, and therefore it is not the cause of concoction (αἰτία πέψεως), as he carefully clarifies, but rather a cause which contributes to good digestion. It may appear rather surprising how much attention Aristotle pays to an obvious fact that chewing is beneficial for digestion, unless we assume, as I suggest, that he closely follows peculiar authorities and silently builds on their accounts as far as he can, while he raises a flag when something important is omitted in their accounts, such as the role of mouth in digestion (about which we read nothing either in *De carn.* or in *De diaeta*), or presented incorrectly, such as the false assumption that heat feeds on cold, which Aristotle elsewhere (*Juv.* 12(6), 473a3–6) explicitly refuses and in our passage discreetly but significantly corrects when saying that "nourishment is in every case from moist and dry".[41]

---

[40] Cf. *PA* II,10, 655b29–32, transl. Lennox: "For in all animals, at least those which are complete, there are two parts that are most necessary, that by which they receive nourishment and that by which the residue departs; for it is impossible to be or to grow without nourishment."

[41] Cf. *Mete.* II,2, 355a5: "moisture is the only food for fire". Nonetheless, nourishment in terms of foods and drinks is obviously not only moist (and dry) but also cold and hot (cf. *de An.* II,3, 414b8–14, transl. Hett: "All animals feed on what is dry or wet, hot or cold, and touch is the sense which apprehends these [...] Hunger and thirst are desire, the former for what is dry

After this clarification concerning the limited but still relevant role of mouth in digestion, Aristotle continues in his account of the digestive system and spells out his argument, again in terms of hypothetical necessity:

> Just as the mouth is a channel for undigested nutrients – and the part continuous with it extending to the stomach, called the oesophagus, in those which have it – so there must also be many other origins, through which the body takes all the nutrients from the stomach and from the nature of the intestines, as from a trough.[42] For while plants take their already worked-up nutrients from the earth by means of their roots (which is also why residues do not come about in plants, since plants use the earth and its heat as a stomach), virtually all animals, and clearly the locomotive ones, have the stomach cavity, like an earth within them. From this they must somehow take the nutrients – just as plants do with their roots – until they reach the end of this continuous concoction. For the operation of the mouth passes its products on to the stomach, and it is necessary for something else to take it from this, which is just what occurs. For the blood vessels extend all through the intestines, beginning beneath the stomach and extending up to it. These things should be studied with the help of the dissections and natural enquiries.[43]
>
> Since there is something receptive of all nutrients and generated residues, and the blood vessels are like a container for blood, it is apparent that blood is the final nutrient for the blooded animals, and its analogue for the bloodless. And the amount of blood decreases on account of not taking nourishment, and increases on account of taking in. And when the nourishment is wholesome the blood is healthy, while when it is bad, the blood is bad [...]
>
> The way in which the parts derive their growth from blood, and the subject of nourishment generally, is more appropriately considered in the works on generation as well as in other works. For now let this much be said (for so much is useful): blood is for the sake of nourishment (τροφῆς ἕνεκα), i.e. nourishment of the parts (τροφῆς τῶν μορίων ἐστίν). (PA II,3, 650a14–b13, transl. Lennox)

---

and hot, the latter for what is cold and wet; favour is a kind of seasoning of these.") See also Aristotle's definition of moist and dry in de Generatione et Corruptione II,2, 329b29–33.

42 Cf. PA IV,5, 680b31, transl. Peck (1937): "growth has its origin from the stomach."

43 Cf. PA III,14, 674a9–21 (transl. Lennox): "Beneath the diaphragm lies the stomach; in animals with an oesophagus it lies where this part ends, while in those without one it lies right next to the mouth; and following the stomach is what is called the intestine. And the cause owing to which each of the animals has these three parts is apparent to everyone. For it is necessary both that the incoming nourishment be received and that the dehydrated nourishment be expelled; and the unconcocted nourishment and the residue must not be in the same location, and there must be a certain location in which the nourishment changes. Indeed, the one part will hold the incoming nourishment, the other the useless residue; and just as it is necessary for there to be a distinct time for each of these, so is it that they be divided in their locations as well. But while the definition of these parts is more appropriate to the works on generation and on nutrition, the differentiation of the stomach and of its contributory parts should be examined now."

The general scheme of internal anatomy, under which Aristotle operates in this particular discussion, is perfectly comparable to the Hippocratic one. In accordance with the Hippocratic author, Aristotle divides the digestive system simply into stomach, upper and lower intestines, and blood vessels, and presupposes a continuous system of blood vessels distributing the nourishment from the stomach and upper intestines to all other parts of the body.

In the concluding summary of his argument Aristotle specifies that when he says "blood is for the sake of nourishment" he means the nourishment of the particular parts of body (τροφῆς τῶν μορίων ἐστίν) which receive the blood distributed via the system of blood vessels.[44] In this sense, nourishment is the final product of concoction that feeds all bodily parts, it is "the material of which the whole body consists". But it is only one of two possible meanings of the term *trophê*, as Aristotle clarifies in his discussion of the nutritive soul in *de Anima*.

In *de An.* II,4 Aristotle builds on the assumption that elements, such as fire and earth, tend to move in contrary directions and, accordingly, there must be something in living organisms which hold the elements together, and concludes that "this is the soul and the cause of growth and nourishment" (416a8–9). As for his predecessors, he reports that according to "some" the nature of fire is the cause of nutrition and growth, and states the most obvious justification of the view: "for it alone of all bodies and elements seems to be nourished and grow of itself" (*de An.* 416a9–18). Aristotle thus pinpoints an essential feature of the tradition upon which he himself draws: fire (or heat) itself needs to be fed, it grows when sufficient nutriment is available and ceases when the sources dry out. Aristotle's principal objection against those who consider fire as the efficient cause (τὸ ἐργαζόμενον) is that fire, on its own, has an unlimited appetite (it burns up whatever the given conditions allow it to burn, "so long as there is something to be burned").[45] And since there must be some limit and proportion in all naturally composed beings, Aristotle proposes, fire is only a "concurrent

---

[44] Cf. *PA* II,6, 652a21–24 (transl. Lennox): "[...] it is also apparent what marrow is – the enclosed, concocted residue of the sanguineous nourishment apportioned to bones and fish-spine."

[45] In comparison with the account in *De carn.*, especially with the image of a flame in a lamp drawn in *De carn.* 6, the account in *De diaeta* seems to be relatively less vulnerable to Aristotle's objection due to its stress on water that counterbalances and restricts the growth of the fire within the mixture (*De diaeta* I,3).

cause" (*sunaition*), while the absolute (*haplôs*) cause is the soul, more precisely the nutritive soul using the fire as its instrument.[46]

Having made this clarification concerning the concept of heat/fire, Aristotle proceeds to the account of nutriment (*trophê*) which starts again with an evaluation of the most relevant views of his predecessors:

> Since the same faculty of the soul is at once nutritive and generative, we must first define nutriment carefully; for the nutritive faculty is distinguished from the others by its function of nutrition. There is a general opinion that contrary is nutriment to contrary; not of course in every case, but among such contraries as have not merely their birth from each other, but their growth as well; for many things arise from each other, but they are not all quantities; e.g., a healthy from a diseased thing. But not even the things mentioned seem to be food for each other in the same way; water feeds fire, but fire does not feed water. It seems, then, that in simple bodies especially the food and the thing fed are contraries. But this presents a difficulty; for some say that like is fed, as also it grows, by like, but others, as we have said, hold the opposite view, that contrary is fed by contrary, on the ground that like is unaffected by like, but that food changes and is digested. But all change is to the opposite, or to an intermediate state. Again, the food is affected by the thing fed, and not *vice versa*, just as the carpenter is not affected by his material, but the material by the carpenter; the carpenter merely changes from idleness to activity.
>
> Now it makes a difference whether "food" means the last or the first form of what is added. If both are food, the one being undigested and the other digested, we might speak of food in both the ways referred to above; for when the food is undigested, contrary feeds on contrary, but when it is digested, like feeds on like. Thus clearly both views are, in a sense, both right and wrong. But since nothing is fed which does not share in life, that which is fed must be the body which has a soul, *qua* having a soul, so that food is related to that which has a soul and that not accidentally. (*de An.* II,4, 416a19–b12, transl. Hett)

Aristotle identifies in this passage the same two patterns of growth as employed in *De carn.*, namely that "like is fed by like" and that "contrary is fed by contrary" (i.e. the heat is fed by the cold). He speaks of two opponent camps ("some say [...] but others hold [...]") and suggests a kind of reconciliation ("both views are, in a sense, both right and wrong"). First, he explains that in some cases ("not of course in every case") contrary is indeed nutriment to contrary, such as in the paradigmatic example of fire and water, and concludes that "in simple bodies especially the food and the thing fed are contraries". Secondly, he admits that this view is in conflict with a no less legitimate assumption that "like is fed

---

[46] As for Aristotle's terminology, cf. the Hippocratic *De flatibus* distinguishing between the cause (αἴτιον) of diseases, which is in all cases the winds (φῦσαι), and all other things that are secondary and subordinate causes (συναίτια καὶ μεταίτια)" (*De flat.* 15, Jones 252,13–15).

by like",[47] and submits a solution which is in remarkable correspondence with the Hippocratic account. He distinguishes between the *trophê* at the beginning of the process (which the Hippocratic author calls "foods and drinks") and at the end (which they both call *trophê*).[48] The ingested food and drink is worked out by the heat in the stomach and upper intestines, and as it travels through these structures it is heated and the most nutritive part, the finest moisture, is driven from these digestive organs into the blood vessels attached to the stomach and intestines, and via these vessels nutriments travel up to the heart and from there to all remaining parts of the body which are all nourished by this nutritive moisture (identified with the blood in blooded animals).

As we have seen, both Aristotle and the Hippocratic author provide relatively simple and schematic yet for their purposes perfectly sufficient expositions of the digestive organs connected with other parts of the body by means of blood vessels, both combine this scheme with the concept of vital heat and assume that the system of blood vessels not only distributes the nourishment around the body, but also actively sucks it from the stomach and intestines and further concocts it. On both accounts this arrangement is adequate to provide a complex explanation of the whole process of digestion, without the need to go into further details. Although Aristotle concludes with a remark that "these things should be studied with the help of the dissections and natural enquiries", in the present account he operates on the same level of abstraction as his Hippocratic forerunner, draws on more or less the same level of anatomical knowledge, and – most importantly – systematically draws on the same first principles of living bodies relevant to material and efficient causality.

Both Aristotle and the author of *De carn.* take for granted that (a) heat and cold are the main opposites in the body, and that (b) the heat/fire actively draws into itself appropriate amount of nourishment, which also explains the continuous flow of the nutrition from the stomach via blood vessels to the terminal parts. Nonetheless, while in the Hippocratic account these two assumptions are connected by the supposition that the heat feeds on the cold, for Aristotle these are two different things: there is an opposition between the heat and the cold in the body on the one hand, which is kept in balance by respiratory (or

---

[47] Cf. Plato, *Phd.* 96c3–d5, transl. G. M. A. Grube: "I thought before that it was obvious to anybody that men grew through eating and drinking, for food adds flesh to flesh and bones to bones, and in the same way appropriate parts were added to all other parts of the body, so that the man grew from an earlier small bulk to a large bulk later, and so a small man became big."

[48] Cf. Aristotle's distinction between two grades of nourishment in *GA* II,6, 744b32–37 and in *HA* I,5, 489b8–10.

an analogous) system (and brain), and, on the other hand, a balance between the degree of the heat and the quality and quantity of its nourishment, which is not the cold but rather the nutritive "moisture" or "water". On this point, Aristotle picks up on the tradition represented by *De diaeta* and presupposes that moisture rather than cold is the main quality of nourishment at all levels of description: at the level of elements fire "feeds on water" (*de An.* II,4, 416a19–b12), in other words "moisture is the only food of fire" (*Mete.* II,2, 355a5), at the level of the homoiomerous parts of blooded animals blood is the final form of nourishment (while plants as well as some bloodless animals can live solely on the moisture provided by their environment).[49] All parts of the body have some portion of the life-bearing heat/fire and therefore also of some natural moisture which feeds the heat/fire. When the innate heat leaves the body (or the particular part of the body), "its natural moisture evaporates, and there is nothing to suck moisture into it (this being the function of its own heat, which attracts and draws moisture in)" (*Mete.* IV,1, 379a23–26).

Both Aristotle and the author of *De carn.* pay remarkable attention to the growth of teeth and its causes,[50] both attempt to explain why the first set of teeth is shed and replaced with a second one, and also why some teeth keep growing when the other bony parts stop growing. They both take for granted that heat is the main moving cause of teeth production and growth[51] and

---

49 Cf. *PA* IV,5, 682a22–27, transl. Lennox: "All the insected animals are light feeders, not so much because of their smallness as because of their coldness (for the hot needs nourishment and concocts nourishment quickly, while the cold needs little nourishment), and most of all the cicada kind; for the moisture left behind by the air is sufficient nourishment for their body [...]."

50 In *Physica* II,8 Aristotle raises the question why nature is to be ranked among final causes and whether we have any reason to regard nature as making for any goal at all. He draws an example with rain which nature drops "not to make the corn growth, but of necessity". Then he takes teeth as his main illustration: "So why should it not be the same with natural organs like teeth? Why should it not be a coincidence that the front teeth come up with an edge, suited to dividing the food, and the back ones flat and good for grinding it, without there being any purpose in the matter? And so with all other organs that seem to embody a purpose" (*Ph.* II,8, 198b23–28, transl. Wicksteed-Cornford, modified). Cf. Macfarlane (2016), p. 274: "It is instructive to understand the medical background that Aristotle incorporates into his account of teeth in terms of final cause (cf. esp. *Physics* 2.8)".

51 Aristotle uses the same assumption also in *GA* V,8 in his refutation of Democritus who allegedly held that animals shed their teeth because the first teeth are formed prematurely due to their suckling. But suckling in itself, objects Aristotle, contributes nothing to the formation of the teeth, though the warmth of the milk makes them come through more quickly. "A proof of this [i.e. that the warmth of the milk makes the teeth to come through more quickly] is that within the actual class of those which suckle, those young ones which get hotter milk

build their answers on the assumption that there is abundance of nutrition supplied by the large veins attached to the jaws.[52] Nevertheless, the Hippocratic author, as we have seen, never suggests any purpose or function of the teeth. Aristotle agrees with the Hippocratic author that teeth are shed and grow second time as a result of necessity, but he would certainly object that it lacks a teleological explanation (as he explicitly does in *GA* V,8 to the account of Democritus) and omits to mention the final cause of their second growth, namely that they grow second time "for the sake of the better" (*GA* 789a9–13), i.e. to replace the worn-down and blunted teeth with a fresh relay of teeth which can serve better its own purpose,[53] which is dividing and grinding food and thus facilitating digestion. The Hippocratic author never uses such a language and actually does not even raise the question of what the teeth are good for. He acknowledges the role of the teeth in the production of articulated sounds but he does not discuss the purpose of speech as such.[54] He rather asks: in which way teeth come into being, or why teeth are formed later than the other bones, and provides explanations on the level of material and efficient causality.[55]

4 Conclusion

In his methodological introduction into zoology in *PA* I Aristotle describes the previous tradition of the inquiry into nature as limited to questions related to material and efficient causality and raises two principal objections to his predecessors. First, while "almost all" of them "attempt to refer their accounts back to necessity", they fail to distinguish "in how many ways the necessary is said", namely they do not differentiate between the "absolute" necessity,

---

grow their teeth quicker, because that which is hot tends to promote growth" (*GA* V,8, 789a4–7, transl. Peck 1942).

52 Aristotle, *GA* V,8, 789b19–20, transl. Peck (1942): "The reason for this [i.e. that the flat teeth grow in adults] is that there is a great deal of nourishment in the wide part of the bones."
53 Cf. *GA* II,6, 745a33–35, transl. Peck (1942): "If life went on for 10 000 or even 1000 years, the teeth would have had to be quite enormous to begin with, and they would have had to grow afresh many times over."
54 Cf. *De carn.* 18, Potter 158,1–4: "The tongue articulates by touching: as the tongue encloses the air in the throat and touches the palate and the teeth, it gives the sound clarity."
55 Cf. Oser-Grote (2004), p. 31, "Den Autor beschäftigte bei der Abfassung daher nicht die Frage, wie die Einzelteile am Ende richtig zusammengesetzt werden, sondern er ging umgekehrt von der Problemstellung aus, wie der Mensch und überhaupt alle Lebewesen, so wie sie ihn umgaben, einmal zustande gekommen sind. Er hat sozusagen die Konstitution ‚zurückgebildet' und wieder ‚von vorne' entstehen lassen."

which belongs to eternal things, and the "hypothetical" necessity, which has to do with generated things, both natural and artificial (639b22–30). And secondly, Aristotle explains that his predecessors could not have arrived at appropriate causal explanations because they lacked the concept of essence (*ousia*). It was not before Socrates' time, Aristotle reports, that the interest in the conception grew, although, at the same time, "research into the natural world ceased, and philosophers turned instead to practical virtues and politics" (642a24–31).

Although natural philosophy stagnated in the period between Socrates and Aristotle, medical literature flourished and most of the texts of the Hippocratic collection were written. Accordingly, in *Parva Naturalia* (*de Sensu et Sensibilibus* 436a17–b3, *Juv.* 27(21), 480b21–30) Aristotle recommends to his readers paying special attention to medical authorities who "practice their discipline in a more philosophical manner" and also "have something to say about nature", since they work in neighbouring and partly overlapping field of study and employ the same first principles as natural philosophers.

In this chapter I have developed the hypothesis that Aristotle in his account of digestion closely draws on a specific medical tradition and that, among the extant pre-Aristotelian sources, the Hippocratic *De carn.* and *De diaeta* provide the most relevant evidence for this tradition.[56] *De diaeta* attests remarkable methodological considerations, peculiar theoretical conceptions, rare technical terms and unprecedented analogies which play important role also in Aristotle's accounts, especially on the level of elements and their mixtures.[57] But regarding the *anhomoiomerous* parts, i.e. the particular organs and structures of the digestive system, the relevance of *De diaeta* is extremely limited. As for the anatomy of the digestive system, Aristotle's accounts are deeply embedded in the tradition attested in *De carn.*[58]

As we have seen, the accounts of these philosophizing physicians suffer from the same drawbacks as those of pre-Socratic philosophers, namely that

---

[56] As for other relevant Hippocratic sources, see especially *De Septimanis* and *De flat.* (cf. Bartoš 2020).
[57] Cf. Smith (1992), p. 270–271, Bartoš (2015), p. 39–40, and Bartoš (forthcoming).
[58] Cf. Freudenthal (1995), p. 100: "[...] among the characteristic ideas Aristotle picked up from *On Fleshes* is the account of the formation of the elastic parts: Aristotle states that 'all bodies depend on something glutinous to hold them together' (*GA* II,3, 737b2), his term, *glischros*, being one which occurs in *De carn.* (5,1) as a synonym for *kollôdês*. Specifically, it is the sinew 'which holds the parts of animals together' (737b4). He also argues, again in conformity with *On Fleshes*, that sinews, skin, blood-vessels and 'all that class of substances' are glutinous, differing only by 'the more or the less' (737b4–5; cf. also II,6, 743b5–12). Beyond the mere dependence of Aristotle on *On Fleshes*, we should note that the elastic parts are here taken to assume physically one of the roles of the nutritive soul, namely to hold together the body."

they lack the notion of *ousia* as the formal and final cause of animals and plants, and the concept of hypothetical necessity. Aristotle therefore in his accounts corrects his predecessors' views on these specific points. The first novelty is mentioned at the beginning as well as at the end of his account of the nutritive soul in *de An.* II,4. Although the Hippocratic authors also employ some conception of soul,[59] their methodological approach is – in a sense – opposite to the one of Aristotle. In their bottom-up approach, they work on the assumption that every animal and plant has some innate fire/heat of such and such capacities, operating under such and such conditions, and they explain the more complex organic structures and life-processes as outcomes of the activity of the fire/heat on various materials and under various conditions. In contrast to their approach, Aristotle in *de An.* proceeds from the top down and insists that in order to understand what the innate heat or fire is, what it actually does in the body and how it is involved in the various physiological processes, one has to understand first of all what the soul and its capacities are.[60] The second novelty consists in rearranging the empirically known facts and accepted assumptions of his predecessors into teleologically structured arguments based on the notion of hypothetical necessity. Aristotle thus binds together the final causes (i.e. the specific functions of each and every bodily organ existing for the sake of nutrition and vegetative life in general), with the material causes identified already in the preceding tradition.[61]

---

[59] As already mentioned, despite the fact that the author of *De carn.* suggests in chap. 1 to explain "what the soul (*psychê*) is", immediately before introducing the heat in chap. 2 as the main cause of practically all vital processes, he never uses the term again. The author of *De diaeta* uses the term *psychê* frequently (almost one hundred times), and although the relation between soul and body is closely similar to the relation between the fire and water on his account (see Bartoš 2015, p. 185–222), fire is never explicitly identified with the soul in the treatise.

[60] Accordingly, whenever Aristotle speaks about innate heat and its works, he presupposes that there must be some nutritive soul using the heat as its instrument. This holds even in borderline cases such as the spontaneously generated animals: "animals and plants come into being in earth and in liquid because there is water in earth, and air in water, and in all air there is vital heat so that in a sense all things are full of souls" (*GA* III,11, 762a18–21, transl. Platt, modified).

[61] After the conference in Berlin, I had the opportunity to present this essay at the History of Philosophy Roundtable (University of California, San Diego). I am very thankful to these audiences for their questions and suggestions as well as to Monte R. Johnson, Colin G. King, Tiberiu Popa, and Giouli Korobili for their helpful comments on previous drafts of the text. It is an outcome of the research project no. 19–07091S supported by the Czech Science Foundation.

Bibliography

Bartoš, Hynek (forthcoming): "Aristotle and Hippocratic Writings". In: Connell, S. (ed.): *Cambridge Companion to Aristotle's Biology*. Cambridge: Cambridge University Press.

Bartoš, Hynek (2020). "Heat, *Pneuma* and Soul in the Medical Tradition". In: Bartoš, H./King, C. G. (eds.): *Heat, Pneuma and Soul in Ancient Philosophy and Medicine*. Cambridge: Cambridge University Press, p. 21–31.

Bartoš, Hynek (2018): "Soul, Perception and Thought in the Hippocratic Corpus". In: Sisko, J. (ed.): *Philosophy of Mind in Antiquity*. Routledge, p. 64–83.

Bartoš, Hynek (2015): *Philosophy and Dietetics in the Hippocratic On Regimen: A Delicate Balance of Health*. Leiden: Brill.

Bartoš, Hynek (2012): "The analogy of Auger Boring in the Hippocratic De Victu". In: *The Classical Quarterly* 62.1, p. 92–97.

Betegh, Gábor (2020). "Fire, Heat and Motive Force in Early Greek Philosophy and Medicine". In: Bartoš, H./King, C. G. (eds.): *Heat, Pneuma and Soul in Ancient Philosophy and Medicine*. Cambridge: Cambridge University Press, p. 35–60.

Craik, Elizabeth (2015): *The 'Hippocratic' Corpus: Content and Context*. London and New York: Routledge.

Deichgräber, Karl (1973): *Pseudhippokrates Über die Nährung: Text, Kommentar und Würdigung einer stoisch-heraklitisierenden Schrift aus der Zeit um Christi Geburt*. Mainz: Verlag der Akademie der Wissenschaften und der Literatur Wiesbaden.

Eijk, Philip J. van der (2005): *Medicine and Philosophy in Classical Antiquity: Doctors and Philosophers on Nature, Soul, Health and Disease*. Cambridge: Cambridge University Press.

Eijk, Philip J. van der (2007): "Les mouvements de la matière dans la génération des animaux selon Aristote". In: Boudon-Millot, V., Guardasole, A., C. Magdelaine, C. (eds.): *La science médicale antique: nouveaux regards*. Paris: Beauchesne, p. 405–424.

Freudenthal, Gad (1995): *Aristotle's Theory of Material Substance: Heat and Pneuma, Form and Soul*. Oxford: Oxford University Press.

Jones, William H. S. (ed./trans.) (1923): *Hippocrates*, vol. II. [Loeb Classical Library 148.] Cambridge: Harvard University Press.

Jouanna, Jacques (1999): *Hippocrates*. Translated by M.B. DeBevoise. Baltimore: John Hopkins University Press.

Kullmann, Wolfgang (2007): *Aristoteles: Über die Teile der Lebewesen*. Berlin: Akademie-Verlag.

Lefèvre, Charles (1972): *Sur l' évolution d'Aristote en psychologie*. Louvain: Éditions de l'Institut supérieur de philosophie.

Lennox, James G. (1996): "Aristotle's Biological Development: The Balme Hypothesis". In: Wians, W. (ed.): *Aristotle's Philosophical Development: Problems and Prospects*. Lanham: Rowman & Littlefield, p. 229–248.

Lennox, James G. (2020): "Why animals must keep their cool: Aristotle on the need for respiration (and other forms of cooling)". In: Bartoš, H./King, C. G. (eds.): *Heat, Pneuma and Soul in Ancient Philosophy and Science: From the Presocratics to Aristotle*. Cambridge: Cambridge University Press, p. 217–242.

Macfarlane, Patrick (2016): "Teeth in the Hippocratic Corpus". In: Dean-Jones, L./Rosen, R.M. (eds.): *Ancient Concepts of the Hippocratic: Papers Presented at the XIIIth*

International Hippocrates Colloquium, Austin, Texas, August 2008. [Studies in ancient medicine 46]. Leiden-Boston: Brill, p. 273–291.

Mendelssohn, Everett (1964): *Heat and Life*. Cambridge.

Ogle, William (1882): *Aristotle on the Parts of Animals*. London.

Oser-Grote, Carolin M. (2004): *Aristoteles und das Corpus Hippocraticum. Die Anatomie und Physiologie des Menchen*. Stuttgart: Franz Steiner.

Peck, Arthur L. (1928): *Pseudo-Hippocrates Philosophus: or, The Development of philosophical and other theories as illustrated by the Hippocratic writings, with special reference to the De victu and the De prisca medicina*. Diss., Cambridge.

Peck, Arthur L. (ed./transl.) (1937): *Aristotle: Parts of animals*. [LCL 323.] Cambridge: Harvard University Press.

Peck, Arthur L. (ed./transl.) (1942): *Aristotle: Generation of animals*. [LCL 366.] Cambridge: Harvard University Press.

Peck, Arthur L. (ed./transl.) (1965): *Aristotle: Historia animalium*. Books 1–3. [LCL 437.] Cambridge: Harvard University Press.

Popa, Tiberiu (2014): "Observing the Invisible Regimen I on Elemental Powers and Higher Order Dispositions". In: *British Journal for the History of Philosophy* 22, p. 1–20.

Popa, Tiberiu (2020): "Aristotle on the Powers of Thermic Equilibrium". In: Bartoš, H., King, C. G. (eds.): *Heat, Pneuma and Soul in Ancient Philosophy and Medicine*. Cambridge: Cambridge University Press, p. 202–216.

Potter, Paul (ed./transl.) (1995): *Hippocrates*, vol. VIII. [Loeb Classical Library 482.] Cambridge: Harvard University Press.

Smith, Wesley D. (1992): "Regimen, κρῆσις, and the History of Dietetics". In: López Férez, J.A. (ed.): *Tratados Hipocráticos*. Madrid: Universidad Nacional de Education a Distancia, p. 263–271.

Wilson, Malcolm (2020): "Heat, Meteorology and Spontaneous Generation". In: Bartoš, H., King, C. G. (eds.): *Heat, Pneuma and Soul in Ancient Philosophy and Medicine*. Cambridge: Cambridge University Press, p. 159–181.

Willerding, Ferdinand (1914): *Studia Hippocratica*. Göttingen (diss.).

Giouli Korobili
Aristotle on the Role of Heat in Plant Life

Abstract: Any modern scholar of Aristotle's natural philosophy would right away admit that, according to Aristotle, all living things, in order to maintain their lives, undoubtedly need, among other factors, a principle of soul and vital heat. Despite this scholarly consensus, so little has been written concerning vital heat in plants, even though Aristotle treats them as ensouled beings endowed with the most basic part of the soul, the nutritive soul. Above all, one of the most crucial questions remains obscure: 'What does this vital heat actually do inside a plant?', especially in the light of the idea that plants manifest a far less complex structure than animals and humans. In this paper, I try to answer this question by offering an interpretation of the role heat plays in the internal processes taking place throughout a plant's life cycle.

1 Introduction

In *de Anima* II,4, 416a13–15, Aristotle calls the nature of fire and the soul a sort of co-cause (συναίτιον) and the unqualified cause of nutrition respectively. The nature of fire is preserved only if the heat is unhinderedly nourished.[1] As long as there is fuel to feed it, fire grows without limit. In living beings, however, the size and growth of their natural fire are subject to the limit and proportion imposed by the (nutritive) soul (*de An.* II,4, 416a10–18). In *de Juventute et Senectute, de Vita*

Note: This paper would not have seen the light of day without the generous support of the Berlin Cluster of Excellence 'TOPOI', as part of the Research Project 'Mapping the Vegetative Soul in Aristotle and Beyond'. I am thankful to Sirio Trentini for his comments on an earlier draft of this paper, as well as to the participants of the Conference Nutrition and Nutritive Soul in Aristotle and Aristotelianism (Berlin, 2017) for the excellent discussions and stimulating exchange of ideas on the subject matter.

1 Cf. *de Partibus Animalium* IV,5, 682a23–25: "for it is heat which requires nourishment (δεῖται τροφῆς), just as it is heat which speedily concocts it; but cold requires no nourishment (ἄτροφον)" (transl. W. Ogle, slightly modified). Translations of *de Juventute et Senectute, de Vita et Morte, de Respiratione* 1–6, are my own. Translations of other texts of Aristotle are taken from Barnes (1995). Occasionally, I have altered translations. Those interventions are marked 'translation slightly modified'.

et Morte, de Respiratione,[2] Aristotle highlights that heat and soul remain inseparable from one another in order to keep the living being alive (4, 469b11–17). At the same time, heat (and usually cold) is always conceived of as an active factor (ποιητικόν), that is, it predominates on the passive factors, such as dry and wet, which suffer a *pathos* (*Meteorologica* IV,1, 378b10–379b9). Thus, when we speak of heat in Aristotle, we always refer to two notions, namely nutrition and change.

On reading these passages, one can find a close relationship between the heat and the soul, which holds *for all living beings*, plants included. In Aristotle's psychology plants possess a soul, even the most basic part of it, while there are also instances where he speaks of their heat, often in discussions based on comparisons with animals and humans. Aristotle distinguishes among different kinds of heat, such as the natural 'innate' heat, the sun's heat or fire's heat, always emphasising how important vital heat is for the maintenance of living beings.[3] Partial lack of vital heat from the body can lead to disease, while total lack thereof results in its death.[4] The interpretation of the role that natural heat plays in a plant's life becomes therefore a *desideratum*, especially when we think that plants are the simplest form of life, but appear at the same time much more mysterious than expected.[5] Unfortunately, Aristotle's treatise on plants, if ever written, has not come down to us (Falcon 2015, p. 75–76, n. 1). If this were not the case, we could probably draw a much clearer picture of the role of natural heat in plants. We can, nevertheless, try to reconstruct his opinion from scattered passages in the works known to us today. In what follows, I will first examine the question of whether there is actually any innate principle of heat in plants, and, if so, where it is located in their bodies, and then present its role in the sub-functions guided by their nutritive soul.

---

[2] In this paper, I take for granted that *de Juventute et Senectute, de Vita et Morte, de Respiratione* comprise a unified treatise, a theme that I explore in more detail in my dissertation.

[3] In the course of this essay, as a qualifier of a living thing's heat I often use the word 'vital'. Of course I do not mean with this word to ascribe accountability to that heat for the processes of generation or animation, since, on Aristotle's account of living things, 'vital' heat is unable to generate on its own without the accompanying presence of soul. In the case of cheese-making, for example, 'vital' heat may set milk into cheese but is not by itself sufficient for the cheese to count as a living thing, since the cheese lacks a soul (*de Generatione Animalium* II,4, 739b20–25). For a discussion of whether Aristotle's picture of living processes is a physical or a vitalistic one, see Connell (2016), p. 221–224.

[4] *GA* V,4, 784b25–28, *Juv.* 23(17), 478b31–33.

[5] What makes them even more attractive targets for investigation is not only that some of their most fundamental functions, such as nutrition and growth, can be observed, recorded or studied, as opposed to the respective functions performed by more complex 'organisms', but also that "they exhibit a form of being that is uncomplicated by animal and rational faculties" (Holmes 2017, p. 360). The studies by Repici (2000) and Hardy/Totelin (2016) provide two very useful guides to plant physiology, yet they are not concerned with the role of vital heat in plants.

2 Is There any Natural Heat in Plants?

Since then everything that lives (πᾶν ζῶν) has a soul, and this [i.e. the soul], as we said, exists not without natural heat (φυσικῆς θερμότητος), in plants the [cooling effected] through nutriment and the surrounding environment furnishes sufficient assistance for the preservation of the natural heat (τὴν τοῦ φυσικοῦ θερμοῦ σωτηρίαν). And indeed the food brings about cooling (κατάψυξιν) when it enters, just as [it does] in man when he first takes it, while fasting causes heat and produces thirst; for when the air is motionless, it always becomes heated, but, moving as the food enters, it is cooled, until [the food] has undergone digestion (πέψιν). And if the environment exceeds in coldness (ὑπερβάλλῃ ψυχρότητι) due to the season of the year, the frost being severe, the force of the heat diminishes (ἐξαυαίνεται ἡ τοῦ θερμοῦ ἰσχύς); but if there are heat-spells and the moisture drawn from the ground cannot produce a cooling effect (μὴ δύνηται τὸ σπώμενον ἐκ τῆς γῆς ὑγρὸν καταψύχειν), the heat is brought to an end by exhaustion, and around such seasons trees are said to be blighted (σφακελίζειν) and star-struck (ἀστρόβλητα). (*Juv.* 6, 470a19–32)

This passage is crucial because it confirms our suspicion that plants, *qua* living things, must have their own principle of vital heat, notwithstanding their simplicity in structure in comparison to animals and humans; it is also crucial because:
(a) it introduces three different sources (and hence kinds) of heat in plants which interact with one another, namely (i) heat stemming from the (incoming) nourishment attracted from the soil by means of their roots, (ii) heat coming from the environment, and (iii) (their) natural heat; and
(b) it provides valuable evidence on how the natural heat in plants can be affected.

Interestingly, (a) seems to represent a view which resonates with the Hippocratic idea that in summer the interior of the earth is cooler than its surface, whereas in winter the opposite is the case;[6] and to which Aristotle remains faithful, as is clear, for instance, from the second book of *Mete.*, where he manifestly distinguishes between the heat of the earth and the heat of the environment (*Mete.* II,4, 360b30–32, 362a5–7). Now as regards (b), Aristotle mentions two factors that are capable of affecting a living thing's internal heat: (1) lack of nourishment and (2) excessive external heat or coldness. I will return to these factors in due course.

---

[6] For example, the author of *De natura pueri* (ch. 24–26) sets out to defend his point that the earth under its surface is cool in summer and warm in winter (as opposed to the opposite conditions on the surface), thus stressing the fact that receiving a due proportion of heat and cold is a necessary precondition for plants' survival; cf. *De humoribus* 11.

Finally, another feature that stands out in this passage – for which evidence is in fact scattered throughout the corpus of Aristotle's writings – is the attempt to build up a clear parallel between humans and plants with respect to their physiology. Since plants are parts of the sublunary world and, like animals, fall into the category of living things (i.e. they possess a soul), it comes as no surprise that their study should form a branch of natural philosophy.[7] Yet there are passages in Aristotle where the way in which he expresses his views on plants may seem to encourage the assumption that he has incidentally incorporated information concerning them into his teachings on nature, sometimes without even preventing himself from falling into contradictions. For example, at some points Aristotle claims that plants do not emit any residue (*PA* II,3, 650a21–23), while at others he goes on to describe fruit production as a process analogous to excrement discharge (*PA* II,10, 655b32–36); again, sometimes he acknowledges that plants do have organs (or parts, *de Caelo* III,1, 298a32), even relatively few in number (*PA* II,10, 656a2) or of a relatively low degree of complexity (*de An.* II,1, 412b1–4), and at others he even outright argues that they are of a comparatively less complex nature (*Physica* II,8, 199b10) and hence keeps his reservations as to whether they should be called 'organs' at all (*de An.* II,4, 416a5).[8] These apparent inconsistencies can, I believe, be straightforwardly resolved by bringing to mind that Aristotle's analogical comparison between plants and other forms of living things is usually intended to serve a twofold purpose: (a) to highlight the structural and functional similarities among different types of living beings; and at the same time (b) to stress the limitations of comparing the things having the lowest kind of soul to those occupying the highest position in the hierarchy of living beings. It seems therefore plausible not only to take into serious consideration what Aristotle has to say about natural heat in plants but, even more, to trace – just as in the case of every other living thing – its overall significance for plant survival.[9]

---

7 Cf. *de Sensu et Sensibilibus* 1, 436a1–6.
8 Cf. *PA* IV,7, 683b17–24; Falcon (2015), p. 83–84.
9 For a different approach, see Freudenthal (1995) and Murphy (2005). Freudenthal (1995) refers to the vital heat of plants in the Appendix of his chapter 1 (p. 70–73). In this, though he appeals to *PA* II,3, 650a2–7 in order to prove that plants do have vital heat, he nonetheless goes on to claim that this kind of heat is identical with the heat they receive from the ground. Murphy (2005) also argues in favour of the identicalness of the plant heat and the heat of the surroundings. Yet he proceeds by making a number of observations that seem to deflect the focus away from plants, in taking as their starting point what in fact applies to more complex 'organisms'. For instance, when Murphy says that "A plant does not have the ability to bring it about that its body returns quickly to its normal temperature. For a plant has no control over its own body temperature" (p. 329), or that "Since for a plant there are no normal conditions, because its

In *Juv.* 23(17), Aristotle refers to the kinds of destruction that affect, without exception, all living beings:

> It is always to some lack of heat that destruction is due, and in perfect creatures the cause is its failure in the part containing the source of the creature's essential nature. This source is sited, as has been said, at the junction of the upper and lower parts; in plants it is intermediate between the root and the stem, in sanguineous animals it is the heart, and in those that are bloodless the corresponding part of their body. (478b31–479a1, transl. G.R.T. Ross, slightly modified)[10]

According to this passage, destruction takes place when a living being's vital principle lacks heat. Both vital principle and heat in blooded animals are located in the heart, while in plants 'in the middle', that is, in the place where both the sprout and the roots meet. Concerning blooded animals, Aristotle has explained elsewhere that the principle of their heat is situated in the heart, which is why the latter is the hottest organ in the body and the richest in blood (*PA* III,4, 665b27–666a3). So, the parallel Aristotle draws between plants and animals seems to suggest that the part of the plant which is analogous to the heart of blooded animals and hosts the principle of the plant's natural heat is to be found in this mid-place, wherein sprout and root join. The destruction of this sort of heat leads plants to death, what Aristotle called a little earlier 'drying out' (αὔανσις, 478b28). This description of the location of a plant's vital principle is admittedly far from infallible, especially if compared with the corresponding description in the case of blooded animals. And while 'the middle place' seems to be more easily identifiable in seeds during the period of the emergence of roots-sprouts, we cannot say the same for, say, a mature tree.

---

own body temperature is not fixed but rather it acquires the temperature of its surroundings [...]" (p. 330), he seems to base his claims, as can be inferred most notably from the use of such words as 'quickly' or 'normal', on what is more or less readily observable in the case of humans. To cite a final example: when Murphy states that "animals can maintain their body temperature to a greater extent than plants" (p. 330), he appears to disregard all those factors allowing an animal to maintain for itself its own 'temperature', such as its ability to handle extreme climatic conditions by moving towards more favourable ones, or to drink/plunge into water more frequently in order to cool itself etc. Plants are incapable of doing so, yet lacking such a skill does not *ipso facto* render them anyhow defective, according to Aristotle. That is why, I think, we need to look more closely at the role ascribed to the heat of a plant inside its body.

10 πᾶσι μὲν οὖν ἡ φθορὰ γίνεται διὰ θερμοῦ τινος ἔκλειψιν, τοῖς δὲ τελείοις, ἐν ᾧ τῆς οὐσίας ἡ ἀρχή. αὕτη δ᾽ ἐστίν, ὥσπερ εἴρηται πρότερον, ἐν ᾧ τό τε ἄνω καὶ τὸ κάτω συνάπτει, τοῖς μὲν φυτοῖς μέσον βλαστοῦ καὶ ῥίζης, τῶν δὲ ζῴων τοῖς μὲν ἐναίμοις ἡ καρδία, τοῖς δ᾽ ἀναίμοις τὸ ἀνάλογον.

In *Juv.* 3, Aristotle resorts to this idea, focusing at the same time on the process of generation. In natural generation, plants are generated from their seeds, and their vital principle lies in the middle (*Juv.* 3, 468b18–23). Yet there is also another way in which they can generate themselves, a type of generation we would call 'artificial generation', or 'transplantation':

> [...] and in both grafting and propagation by slips or cuttings, this [i.e. generation] takes place, in most cases, about the eyes. For the eye is a sort of starting point of the branch, and at the same time [it is its] middle [part], therefore they either remove this, or insert [something] into this, in order that either a branch or roots may be generated from it, in the belief that the origin [of growth] springs from the middle between branch and root. (468b23–28)[11]

According to Aristotle, transplantation of a part/parts of a plant takes place, in most cases, about the eyes. The eye serves at the same time as a sort of starting point of the branch, and as its middle part. It goes without saying that since plants have many eyes, their bodies abound in potential locations in which the vital principle may be hosted. This view is consistent with the one found in *de Longitudine et Brevitate Vitae* 6, according to which plants possess potentially the life principle in every part.[12]

After ascertaining the presence of the principle of vital heat in plants and localising it in their bodies, let us turn to the main question of this paper, namely 'What role does heat play in a plant's life?'. In order to critically assess the importance of the natural heat's presence and maintenance for the life and growth of a plant, we should examine whether it actually contributes to the functioning

---

11 [...] ἔν τε ταῖς ἐμφυτείαις καὶ ταῖς ἀποφυτείαις μάλιστα συμβαίνει τοῦτο περὶ τοὺς ὄζους· ἔστι γὰρ ἀρχή τις ὁ ὄζος τοῦ κλάδου, ἅμα δὲ καὶ μέσον, ὥστε ἢ τοῦτο ἀφαιροῦσιν ἢ εἰς τοῦτο ἐμβάλλουσιν, ἵνα ἢ ὁ κλάδος ἢ αἱ ῥίζαι ἐκ τούτου γίνωνται, ὡς οὔσης τῆς ἀρχῆς ἐκ τοῦ μέσου κλάδου καὶ ῥίζης.

12 "For the plant possesses potentially both root and stem in every part of it (πανταχῇ). Hence it is from this source that issues that continued growth when one part is renewed and the other grows old; it is practically a case of longevity. The taking of cuttings furnishes a similar instance; for we might say that, in a way, when we take a cutting the same thing happens; the shoot cut off is part of the plant. Thus, in taking cuttings this perpetuation of life occurs though their connexion with the plant is severed, but in the former case it is the continuity that is operative. The reason is that the life principle potentially belonging to them is present in every part (αἴτιον δ' ὅτι ἐνυπάρχει πάντῃ ἡ ἀρχὴ δυνάμει ἐνοῦσα)." (467a22–30, transl. G.R.T. Ross, slightly modified). Cf. *GA* III,2, 752a18–23: "The same is the case also in the seeds of plants; the principle of the seed is attached (προσπέφυκε) sometimes to the twig, sometimes to the husk, sometimes to the pericarp. This is plain in the leguminous plants, for where the two cotyledons of beans and of similar seeds are united (συνῆπται), there is the seed attached to the parent plant, and there is the principle of the seed."

of its ensouled body. Besides nutrition, other 'works' or *pathê*, as sometimes Aristotle calls them, that take place during a plant's lifetime are: flowering, ripening and leaf-shedding. Let us start with nutrition.

3 Nutrition

Nutrition is the prime function of the nutritive soul thanks to which a living thing succeeds in remaining alive. In *de An.* II,4, which has been extensively discussed in this volume, Aristotle says that the nutritive soul has two functions: "reproduction and the use of food" (γεννῆσαι καὶ τροφῇ χρῆσθαι, 415a26). But how can plants use their food? By means of what organs? And what kind of relationship is established between user and used in such a case? That is, do plants have absolute authority over what feeds their body? In *GA* II,4, Aristotle makes it emphatically clear that in both animals and plants, the nutritive-generative soul *is* the active agent; yet, as a user, it is able to accomplish its task of inducing change in the material substrate upon which it acts only by means of its tools, namely heat and cold.[13] In what way does the soul, with the aid of heat, act upon food (i.e. matter) in the case of plants? And how can this action be explained, especially in view of the fact that plants themselves do not actually concoct the food they take from the earth by means of their roots, but receive it already concocted?

Again, dwelling on a profound similarity between animals and plants, a passage from *PA* seems to have this latter reasonable question as its background:

> Now since everything that grows must take nourishment, and nutriment in all cases consists of moist and dry substances, and since it is through the potency of heat that these are concocted and changed, it follows that all living things, animals and plants alike, must on

---

[13] "[...] As later on in the case of mature animals and plants this soul causes growth from the nutriment, using heat and cold as its tools (οἷον ὀργάνοις) (for in these is the movement (κίνησις) of the soul and each comes into being in accordance with a certain formula (λόγῳ τινί)), so also from the beginning does it form (συνίστησι) the product of nature. For the material by which this latter grows is the same as that from which it is constituted at first (συνίσταται τὸ πρῶτον); consequently also the power which acts upon it (ἡ ποιοῦσα δύναμις) is identical with that at the beginning (but greater than it); thus if it is the nutritive soul, it is also the generative soul, and this is the nature of every being, existing in all animals and plants. But the other parts of the soul are present in some living things and not in others" (740b29–741a3, transl. A. Platt, slightly modified).

this account, if on no other, have a natural source of heat; and this, like the working of the food, must belong to many parts. (*PA* II,3, 650a2–8, transl. W. Ogle, slightly modified)[14]

The message is clear: If for no other reason, everything that grows, animals and plants alike, has a natural principle of heat, precisely because it is by the power of heat that food/matter is concocted and changed. Similar information is given in *de An.* II,4, 416b28–29 and *Sens.* 4, 442a6–8. In what follows, I will offer a model of interpretation that centres on the role of a plant's internal source of heat by drawing on Aristotle's typical manner of reasoning when discussing plants and their characteristic features, namely by analogy with animals.

It seems that the route the incoming food follows until its assimilation by the plant consists of two phases. Under normal circumstances, the food drawn by means of the roots is already concocted, that is heated and liquefied.[15] Often likened to veins, the roots provide the means by which nourishment is drawn towards the middle of the plant (i.e. wherein its soul and hearth of heat are situated), and is subsequently distributed along its entire length.[16] They constitute the upper part of the plant which, conceived in analogy to that of an animal's body, must be the hottest part in comparison to the rest of their body, since it hosts the principle of life, as we have seen earlier.[17] Thus, as we move from the hearth towards the edges of the plant, it is very likely that its nature becomes colder. This means that, when distributed to the branches, the nourishment, which has already liquefied on concoction in the ground, is now at risk of becoming congealed due to the fact that the active agent that brought about its change can no longer act upon it.[18]

It is with reference to 'the middle of the plant' that we are encouraged to speak of the 'active' role of the plant's internal heat, seeing that in the middle, on my interpretation, the heat of the ground appears to turn into the plant's vital heat. The source of heat, on this view, is expected to preserve the heat of the liquefied food drawn from the soil and to send it increasingly further away from the roots. At the same time, it is expected to be unable to effect concoction of the

---

14 Ἐπεὶ δ' ἀνάγκη πᾶν τὸ αὐξανόμενον λαμβάνειν τροφήν, ἡ δὲ τροφὴ πᾶσιν ἐξ ὑγροῦ καὶ ξηροῦ, καὶ τούτων ἡ πέψις γίνεται καὶ ἡ μεταβολὴ διὰ τῆς τοῦ θερμοῦ δυνάμεως, καὶ τὰ ζῷα πάντα καὶ τὰ φυτά, κἂν εἰ μὴ δι' ἄλλην αἰτίαν, ἀλλὰ διὰ ταύτην ἀναγκαῖον ἔχειν ἀρχὴν θερμοῦ φυσικήν, καὶ ταύτην ὥσπερ † [...] αἱ ἐργασίαι τῆς τροφῆς πλειόνων εἰσὶ μορίων.
15 Cf. *PA* II,10, 655b34–36; *GA* III,2, 753b23–29; *de Generatione et Corruptione* II,8, 335a11–14.
16 *PA* IV,4, 678a9–15; *GA* II,4, 740a33–35; *IA* 4, 705a32–b1; cf. *PA* III,5, 668a19–22, *de An.* II,1, 412b3–4.
17 See p. 157 above and cf. *Long.* 6, 467a30–b5; *IA* 4, 705a29–b8.
18 Cf. *Juv.* 14(8), 474b15–19.

incoming nourishment, an idea to which Aristotle is expressly opposed in the case of plants. Thus, the natural principle of heat seems to serve as a sort of maintainer of heat against the natural tendency of plants' material constitution to be cold owing to the fact that they are composed predominantly of earth;[19] that is to say, though it lacks the capacity to change the quality (through πέψις) of the food received, yet it is able to preserve its heat, so that the nutritive soul can make use of it. The source functions therefore as a kind of heat receptor or accumulator, which, notwithstanding the fact that it brings about no change at all in the quality of the incoming food, is responsible for maintaining its heat at every part of the plant, providing at the same time a counterbalance to the effects of the external heat of the environment.[20] Accumulation of heat remains unthreatening so long as the internal heat is properly cooled (*Juv.* 5), whereas cooling in plants is *also* effected through the movement that the intake of food occasions within the living body (*Juv.* 6, 470a23–27).

The end of phase 1 is signalled, as I wish to stress, by the arrival of the concocted (by the earth) food in the hearth of a plant's body. Similarly in animals the nutriment, after passing through the various stages of elaboration along the digestive tract, proceeds 'towards the principle', the heart, in order to take its ultimate form (*de Somno et Vigilia* 3, 456a32–b5). This view is further reinforced by *PA* II,3, 650a2–31. Here Aristotle does not embark on a direct comparison *of all* the stages of food transformation in animals and plants; rather, he launches into a comparative appraisal of the stages occurring, before the concocted food reaches the middle region, the region of the heart.[21] Phase 2 begins, once the food passes through the psychic heat's hearth; in this phase, the food moves toward the extremities of the plant thanks to the natural tendency of heat to move upwards, that is, as we shall see next, thanks to the evaporation of the

---

19 *Juv.* 19(13), 477a28; *de An.* III,13, 435b1.
20 Cf. *Ph.* VII,3, 246b4–6, which, however, is not referring to plants but to living bodies in general: τὰς μὲν γὰρ τοῦ σώματος, οἷον ὑγίειαν καὶ εὐεξίαν, ἐν κράσει καὶ συμμετρίᾳ θερμῶν καὶ ψυχρῶν τίθεμεν, ἢ αὐτῶν πρὸς αὑτὰ τῶν ἐντὸς ἢ πρὸς τὸ περιέχον ("Thus bodily excellences such as health and fitness we regard as consisting in a blending of hot and cold elements in due proportion, in relation either to one another within the body or to the surrounding"). A description of the role of homeostasis in Aristotle's ethical thought in relation to the doctrine of the mean has been already given by Terzis (1995). Cf. also King's remark: "It is noticeable that Aristotle devotes almost all of his attention to the case of the loss of balance, rather than to the proper functioning of the plant" (King 2001, p. 107).
21 650a20–23. Similarly in *PA* IV,4, 678a11–15: Τὰ μὲν οὖν φυτὰ τὰς ῥίζας ἔχει εἰς τὴν γῆν (ἐκεῖθεν γὰρ λαμβάνει τὴν τροφήν), τοῖς δὲ ζῴοις ἡ κοιλία καὶ ἡ τῶν ἐντέρων δύναμις γῆ ἐστιν, ἐξ ἧς δεῖ λαμβάνειν τὴν τροφήν· διόπερ ἡ τοῦ μεσεντερίου φύσις ἐστίν, οἷον ῥίζας ἔχουσα τὰς δι' αὑτῆς φλέβας.

nutritive moisture that takes place due to excessive accumulation of heat in the 'middle' of the plant.

The assumption that the hearth of the plant functions as a heat receiver/emitter could be buttressed by Aristotle's own views on fig-juice. In *GA* IV,4, 771b33–772a25, for example, Aristotle explains that the number of offspring produced is not merely dependent upon the seminal material from which each animal is formed, but also upon the limit of size imposed upon animals when being perfected. And this recalls the idea found earlier in *GA* IV,4, that large trees bear less fruit. In the course of this discussion, Aristotle notes:

> *So fire also does not continue to make water hotter in proportion as it is itself increased, but there is a fixed limit to the heat of which water is capable; if that is once reached and the fire is then increased, the water no longer gets hotter but rather evaporates and at last disappears and is dried up.* Now since it appears that the secretion of the female and that from the male need to stand in some proportionate relation to one another (I mean in animals of which the male emits semen), what happens in those that produce many young is this: from the very first the semen emitted by the male has power, being divided, to form several embryos, and the material contributed by the female is so much that several can be formed out of it. (*The parallel of curdling milk, which we spoke of before, is no longer in point here, for what is formed by the heat of the semen is not only of a certain quantity but also of a certain quality, whereas with fig-juice and rennet quantity alone is concerned.*) (*GA* IV,4, 772a12–25, my emphasis)[22]

Two things stand out in this passage: first, that the reference to the semen's heat to give shape to something occurs at the level of the nutritive/generative soul, that is, the basic part of the soul that plants also possess; second, that the last two lines of this passage emphasise that the acid juice of the fig-tree acts upon that into which it is put only in respect of quantity, not quality. By transferring its natural heat to the material it is poured into, fig-juice causes milk to curdle, or, to put it differently, varies its amount. Under conditions of excessive heat, as stated at the beginning of the passage, the water undergoes evaporation. Indeed, this small detail could be very helpful in explaining why the plant's source of heat, during its 'charging' process – if I may use that expression –,

---

[22] οὐδὲ γὰρ τὸ πῦρ θερμαίνει τὸ ὕδωρ μᾶλλον, ὅσῳπερ ἂν ᾖ πλεῖον, ἀλλ' ἔστιν ὅρος τις τῆς θερμότητος, ἧς ὑπαρχούσης ἐὰν αὔξῃ τις τὸ πῦρ, θερμὸν μὲν οὐκέτι γίγνεται μᾶλλον, ἐξατμίζει δὲ μᾶλλον καὶ τέλος ἀφανίζεται καὶ γίγνεται ξηρόν. ἐπεὶ δὲ φαίνεται συμμετρίας δεῖσθαί τινος πρὸς ἄλληλα τό τε περίττωμα τὸ τοῦ θήλεος καὶ τὸ παρὰ τοῦ ἄρρενος (ὅσα προΐεται σπέρμα τῶν ἀρρένων), τὰ πολυτόκα τῶν ζῴων εὐθὺς ἀφίησι τὸ μὲν ἄρρεν δυνάμενον πλείω συνιστάναι μεριζόμενον, τὸ δὲ θῆλυ τοσοῦτον ὥστε πλείους γίγνεσθαι συστάσεις. τὸ δ' ἐπὶ τοῦ γάλακτος παράδειγμα λεχθὲν οὐχ ὅμοιόν ἐστιν· ἡ μὲν γὰρ τοῦ σπέρματος θερμότης οὐ μόνον συνίστησι ποσόν τι ἀλλὰ καὶ ποιόν τι, ἡ δ' ἐν τῷ ὀπῷ καὶ τῇ πυετίᾳ τὸ ποσὸν μόνον. Cf. *GA* I,20, 729a9–14.

can at the same time withhold itself from depriving the incoming food of the heat necessary for the survival of the plant extremities. These (re)charging needs, it seems, *grow in proportion to the heat consumption* taking place during that state in which the heat emitter is in charge of maintaining the life of the plant in the absence of food intake. And this leads us to the following reasonable question: what happens in the case of extreme weather conditions?

According to *Juv.* 6, 470a19–32 (p. 155 above), when changes in the heat of the environment occur due to extreme seasonal conditions, the hearth's heat is not powerful enough to meet the needs of the plant, and even if it has not entirely lost its ability to feed itself, it eventually remains helpless, as the nutritive mixture drawn from the ground proves inadequately salutary. Under such circumstances, it seems, the heat emitter ultimately finds no way either to recharge itself (in the case of excessive cold) or to get the refrigeration it needs to conserve itself (in the case of excessive heat).

The need for a balance between internal and external heat in plants is also highlighted by the criticism Aristotle levels against Empedocles in chapter 20(14) of *Juv.*[23] According to Aristotle, Empedocles claimed that the material of living things may be of a different nature than the environment in which they exist; and that such differentiation in practice proves beneficial to them: one needs only observe the aquatic animals, which live in water precisely because they are of a hot nature (477a32–b4). Against this, Aristotle raises the objection that the nature of the material can be preserved only within an environment of the same nature. Deviations from the rule are simply due to the dispositions of matter, which account for the need of an opposing power to counterbalance and subsequently prevent excess at the opposing extremes (477b14–17).

---

23 διὰ τοῦτο τὰ δένδρα οὐκ ἐν ὕδατι φύεται, ἀλλ' ἐν γῇ. [...] αἱ μὲν οὖν φύσεις τῆς ὕλης, ἐν οἵῳπερ τόπῳ εἰσί, τοιαῦται οὖσαι τυγχάνουσιν, αἱ μὲν ἐν ὕδατι ὑγραί, αἱ δ' ἐν τῇ γῇ ξηραί, αἱ δ' ἐν τῷ ἀέρι θερμαί· αἱ μέντοι ἕξεις αἱ μὲν ὑπερβάλλουσαι θερμότητι ἐν ψυχρῷ, αἱ δὲ τῇ ψυχρότητι ἐν θερμῷ τιθέμεναι σῴζονται μᾶλλον· ἐπανισοῖ γὰρ εἰς τὸ μέτριον ὁ τόπος τὴν τῆς ἕξεως ὑπερβολήν. τοῦτο μὲν οὖν δεῖ ζητεῖν ἐν τοῖς οἰκείοις τόποις ἑκάστης ὕλης καὶ κατὰ τὰς μεταβολὰς τῆς κοινῆς ὥρας· τὰς μὲν γὰρ ἕξεις ἐνδέχεται τοῖς τόποις ἐναντίας εἶναι, τὴν δ' ὕλην ἀδύνατον. (*Juv.* 20(14), 477b26–478a7) ("Thus trees grow not in water but on dry land [...] Thus the natural character of the material of objects is of the same nature as the region in which they exist; the liquid character is found in water, the dry on land, the warm in air. But the states of body which are excessively hot are rather preserved when they are placed in the cold; and those that are excessively cold when they are placed in the warm; for the region reduces to a mean the excess in the bodily condition. This must be sought in the regions appropriate to each type of matter, and according to the changes of the seasons which are common to all; for, while states of the body can be opposed in character to the environment, the material of which it is composed can never be so." Transl. G.R.T. Ross, slightly modified).

The model of interpretation offered above conforms with this objection and seems capable of explaining why plants begin to dry up when exposed to extreme environmental conditions, that is, when they are no longer 'familiar' with the environment in which they live. Since the source of their internal heat can only maintain but not bring about any change of quality on the heat inside them, plants are directly affected by extreme external conditions. The situation is even further complicated by the fact that it is also impossible for them to avoid intake of nourishment already pre-heated in the external environment. Moreover, as the stage of fruit-bearing comes to an end, which in many cases happens during the cold months of the year, plants suffer additionally from constant loss of vital heat, seeing that, as mentioned earlier, fruit is a kind of residue, and loss of residue entails loss of vital heat.[24] For all the above reasons therefore it seems quite reasonable to infer that plants are in need of an internal mechanism to preserve their natural heat.

4 Ripening

Seed-bearing is preceded by flowering, as Aristotle informs us,[25] on which, however, the material we find in the Stagirite's extant works is too scanty to allow us to draw a fairly coherent picture. Seed-bearing results from ripening (πέπανσις), a kind of concoction of the nutritive material. Ripening, in turn, when completed, will yield the fruit of the plant – actually, it takes place inside that part of the plant which surrounds and protects the future fruit, namely the pericarp.[26] As a kind of concoction, πέπανσις is a natural process that occurs due to the internal source of plant heat, yet not without the assistance of the external heat.[27] One

---

[24] See *Long.* 5, 466a29–b9 and *GA* V,3, 783b17–18, which will be discussed under the section 'Leaf-shedding', *GA* III,1, 750a21–26 and p. 166 below.

[25] Following Alcmaeon of Croton: "and at the same time hair appears upon the pubes, in like manner, so Alcmaeon of Croton remarks, as plants first blossom (ἀνθεῖν) and then seed (σπέρμα φέρειν)" (*Historia Animalium* VII,1, 581a14–16).

[26] πέπανσις δ' ἐστὶ πέψις τις· ἡ γὰρ τῆς ἐν τοῖς περικαρπίοις τροφῆς πέψις πέπανσις λέγεται (*Mete.* IV,3, 380a11–12) ("Ripening is a sort of concoction; for we call it ripening when there is a concoction of the nutriment in the pericarp." Transl. E.W. Webster, slightly modified).

[27] πέψις μὲν οὖν ἐστι τελείωσις ὑπὸ τοῦ φυσικοῦ καὶ οἰκείου θερμοῦ ἐκ τῶν ἀντικειμένων παθητικῶν· ταῦτα δ' ἐστὶν ἡ οἰκεία ἑκάστῳ ὕλη. ὅταν γὰρ πεφθῇ, τετελείωταί τε καὶ γέγονεν. καὶ ἡ ἀρχὴ τῆς τελειώσεως ὑπὸ θερμότητος τῆς οἰκείας συμβαίνει, κἂν διά τινος τῶν ἐκτὸς βοηθείας συνεπιτελεσθῇ, οἷον ἡ τροφὴ συμπέττεται καὶ διὰ λουτρῶν καὶ δι' ἄλλων τοιούτων· ἀλλ' ἥ γε ἀρχὴ ἡ ἐν αὐτῷ θερμότης ἐστίν (*Mete.* IV,2, 379b18–25) ("Concoction is maturity, produced by a thing's natural and proper heat out of the opposite, passive qualities, which are the proper

should bear in mind that πέπανσις concerns a process taking place away from the region of the natural heat's hearth, so that the parts of the plant being involved in this process are in greater need of the heat of the environment for their conservation. The sun's heat acts upon that point out of which fruits spring, and thus activates the process of ripening. And it is only under circumstances like those occurring inside the pericarp (as part of the whole plant) that the sun's heat, in cooperation with the internal heat of the plant, can accomplish the ripening of the plant's seeds. For, as Aristotle points out in *Sens.*, the result would not have been the same had a pericarp, after being separated from the plant, been exposed in the sun.[28]

During ripening, the underlying matter passes through three distinct stages of qualitative change:

> So everything that ripens turns from an airy into a watery state, and from a watery into an earthy state, and in general from being rare becomes dense. In this process nature incorporates some of the matter in itself, and some it rejects. (*Mete.* IV,3, 380a23–26)[29]

As the text makes clear, the moisture of the nutritive material that is present inside the pericarp gradually begins to congeal, and the result is easily observable in the seed, which becomes denser. In such a change (or course of a change), the active factor is, as Aristotle mentioned immediately before (while explaining that

---

matter of any given object. For when concoction has taken place we say that a thing has been perfected and has come to be itself. It is the proper heat of a thing that sets up this perfecting, though external influences may contribute in some degree to its fulfilment. Baths, for instance, and other things of the kind contribute to the concoction of food, but the primary source, at any rate, is the proper heat of the body." Transl. E.W. Webster, slightly modified). See also *Juv.* 6, 470a19–22, cited above.

28 "For we see that when pericarpal fruits are plucked and exposed in the sun, or subjected to the action of fire, their savours are changed by the heat, which shows that their qualities are not due to their drawing anything from the water in the ground, but to a change which they undergo within the pericarp itself; and we see, moreover, that these juices, when extracted and allowed to lie, instead of sweet become by lapse of time harsh or bitter, or acquire savours of any and every sort; and that, again, by the process of boiling they are made to assume almost all kinds of new savours. It is likewise impossible that water should be a material qualified to generate all kinds of savour germs; for we see different kinds of taste generated from the same water, having it as their nutriment" (4, 441a11–20).

29 ἐκ μὲν οὖν τῶν πνευματικῶν ὑδατώδη, ἐκ δὲ τῶν τοιούτων τὰ γεηρὰ συνίσταται, καὶ ἐκ λεπτῶν ἀεὶ παχύτερα γίγνεται πεπαινόμενα πάντα. καὶ τὰ μὲν εἰς αὑτὴν ἡ φύσις ἄγει κατὰ τοῦτο, τὰ δὲ ἐκβάλλει.

boils and phlegm are subject to ripening too),[30] the natural heat inherent in the moisture of matter, while the matter undergoing ripening is nothing other than the nutritive material that entered through the roots. For the most part, this material has been distributed to the parts of the plant in order to cover its nutritional needs; yet a small amount thereof has been stored to constitute the future seed.[31] In general, the amount of nutritive material that flows in, and is consumed by, the plant body is inversely proportional to the quantities in which fruits grow; as a result, the largest plants bear less fruit (*GA* IV,4, 771a27–b14).

5 Leaf-Shedding

In *Analytica Posteriora*, Aristotle, in order to illustrate what he means by the concept of commensurateness between cause and effect, provides the example of leaf-shedding, which is considered as being caused by solidification of the moisture of a tree's leaves (II,16, 98b32–38). The same idea is further developed in *GA* V, where, in a process of reasoning resting again on argument from analogy between plants and animals, leaf-shedding is presented as a condition analogous to going bald (φαλακροῦσθαι) and moulting (πτερορρυεῖν):

> Similar to this is the condition of baldness in those human beings to whom it is incident. For leaves are shed by all plants, from one part of the plant at a time, and so are feathers and hairs by those animals that have them, it is when they are all shed together that the condition is described by the terms mentioned, for it is called 'going bald' and 'the fall of the leaf' and 'moulting'. The cause of the condition is deficiency of hot moisture, such moisture being especially the greasy and hence greasy plants are more evergreen. (However, we must elsewhere state the cause of this – for other causes also contribute to

---

[30] ἔστι δὲ ἡ φυμάτων καὶ φλέγματος καὶ τῶν τοιούτων πέπανσις ἡ ὑπὸ τοῦ φυσικοῦ θερμοῦ τοῦ ἐνόντος ὑγροῦ πέψις (*Mete*. IV,3, 380a20–22) ("In the case of boils and phlegm, and the like, the process of ripening is the concoction of the moisture in them by their natural heat").

[31] ὥσπερ γὰρ καὶ ἐκ τῆς πρώτης τροφῆς ἐκ πολλῆς ὀλίγον ἀποκρίνεται τὸ χρήσιμον ἐν ταῖς περὶ τοὺς καρποὺς ἐργασίαις, καὶ τέλος οὐθὲν μέρος τὸ ἔσχατον πρὸς τὸ πρῶτον πλῆθός ἐστιν, οὕτω πάλιν καὶ ἐν τῷ σώματι διαδεχόμενα τὰ μέρη ταῖς ἐργασίαις τὸ τελευταῖον πάμπαν μικρὸν ἐξ ἁπάσης γίγνεται τῆς τροφῆς. τοῦτο δὲ ἐν μέν τισιν αἷμά ἐστιν ἐν δέ τισι τὸ ἀνάλογον (*GA* IV,1, 765b28–35) ("For consider the production of fruit; the nutriment in its first stage is abundant, but the useful product derived from it is small, indeed the final result is nothing at all compared to the quantity in the first stage. So is it with the body; the various parts receive and work up the nutriment, from the whole of which the final result is quite small. This is blood in some animals, in some its analogue."); cf. *GA* I,23, 731a5–9; II,4, 740b2–8.

it.) It is in winter that this happens to plants (for the change from summer to winter is more important to them than the time of life) [...] (*GA* V,3, 783b12–23)³²

It is worth noting that hot-greasy moisture is also mentioned in *Long.* 6, where it is viewed as a vital factor and one of the main causes of longevity in plants.³³ Changes in the time of life are regarded as highly important to animals; in plants, however, it is seasonal change that is of paramount significance. As the weather grows colder from summer to winter, plants undergo a period of leaf-drop: they become deficient with respect to natural hot moisture, precisely because the external heat does not provide useful assistance to the internal heat source, at least as much as it does during the summer months. In the hot months, however, plants, unlike humans, can sprout new leaves. In other words, a plant's internal heat 'mechanism', owing to the cold of winter, is not rendered completely destroyed but merely inoperable, in the sense that it fails to maintain the heat in the most extreme parts of the plant body.

In the subsequent chapter of *APo.* II, when the discussion arrives at the issue of how cause, effect and subject are related to one another, Aristotle raises anew the question of what leaf-shedding is. Then, he offers a somehow clearer answer, at least as regards the location in which leaf-shedding takes place: τὸ πήγνυσθαι τὸν ἐν τῇ συνάψει τοῦ σπέρματος ὀπόν (II,17, 99a28–29). That is, leaf-shedding is the solidifying of the plant juice *at the junction of the seed*, i.e., at the point where the seed is connected with the rest of the plant. This, I believe, shows that the tiny point connecting the leaf to the tree, owing to its size and hence to small quantity of heat, fails to preserve through heat-transferring the remaining part of the leaf, particularly given that it gradually receives less and less support by the heat of the environment.

---

32 τοιοῦτον δέ τι πάθος καὶ ἡ φαλακρότης ἐστὶν ἐπὶ τῶν ἀνθρώπων ὅσοις συμβαίνει φαλακροῦσθαι· κατὰ μέρος μὲν γὰρ ἀπορρεῖ καὶ τὰ φύλλα τοῖς φυτοῖς πᾶσι καὶ τὰ πτερὰ καὶ αἱ τρίχες τοῖς ἔχουσιν, ὅταν δ' ἀθρόον γένηται τὸ πάθος λαμβάνει τὰς εἰρημένας ἐπωνυμίας· φαλακροῦσθαί τε γὰρ λέγεται καὶ φυλλορροεῖν (καὶ πτερορρυεῖν). αἴτιον δὲ τοῦ πάθους ἔνδεια ὑγρότητος θερμῆς, τοιοῦτον δὲ μάλιστα τῶν ὑγρῶν τὸ λιπαρόν· διὸ καὶ τῶν φυτῶν τὰ λιπαρὰ ἀείφυλλα μᾶλλον. ἀλλὰ περὶ μὲν τούτων ἐν ἄλλοις τὸ αἴτιον λεκτέον· καὶ γὰρ ἄλλα συναίτια τούτου τοῦ πάθους αὐτοῖς. γίγνεται δὲ τοῖς μὲν φυτοῖς ἐν τῷ χειμῶνι τὸ πάθος (αὕτη γὰρ ἡ μεταβολὴ κυριωτέρα τῆς ἡλικίας) [...].

33 Ἐν δὲ τοῖς φυτοῖς ἐστι τὰ μακροβιώτατα, καὶ μᾶλλον ἢ ἐν τοῖς ζῴοις, πρῶτον μὲν ὅτι ἧττον ὑδατώδη, ὥστ' οὐκ εὔπηκτα· εἶτ' ἔχει λιπαρότητα καὶ γλισχρότητα, διὸ καὶ ξηρὰ καὶ γεώδη ὄντα ὅμως οὐκ εὐξήραντον ἔχουσι τὸ ὑγρόν (467a6–9).

6 Concluding Remarks

Let us sum up what has been said so far about vital heat in plants:
- First of all, it is inseparable from the nutritive soul, serving as its instrument inside the body of the plant in order for life to be preserved;
- second, its principle exists in actuality in the middle part of the plant (in analogy with the animal), but is in potentiality present in every part;
- third, its principle interacts continuously with the heat of the environment;
- and finally, it preserves itself, in this way preserving the plant's life.

Now the above remarks lead to the following question: If heat preserves itself by functioning as a kind of heat receptor or accumulator, is it not the case also that it should somehow preserve its aggregated quantity each time there is some failure of heat inside the plant's body? In other words, is the principle of the plant's heat capable of increasing or decreasing the heat inside the plant in accordance with the external environmental conditions? I am inclined to think that, rather, the nutritive soul achieves this goal, and it does so by continuously feeding the source of heat, while at the same time accomplishing the purpose of determining the plant's size (and increase) according to a certain limit and ratio. This, it seems, suffices to explain why the principle of heat and the nutritive soul are considered inseparable; because only under these circumstances is vital heat capable of distributing itself over the entire length of the plant in the appropriate ratio.

This view prompts a last remark regarding Aristotle's methodological approach to plants *qua* living beings in the surviving corpus. What has, hopefully, become clear from my analysis is that Aristotle's customary way of discussing, describing and accounting for the various characteristics of plants is by analogy with the corresponding attributes in animals. The animal body and its functions stand, thus, as an exemplary model illustrating the necessary conditions for successful sustenance of plant life. In terms of methodology, it seems to me, Aristotle is fully in accordance with his protreptic to the study of animals and plants at *PA* I,5, read from a very specific perspective: here, though the contrast is between imperishable and perishable things with respect to the worthiness of their study, we are clearly told that better means of information are provided by objects which are closer to sense perception. It is to be expected, however, that all things that are more knowable to us, the perishable things, will not be in the same proximity to sense perception.[34] Admittedly, the internal structure of plants is far less obvious, and complex, than the structure(s) of the animal body, which proves to be a disadvantage when it comes

---

34 Cf. *Ph.* I,1; *APo.* I,2.

to causal explanations. Nevertheless, since plants exhibit signs of life, in the sense that they perform such functions as growth, nutrition, or generation, and hence are included by Aristotle among living things, it comes as no surprise that in many places Aristotle launches into a comparative discussion and hierarchical assessment of different forms of life, including plants; and that plants are frequently inferred to share properties with even the most complete animals. Interestingly enough, the process of nutrition seems to serve as the paradigmatic case which helps him, first and foremost, to establish the structural affinity between animals and plants.

Bibliography

Barnes, Jonathan (1995): *The Complete Works of Aristotle. The Revised Oxford Translation. One volume digital edition.* [Bollingen Series LXXI.2.] Princeton: Princeton University Press.

Connell, Sophia M. (2016): *Aristotle on Female Animals. A Study of the* Generation of Animals. Cambridge: Cambridge University Press.

Drossaart Lulofs, Hendrik J. (ed.) (1965): *Aristotelis* De Generatione Animalium *recognovit breviaque adnotatione critica instruxit.* Oxford: Clarendon Press.

Falcon, Andrea (2015): "Aristotle and the Study of Animals and Plants". In: Holmes, Brooke/Fischer, Klaus-Dietrich (eds.): *The Frontiers of Ancient Science. Essays in Honor of Heinrich von Staden.* [Beiträge zur Altertumskunde Band 338.] Berlin/München/Boston: De Gruyter, p. 75–91.

Freudenthal, Gad (1995): *Aristotle's Theory of Material Substance. Heat and Pneuma, Form and Soul.* Oxford: Clarendon Press.

Hardy, Gavin/Totelin, Laurence (2016): *Ancient Botany.* London/N. York: Routledge.

Holmes, Brooke (2017): "Pure Life: The Limits of the Vegetal Analogy in the Hippocratics and Galen". In: John Z. Wee (Ed.): *The Comparable Body. Analogy and Metaphor in Ancient Mesopotamian, Egyptian, and Greco-Roman Medicine.* Studies in Ancient Medicine 49. Leiden/Boston: Brill, p. 358–386.

King, Richard A. H. (2001): *Aristotle on Life and Death.* London: Duckworth.

Louis, Pierre (ed.) (1982): *Aristote*, Météorologiques *Tome II (Livres III et IV), Texte établi et traduit.* Paris: Les Belles Lettres.

Louis, Pierre (ed.) (1993[2]): *Aristote,* Les parties des animaux, *Texte établi et traduit.* 2ème édition revue et corrigée. Paris: Les Belles Lettres.

Murphy, Damian (2005): "Aristotle on Why Plants cannot Perceive". In *OSAPh* 29, p. 295–339.

Repici, Luciana (2000): *Uomini capovolti. Le piante nel pensiero dei Greci.* Bari: Laterza.

Ross, William D. (ed.) (1950): *Aristotelis* Physica *recognovit breviaque adnotatione critica instruxit.* Oxford: Clarendon Press.

Ross, William D. (ed.) (1964): *Aristotelis* Analytica Priora et Posteriora *recensuit breviaque adnotatione critica instruxit.* Oxford: Clarendon Press.

Terzis, George N. (1995): "Homeostasis and the Mean in Aristotle's Ethics". In *Apeiron* 28.4, p. 175–189.

Aristotelianism

Robert Mayhew
Reading and Sleep in Pseudo-Aristotle, *Problemata* XVIII,7
On the Nutritive Soul's Influence on the Intellect, and *Vice Versa*

Abstract: After brief discussions of Aristotle's *De somno et vigilia* and the fragmentary evidence for Strato of Lampsacus' *De somno*, which together provide the relevant Peripatetic context, this essay is in effect a commentary on pseudo-Aristotle, *Problemata* XVIII,7. This neglected text discusses the purported relationship in different individuals between reading and sleep (or insomnia), which in turn involves the unknown author's conception of the connection between the nutritive part of the soul and the rational part. Discussion and debate in the Lyceum in the generation after Aristotle was, it is argued, the likely intellectual soil from which *Pr.* XVIII,7 emerged.

1 Introduction

One major concern of the *Problemata physica* in the *corpus Aristotelicum* are the problems that arise in attempting to understand the interaction between body and soul, or more specifically (as we might put it, though the authors of the *Problemata*[1] never did) the nutritive part of the soul and the two other parts (the perceptual-appetitive part, and the rational part). For instance: *Pr.* XIV focuses on the relationship between the climate of a particular region, and its effects on the bodies and so on the temperament and character of the people who live there. The focus of most of the chapters in *Pr.* XXVII is the physiological causes of the external manifestations of fear: trembling, loose bowels, rapid heartbeat, pallor, etc. The longest and most famous chapter in the entire work – *Pr.* XXX,1 – is concerned with answering the question: "Why is it that all those men who

---

[1] Although the extant *Pr.* was attributed in antiquity and beyond to Aristotle, who the ancient biographical tradition reports wrote more than one work on 'problems' (see e.g. Diogenes Laertius V,23; V,26), and although there may be remnants of such lost works in the extant *Pr.*, the bulk of its chapters were likely written later (though not necessarily much later), by other Peripatetics. See Flashar (1962), p. 303–358, Mayhew (2011), p. xvi–xxiv, and Bodnár (2015), and note 8 below. I return later to the question of when *Pr.* XVIII,7 was likely written.

https://doi.org/10.1515/9783110690552-011

have become extraordinary (περιττοί) in philosophy, politics, poetry or the arts are obviously melancholic (μελαγχολικοί)?"[2] My interest here, however, is in *Pr.* XVIII,7. Like *Pr.* XVIII as a whole, this chapter has received virtually no scholarly attention.[3] It is concerned with the interaction between the nutritive part of the soul and the rational part.

The title of *Pr.* XVIII that has come down to us in the manuscripts is ὅσα (sc. προβλήματα) περὶ φιλολογίαν. Nuchelmans, in his study of φιλολογία and its cognates, writes (1950, p. 58): "Unter diesem Titel werden Fragen allgemeiner Bildung (Stil, Rhetorik, Lesen, Geschichte) zusammengefasst." In fact, the most prominent topic of *Pr.* XVIII (eight of its ten chapters) is rhetoric, broadly understood.[4] The only exceptions are 1 & 7, which deal with reading and sleep. This is the topic that concerns me here. There is a great deal of overlap between *Pr.* XVIII,1 & 7, but in what follows I focus on the more developed and likely earlier of the two (namely XVIII,7), occasionally providing notes indicating where it diverges from XVIII,1.[5]

2 The Peripatetic Context: Aristotle and Strato on Sleep

I next present a sketch of Aristotle's *De somno et vigilia*, as well as the fragmentary evidence for Strato of Lampsacus' *De somno*, which together provide some of the background needed to understand *Pr.* XVIII,7 and its relationship to the work of Aristotle and his school.

2.1 Aristotle, *De Somno et Vigilia*[6]

In chapter 1, Aristotle says that sleep and waking belong to the same element or part of an animal, and that they are opposite states, sleep being a privation

---

[2] Unless otherwise indicated, translations from the Greek are my own. Translations from the *Problemata* come from Mayhew (2011), in many cases revised.
[3] See however Barthélemy-Saint Hilaire (1891), p. 26–28, 32–34, Flashar (1962), p. 591–594, and Louis (1993), p. 85–87.
[4] See Mayhew (forthcoming).
[5] It should become clear that *Pr.* XVIII,1 is inferior to (and perhaps a later revision of) XVIII,7. The differences are such that I do not think that XVIII,1 is a mere abridgement of XVIII,7. On how *Pr.* XVIII,1 & 7 might be thought to be connected to the topic of φιλολογία, see below note 26.
[6] This summary is necessarily terse and selective, especially in its coverage of chapter 3.

(στέρησις) of waking. Waking is closely associated with perception (αἴσθησις), as one who is awake just is one who is perceiving, so that that in virtue of which animals perceive is that in virtue of which they are awake or asleep. Further, like perception, waking and sleep are neither peculiar to body alone nor to soul alone but involve both. Of the parts of the soul, sleep and waking involve the nutritive and perceptual parts (which is why *animals* are said to sleep, but not plants).[7] He goes on to argue that all animals sleep, and no animal is always awake or always asleep. He offers a provisional account of what sleep is (which makes clear why no animal can be asleep always): "Sleep is an affection – that is, a certain binding or immobility – of the perceptual part" (ὁ γὰρ ὕπνος πάθος τι τοῦ αἰσθητικοῦ μορίου ἐστίν, οἷον δεσμός τις καὶ ἀκινησία) (454b9–11).

In chapter 2, Aristotle investigates "why it is that one sleeps or wakes, and owing to what sort of sense – or what sorts, if more than one" (455a4–5). He argues at some length that sleep does not involve the cessation of any one particular sense or set of senses, nor is it the result of just any kind of impairment of or incapacity in the senses (for instance fainting, or unnatural causes of unconsciousness like constricting the veins in the neck). Rather, sleep occurs in the primary organ by which one perceives everything (ἐν τῷ πρώτῳ ᾧ αἰσθάνεται πάντων, 455b10) – i.e., in the heart. He next turns to the causes of sleep, beginning with its final cause: Sleep is (to speak metaphorically) rest or repose (ἀνάπαυσις), in animals that perceive and move, for the sake of the preservation of the animal – keeping in mind, of course, that being awake (not sleep) is the end or goal (ἡ [...] ἐγρήγορσις τέλος) (455b16–25). Next, he briefly indicates "from what sort of motion or action (ποίας κινήσεως καὶ πράξεως) occurring in their bodies waking or sleeping arises in animals" (455b28–30). In sanguineous animals, sleep and waking, like perception (and movement and respiration), originate in the region around the heart (in bloodless animals, in the region around the analogue of the heart); and owing to respiration, cooling takes place in the heart (or its analogue) – which likely has a role in sleep, though much is unclear (456a4–11).[8]

---

[7] "Having determined earlier in other works what are called parts of the soul (μορίων τῆς ψυχῆς), and that the nutritive part (τοῦ θρεπτικοῦ) in [some] bodies possessing life exists separately from the others, whereas none of them exists without this, it is clear that neither sleep nor waking belongs to living things that partake only of growth and decay, i.e. to plants, for they do not possess the perceptual part (τὸ αἰσθητικὸν μόριον)" (*Somn. Vig.* 1, 454a11–17).
[8] The second chapter ends with an intriguing reference to Aristotle's (lost) *Pr.*, which indicates that a set of problems on sleep likely originated in the Lyceum under the direction of Aristotle: Having mentioned that "some people move in their sleep and perform many awake-like activities" – I assume he is referring to sleep-walking – he states: "Why people when awakened re-

In chapter 3, Aristotle describes the specific natural processes (involving the nutritive part of the soul) from which sleep and waking originate. Once eaten, food in the body is transformed into blood (or the analogue of blood in bloodless animals), which goes to the heart (or its analogue) and then to the veins. During this process, exhalation (ἀναθυμίασις) is produced and enters the veins.[9] As it is hot it moves upward, but then it turns back again (as it has no place else to go, and perhaps too because it has been cooled by the brain) and travels downward in a mass. The (material-efficient) cause of sleep is this mass of (still hot, but cooler) material reaching the heart, which as we have seen is the common organ of perception. Presumably the exhalation binds or immobilizes the heart, at least *qua* common organ of perception. Here are Aristotle's three (similar) descriptions of the nature or cause of sleep:

> So, it is manifest from what has been said that sleep is a certain concentration inward of the heat and a natural reflux (ὁ ὕπνος ἐστὶ σύνοδός τις τοῦ θερμοῦ εἴσω καὶ ἀντιπερίστασις φυσική) (457a33–b2).

---

member their dreams, but do not remember their awake-like activities, has been explained in the *Pr.*" (ἐν τοῖς Προβληματικοῖς εἴρηται) (456a24–29). There is no such discussion in the extant *Pr.*

[9] The precise composition of this exhalation (and especially whether it contains *pneuma*) is unclear. In three passages, Aristotle seems to claim that the exhalation is liquid and corporeal alone (and especially the latter): see 3, 456b17–26 (τό τε ὑγρὸν καὶ τὸ σωματῶδες), 457b20–23, and 458a25–28; see also Bubb (forthcoming). But Aristotle also implies that what is consumed in eating and drinking contains *pneuma*, that the abundance of *pneuma* in what is consumed can be problematic (more on this below, in connection with melancholy), and that one problem that some people have with wine is its pneumatic nature (πνευματῶδες γὰρ ὁ οἶνος): see 3, 457a9–25, and Meeusen (2020). Cf. Theophrastus, *de Vertigine* 1,1–5, where he claims that one cause of dizziness is alien *pneuma* from certain nourishment (οἱ ἴλιγγοι γίνονται ὅταν ἢ πνεῦμα ἀλλότριον περὶ τὴν κεφαλὴν ἔλθῃ ἢ ὑγρότης περιττωματική [ἢ] ἀπὸ τροφῆς ἐνίας), and see Sharples (2003), p. 199. It is also unclear what relation, if any, the exhalation in Aristotle's account of sleep has to the exhalations so central to his explanation of meteorological phenomena. See especially *Meteorologica* I,4, with Wilson (2013), chapters 2–3. Here is his basic description (*Mete*. I,4, 341b6–12):

> When the earth is heated by the sun the exhalation (τὴν ἀναθυμίασιν) that necessarily arises is not of one kind, as some think, but of two: the one more vapor-like, the other more *pneuma*-like (τὴν μὲν ἀτμιδωδεστέραν τὴν δὲ πνευματωδεστέραν). The one from the moisture in the earth and on the earth is vapor (ἀτμίδα), while the one from the earth itself, being dry, is smoke-like (καπνώδη). And of these the *pneuma*-like one rises to the top owing to its heat, whereas the moister one sinks below owing to its weight.

As we shall see, whatever Aristotle himself thought, some of the authors of the *Pr.* regarded the exhalation produced in the concoction of nourishment as *pneuma*-like (to use the language of the *Mete.*).

For sleep occurs [...] when the solid [matter] is carried upwards by the heat, along the veins, to the head (τοῦ σωματώδους ἀναφερομένου ὑπὸ τοῦ θερμοῦ διὰ τῶν φλεβῶν πρὸς τὴν κεφαλήν); and when what has been carried up can no longer do so, but is too great in quantity, it is pushed back again and flows downwards (πάλιν ἀνταπωθεῖται καὶ κάτω ῥεῖ) (457b20–23).

What the cause of sleeping is has now been stated: the reflux of the solid [matter] carried up by the connate heat in a mass [back down] to the primary sense-organ (ἡ τοῦ σωματώδους τοῦ ἀναφερομένου ὑπὸ τοῦ συμφύτου θερμοῦ ἀντιπερίστασις ἀθρόως ἐπὶ τὸ πρῶτον αἰσθητήριον) (458a25–28).

A couple of additional points in concluding this account of Aristotle on sleep, with a view to better understanding *Pr.* XVIII,7: (1) Note that the brain is not the organ of perception or cognition, the heart is. Aristotle claims in this work and elsewhere that the brain is the coldest organ, as its purpose is to regulate (by cooling) the temperature of the body, and especially the heat in the heart (see *Somn. Vig.* 3, 457b26–31 and *de Partibus Animalium* II,7). (2) Though Aristotle does not refer to *pneuma* in presenting his theory of sleep, he does at one point (3, 457a8–14) refer to it in comparing sleep to epilepsy, which he says are similar (ὅμοιον [...] ὁ ὕπνος ἐπιλήψει). He explains that in epilepsy, "when a great deal of *pneuma* is carried upward (ὅταν [...] πολὺ φέρηται τὸ πνεῦμα ἄνω), in going down again (καταβαῖνον πάλιν) it distends the veins and compresses the passage through which respiration (ἡ ἀναπνοή) occurs." So *pneuma* may be one component of the exhalation (or an accompanying material or by-product) that is central to Aristotle's theory of sleep, though he does not explicitly say so.[10]

2.2 Strato of Lampsacus, *De Somno*

According to Diogenes Laertius, Strato of Lampsacus (scholarch of the Lyceum from 286 to 268 BC) wrote a work entitled Περὶ ὕπνου (V,59). The fragmentary

---

[10] See the previous note. According to Aristotle in *de Generatione Animalium*, the male contribution to generation (semen, ἡ γονή) is fully concocted blood, consisting of water and *pneuma* – a special kind of hot air (II,2, 736a1), which is or contains soul-heat (θερμότητα ψυχικήν, III,11, 762a20). This seems to support the view that the blood that results from the concoction of nutrition, referred to in his account of sleep, contains *pneuma* as well. Or at the very least, that account adds to the confusion. There is an abundant literature on Aristotle on *pneuma*; see for instance Peck (1942), p. 576–593, and Nussbaum (1978), p. 143–164. Nussbaum refers to Aristotle's *pneuma* as "a hypothetical gap-filler whose workings cannot be scrutinized too closely" (1978, p. 163).

remains are paltry: there are four relevant texts (fr. 66–69 Sharples), two on sleep and two on dreams. The most important is fr. 66 (= pseudo-Plutarch, *Placita* 5.24 [*Mor.* 909E–F]), though it is in a wretched state:[11]

> [Πλ]⟨Στρ⟩άτων[12] οἱ Στωικοὶ τὸν μὲν ὕπνον γίνεσθαι ἀνέσει τοῦ αἰσθητικοῦ πνεύματος οὐ κατ' ἀναχαλασμόν, καθάπερ ἐπὶ τῆς [γ]⟨μέθ⟩ης,[13] φερομένου δ'ὡς ἐπὶ τὸ ἡγεμονικὸν μεσόφρυον ⟨...⟩[14] ὅταν δὲ παντελὴς γένηται ἡ ἄνεσις τοῦ αἰσθητικοῦ πνεύματος, τότε γεγενῆσθαι θάνατον.

> Strato [and][15] the Stoics [claim] that sleep comes about through an abatement of the sensory *pneuma*, not through a relaxation as in the case of drunkenness, whereas when it is carried towards the ruling space between the eyebrows[16] [...] But when the abatement of the sensory *pneuma* is total, death occurs.

I believe that somewhere along the way the ideas presented in pseudo-Plutarch's ultimate source were condensed to the point of confusion or corruption. First, the name of Plato appears in the text (see note 12) either because a description of his view of sleep was originally there, but dropped out, or simply as a result of scribal error (Πλάτων being a mistake for Στράτων). Either way, the reference to the space between the eyebrows is a fairly certain indication that Strato belongs here.[17] Next, I suspect (based on fr. 67 = Tertullian, *De anima* 43,1–2 [part of which is quoted below]) that there is some kind of confusion involving the relationship between Strato and the Stoics, namely, that Strato and the Stoics did not have the same view (or did not use the same terminology), but that Strato viewed sleep as an abatement (ἄνεσις) of, i.e. a reduction in, the sensory *pneuma* (in the

---

11 I use Sharples's text (2011), p. 152 (with modifications, discussed below). Sharples marks fr. 66 as dubious, but I think it does provide evidence (however imperfect) for Strato's views.
12 Sharples (2011), p. 153 n. 1: "The MSS, and pseudo-Galen, have 'Plato.' 'Strato' is an emendation by Corsinus, accepted by Diels because of the reference to the space between the eyebrows (cf. 57 and 58)." See below note 21.
13 The manuscripts have γῆς ("on the *earth/ground*", which is meaningless in context). Sharples follows Mau's conjecture and prints γη⟨ράσεω⟩ς ("in *old age*"); I prefer Diels's μέθης ("in the case of *drunkenness*"). But which is correct makes little difference in the present context.
14 I mark a lacuna here, the reason for which I provide shortly.
15 Though one might have expected a καί or τε here, asyndeton is not uncommon in doxographical accounts, such as pseudo-Plutarch's *Placita*. (See below note 21, e.g. Πλάτων Δημόκριτος.)
16 I take this to be shorthand for: the ruling part of the soul, which is located in the space between the eyebrows.
17 See fr. 57 (= ps.-Plut. *Placita* 4,5 [*Moralia* 899 A]), quoted below (note 21).

relevant region), whereas the Stoics viewed it as a relaxation or slackening (ἀναχαλασμός) of sensory *pneuma* (whatever precisely that means).[18]

I must now justify my reading of the μέν – δέ construction in our text, and especially my insertion of a lacuna. I think the best way to do this is to present, by way of contrast, the interpretation of Sharples (2011, p. 152–153). He takes the μέν – δέ construction to be marking a contrast between different ways of understanding ἄνεσις (which he renders 'relaxation'), and therefore he had no reason to posit a lacuna here. He translates the first sentence:

> Strato [and] the Stoics [say] that sleep comes about through a relaxation of the sensory *pneuma*, not through a slackening as in growing old (γη⟨ράνσεω⟩ς),[19] but when it is carried towards the ruling [principle in] the space between the eyebrows.

This has the advantage of sticking more closely to the manuscript tradition. But it is, I believe, a less natural way of understanding the μέν – δέ construction. I think it more natural to take the line to be contrasting the ἄνεσις of sensory *pneuma* (which produces sleep), with sensory *pneuma* being carried "towards the ruling space between the eyebrows" (which produces some other state, not mentioned in the text – hence the lacuna). In my view, the absence of this other state is just one more sign that pseudo-Plutarch's source has been abridged to the point of inaccuracy or corrupted in some other way. So, putting aside the clause on drunkenness (or old age), I think this text is saying the following: on the one hand, sleep comes about through an abatement of the sensory *pneuma*; on the other, when the sensory *pneuma* is carried towards the ruling part located in the space between the eyebrows... The statement of what happens as a result of the latter is missing: I would speculate it is that one *awakens*. If "sleep comes about through an abatement of the sensory *pneuma*", then it makes sense that one would awake when the sensory *pneuma* is carried towards the ruling part. On the other end of the spectrum, a total or final abatement means death. Drunkenness would be some state between the presence and absence of *pneuma*, namely its relaxation.

I think my interpretation gets some support from fr. 67 (= Tertullian, *De anima* 43,1–2), which reports on different accounts of the nature of sleep. I quote merely the two relevant opinions:

---

18 That ἄνεσις and ἀναχαλασμός can be synonymous likely contributed to the confusion. (LSJ s.v. ἄνεσις [*loosening, relaxing; remission, abatement*], s.v. ἀναχαλασμός [*relaxation*]). Perhaps the source originally claimed the Stoics held that *sleep* is a relaxation of the soul, whereas Strato said *drunkenness* (or old age, see note 13) was.

19 See above note 13.

The Stoics claim that sleep is a slackening of the strength of sensation (*Stoici somnum resolutionem sensualis uigoris affirmant*) [...] , Strato that it is a separation of the connate *pneuma* (*Strato segregationem consati spiritus*) [...].

This confirms or supports aspects of my speculations concerning fr. 66: that the Stoics viewed sleep as a relaxation of the sensory *pneuma*; and further, that in the case of Strato, if the abatement of the sensory *pneuma* in the ruling part produces sleep, then sleep could (loosely) be considered a separation of *pneuma* from the ruling part.[20]

Whatever else might be inaccurate in the above presentation of (and speculations about) Strato's conception of sleep, two points should be kept in mind going forward, both of which I think we can be fairly certain are true: (1) Strato's account of sleep and awakening involves some conception of sensory *pneuma* (and its movement) interacting with and affecting the perceptual part of the soul; (2) the part of the body with which the sensory *pneuma* interacts to cause sleep is *not* the heart, but the space between the eyebrows.[21]

3 Pseudo-Aristotle, *Problemata* XVIII,7[22]

The most prevalent (and basic) format of the over 900 problems in the *Pr.* (though there are many exceptions) consists of three parts:
1. The statement of the problem (beginning Διὰ τί, 'Why ... ?').

---

[20] Tertullian goes on to say that according to Aristotle, sleep is "a weakening of the heat around the heart" (*Aristoteles marcorem circumcordialis caloris*). This is fairly accurate, if the heat descending from above to the heart has been cooled somewhat by the brain (a possibility I mentioned above). I owe this point to Michiel Meeusen.

[21] Perhaps this refers to the frontal lobe. In any case, in fr. 57 (= ps.-Plut., *Placita* 4,5 [*Moralia* 899 A]), Strato is included in the group of thinkers who locate the ruling part of the soul in (some part of) the head:

What is the ruling (ἡγεμονικόν) [part] of the soul, and in what is it.

Plato [and] Democritus [say it is] in the head as a whole (Πλάτων Δημόκριτος ἐν ὅλῃ τῇ κεφαλῇ).

Strato in [the space] between the eyebrows (Στράτων ἐν μεσοφρύῳ).

Erasistratus in the membrane around the brain, which he calls ἐπικρανίδα.

Herophilus, in the cavity of the brain, which is also its base.

This is Sharples's translation, modified. For context, I have added the line on Herophilus (not included by Sharples). After Herophilus, pseudo-Plutarch turns to those who locate the ruling part of the soul elsewhere (e. g. in the heart).

[22] For the text of *Pr.* XVIII,7, I have used Louis (1993), p. 90–91.

2. The follow-up question (almost always beginning ἤ ὅτι/διότι, 'Is it that/because ... ?'), which presents a tentative (or perhaps respected or widely held) solution to the problem (not necessarily the author's committed view).
3. Commentary on or discussion of that solution.[23]

Both *Pr.* XVIII,1 & 7 have (for the most part) this basic format.[24] Their statements of the problem are virtually identical, and their follow-up questions *are* identical. They differ most in their respective (and relatively lengthy) discussions following the follow-up question (though even here there are some similarities).

3.1 The Statement of the Problem: *Pr.* XVIII,7, 917a18–20

Here is the problem to be solved:

> Why does sleep overtake some people, if they begin to read, even though they do not want [to sleep], whereas others who want to, it does not make able [to sleep] when they take up a book?[25]

---

[23] There are more complex formats, involving more than one solution, and/or more than one follow-up question with discussion. These are not relevant in the present case (though see the following note).

[24] I add the parenthetical qualification, because (1) there are scholarly disagreements about whether this is true of the second part in *Pr.* XVIII,1 & 7 (see note 28), and (2) the third part in *Pr.* XVIII,7 especially is lengthy and complex.

[25] Διὰ τί τοὺς μέν, ἂν ἄρξωνται ἀναγινώσκειν, ὕπνος λαμβάνει καὶ μὴ βουλομένους, τοὺς δὲ βουλομένους οὐ ποιεῖ δύνασθαι, ὅταν λάβωσι βιβλίον; Following τοὺς δέ, the manuscripts have οὐ βουλομένους, which Ross (*apud* Forster 1927, n.p. *ad loc.*) recommended transposing, as I think is necessary (though admittedly the result is somewhat awkward). Richards's suggestion (1915), p. 141, accepted by Flashar (1962), p. 593, yields a similar (and arguably improved) result, but is unnecessarily complex: τοὺς δὲ αὖ βουλομένους ⟨μὴ⟩ ποιεῖ δύνασθαι κτλ. Ruelle et al. (1922), p. 162 bracket οὐ, pointing for support to *Pr.* XVIII,1, 916b2 (which has βουλομένους alone). But if this were correct, the problem would be about why reading causes sleep, whether or not the person wants to sleep, which is not supported by the remainder of XVIII,7. There are two other differences in the statement of the problem between XVIII,1 (916b2–4) and XVIII,7: ἐάν for ἄν (not significant); and, προ⟨σ⟩εγρηγορέναι for οὐ ποιεῖ δύνασθαι ('are kept awake' instead of 'does not make able [to sleep]'). Re. προ⟨σ⟩εγρηγορέναι: The manuscript reading is προεγρηγορέναι, which Bekker easily emended. It is most natural to take this to be the perfect infinitive active of προσεγείρω (LSJ s.v. *lift up, stimulate, excite*). As the infinitive does not work, however, and because of the reading of the parallel line at *Pr.* XVIII,7, 917a19, some scholars (e.g. Forster 1927, n.p. *ad loc.*, Mayhew 2011, p. 516) accepted Bussemaker's conjecture ποιεῖ ἐγρηγορέναι (1869, p. v). (I now regret having done this in my Loeb edition.) Other scholars, however (e.g. Louis 1993, p. 89, Ferrini 2002, p. 268) follow Bekker, I assume on the grounds that προσεγρη-

Judging by this opening question alone, one might conclude that the author is concerned with the interaction between the nutritive part of the soul and the perceptual part, with the aspect of reading referred to here specifically being the *perception* of the words being read. But later references (see e.g. 917a23) to intellect (ἡ διάνοια) and thinking (νοέω) make it clear that that is not the case. This purported connection between the nutritive part of the soul and the rational part may in fact have been what in part motivated this problem (for as we have seen, the central interaction in Aristotle's account of sleep is between the nutritive and perceptual parts) – that is, unless the author came to make this connection only in attempting to solve the problem.²⁶

In fact, I suspect what primarily motivated this problem is that it seems to be the case (though it cannot be) that the same cause (i.e. the activity of reading) in the same kind of entity leads to two opposite effects:
1. reading → sleep (in people who do *not* want to sleep);
2. reading → insomnia (in people who *do* want to sleep).

It remains to be seen whether the reference to people wanting or not wanting to sleep merely acts as a control of sorts, focusing the attention on the activity of reading as a cause, or whether wanting or not wanting to sleep is an additional or contributing causal factor. In any case, the problem seems to be the following: Reading can produce sleep *even in those who don't want to sleep*; and, reading can prevent sleep (or fail to produce it) *even in those who want to sleep*. This apparent contradiction or anomaly is what must be explained – the problem to be solved.

3.2 The Follow-up Question: *Pr.* XVIII,7, 917a20–23

I take the next line to be a (somewhat) standard follow-up question, as described above:

γορέναι can also be (and in this case is) a third person plural perfect indicative active (see LSJ s.v. προσεγρήγορα). Now as Louis (1993), p. 249 n. 2 points out: *Le verbe προσεγείρειν [...] n'appartient pas à la langue classique*. This suggests that *Pr.* XVIII,7's ποιεῖ δύνασθαι is more likely (closer to) the original text.

26 Another possible motivation, suggested to me by Giouli Korobili, connecting *Pr.* XVIII,7 to the general topic of *Pr.* XVIII (φιλολογία, in the sense of a love of or interest in literature and language) is (my words, not hers) to show that physiological explanation has a role to play even in understanding certain aspects of literature.

Are these people in whom there are pneumatic movements owing to coldness from nature or from melancholic humors, through which [sc. humors or nature]²⁷ pneumatic residues are unconcocted owing to coldness?²⁸

In a sense, or broadly speaking, the tentative solution offered here is straightforward: It cannot be the case that the same basic activity in the same kind of entity leads to two opposite effects. So these must not in fact be the same kind of entity (i.e. the same kind of people): one is in a normal state, the other (introduced here) in an abnormally cold one (whether through nature or through an illness involving melancholic humors [μελαγχολικοὶ χυμοί], i.e. black bile). For the rest, this passage is unclear and raises the following questions: Who are *these people*? What are these cold pneumatic movements and unconcocted pneumatic residues? What is meant by "coldness from nature or from melancholic humors"?

---

27 The antecedent of οὕς is μελαγχολικῶν χυμῶν, though the author must be referring to whichever of the two (melancholic humors, or nature) is the cause of coldness in a given case (and so in effect to the coldness itself, as the second διὰ ψυχρότητα was likely meant to make clear).
28 ἢ ὅσοις μέν εἰσι πνευματικαὶ κινήσεις διὰ ψυχρότητα φύσεως ἢ μελαγχολικῶν χυμῶν, δι' οὕς περίττωμα γίνεται πνευματικὸν ἄπεπτον διὰ ψυχρότητα; What I am calling the follow-up question is identical in *Pr.* XVIII,1 (916b4–7) and XVIII,7. I follow Louis (1993, p. 90) here, in placing a question mark after the second occurrence of διὰ ψυχρότητα. But most editors and translators (Sylburg, Septalius, Bekker, Bussemaker, Ruelle, Flashar), place a comma after the second occurrence of διὰ ψυχρότητα, but do not insert another question mark at all, ending the sentence somewhere else with a period. Barthélémy-Saint Hilaire and Forster place a comma here as well, but insert a question mark further down in the chapter. There are a couple of noteworthy oddities in this text: (1) ἢ without ὅτι or διότι. (See above on the basic format of an Aristotelian problem.) I think that whether ὅτι/διότι was intentionally omitted, or dropped out in the transmission of the text, ultimately makes little difference: If added or restored, the translation should be changed to "Is it that/because these are people in whom" etc. (That having been said, the absence of ὅτι/διότι may explain why some scholars do not consider this line to be a question.) (2) μέν without δέ (the next section beginning with a μέν – δέ construction). Either a clause with δέ answering this μέν has dropped out (though there is no evidence of a specific alternative explanation in what follows), or this first μέν is the result of scribal error, or there is meant to be some contrasted idea left to implication (making this an instance of μέν *solitarium*): e.g. 'Are these people in whom there are cold pneumatic movements, *or is there some other explanation?*' (See Denniston 1950, p. 380 [s.v. μέν III. Preparatory, (5) Contrasted idea not expressed].) For I doubt that this μέν goes with the first one in the next passage (see Denniston 1950, p. 384 [IV. Duplication of μέν, (1) Resumption of clause]), though I suspect that is how some translators have understood the text. As these oddities have full manuscript support in both XVIII,1 and XVIII,7, I doubt they are the result of textual corruption. (One might add a third oddity: the repetition of διὰ ψυχρότητα. This could be the result of dittography, but I suspect it was more likely included for clarity. See the previous note.)

This first question is the easiest to answer. Though it is impossible to tell just from this passage which people ὅσοις refers to – those for whom reading causes sleep or for whom it causes insomnia – it eventually becomes clear that the author is referring to the latter. As for the second question: According to Aristotle (especially in *De generatione animalium*), nourishment (food and drink) is concocted (or cooked) in the stomach, liver, and heart (the heat from the heart is especially important), until (in the form of blood) it is sufficient to perform the nutritive functions of the soul: namely, growth and maintenance. After puberty, however, when growth stops, there is an excess of blood, and it is further concocted into seed – though the female's seed is not fully concocted.[29] Assuming the author of *Pr.* XVIII,7 had something like this in mind, he of course cannot be saying that there is no properly concocted food in these abnormally cold people, or else they could neither grow nor stay alive (nor contribute to reproduction). Nor would it make sense to claim that pneumatic movements are the cause of pneumatic residues being unconcocted. One would rather expect that, owing to the coldness of these people, their nourishment is not sufficiently concocted, and consequently there is less residual *pneuma* and so fewer or weaker (and in a sense colder) pneumatic movements (rising up in the body, as we learn in the next section).

Turning now to the third interpretive question, I should specify in what people "there are pneumatic movements owing to coldness (a) from nature or (b) from melancholic humors." Re. (a): There is an indication of the identity of these people in some of the chapters in *Pr.* XIV (esp. 8, 15, 16). People being cold by nature or hot by nature (i.e. colder or hotter than is normal for humans) is connected to their living in very hot or very cold regions: people are cold by nature in hot regions, and hot by nature in cold regions (in order to produce a livable balance, as nature does nothing in vain). These differences produce different characteristics in people.[30] Take for instance *Pr.* XIV,8, 909b9–13 (≈ XIV,16, 910a38–b3):

> Why are people living in hot places cowardly, while people living in cold ones are courageous? Is it because nature is the opposite in locations and in seasons (ἢ ὅτι ἐναντίως τοῖς τόποις καὶ ταῖς ὥραις ἡ φύσις ἔχει), since, if they were the same, people would of necessity be quickly destroyed? Now those who are hot with respect to their nature are courageous, whereas those who have been cooled are cowardly (ἀνδρεῖοι δέ εἰσιν οἱ τὴν φύσιν θερμοί, δειλοὶ δὲ οἱ κατεψυγμένοι).

---

29 See Peck (1942), p. lxiii–lxvii, Dean-Jones (1994), p. 60–61, 184–87, and Sophia Connell's chapter in this volume.
30 On *Pr.* XIV, see Leunissen (2015).

Further, *Pr.* XIV,15 claims that these colder people are cowardly but also wiser.[31] So one can speculate that the author of *Pr.* XVIII,7 may be referring to people from hot climates (hotter than Greece), who are therefore colder by nature. And as they are wiser, and so probably much more likely to read, how reading tended to affect their sleep may have been observed (or was thought to have been).[32] Now I am not suggesting that *Pr.* XVIII and *Pr.* XIV had the same author, but that the same views about climate and temperament were in the air, so to speak, in the Lyceum, when the above mentioned problems were composed. The same is true for *Pr.* XVIII,7 and *Pr.* XXX,1,[33] which I turn to next. Re. (b): According to *Pr.* XXX,1, 954a11–26, melancholy or black bile is in one sense by nature a mixture of hot and cold, in that it can, like water, become either very hot or very cold. But in another (primary) sense, it is cold by nature (ἡ χολὴ δὲ ἡ μέλαινα φύσει ψυχρὰ [...] οὖσα).[34] Now when there is an excess of black bile in the human body, it produces different diseases or symptoms depending on whether it is cold or hot. When it is cold, "it produces apoplexy or torpor or lack-of-spirit or fear" (ἀποπληξίας ἢ νάρκας ἢ ἀθυμίας ποιεῖ ἢ φόβους). Compare this to what Aristotle says about sleep and melancholic people (*Somn. Vig.* 3, 457a27–29): They are not prone to sleep, "for the inner region is cooled, so that the quantity of exhalation in them is not great."[35] Such cooling of the relevant region (though not the effects of this, and especially not torpor) may be what the author of *Pr.* XVIII,7 is referring to: As philosophers, poets, etc. tend to be melancholic (see the beginning of *Pr.* XXX,1, quoted above), the effect of reading on their sleep may have been observed (or was thought to have been).

---

31 Cf. Arist. *Politica* VII,7 and *de Juventute et Senectute, de Vita et Morte, de Respiratione* 20(14). (I owe the latter reference to Giouli Korobili.)
32 Determining the percentage of people who could read in antiquity (including the Classical and Hellenistic periods) is difficult. In any case, it must be low – particularly the ability to read at the level required for philosophic and scientific texts. (See the entry on literacy in the *OCD*[3].) This does not I think matter for the point I am making, though it may explain why the author of *Pr.* XVIII,7 regards reading as an unusual or even unnatural intellectual activity.
33 *Pr.* XXX,1 is the *locus classicus* for the Aristotelian conception of melancholy. See e.g. Flashar (1962), p. 711–722, van der Eijk (1990), Centrone (2011), and Schütrumpf (2015).
34 Cf. Arist. *Somn. Vig.* 3, 457a31: ἡ δὲ μέλαινα χολὴ φύσει ψυχρὰ οὖσα [...].
35 οὐδ' οἱ μελαγχολικοί (sc. ὑπνωτικοί)· κατέψυκται γὰρ ὁ εἴσω τόπος, ὥστ' οὐ γίγνεται πλῆθος αὐτοῖς ἀναθυμιάσεως.

3.3 The Solution, pt. 1: *Pr.* XVII,7, 917a23–28

The solution (the third and longest section of *Pr.* XVIII,7) itself divides naturally into three parts: (1) the abnormal case; (2) the normal case; (3) a final comment or consideration. Here is part 1:

> In these people, when the intellect is moved but does not think while focusing on something, the other [i.e. pneumatic] movement is checked,[36] which is why, as the intellect undergoes a great change, they are more inclined to sleep. For the pneumatic [movement] is defeated. Whereas when they fix the intellect on something, which is just what reading does, they are moved by pneumatic movements, which are not checked by anything, so that they are not able to sleep.[37]

The author begins by treating the abnormal case first (τούτοις picking up ὅσοις from the previous section): those with unconcocted pneumatic residues and fewer or weaker pneumatic movements. And he contrasts the effects of both reading and not reading on such a person. Unfortunately, much of what he says is unexplained and so unclear.

Behind the solution to this problem seems to be the author's conviction that there are two kinds of processes involved, capable of interacting: changes to the intellect (ἡ διάνοια), and pneumatic movements. At work here then is the notion that the rational part of the soul undergoes changes that have physical manifestations or corresponding physical actions, which can thereby interact with movements associated with the nutritive part of the soul. To understand the abnormal case, we need to know what the author means by cold(er) pneumatic movements

---

[36] Where *Pr.* XVIII,7, 917a24 has ἐκκρούεται ἡ ἐτέρα κίνησις ("the other movement is checked"), *Pr.* XVIII,1, 916b8 has τῇ ἐτέρᾳ κινήσει οὔσῃ καταψυκτικῇ (the intellect "is checked by the other movement, which cools"). Forster (1927, n.p. *ad loc.*), followed by Flashar (1962, p. 593), reads τῇ ἐτέρᾳ κινήσει in place of ἡ ἐτέρα κίνησις in *Pr.* XVIII,7. This may be smoother Greek; but the manuscript reading is fine, and in fact is supported by ἡττᾶται γὰρ ἡ πνευματική (the sentence that immediately follows) – if, as I believe is the case, the other movement (ἡ ἐτέρα κίνησις) *is* the pneumatic movement (ἡ πνευματική). This movement is checked and so defeated (ἐκκρούεται, ἡττᾶται).

[37] τούτοις ὅταν μὲν κινῆται ἡ διάνοια καὶ μὴ νοῇ ἐπιστήσασά τι, ἐκκρούεται ἡ ἐτέρα κίνησις, διὸ μᾶλλον μεταβάλλοντες πολὺ τὴν διάνοιαν καθεύδουσιν. ἡττᾶται γὰρ ἡ πνευματική. ὅταν δὲ ἐρείσωσι πρός τι τὴν διάνοιαν, ὅπερ ἡ ἀνάγνωσις ποιεῖ, κινοῦνται ὑπὸ τῆς πνευματικῆς κινήσεως, οὐκ ἐκκρουομένης ὑπ' οὐδενός, ὥστε οὐ δύνανται καθεύδειν. *Pr.* XVIII,1, 916b7–12 has virtually the same beginning and ending as *Pr.* XVIII,7, 917a23–28. In between, however, there are a number of variations, but two are especially noteworthy: (1) ἡττᾶται γὰρ ἡ πνευματική (XVIII,7, 917a25–26) is missing in *Pr.* XVIII,1, and (2) where *Pr.* XVIII,7 has ὑπὸ τῆς *πνευματικῆς* κινήσεως (917a27), XVIII,1 has ὑπὸ τῆς *θερμαντικῆς* κινήσεως (916b10–11). I believe the text of *Pr.* XVIII,7 is superior in both cases. See the preceding note for a third variation.

and unconcocted pneumatic residue. All one can do is speculate, which I proceed to do.

Whatever Aristotle thought about the presence of *pneuma* in the exhalation resulting from concoction, and the role of *pneuma* (if any) in sleep, the author of *Pr.* XVIII,7 clearly thought *pneuma* had a central role to play, and almost certainly held that it was one result of concoction.[38] In fact, I suspect he has replaced Aristotle's exhalation with *pneuma*. He holds that in normal people there is a quantity of hot *pneuma* that results from the concoction of nutrition, but that this is much less abundant in the abnormally cold people.[39] As a result, there is not in these people a great deal of *pneuma* rising in a mass to, and interacting with, the physical part of the body that houses the intellect (on this view) – somewhere in the head. Not only is that location made clear in part 2 of the solution, there is no mention of the heart in *Pr.* XVIII,7, nor of any reflux or recoil of pneumatic movement with a resulting downward motion.

I turn now to the non-reading case in abnormally cold people. It becomes clear that the intellect thinking without focusing on something refers to a person being intellectually active or awake, but not reading. The implication is that in this case, the intellect is active all over, and so causing movement all over the physical part of the body that houses it, and not merely in one small spot. So, when the other movement – the weaker or less abundant movement of *pneuma* – reaches the seat of the intellect in the head, this pneumatic movement is checked and defeated, and so it does not affect and thereby excite the intellect. (Perhaps if unchecked it excites the intellect precisely because it is cold – i.e. not as hot – whereby a quantity of hot *pneuma* has a soporific effect.) Now the author refers to the intellect (ἡ διάνοια) undergoing a great change, which inclines the person to sleep. In part 2 of the solution (more on this shortly), he says that when thought is fatigued it undergoes a change (ὅταν κοπιάσῃ, ὁ νοῦς μεταβάλλει);[40] and I take him to be referring to fatigue here as well. It is especially likely that fatigue is the reference if we recall that the author is here referring

---

38 On *pneuma* in the *Pr.* (with a focus on *Pr.* IV), see Meeusen (2020). He concludes: "the *Problems* seem to employ the concept [of *pneuma*] in a more liberal way than Aristotle's texts allow, and one can only assume that this procedure was deliberate, in view of re-opening these texts for debate, for testing alternative [...] approaches, for venting criticism etc."

39 Cf. Aristotle on melancholic people and sleep: they are not prone to sleep because they are cooled inside, "so that the quantity of exhalation in them is not great" (*Somn. Vig.* 3, 457a27–29). The author of *Pr.* XVIII,7 is likely using different language (*pneuma* rather than an exhalation) to refer to the same thing.

40 It is unclear what distinction the author of *Pr.* XVIII,7 sees between ἡ διάνοια (917a23, 25, 26, 29) and ὁ νοῦς (917a23, 35, 38, 39). Michiel Meeusen suggested to me the possibility that διάνοια is the activity of νοῦς.

specifically to people who *want* to sleep. Now the nature of this great change to the intellect is not specified, and once again one must speculate. I take it to be thought or intellect ceasing to operate (as happens when one sleeps) – in effect, thought ceases to think. If I am right, then this is the 'psychic' change, so to speak, corresponding to physical changes in the part of the soul that houses the intellect.[41]

The reading case in abnormally cold people is now fairly straightforward: When such a person reads, his intellect is focused and so fixed on one thing, and consequently this causes physical changes or movements in only one small spot in the seat of the intellect. As a result, there are no widespread physical movements in this region to counteract and check any cold pneumatic movements. Consequently, these pneumatic movements – though not massive – affect and so stimulate this region, and that prevents the person from sleeping.[42]

3.4 The Solution, pt. 2: *Pr.* XVIII,7, 917a28–36

The author turns now to the normal case, and unlike his discussion of the abnormal one, he does not contrast the effects of both reading and not reading, but discusses only how reading produces sleep in such a person (even when the person does not want to sleep).

> Now for those who are in a natural state, when the intellect stands at one point and does not undergo change in many places, all the others [i.e. parts][43] around the region stand still as well, and their calm is sleep.[44] For when a single leader stands still, as in a rout, the

---

41 I think this gets some confirmation from *Pr.* XVIII,7, 917a37–b3, discussed below in § 3.5.
42 It should be clear that cold pneumatic movements are cold relatively. *Pneuma* is hot by nature, and even in these cases, it rises in the body because of its heat. But it is colder than normal owing to melancholy or nature (as described). Consider the following analogy: One might find hot showers before bed soporific, whereas a somewhat cooler shower (though still warm, a mix of hot and cold water) might have the opposite effect.
43 A translator needs to fill out τὰ ἄλλα ὅσα here (see below note 47). Forster (1927, n.p. *ad loc.*) translates this "every function", Flashar (1962, p. 155) *alle anderen (Tätigkeiten)*, Louis (1993, p. 90) *autres facultés*. The author is likely referring to parts (looking ahead to τὰ ἄλλα μόρια), though he may simply have in mind *everything else* (in that region): that is, activities in the rational part of the soul, and any corresponding physical movements in the seat of that part in the body.
44 As we have seen, Aristotle says that sleep is or involves the immobility (ἀκινησία) of the perceptual part (*Somn. Vig.* 1, 454b11), and that metaphorically it is rest (ἀνάπαυσις) in animals that perceive and move (2, 455b18). These are different ways of saying the same thing, as is (*mutatis*

other parts naturally come to a standstill as well. For by nature what is light travels upward, and what is heavy downward.[45] Therefore, when the soul moves according to nature, it does not sleep; for this is its condition.[46] And when it is stationary and as it were fatigued, thought undergoes change, and the bodily elements rising to the head produces sleep.[47]

The author is here referring to normal (i.e. not cold) people, and in this case reading produces sleep. The reason is that such people do not generate unconcocted pneumatic residues. So when they read, the intellect focusing on one point does not allow cold pneumatic movements to go unchecked and thus excite the part of the body that houses the intellect (for there are no such cold movements). Rather, it leads to a general immobility in the rational part of the soul and the region in the body that houses it – an immobility that does not occur when the intellect is functioning in its normal way (i.e., in contrast to when one is reading). And this immobility or calm just is sleep (with a corresponding cessation of thought, as described above). Where Aristotle claimed that sleep is in a sense a certain immobility of the *perceptual* part of the soul,[48] the author of *Pr.* XVIII,7 claims it is (in the normal case) a certain immobility of the *rational* part (or of the intellect at any rate). For this immobility is not stirred, so to speak, by cold pneumatic movements. I assume that according to the author of *Pr.* XVIII,7, in normal (i.e. non-reading) cases this immobility

---

mutandis) the claim of the author of *Pr.* XVIII,7 here that sleep is calm (ἠρέμησις); for ἀνάπαυσις and ἠρέμησις are, in the present context, pretty much synonymous.

45 I agree with Forster (1927 n.p., *ad loc.*, n. 2) that this sentence "seems out of place here". (It is however orthodox Aristotelianism: see e.g. *Physica* IV,4, 212a24–26, *de Caelo* IV,1, 308a29–31.) I suspect it was a marginal gloss that made its way into the text at the wrong location, and that it actually belongs with the final sentence. (See note 47.)

46 *Pr.* XVIII,1 & 7 diverge fairly considerably at this point. First, where XVIII,7 has "Therefore, when the soul moves according to nature, it does not sleep; for this is its [sc. natural?] condition" (ὅταν οὖν ἡ ψυχὴ κινῆται κατὰ φύσιν, οὐ καθεύδει· οὕτω γὰρ ἔχει [Flashar 1962, p. 155, 593 obelizes *so hat sie*, his translation of οὕτω γὰρ ἔχει]), XVIII,1 has "But when the soul moves according to nature, it does not sleep; for then especially it is alive" (κινουμένης δὲ τῆς ψυχῆς κατὰ φύσιν οὐ καθεύδει· ζῇ γὰρ τότε μάλιστα). Second, whereas XVIII,7 continues with over seven more lines, XVIII,1 concludes at this point with the brief remark: "But being awake is the cause of life rather than being asleep" (τὸ δ' ἐγρηγορέναι τοῦ ζῆν αἴτιόν ἐστιν ἢ τὸ καθεύδειν).

47 τῶν δὲ κατὰ φύσιν ἐχόντων ὅταν στῇ πρὸς ἓν ἡ διάνοια καὶ μὴ μεταβάλλῃ πολλαχῇ, ἵσταται καὶ τὰ ἄλλα ὅσα περὶ τὸν τόπον, ὧν ἠρέμησις ὁ ὕπνος ἐστίν. ἑνὸς γὰρ κυρίου στάντος, ὥσπερ ἐν τροπῇ, καὶ τὰ ἄλλα μόρια ἵστασθαι πέφυκεν. φύσει γὰρ ἄνω τὸ κοῦφον φέρεται, τὸ δὲ βαρὺ κάτω. ὅταν οὖν ἡ ψυχὴ κινῆται κατὰ φύσιν, οὐ καθεύδει· οὕτω γὰρ ἔχει. ὅταν δὲ στῇ καὶ οἷον κοπιάσῃ, ὁ μὲν νοῦς μεταβάλλει, καὶ ἄνω τὰ σωματώδη πρὸς τὴν κεφαλὴν ἰόντα ποιεῖ τὸν ὕπνον.

48 *Somn. Vig.* 1, 454b9–11, quoted above.

and thus sleep comes at the end of the day, from fatigue. But it can come from reading, too, even when one does not want to go to sleep.

I find the last clause of this passage (καὶ ἄνω τὰ σωματώδη πρὸς τὴν κεφαλὴν ἰόντα ποιεῖ τὸν ὕπνον) somewhat puzzling, on two counts: First, the reference to the bodily elements (τὰ σωματώδη) rising to the head is surprising, as this sounds more like the exhalations in Aristotle's account of sleep, and less like a description of *pneuma*; though to make sense, it must refer to this latter. Second, if the region of the body that houses thought comes to a standstill, and this calm state just is sleep, why does the author then say that the *pneuma* "rising to the head produces sleep"? The *pneuma* being its normal (hot) temperature certainly contributes to sleep in a negative way: it does not excite this part of the soul (as colder *pneuma* does in the abnormal case). But this is not the same as *producing* sleep. So perhaps the author holds (and is here making explicit for the first time, however ambiguously) that in normal cases, sleep comes at the end of the day, from fatigue, in combination with the (normally hot) *pneuma* rising to the head.

The author was here directly influenced by two earlier (but perhaps not all that earlier) Peripatetics (more on the dating of *Pr.* XVIII,7 shortly). First, the central claim of the opening line of our passage bears a striking resemblance to a line in *De vertigine* 9, where Theophrastus is discussing the causes of a certain type of dizziness (ἴλιγγος). One cause is staring fixedly at a single object (*Vert.* 9, 69–70 Sharples). Dizziness occurs, because "when the sight in a single part[49] stands still, the other parts in the brain stand still as well" (τῆς ὄψεως δὲ στάσης ἑνὸς μορίου, καὶ τἆλλα τὰ συνεχῆ ἐν τῷ ἐγκεφάλῳ ἵσταται) (9,75–76).[50] The author of *Pr.* XVIII,7 transfers and applies this same phenomenon to a faculty of the rational part of the soul, to explain not dizziness but sleep: "when the intellect stands at one point and does not undergo change in many places, all the others around the region as well stand still." The region he is referring to is likewise the brain or some part of it.

Second, the author combines this Theophrastean phenomenon with an analogy used by Aristotle in *Analytica Posteriora* II,19 (in a very different context, dis-

---

[49] I assume this one part is a combination of the eyes and the region of the brain to which they are connected, treated together (or perhaps just the latter) – the physical part that is the seat of sight or vision. The eyes are (Theophrastus knew) continuous with the brain.

[50] It is somewhat surprising that the explanation involves the brain, though this should not be taken to mean that Theophrastus disagreed with Aristotle about the heart being the seat of perception. (See note 58 below.) Cf. Sharples (2003), p. 175: "The fact that the importance of the brain was not generally realized by Aristotle or by Theophrastus makes the role given to the head in the explanation of dizziness all the more striking."

cussing universals and the acquisition of first principles). Aristotle writes (100a12–13): "as in battle when a rout has occurred, one stops, then another [...]."[51] Compare *Pr.* XVIII,7, 917a31–32: "For when a single leader stands still, as in a rout, the other parts [i.e. the soldiers under his command] naturally come to a standstill as well." The immobility of the leader (presumably the intellect, the leader of the rational part of the soul) causes the rest of that part of the soul – as well as any movements in the part of the body that houses it – to stop moving or acting.

What the author of *Pr.* XVIII,7 describes must happen in the abnormal case of reading as well. The difference is that in that case the immobilized army (so to speak) allows cold pneumatic movements to pass by unchecked so as to excite the intellect, whereas in the normal case, the pneumatic movements that pass by are more massive and hot, and that apparently has a soporific effect – even when the reader does not want to sleep.

3.5 The Solution, pt. 3: *Pr.* XVIII,7, 917a37–b3[52]

Pr. XVIII,7 ends by considering a puzzle that likely arose in the Lyceum in discussions of the relationship between reading and sleep:

> Now reading might seem to prevent sleep. But wakefulness is not due to the thinking – for then[53] the soul is more concentrated – but to the changing, since in fact such thoughts[54] are wakeful in which the soul investigates and questions, and not those in which it continuously contemplates; for the former cause a lack of concentration, whereas the latter do not.[55]

---

51 οἷον ἐν μάχῃ τροπῆς γενομένης ἑνὸς στάντος ἕτερος ἔστη, εἶθ' ἕτερος [...]. For a recent discussion of this passage, and how properly to understand it in context, see McKirahan (2018). The author of *Pr.* XXVI,8 makes use of this same analogy in a meteorological context (941a11–13): "for just as in a rout, when one man resists the others also remain, so also in the case of air" (ὥσπερ γὰρ ἐν τροπῇ ἑνὸς ἀντιστάντος καὶ οἱ ἄλλοι μένουσιν, οὕτω καὶ ἐπὶ τοῦ ἀέρος).
52 As I indicated earlier, this passage has no parallel in *Pr.* XVIII,1.
53 I take τότε here to refer specifically to when a person is reading.
54 Louis (1993), p. 91, points out: "L'emploi du mot νοήσις au pluriel est exceptionnel chez Aristote. Voir toutefois *De l'Ame* I 3, 407a24." He translates νοήσεις straight here, "les pensées", as do I, though Forster's (1927, n.p. *ad loc.*) "intellectual activities" may be closer to the mark.
55 δόξειε δ' ἂν ἡ ἀνάγνωσις κωλύειν καθεύδειν. ἔστι δὲ οὐ διὰ τὸ νοεῖν, ὥρισται γὰρ ἡ ψυχὴ τότε μᾶλλον, ἀλλὰ διὰ τὸ μεταβάλλειν ἡ ἀγρυπνία, ἐπεὶ καὶ νοήσεις αἱ τοιαῦται ἄγρυπνοί εἰσιν, ἐν αἷς ζητεῖ ἡ ψυχὴ καὶ ἀπορεῖ, ἀλλ' οὐκ ἐν αἷς ἀεὶ θεωρεῖ· ἐκεῖναι μὲν γὰρ ἀοριστεῖν ποιοῦσιν, αὗται δὲ οὔ.

Sleep is an absence of thought or intellectual activity, so (one might ask) should not reading, which is certainly an intellectual activity, keep the reader awake? The puzzle is that in fact reading does the opposite (in all but the abnormally cold people). The author's solution – which flows out of what he presented earlier – is that it is not thinking *per se* that causes wakefulness or prevents sleep, but a certain kind of thinking: the more normal kind, in which the objects of thought change, as the person investigates and raises questions etc. The sort of concentrated thinking involved in reading, however, it is claimed, does not prevent sleep, but rather induces it, for the reasons presented earlier.

4 Postscript

Aristotle was head of the Lyceum, the school he founded, till 323/2 BC, when he left Athens (dying not long thereafter). Theophrastus took over and served till his death in 286; and he was followed by Strato, who served as scholarch till 268. I would speculate that the relevant discussion and debate in the Lyceum in the generation after Aristotle was the intellectual soil from which emerged *Pr.* XVIII,7. Although it is possible that it was written long after Strato, I think that unlikely – or at least I see no evidence (linguistic or otherwise) that points to or favors that conclusion specifically.[56]

For Aristotle, as we have seen, it is the heart that plays the central role in sense perception and thus in sleep. He also held that although all of the sensory faculties have corresponding bodily organs, there is no organ in the body that is the seat of thought (ὁ νοῦς) – and perhaps of related faculties, like intellect (ἡ διάνοια).[57] And though *pneuma* may be one constituent or byproduct of the exhalation central to Aristotle's account of sleep, it is not explicitly said to be, and in any case does not play a central role. Theophrastus appears to have agreed with Aristotle about the role of the heart and about there being no bodily

---

[56] Louis (1993), p. 86–87, argues (largely on linguistic grounds) that *Pr.* XVIII is very late – in fact he claims that these problems may be from the first or second century AD, introduced into the last edition of the *Problemata* collection! He refers to the word προσεγείρειν (see above note 25), as well as to *constructions insolites* in *Pr.* XVIII,1 (916b9, 16, 18) and 7 (917a30). I find his arguments for such a late date completely unconvincing, as all that these passages show is that the authors of *Pr.* XVIII,1 & 7 have a different conception of sleep than that of Aristotle. In Mayhew (forthcoming) I argue that *Pr.* XVIII,4–6 are in fact relatively early – likely from the Lyceum of Aristotle and Theophrastus.

[57] See *De Anima* II,1, 413a3–7; II,2, 413b24–29; III,5, 430a17–19; cf. *GA* II,3, 736b21–29. See, however, van der Eijk (1997).

organ serving as the seat of thought (though the evidence is fragmentary).[58] Theophrastus' view of the role played by *pneuma* in sleep, if any, cannot be determined; but note that *pneuma* plays no role in his *De lassitudine*.

Unlike Aristotle and Theophrastus, but like Strato, the author of *Pr.* XVIII,7 held that the brain or some part thereof is the central organ of human cognition. And unlike Aristotle and Theophrastus, but probably like Strato, he held that there *is* an organ in the body that is the seat of thought (ὁ νοῦς) and intellect (ἡ διάνοια), namely the brain or some part thereof.[59] Further, unlike Aristotle (and perhaps Theophrastus, too), but like Strato, he believed *pneuma* played a central role in explaining sleep. Finally, unlike Aristotle, Theophrastus, *and* Strato, he held that sleep is to be defined or explained not in connection with sense perception and the perceptual part of the soul, but in connection with intellect or thought (ἡ διάνοια, ὁ νοῦς) and the rational part – interacting with the body and the nutritive part of the soul. This is the most original feature of *Pr.* XVIII,7.

I offer one last possible bit of evidence for dating *Pr.* XVIII,7 roughly to around the time of Strato: fr. 62 Sharples (= Plut. *de Sollertia Animalium* 3,960E–961 A). Strato is here said to have discussed the connection between thought and perception in the act of reading (among other things):

> And indeed there is an argument (λόγος) of Strato the naturalist showing that not even perceiving is present at all in the absence of thinking (οὐδ' αἰσθάνεσθαι τὸ παράπαν ἄνευ τοῦ νοεῖν ὑπάρχει). For in fact we frequently fail to notice letters (γράμματα) when we go over them with our sight, and words (λόγοι) when they strike our ears, because we have our thought [directed] at something else (πρὸς ἑτέροις τὸν νοῦν ἔχοντας) [...]. (961 A)[60]

This is the only other early Peripatetic text (besides *Pr.* XVIII,1 & 7) I am aware of that discusses (or mentions) reading in the context of the connection between the rational part of the soul, and the other parts – though Strato is concerned, not with the nutritive part of the soul, but with the perceptual-appetitive part.

As is true for many of the chapters in the extant *Pr.* attributed to Aristotle, *Pr.* XVIII,7 is not Peripatetic philosophy and science at its best. But it is interesting,

---

58 On the centrality of the heart, see fr. 330 FHS&G (= Galen, *De placitis Hippocratis et Platonis* VI,1,1–2, De Lacy p. 360,4–12). It is surprising, however, that the heart is not mentioned in his *de Sensu et sensibilibus*. On ὁ νοῦς, see especially fr. 307 A FHS&G (= Themistius *in libros Aristotelis de Anima*, CAG 5,3, p. 107,30–108,18 Heinze), which ends: "And going forward, he [sc. Theophrastus] claims that the senses are not without body, whereas thought is separate" (καὶ προϊὼν φησι τὰς μὲν αἰσθήσεις οὐκ ἄνευ σώματος, τὸν δὲ νοῦν χωριστόν).
59 On Strato, see perhaps fr. 61 Sharples (= Sextus Empiricus, *adversus Mathematicos* 7,348–350).
60 This is the translation of Sharples (2011), p. 147, but modified significantly.

as an example of Peripatetic 'psychology' in action – solving a problem – likely in the generation after Aristotle, and not constrained or limited by concerns of orthodoxy, and with a fairly unique interest in the interaction between the rational part of the soul, and the body and its nutritive functions.[61]

Bibliography

Barthélémy-Saint Hilaire, Jules (1891): *Les Problèmes d'Aristote*. Vol. 2. Paris.
Bekker, Immanuel (ed.) (1831): *Aristotelis Opera*, vol. 2, Berlin.
Bodnár, István (2015): "The *Problemata physica:* An Introduction". In:Mayhew, Robert (ed.): *The Aristotelian* Problemata Physica: *Philosophical and Scientific Investigations*. Leiden, 1–9.
Bubb, Claire (forthcoming): "Blood Flow in Aristotle". In: *Classical Quarterly.*
Bussemaker (ed.) (1869): Ulco Cats Bussemaker, *Problemata*, in: Friedrich Dübner et al. (eds.), *Aristotelis Opera Omnia Graeca et Latine*, vol. 4. Paris.
Centrone, Bruno (2011): "ΜΕΛΑΓΧΟΛΙΚΟΣ in Aristotele e il *Problema* XXX 1". In: Idem (Ed.): *Studi sui* Problemata Physica Aristotelici. Naples, 309–39.
Dean-Jones, Lesley Ann (1994): *Women's Bodies in Classical Greek Science*. Oxford.
Denniston, John Dewar (1950): *The Greek Particles*. 2nd ed., revised by Kenneth J. Dover. Oxford.
Ferrini, Maria Fernanda (2002): *Aristotele,* Problemi. Milan.
Flashar, Helmut (1962): *Aristoteles: Problemata Physica*. Berlin.
Forster, Edward Seymour (1927): *The Works of Aristotle Translated into English Under the Editorship of W.D. Ross.* Vol. 7: *Problemata*. Oxford.
Leunissen, Mariska (2015): "The Ethnography of *Problemata* 14 in (Its Mostly Aristotelian) Context". In: Mayhew, Robert (Ed.): *The Aristotelian* Problemata Physica: *Philosophical and Scientific Investigations*. Leiden, 190–213.
Louis, Pierre (1993): *Aristote: Problèmes*. Vol. 2: *Sections XI–XXVII*. Paris.
Mayhew, Robert (ed./transl.) (2011): *Aristotle: Problems*. Vol. 1: *Books 1–19*. Cambridge, MA.
Mayhew, Robert (forthcoming): "Pseudo-Aristotle, *Problemata* 18.4–6 on the status of rhetoric". In: Johnson, Monte/Destrée, Pierre (Eds.): *Protreptic Rhetoric in the Aristotle Corpus.*
McKirahan, Richard (2018): "'As in a Battle When a Rout has Occurred'". In: Sfendoni-Mentzou, Demetra (Ed.): *Aristotle–Contemporary Perspectives on His Thought*. Berlin, 297–322.
Meeusen, Michiel (2020): "Aristotle's Second Breath: Pneumatic Processes in the *Natural Problems* (on Sexual Intercourse)". In: Coughlin, Sean, et al. (Eds.): *The Concept of Pneuma After Aristotle*. Berlin, 63–90.

---

[61] I would like to thank Michiel Meeusen and Claire Bubb for their helpful comments on an earlier draft of this essay, and for sharing relevant papers of theirs prior to publication (see the bibliography for details). Thanks as well to Giouli Korobili, for her helpful comments on the penultimate draft of this essay.

Nuchelmans, G.R.F.M. (1950): *Studien über φιλόλογος, φιλολογία und φιλολογεῖν*. Dissertation, University of Nijmegen.
Nussbaum, Martha Craven (1975): *Aristotle's* De Motu Animalium. Princeton.
Peck, Arthur Leslie (1942): *Aristotle:* Generation of Animals. Cambridge, MA.
Richards, Herbert (1915): *Aristotelica*. London.
Ruelle, Carolus, et al. (ed.) (1922): *Aristotelis Quae Feruntur Problemata Physica*. Leipzig.
Schütrumpf, Eckart (2015): "Black Bile as the Cause of Human Accomplishments and Behaviors in *Pr.* 30.1: Is the Concept Aristotelian?". In: Mayhew, Robert (Ed.): *The Aristotelian* Problemata Physica: *Philosophical and Scientific Investigations*. Leiden, 357–380.
Sharples, Robert W. (2003): "Theophrastus, *On Dizziness*". In: Fortenbaugh, W. W., et al. (Eds.): *Theophrastus of Eresus*, On Sweat, On Dizziness *and* On Fatigue. Leiden, 169–249.
Sharples, Robert W. (2011): "Strato of Lampsacus: The Sources, Texts and Translations". In: Desclos, M.-L./Fortenbaugh, William W. (Eds.): *Strato of Lampsacus: Text, Translation and Discussion*. New Brunswick, NJ, 5–230.
Septalius, Ludovicus (ed.) (1632): *In Aristotelis problemata commentaria*. Lyon.
Sylburg, Friedrich (ed.) (1585): *Aristotelis, Alexandri, et Cassii Problemata etc*. Frankfurt.
van der Eijk, Philip J. (1990): "Aristoteles über die Melancholie". In: *Mnemosyne* 43, 33–72. English version in van der Eijk (2005), chapter 5.
van der Eijk, Philip J. (1997): "The matter of mind: Aristotle on the biology of 'psychic' processes and the bodily aspects of thinking". In: Kullmann, Wolfgang/Föllinger, Sabine (Eds.): *Aristotelische Biologie. Intentionen, Methoden, Ergebnisse*. Stuttgart, 221–258. Reprinted in van der Eijk (2005), chapter 7.
van der Eijk, Philip J. (2005): *Medicine and Philosophy in Classical Antiquity: Doctors and Philosophers on Nature, Soul, Health and Disease*. Cambridge.

Gweltaz Guyomarc'h
Dividing an Apple
The Nutritive Soul and Soul Parts in Alexander of Aphrodisias

Translated by Jeanne Allard

Abstract: The nutritive soul provides a relevant test case to examine Alexander of Aphrodisias' conception of the parts of the soul, since it appears in Alexander's *De anima* along with methodological considerations, especially an analogy with the division of an apple. I examine here the unity of the powers of the soul, focusing especially on the case of the vegetative soul. If the division of soul parts and soul powers is neither local, nor numerical, what is it? I put forward three correlated hypotheses: 1) Even if there is no lexical distinction in Alexander between "powers of the soul" and "parts of the soul", Alexander nonetheless comes up with criteria which distinguish a soul power from a soul part, or from a soul of its own. The difference between his position and Aristotle's is found chiefly in Alexander's effort to clarify these criteria. 2) As will become clear in the case of the vegetative soul, even the powers that do constitute a soul or a soul part (vegetative/animal/human) are objectively distinct (in a sense that remains to be clarified) and are not simply the result of a change in perspective. 3) The main criterion by which one can account for the organization and the unification of soul parts is the teleological criterion.

1 The Soul and the Apple

Alexander's *de Anima* must be distinguished from his lemmatic commentary on the Aristotelian treatise, which is only known to us through testimonies.[1] The *de*

Note: First, I am grateful to G. Korobili and R. Lo Presti for their invitation to take part in the conference which led to the publication of this volume, and to all the conference participants from whom I learned greatly. Thanks are also due to V. Caston for communicating to me his unpublished translation of the second part of Alexander's *De anima*, as well as to A. Hangai for sharing with me his excellent dissertation, especially the passages concerning Alexander on the soul. Finally, I am very grateful to Jeanne Allard for her translation and our discussions about it.

1 For a reconstruction of this commentary, see Kupreeva (2012).

An. is described as "personal",[2] has a polemical aim and targets a wider, less-specialized audience than commentaries typically do. Nonetheless, in many cases, Alexander's *de An.* conforms to the agenda and arrangement of its Aristotelian source. The examination of the nutritive soul is one such case: the discussion offers Alexander the opportunity to introduce methodological distinctions, as Aristotle himself did in *de An.* II,4. But, whereas Aristotle, at the beginning of *de An.* II,4, stresses that in order to determine the essence of a soul power, it is required to look into the activity of this power and, ultimately, into its object,[3] Alexander uses an analogy that does not appear in Aristotle's text:

> **T.1.** For we do not divide the soul as though it were composed from the parts into which we divide it as separate things. Rather, we divide the soul by enumerating the powers it has and by ascertaining the differences between them, just as if one were to divide an apple into its fragrance (εἴς τε εὐωδίαν), lustre, shape, and flavour. For dividing an apple in this way is not dividing it as a body (ὡς σώματος), even though the apple is certainly a body, nor as a number. (Alexander, *de An.* 31,1–6, transl. Caston 2012, slightly modified)

The origin of the analogy between soul and apple is an open question and has been discussed. In the literature, the two likely candidates are either the Stoics or other Peripatetics, like Nicolaus of Damascus.[4] The Stoic origin of the analogy can be traced back to Iamblichus. In a passage of his *de Anima*, Iamblichus likens the functions of the leading part of the soul (impression, assent, impulsion and reason) to the sweetness and the aroma of the apple (τὴν εὐωδίαν), which both inhere in the same body and differ only in quality.[5] In the Stoic view, the functions of the leading part of the soul are corporeal insofar as they are qualities, but they are not separated in place. If Alexander had something like the Stoic view in mind, i.e. if the passage T.1 had a polemical aim,[6] one must admit that his refutation would remain entirely implicit and, even, that it would miss its intended target. For, precisely because of what is commonly called their "monopsychism", the Stoics (or most of them) *do not* divide the leading part of the soul as they would apple wedges.

In fact, it was common in Alexander's time to use the apple and its qualities as an *analogon*. One occurrence is found in a text by the Pseudo-Galen on the incorporeality of qualities, and many others in Sextus Empiricus, where they

---

[2] For problems arising from the designation "personal" and from the distinction between personal treatises and commentaries, see Rashed (2007), p. 3.
[3] Aristotle, *de Anima* II,4, 415a14–23.
[4] Caston (2012), p. 126, following Inwood (1985), p. 30–32.
[5] Iamblichus, *de Anima*, apud Stobaeus I,17(18 H.), 368,12–20 (SVF II,826; LS 53K).
[6] As Caston (2012), p. 126 claims.

refer to cases where one object appears in various ways to the senses.[7] Alexander also uses it in this latter sense.[8] The apple and its qualities are a model, either of accidents inhering in a subject,[9] or of mereological relations. As a result, the meaning of the analogy in Alexander becomes clearer. There is an implicit premiss: that the most usual way to divide an apple is to cut it into wedges. But soul parts are not apple wedges. They are rather analogous to its "fragrance, lustre, shape, and flavour". Since these sensible qualities are coextensive with the entire apple, they cannot be divided spatially. This is also how the analogy was read by Themistius and Pseudo-Philoponus (both of whom may have come into contact with this specific use of the analogy in Alexander's *de An.* or in his lemmatic commentary).[10] Pseudo-Philoponus, in particular, explicates the idea underlying the analogy: "this is how a power differs from a part, that a power, as we said, runs through the whole substance (δι' ὅλης τῆς οὐσίας κεχώρηκε), whereas a part is not in the whole substance but in something of it."[11]

Not only are soul parts not divided locally – they are not divided numerically either, Alexander adds. But this comes as a surprise: in the passage quoted above, Alexander maintained that, in order to divide the soul, one must first "enumerate" its powers (καταριθμήσει, 31,2–3). In this case, however, as V. Caston notes, Alexander is most likely arguing against those who conceive the soul as an aggregate of discrete parts (Caston 2012, p. 125). Furthermore, Alexander sometimes combines the spatial and numerical divisions or even conflates one with the other, with the thought that any particular thing occupies a determined place.[12]

---

[7] Ps.-Galen, *Quod qualitates incorporea sint* 3, K. XIX,469,15–472,2 (see Alexander, *Mantissa* §6, 123,23–34); Sextus Empiricus, *P.* I,94–97, 99; *M.* VII,103.
[8] *In librum de Sensu et Sensilibus,* 165,26–166,2. On this, see Diogenes Laertius, IX,81. Concerning the apple argument in the Sceptics, cf. Annas and Barnes (1985), p. 71–74.
[9] The apple analogy will be used in this way in Neoplatonic commentaries on the *Categoriae.*
[10] Themistius, *in de An.* 37,21–23 (concerning *de An.* I,5, 411b6–19) and 117,3–4 (about III,9, 432a22–30); [Philoponus], *in de Anima* 571,11–13 (also about *de An.* III,9).
[11] [Philoponus], *in de An.* 571,14–16, transl. Charlton 2005.
[12] In the passage immediately prior to T.1, Alexander uses number as a counter-example to the thesis that all which possesses parts is a magnitude (30,26–29). Our T.1 then aims to say that the soul, even though it possesses parts, is not a magnitude or a body, but also that it cannot be like a number. See also *Quaestio* 3,9, which is all the more important since it comments on *de An.* III,2, 427a9–14. For the division τόπῳ δὲ καὶ ἀριθμῷ, see 94,15 or 95,31. 95,31 is interesting: the phrase τουτέστιν τῷ ὑποκειμένῳ is explanatory of local and numerical indivisibility. Likewise, in his own *de An.*, Alexander rarely talks of a τόπῳ division of the parts of the soul and instead uses phrases like κατὰ τὸ ὑποκείμενον (e.g. 75,8; 99,9). For a contrasting passage, see 94,9–10.

In what follows, I would like to examine the unity of the powers of the soul, focusing especially on the case of the vegetative soul. If the division of soul parts and soul powers is neither local, nor numerical, what is it? In Aristotle's account, the other option is a "logical" division (λόγῳ[13]). But what does that mean? Aristotle himself raises this important question when he lists the aporias of a science of the soul in *de An.* I,1, then again in II,2 when he asks whether the nutritive, sensitive and cognitive powers are a soul of their own or a part of a soul.[14]

For Alexander, as we have mentioned, the apple *analogon* appears in the discussion of the vegetative soul. The vegetative soul is in fact a unique case study and may help us answer our question: 1) it can be a part of the soul in animals; 2) the vegetative part of the soul contains many powers; 3) in plants, it can be a soul of its own. It is the only soul power, Aristotle says, which can be "separated". In Alexander's terminology, it is the only soul power which can "subsist apart" from the other powers (χωρὶς τούτων ὑφίσταται).[15]

Yet, when he begins to discuss the composition of the vegetative soul, Alexander claims:

> **T.2.** The first [power] of the soul, then, in animate things which are subject to coming-to-be and perishing is the [power] for nourishing [oneself], to which the [powers] for growing and for reproducing are both linked (συνέζευκται). (*de An.* 29,1–3, transl. Caston 2012)

The key issue in this passage is the meaning of the verb συζεύγνυμι (yoke together, couple). According to Accattino and Donini, the verb indicates that there is only a difference in perspective between the powers of nourishment, growth and reproduction, such that the three powers of the vegetative soul are really identical, and are one sole soul, one sole power. In this view, the three powers are not really different, but are merely different accounts of the same phenomenon.[16] In T.2, συνέζευκται does not simply mean that the three powers are connected, but that they are, *stricto sensu*, united.[17]

---

13 Aristotle, *de An.* II,2, 413b15. In Alexander, see especially 94,9–10, where the question discussed is whether the entire soul can be found in a substrate that is numerically one, having differences only "according to its powers and its definition" (κατὰ τὰς δυνάμεις μόνον καὶ κατὰ τὸν λόγον ἔχουσα τὰς διαφοράς), or if these powers are separated according to place. This same disjunct is also, expectedly, found in *Quaest.* 2,27, commenting on *de An.* II,2, 413a20 and explicitly referencing 413b15.
14 Respectively at *de An.* I,1, 402b1–3 and II,2, 413b10–15.
15 Aristotle, *de An.* II,2, 413a31–32; Alexander, *de An.* 29,13 and *Quaest.* 2,27; 77,12. See also Gregorić (2007), p. 22.
16 Alexander, they say, "does not consider as really distinct faculties the powers for nourishing oneself, for growing and for reproducing. Rather, he sees them as various explanations of one

Against this, I will put forward three correlated hypotheses: 1) Even if there is no lexical distinction in Alexander between "powers of the soul" and "parts of the soul", Alexander nonetheless comes up with criteria which distinguish a soul power from a soul part, or from a soul of its own. The difference between his position and Aristotle's[18] is found chiefly in Alexander's effort to clarify these criteria. 2) As will become clear in the case of the vegetative soul, even the powers that do constitute a soul or a soul part (vegetative/animal/human) are objectively distinct (in a sense that remains to be clarified) and are not simply the result of a change in perspective. 3) The main criterion by which one can account for the organization and the unification of soul parts is the teleological criterion.

At this point, for clarificatory purposes, two issues must be distinguished:[19] the internal organization of a soul part, on the one hand; and the unity of the different soul parts in the complex cases of animals and in the yet more complex cases of human beings, on the other hand. In this paper, I will focus on the first issue, but the two are intimately connected. This is already true in Aristotle, and it is still the case in Alexander. As we will see below, the impossibility for a given power to meet the criteria that characterize a genuine soul part gives us clues about the internal unity of a given soul. In other words, the reason why the reproductive power, for instance, is not a genuine soul part is grasped together with the correlate reason, i.e. the reason why the vegetative soul can unify the powers of nutrition, growth and reproduction.

2 The nutritive soul as a complex form

On the face of it, Accattino and Donini's unitarian interpretation seems able to federate other passages. An instance of this would be the passage where Alexander introduces the nutritive soul, one page after T.2. Alexander has shown that the soul is a condition of life and that the plants must also be endow-

---

and the same fundamental *prôte dynamis* or *psychê*." Accattino and Donini (1996), p. 155–156 (my translation).
17 Accattino and Donini ascribe to Alexander the standard view of the unity of the soul and of its divisibility in parts. This view is spelled out in Corcilius and Gregorić (2010), p. 82–83, *esp.* 82: "the capacities are merely logical parts or aspects of the soul, which does not imply that they can be detached from the whole so as to exist separately from one another or from the whole."
18 At least in the reading of Aristotle by Corcilius and Gregorić (2010). See also Menn (2002).
19 They are carefully distinguished in Johansen (2012), p. 47–48 and (2014), p. 40.

ed with a soul (more on this below). This soul must account for nutrition and growth, as well as for reproduction. Alexander adds:

> **T.3.** Given, then, that the power for nourishing [oneself] is both a soul and first among the soul's powers, we should speak about it first and show just what its essence and nature is. The account of the soul for reproducing is linked (συνέζευκται) to the account (λόγῳ) of the soul for nourishing [oneself]. For just as being nourished and growing are activities that belong properly (οἰκεῖα ἔργα) to the soul for nourishing [oneself], so too does reproducing something similar to what sows seed. (Alexander, de An. 32,6–11, transl. Caston 2012)

The meaning of λόγῳ, however, is problematic.[20] It can refer both to the formula of the powers of the nutritive soul, and to the discourse about them. Furthermore, the meaning of λόγῳ will impact the meaning of συνέζευκται in the same passage. Should one say that the formulas of the powers of nutrition, growth and reproduction are united, or that, in the inquiry on the soul, the discourses on these three powers are intertwined? The first option would result in an attenuation of the logical difference between the three powers. We can reject it, if we consider that Alexander's T.3 is a reading of Aristotle's *de An.* II,4:

> **T.3'.** Since the same capacity of soul is both nutritive and generative, it is necessary to determine what concerns nutrition first; for it is in virtue of this function that it is marked off (ἀφορίζεται) from the other capacities. (Aristotle, *de An.* II,4, 416a18–21, transl. Shields)

In chapter II,4, T.3' comes after a general introduction on the method of inquiry suited for soul powers (415a14–23). This introduction is followed by a broad account of the nutritive soul (a23–b7) and a general discussion of causality in the soul (b7–28), at the end of which the claim that nourishment and growth do belong to life can be stated (b27–28), this claim implying that nourishment and growth require the possession of a soul. Aristotle then refutes Empedocles and other philosophers (perhaps Heraclitus) who have attempted to explain nourishment and growth with no psychic principle (415b28–416a18). Our T.3' passage justifies the necessity to start with nourishment and, in so doing, inaugurates a positive treatment of the nutritive soul.[21] If T.3' is indeed the Aristotelian passage that Alexander has in mind, then λόγῳ at T.3 refers to the discourse.[22]

---

[20] In his translation, Caston neutrally chooses "account" which suits both meanings. Accattino and Donini (*discorso*) and Bergeron and Dufour (*exposé*) choose the second meaning, which seems preferable to me.

[21] The word τροφή can mean the food one feeds on, as well as the activity of nourishment itself. Even if, in the next clause (416a21–22), τροφή can mean food, it seems better to understand it as the activity of nourishment based on the τῷ ἔργῳ τούτῳ (a21). On this, see Polansky (2007),

But, concerning the nutritive soul, T.3' can be used to defend both a strong and a weak reading. In the strong reading, as C. Shields says, nutrition and reproduction are "twin aspects of the same overarching function" (Shields 2016, p. 201, about *de Generatione Animalium* II,5). This reading, however, encounters a problem further in II,4, where Aristotle points out, contrarily to what the strong reading maintains, that growth and reproduction do not operate in the same way on food.[23] This ambiguity explains why Alexander never goes so far as to describe the three powers of the nutritive soul as identical. On the contrary, the use of συνέζευκται in T.2 signals the intention to articulate the three powers without reuniting them into one power. Whereas Aristotle is ambiguous when he says that "the same capacity of soul is both nutritive and generative", Alexander only maintains that the actions to nourish oneself, grow and reproduce are the proper activities (οἰκεῖα ἔργα) of the nutritive soul.

The endeavor to articulate rather than to identify the three powers is not only the result of Aristotelian exegesis in Alexander – its aim is also polemical. In the *de An.* as in the *Mantissa*, Alexander targets the Stoic doctrine according to which the powers of nutrition and growth are distinct from the power of reproduction. It is based on this distinction that the Stoics deny to plants the possession of a soul, and confine them to the φύσις.[24] Against this view, Alexander strives to show that being alive necessarily requires the possession of a soul. If plants are alive, they are ensouled and, therefore (skipping some premises), the "first power", i.e. the nutritive one (nutritive being taken here in its broader sense of vegetative) is "psychic" (ψυχική).[25] As he often does, Alexander retrieves the Aristotelian critique of the *physikoi* and uses it against the Stoics.

---

p. 213–214 (who nonetheless chooses food to translate τροφή in both clauses, unlike C. Shields) and Johansen (2012), p. 106–107.

22 The rest of Alexander's text after T.3. also supports this reading. Following T.3, after a sentence in which Alexander justifies the inclusion of reproduction in the nutritive soul (32,11–23), the text returns to its main line of argument with the methodological "Object – Activity – Power" rule, which is said to guide the inquiry (ἐπὶ τῶν προκειμένων οὕτω χρὴ ποιεῖν, 32,24). This also encourages us to read λόγῳ at T.3 as referring to the discourse.

23 *De An.* II,4, 416b11–20. See also Johansen (2012), p. 107–108.

24 At least according to Crysippus (but not Panaetius), cf. Inwood (1985), p. 35 and Tieleman (1996), p. 97–99.

25 See, mainly, *de An.* 31,25–32,19 and *Mantissa* §4, 118,16–26. V. Caston has argued that Alexander could not be targeting the Stoics in the *de An.* passage: if that were the case, Alexander's refutation "would misfire", because the Stoic argument would not be impacted by the objection that simple bodies are not alive (*de An.* 31,27–28; Caston (2012), p. 126). Indeed, for the Stoics, simple bodies are not part of "nature" in the strict sense. Rather, the divine *pneuma* creates cohesion, ἕξις, in them. But, if we look at other texts in which the Stoic target is undeniable (most notably *Mantissa* §4), we can see that Alexander does not target the tripartition

However, Alexander's argument against the Stoics would be much more significant if it involved the unification of the three powers of the vegetative soul into one sole power, i.e. if Alexander had offered a strictly unitarian account of the vegetative soul. But Alexander never goes that far.

The reason why Alexander never goes so far as to identify the three powers of the nutritive soul is to be found in his own interpretation of the Aristotelian doctrine. Alexander's *de An.* begins with a general picture of the sublunary substances, which starts from the elements and leads to the souls.[26] Doing so, his aim is to establish that this *scala naturae* is also a *scala formarum*, since Alexander extends and systematizes the hylomorphic model, so that it applies to every sublunary substance, including the elements.[27] He can then make use of arguments from analogy or *a fortiori* like: if x is the case for the forms of elements, x will also be the case for the soul.[28] There is more to these arguments, of course, than an analogy, since the gradation of forms is also a description of the living body from basic components to soul, which is a "form of forms and a kind of culmination of culminations".[29]

Now, Alexander must also distinguish the degrees of this scale of forms. To introduce these distinctions, he uses the criterion of the plurality of powers. In simple bodies, form is a unique power emerging from a pair of properties. For instance, in the case of fire, lightness emerges from the dry – hot pair.[30] But, unlike the simple matter of simple bodies, the matter underlying living bodies, Alexander says, is "a compound body with distinctive parts useful for different activities" (*de An.* 10,28–29). This is why Alexander interprets the ὀργανικόν of Aristotle's famous sketch of a definition of soul in *de An.* II,2 to refer to the ma-

---

ἕξις / φύσις / ψυχή, but rather that he reduces it to the couple φύσις / ψυχή, since he is most concerned with the distinction between animate and inanimate beings.

26 *De An.* 2,25–15,29 (and more specifically 7,14–10,10).
27 Accattino (1995), p. 186–187; Guyomarc'h (2015), p. 246–248.
28 For instance, *de An.* 11,6–7. See also *de An.* 5,12–18 and 21,24–22,23.
29 *De An.* 8,12–13: εἶδος […] εἰδῶν καὶ τελειότης τις τελειοτήτων.
30 *De An.* 5,4–6: "In the case of fire, which is a simple, natural body, heat and dryness, as well as the lightness that emerges from them and above them (ἡ ἐκ τούτων τε καὶ ἐπὶ τούτοις γεννωμένη κουφότης), are its form." On this difficult text, see Accattino (1995), p. 186–189 and Caston (2012), p. 79. The difficulty does not dissipate in what follows, where it is said simultaneously both that lightness is what *proceeds from* the form of the fire (παρὰ τῆς φύσεως καὶ τῆς κατὰ τὸ εἶδος οὐσίας, 5,10) and that lightness *is* the form of fire (ἥτις κουφότης εἶδός τε καὶ φύσις οὖσα τοῦ πυρός, 5,11–12). More work remains to be done in this matter. I am of the opinion that Alexander brings together two claims: 1) elements have a unique power (i.e. a principle for only one type of motion); 2) the form of the elements is a combination of differences (see Alexander *apud* Simplicius, *in Aristotelis Physica commentaria* 282,21–24, and Alexander, *de An.* 8,17–22; *Quaest.* 3,14; *Mantissa* §3, 115,24–25).

terial condition of life. The word qualifies a body "that has many different parts able to subserve the soul powers" (*de An.* 16,11–12).³¹

In other words, life is complex. The material complexity of bodies will be matched by the complexity of powers. Alexander opposes the simple motion of simple bodies to trees and plants, which possess the principle of nutrition, growth and reproduction (*de An.* 8,25–9,4). Numerous texts confirm this proportional relation (ἀναλογία) between the complexity of bodies and the plurality of powers possessed by the psychic form.³² As was shown by M. Rashed, a passage of *GA* II,3 can be used as ground for Alexander's claim. In this passage, Aristotle draws an explicit analogy between the differences in dignity and indignity among souls and the corporeal nature that underlies these souls.³³ In Alexander, this claim will develop as a cosmological one, since the works of the "divine power" are proportionate to the matter on which this power acts.³⁴ These cosmological elements are not explicitly involved in the *de An.*, but the result of Alexander's analogy is – namely a strong pressure to grant living beings a plurality of powers,³⁵ which cannot be only apparent.

In a nutshell, there is a hierarchy to Alexander's polemical aims. Against the Stoics, it is crucial to unify the powers of growth and nutrition with the power of reproduction, and to make them all soul powers. But what matters most is to counteract Stoic monopsychism: if not, the organizational criterion of the Peripatetic *scala naturae* itself falls apart. This is why the differences between the psychic powers must be exacerbated as much as feasible, while avoiding the local separation endorsed by the Platonists and renewed by Galen.³⁶ Hence, Alexander states clearly that the soul powers are multiple and different, and that it is not "just one power which seems to be several" (*de An.* 27,5). He can even enjoy his mediate position and use Platonist (or Platonist-like) arguments (e.g. moral conflicts) against the Stoics' monopsychism. This must at least mean that the plural-

---

31 See also 75,26–30.
32 *De An.* 11,5–13; *de Providentia* 83,6–87,4 Ruland, etc.
33 *GA* II,3, 736b29–33; Rashed (2007), p. 156–157 and 286–291.
34 See chiefly *Quaestio* 2,3; *de Providentia* 77,12 Ruland.
35 Confirmed in Simplicius, *in Ph.* 265,1–3.
36 *De An.* 27,4–8: "To show that the soul has several powers and not just one power which seems to be several – as Democritus and various others thought, because of the changes and activities [it undergoes] at different times, in relation to different things, and by different means – it is sufficient to point to the conflict of powers with one another, which occurs both in people who control themselves and in those who do not." (transl. Caston 2012) Concerning the reference to Democritus: see the discussion between Accattino and Donini (1996), p. 152 and Caston (2012), p. 118, to determine whether Democritus is merely, as Caston maintains, "a beard for the Stoics", or if the ascription of this view to Democritus can be substantiated.

ity of the soul powers cannot be only apparent, and that the difference λόγῳ between the three powers of the vegetative soul cannot be only a difference in perspective.

3 The distinguishing criterion for powers

To the question that asks what differentiates the vegetative powers, a first answer is found in the general rule according to which a power (that is, an irrational power[37]) can only have one type of corresponding activity. The rule is set down in *Metaphysica* Θ and comes from Plato. It has been given many names: power monovalence or "one-capacity one-function thesis".[38] This kind of rule acts as a premiss in the argument sketched by Alexander in chapter 4 of the *Mantissa*:

> **T.4.** Rather, the capacities of the soul themselves differ from one another, and it is not possible [to do different things] with the same capacity, for example to think with the [capacity] of sense-perception or to perceive by sense with the [capacity] of thinking. For if the vegetative [part] too belongs to soul, and it is impossible to act in several ways simultaneously with the same capacity, and the nutritive [part] is always active in living things, then either we will perform no other activity in respect of our soul, if the capacity of the soul is single, or else, if we do perform other activities, for example perceiving and feeding at the same [time], then the capacity of the soul is not single [...] (Alexander, *Mantissa* §4 [viii–ix], 118,28–35, transl. Sharples 2004)

This passage is inserted in a series of arguments aimed at Stoic monopsychism. The monovalence of power here may look less finely-grained than its Aristotelian version at Θ,2, 1046a36–b29.[39] Alexander treats a rational power (the intellect) like irrational powers: all are specified by *one* proper activity. What Alexander adds is a temporal criterion (ἅμα), which is relevant in this particular case, because nutrition is a condition of life and that it must thus be active until the animal's death. Having stated this set of premises (the vegetative power's psychic character, the monovalence of powers and the temporal criterion), Alexander moves to the following dilemma: either all ensouled beings are plants, since only the vegetative power would be constantly active in them; or there would be many powers of the soul, at least in animals. But the monovalence premiss

---

[37] See D. Lefebvre's interrogations about rational powers in Lefebvre (2003).
[38] Lefebvre (2003). Cf. also Hintikka (1974), p. 8.
[39] The following passage (§4, 118,35–38), about the division of the arts according to their proper power, confirms that Alexander is indeed thinking of Θ,2 here.

could still seem to beg the question here.⁴⁰ At the very least, this premiss states as a *fact* the difference between powers. But Alexander has still not definitely explained what makes an activity singular in such a way that it could not be produced by a different power.

The *de An.* elaborates on this topic. Alexander's explanation, in agreement with Θ,8, is that powers are specified by their ends. This is what we learn from the distinction of powers within the vegetative soul:

> **T.5.** To grow and to be nourished are not the same, however⁴¹: they do not occur at the same times, nor does each of them have the same function (σκοπός). (1) For an animal is always being nourished for as long as it exists – it is for just this reason the most continuous of soul activities. But things which are nourished are not always growing, since they are nourished as long as they exist, but some also shrink as they age. (2) The function (σκοπός) of what nourishes is to preserve (σωτηρία) what is nourished, but [the function] of what causes growth is to add to the magnitude of what is nourished. So among powers there is accordingly one that is able to preserve (τηρητική) the being and essence [of a living thing] – since the power for nourishing [oneself] is of this kind – while the [power] for growing is primarily able to produce an increase in size. (*de An.* 35,9–17, transl. Caston 2012)

Shortly before this passage, Alexander has restated his methodological rule "Object – Activity – Power". After a discussion of the objects of the different powers of the vegetative soul, what is now at stake is to distinguish these powers according to their activities. To accomplish this, first, there is the temporal criterion⁴² (1) which is of use here because of the functional independence it implies. Even if nutrition and growth are conjoined as soul parts, they are disjoined in time, and therefore functionally distinct. But this temporal criterion is only the sign or the result of a more intrinsic difference. The monovalence criterion is then introduced and explicated: the definition – and thus the specificity – of a power de-

---

40 In the sense that the monovalence argument *immediately* implies a plurality of powers of the soul, which is the view the argument defends. Alexander's argument seems to target Stoic monopsychism more directly in the negative sentence: "it is not possible [to do different things] with the same capacity." However, this reading requires that we understand Stoic monopsychism as a stronger doctrine, asserting not only the lack of partition in the soul, but also the identity and unity of its powers (e. g., impression, assent, impulse), which is doxographically debatable. For the Stoic notion of "powers of the soul", see the passage of Iamblichus *apud* Stobaeus, SVF II,826, mentioned above, which also (and perhaps more faithfully) calls the powers "qualities". Cf. Inwood (2014). It goes without saying that this case concerns mainly the soul as a *hegemonikon*, following the distinction recounted by Sextus Empiricus, *adversus Mathematicos* VII,234.
41 See Aristotle, *de An.* II,4, 416b11–12.
42 See already in Aristotle, *de Generatione et Corruptione* I,5, 322a24–26.

pends on the σκοπός it aims for (2). For instance, the function of the nutritive power is to "preserve the being and essence". The words used here, τοῦ εἶναί τε καὶ τῆς οὐσίας [...] τηρητική, or σωτηρία, have an ontological echo. The same words are used by Alexander in other texts to qualify the function of a form in general, when it is taken as the "completive part" (συμπληρωτικόν μέρος) of the compound (Rashed 2007, p. 158–159). It is because the nutritive power has for an end to ensure the conservation of an animal's being that it must last for as long as the animal lives: the temporal criterion is indeed derived from the σκοπός criterion.

The question is now: can we identify a more fundamental criterion, for instance a distinction between objects? This would follow from the "Object – Activity – Power" rule. Yet, it is the opposite that is true: the specification of powers according to their objects is also – like the distinction of powers according to their activities – derived from the teleological criterion.

If the specification of vegetative powers was done according to their objects, we would have for each power a certain state of food. For instance, to food that is not yet transformed and contrary to what is being nourished would correspond the nutritive power in its strictest sense; to food that is transformed and assimilable to what is being nourished would correspond the power of growth; finally, the useful residue of digested food would be used for reproduction since, as Alexander explicitly says (following Aristotle), the generative power "uses nourishment itself in a way too" (de An. 36,3).

This reading is attractive, but it must be rejected. A first clue in this sense is the fate of the "Object – Activity – Power" rule in Alexander's de An. While Aristotle introduces the rule at the beginning of II,4, Alexander does so only at de An. 32,23–33,10 – it appears as a simple transitional passage in the middle of the study of the vegetative soul. Alexander again construes Aristotle's "logical" priority (κατὰ τὸν λόγον, de An. II,4, 415a19–20) as applicable only to the "order of inquiry" (33,8). The sequence "Object – Activity – Power" is only valid regarding our knowledge, setting the inquiry to start with what is more known by us. For the order of being, however, it is the opposite: in fact, the power of nutrition "does not require its activity in order to be (πρὸς τὸ εἶναι)" (33,1–2). Alexander's interpretation does not follow the ontological claim of Metaph. Θ concerning the priority of activity.[43] With this context as background, we can better understand the explicit claim in the Mantissa, that the objective criteria cannot differentiate the powers. This claim appears just before our T.4: "the differentiae of the soul

---

[43] Contrarily to what M. Bergeron and R. Dufour maintain in Bergeron and Dufour (2008), p. 267.

are not [a matter of] the things concerning which soul is active; rather, the capacities of the soul *themselves* differ from one another [...]."⁴⁴

This intrinsic differentiation of the powers of the soul is confirmed in the case of the vegetative soul. It would be a mistake to maintain that the nutritive power would have as an object food not yet transformed, while the power of growth would have as an object assimilable food. Both these powers have as their object food in both these states. Alexander clearly says that nutrition consists in the assimilation of food, i.e. that it consists in the passage from food that is contrary to what is nourished to food that is similar to what is being nourished, that passage being identical with the process of digestion.⁴⁵ In the *De mixtione*,⁴⁶ Alexander says that the *same* food can result in nourishment and in growth, depending on the amount of it absorbed into the "flux" (the continuous movement of dissociation and adjunction that affects the flesh throughout its life).⁴⁷ If, Alexander explains, food is inferior or equal in amount to the matter absorbed into the flux, it is only source of nourishment in the strict sense. But if its amount is greater than the amount of matter, it will also be source of growth.⁴⁸ This difference in quantity, however, contains in fact an ontological difference – here, Alexander is thinking of *GC* I,5, where Aristotle says that the food which nourishes and the food which fuels growth are indeed the same thing, but differ in being (I,5, 322a24 – 26). Following Aristotle, Alexander indicates this difference in being using the *qua*-operator: "Nutriment, then, only preserves the substrate when it acts *qua* nutriment (καθὸ τροφή), but when it also acts as quantity (ὡς ποσή) it contributes to growth as well as preservation."⁴⁹

It is then not food in itself, unqualifiedly, which is the object of a given power. The distinction between the power of nourishment and the power of growth is not found in their object *simpliciter*, nor even in a given state of that object, but in what medieval philosophers would later call a formal object (distinct from a material object, i.e. an extensional object). This formal object is the one that is considered *qua* x or y, in its relation with a given power of the soul.⁵⁰ This is why, until I become fully grown, the *bœuf bourguignon* I have for lunch

---

44 *Mantissa* §4, 118,27–29, transl. Sharples 2004: ὅτι οὐκ εἰσὶν αἱ ψυχικαὶ διαφοραί, περὶ ἃ τῇ ψυχῇ ἡ ἐνέργεια, ἀλλ' ὅτι αὐταὶ αἱ δυνάμεις τῆς ψυχῆς διαφέρουσιν ἀλλήλων. See also *de An.* 27,8–15.
45 See *de An.* 34,27–35,2 and 35,8, noting the importance of the καί.
46 *Mixt.* XVI, 236,18–26 Bruns; 43,12–23 Groisard.
47 For this definition of the flux, cf. *Mixt.* 235,22 Bruns; 41,21–22 Groisard.
48 As J. Groisard aptly notes, this idea is already found in Plato, *Timaeus* 81b.
49 *Mixt.* XVI, 236,24–26 Bruns; 43,21–23 Groisard, transl. Todd.
50 Cf. Rashed (2005), p. cxv.

can be *simultaneously* a source of nutrition and a source of growth. This relation between object and power is qualified following the end of that power: in the case of nutrition, food as such aims to "contribute to being as well as preservation of what is nourished"[51], a wording which Alexander also uses in his own *de An.*[52]

In other words, the σκοπός of a power is what constitutes, in the strongest sense of the word, the object of that power. The σκοπός transforms an object *simpliciter* (food as an extensional entity) into a formal and relational object (food *qua*…). This teleological differentiation also involves corporeal operations that are partially distinct (e.g. following a proportional quantity of food) and ontologically independent (T.5). As will be shown in more detail below, the objective criterion is ambiguous. When it concerns the extensional object, Alexander uses it to highlight the unity of the three vegetative powers.[53] But when it concerns the *qua*-object, the criterion indicates the difference of the same three powers. In both cases, however, the objective criterion is derived from the teleological criterion.

In any case, it seems irrevocably impossible to maintain that nutrition, growth and reproduction would be mere perspectives on a reality which is essentially one. There is a real difference (but not in place) between the three powers of the vegetative soul, and this difference is teleological.

Yet, even if the soul is a cluster of powers distinct in reality, its unity must not be compromised. Alexander repeats more than once that it is impossible for a living being to have many souls.[54] Since this claim concerns animals, it is legitimate to maintain that the unity of the soul must be even stronger in plants, which have less powers. On the other hand, in human beings, Alexander says, the inferior powers are "expanded and developed" by the presence of the superior powers (*de An.* 30,5–6). When Alexander sketches a general and typical definition of the soul at the beginning of the treatise, he calls it a "form of forms and a kind of culmination of culminations," a common form which stands above the different forms of organic bodies (*de An.* 8,11–13). The question is ultimately to

---

[51] *Mixt.* 236,23–24 Bruns; 43,19–20 Groisard: εἰς τὸ εἶναί τε καὶ σώζεσθαι συντελούσης τῷ τρεφομένῳ.
[52] See T.5 above and 35,26–36,2.
[53] It should be noted that the study of food in the *de An.* is focused on the unity of the powers of nutrition and growth (33,12–34,26), but that, when Alexander wants to differentiate these powers, he uses the criterion of activity and aim (as in T.5), and does not detail the role of food *qua*. This is perhaps because the distinction is rather technical, which might conflict with the interests of the broader audience this treatise is intended for.
[54] E.g. *de An.* 30,2–3; 96,12; 99,10, etc.

know whether these phrases must be taken literally: is the soul a principle of unity, distinct from its different parts or powers, or is it the unity itself of these powers?

4 The Unity of the Vegetative Soul

What produces the unity of the vegetative soul's powers? According to the text, they are conjoined, συνέζευκται, which means that they must have a symmetrical relation: what has one of the three powers has the two others as well. From the start, it seems then excluded that the soul powers form a serial unity.[55] Of course, Alexander sometimes uses the verb συζεύγνυμι about things that have an asymmetrical relation (de An. 29,16–17). And, of course, the reproductive power develops last and is a "perfection" of the two previous ones (de An. 35,17–22). This is why the entire vegetative soul can be named reproductive, since it is according to their end (τέλος) that things are named (de An. 36,13–15).

Yet, relying on Alexander's commentary on Aristotle's Metaph., we can specify the conditions which must be met by a group of realities for a serial unity (ἐφεξῆς) to justifiably connect them. Elements in a group have a serial unity if 1) there is an asymmetrical dependency between them, in such a way that the prior element can obtain without the posterior element, but not conversely; 2) the order of the asymmetrical dependency is one of ascending perfection. Concerning things that stand in a serial unity, Alexander says, "the posterior things are more perfect whereas, in the case of things said with reference to one thing, the unity to which the other things are referred [is more perfect]."[56] This ascending perfection is due to the fact that the prior element "contributes" (συντελεῖν) to the posterior element, so that the order is not only hierarchical but also teleological, the inferior terms being for the sake of the superior terms.[57]

This second condition is explicitly applied to the animal or the human soul in the de An., and with the same wording as we find in the commentary on the Metaph.:

> **T.6.** In those things in which the powers of the soul are all present, they are related to each other in such a way that the earlier powers are for the sake of the later ones and make a

---

55 *Pace* Bergeron and Dufour (2008), p. 36.
56 *In Aristotelis Metaphysica commentaria* 263,32–33. Concerning textual issues and the translation of this passage, I refer the reader to Guyomarc'h (2015), p. 225.
57 See also *de An.* 75,31–76,6; Caston (2012), p. 121. On the nutritive part in the teleological order, see 28,20–25; 94,11–17.

contribution (συντελεῖν τι) to them, analogous to the parts of the animate body. [...] For the vegetative soul in animals is for the sake of preservation and being, and apart from it the part for perception could not be. (*de An.* 75,24–76,1, transl. Caston)

Even if the nutritive soul's powers satisfy the second condition, they do not satisfy the first. Alexander concedes that some living beings may not have a reproductive power, but this is visibly only the result of a particular accident (i.e. sterility), in accordance with the physical regularities ὡς ἐπὶ τὸ πολύ (*de An.* 35,19–20).[58] Alexander makes similar remarks when he goes over the development of the rational power in men "who are not mutilated" (*de An.* 82,12).[59] It is because plants in general possess the power of reproduction as well as the power of nutrition, that reproductive power is "part of the power and soul for nourishing [oneself]" (*de An.* 32,21–22). If Alexander does admit that there is a serial order within sensitive power (*Quaest.* 2,27), he never claims there is such an order in the case of vegetative power.

But if the unity of the vegetative soul's power is not serial, what is it? Alexander offers some indications about this, shortly before T.6, when he discusses the distinction between the different souls or soul parts in the proper sense.

> **T.7.** These powers differ from each other not only in definition, but can at this stage be separated by underlying subject and activity as well. For there is a difference between them in definition, because what it is to be the part for nourishing and what it is to be the part for perceiving are not the same, since in so far as they are such the definition of each is different. There is a difference in activity, because it is not the case that the part for perceiving is always active when the part for nourishing is. For the part for nourishing is always active, whereas the part for perceiving is inactive when we are asleep. And there is a difference in their underlying subject because the part for nourishing is in every part [of the body], while the part for perceiving is not, and the part for nourishing is in plants as well, while the part for perceiving is not to be found there any longer. (*de An.* 75,2–10, transl. Caston)

This passage is found in a long discussion of soul parts, starting from the study of the impulsive part. Namely, it concerns the difference between the vegetative and sensitive souls. This entire discussion shows that there is an objective difference between souls and that this difference is not a mere *façon de parler*. Immediately before T.7, Alexander reminds the reader of the following: that the vegetative and sensitive parts are not found in the same beings; that the vegetative part is present at the start of the embryo's life; that we are continually active

---

[58] Caston (2012), p. 132.
[59] He is using the same words in 35,20.

where the vegetative part is concerned, but that we do not perceive while sleeping; that the vegetative part is productive whereas the sensitive part is theoretical; and, finally, that all parts of an animal have a share in the vegetative part, while it is obvious that we could not perceive using hair, nails or bones.

Then (at T.7) Alexander summarizes his claims with three criteria that will be also applied next to the difference between nutritive soul and impulsive soul, and to the one between sensitive soul and impulsive soul: in definition, in activity, in their underlying subject. The first proposition of the passage (i.e. that the difference between the soul parts cannot simply amount to a difference in definition) is interesting, since Alexander uses it in other occasions, for instance in the cases of φαντασία: Alexander refuses to make imagination a proper part of the soul precisely because it differs from sensation *only* in being or in definition, but not in its underlying subject (*de An.* 69,5–6). In this passage, "underlying subject" refers to the traces left in the soul by sensible objects, but it can also refer to the corporeal organs the soul depends on. In this case, the extensional difference (some animals do not have φαντασία) is not the sign of a difference that could ground a soul part in the proper sense.

Likewise, the three powers within the vegetative soul do not meet the three criteria and, thus, are not different parts of an animal's soul. Reproduction, growth and nutrition differ, at least partially, in being or in definition. They also differ in activity (as the temporal criterion signals). But they do not differ sufficiently in their underlying subject, since their operations have some organs in common as well as an object (food) which, while different as *qua*-object, is identical as extensional object.[60] Hence, in an animal, they are not soul parts in the proper sense and can be unified in the vegetative soul.

I have mentioned that the vegetative soul's three powers differ at least partially in definition: with this, Alexander draws a subtle distinction, which leads to a more positive approach of the powers' unification. When Alexander summarizes the three criteria, he retains only one, the criterion of definition. The sensitive and rational parts are different from the others on account of being κριτικά, "for cognition"; the impulsive part is practical; the vegetative part is productive (ποιητικόν, *de An.* 75,12–13). I will not discuss here the obvious systematization created by this distinction, which borrows the division of the sciences to apply it to the partition of the soul.[61] More important for us is the reference to an end,

---

[60] For the seed as "the end product of the final stage of nourishment", see *de An.* 36,4 and 92,20–21 (Aristotle, *inter alia, GA* I,18, 725a11–13).

[61] Alexander is, concerning the parts of the soul, as flexible as Aristotle is concerning the different sciences. The latter occasionally divides the sciences only between theoretical ones and practical, or productive ones, and the former, likewise, sometimes puts the nutritive soul togeth-

which encompasses the criteria previously listed. The priority of the end has been made explicit before, in a passage already cited at T.6: "For the vegetative soul in animals is for the sake of preservation and being, and apart from it the part for perception could not be." Of the three criteria – definition, activity and underlying subject – the first, because it refers to an end, is the most crucial and the one which determines the following criteria.

As we have seen, the vegetative soul's powers can be distinguished according to their σκοπός, i.e. according to what for the sake of which they are. But this distinction must not conceal a superior unity: the three powers have in common to be for the sake of life itself (*de An.* 75,31). This is why Alexander also calls the vegetative soul the "vital" soul (ζωτική, *de An.* 38,12–13). Each of the vegetative soul's three powers involves life. Nutrition is the condition of life itself and, without it, there is no ensouled being. The need to nourish oneself is the first element that defines life (*de An.* 29,3–10; 31,9; 78,5–6; 92,18). Its function of "preservation" (of life or of being[62]) is often extended to the entire vegetative soul. Growth, on the other hand, generates an "increase in size", that is also part of the definition of life (*de An.* 29,10; 31,9; 35,16–17; 78,5–6; 92,18). Finally, the reproductive capacity is a general feature of all living beings (*de An.* 32,11–20). Even if Alexander often discusses it separately from the two other vegetative powers, it is an end for every living being *qua* living being (*de An.* 35,22–23). But, differently from the two other vegetative powers, it does not cause solely the preservation or the development (σωτηρίαν τε καὶ τελείωσιν) of living beings. Rather it brings completion and "a share in immortality" (*de An.* 36,5–9; 92,19–20) in a living being who is already developed. This triple function – to produce, develop and reproduce life – suffices to indicate the vegetative powers' unity. It provides support to the empirical fact that all non-mutilated living beings must possess all three of them.[63]

As a result, the teleological criterion is adaptable enough to account both for the soul powers' difference and for their unity, depending on the focus of analysis. This model can bring to mind the interconnection of instrumental ends and ends *per se* in moral conduct. Analyzed in relation with each other, in a plant, the vegetative soul's powers are different, because each of them, in their own way, contributes to the plant's life. Analyzed in a more complex lifeform, like

---

er with the practical part, cf. *de An.* 99,21–22; cf. also 75,14–15 on the impulsive part as both "productive and practical"; 73,22–23 and 99,18 on the mere bipartition of the soul into critical and practical parts.

62 Which are the same thing. On this, see Aristotle, *de An.* II,4, 415b13.

63 Alexander does not appear hindered by the plurivocity of life. See Accattino and Donini (1996), p. 263 and Caston (2012), p. 122–123.

an animal, they serve one common end, and are therefore united into the same part of the soul.

But this teleological schema subordinating means and ends is not as systematic as it seems. For, rigorously, in line with Alexander's view of serial unity, we would be justified to expect the vegetative soul to be for the sake of the sensitive soul, which follows it in the order of souls. Yet that is not the case: for Alexander, the vegetative soul simply produces life as the condition of possibility of sensation. By contrast, the teleological criterion applies more directly to the relation between sensation and impulsion: the discrimination of sensible objects has action for an end.[64] In turn, the impulsive part aims at the rational part[65] – which opens way to an intellectualist reading of the *de An*. But this is another story to tell.

5 The Parts of the Form

In conclusion, I want to take a step back and consider the unity of the vegetative soul and the distinction of its powers from another angle. In a short passage of his commentary to *Metaph*. Δ,25, concerning the meaning of "part",[66] Alexander mentions the "parts of the form":

> **T.8.** 'Or the form' again indicates the division [of a thing] as form. For there are parts of the form as form, for instance as the powers of the soul, which is a form, are parts into which the soul is divided. (Alexander, in *Metaph*. 424,28–31, transl. Dooley 1993)

Chapter 25 follows the possible objects of the division, which produces distinct types of parts. The development at T.8 is found in what Alexander identifies as the fourth sense of "part:" in the case of the hylomorphic compound, x is a part of y if y is divided into x, or if y is composed of x's. Aristotle gives a number of examples of the partition of hylomorphic compounds[67] and Alexander, here, comments on what is for him the second of these examples, the one of form "as form" (which probably refers to form taken in itself, in opposition to form taken in some matter).

---

64 See also *Mantissa* §1, 105,29–30.
65 See mainly *de An*. 76,1–16.
66 V. Caston has already pointed out the relevance of this passage for the interpretation of *de An*. 31,2–6 (2012, p. 126, n. 275).
67 Alexander is commenting 1023b20 τὸ ὅλον ἢ τὸ εἶδος ἢ τὸ ἔχον εἶδος. With Bonitz's punctuation (τὸ ὅλον, ἢ τὸ εἶδος ἢ τὸ ἔχον εἶδος), "whole" is a general term covering two cases: form itself and whatever possesses a form. For Alexander, there are three distinct cases.

The idea that the parts of the soul are "parts of the form" is of course not Alexander's invention.[68] But, in the rest of the Aristotelian corpus, the phrase "parts of the form" is logical, and is usually understood as an ellipsis of the more complete phrase "parts of the account of the form," i.e. of the definition.[69] Alexander is well aware of this usage, as the rest of his commentary to Δ,25 shows. But he makes a clear distinction between the logical usage and the one he has here in mind: the parts of the definition mentioned at the end of Δ,25 are indeed parts, but they are, Alexander says, παρὰ τὰ τοῦ εἴδους μέρη (in Metaph. 425,1), distinct of the parts of the form. The definition is formulated in connection to the form, but also includes the matter, here, the genus. There is then a difference between the parts of the form in a logical sense and the 'physical' parts of form as form.

From this point on, the problem unfolds predictably: if form is, in the wording Alexander often uses, the cause of being and thus of unity for any substance,[70] it seems risky to no longer think of the form as a mereological atom, i.e. as a fundamentally simple and indivisible thing. If indeed the form has parts, what is the glue that holds them together? There are two likely solutions: either form is the combination of its own parts, that is, in the case of the soul, a mere bundle of powers; or soul is the cause of the unity of its parts, in such a way that this cause is distinct from these parts. In this latter case, given that Alexander lays much more emphasis on power than on activity in the definition of the soul, the soul itself would be a power possessing various powers.

But, following what we have said above, it seems to me that Alexander leans towards the former case. The soul's unity is – if I can put it this way – self-sufficient: the unification of the parts of the soul does not require a cause other than the teleological schema which structures the embedding of means and ends. The soul is not a mysterious fastening, an additional "yoke" (to recall the meaning of the verb συνέζευκται) that would unify from above its different powers or parts.

This is how we must understand Philoponus' testimony on Alexander. According to Philoponus, Alexander claimed that the soul is not the principle of its different powers, but the principle of the activities resulting from these pow-

---

[68] He could have taken it, for instance, from Metaph. Z,10, 1035b14–21. On this passage, see Morel (2010).
[69] See the next passage at Δ,25, 1023b23–25; see also, among other passages, Z,10, 1035a4 and 21.
[70] For instance, de An. 15,28–29; in Metaph. 373,25; 375,10–11, etc. For being and unity, see e.g. in Metaph. 249,5–14. On this claim, see also Rashed (2007), especially p. 31, and Guyomarc'h (2015), p. 256–265.

ers. This amounts to say that the psychic form is only the combination of its powers.

> **T.9.** Those who want to make all soul immortal say that that which nourishes (τὸ μὲν θρεπτικόν), that which augments and the like are activities of soul which, they say, Aristotle too says are inseparable, but the soul and the powers from which these activities proceed (προέρχονται), these are separable. They claim, then, that he says, that the soul is cause and source (αἰτίαν καὶ ἀρχήν) of these activities, the nourishing, the perceiving and the rest. But that Aristotle does not think this has been stated many times. Alexander interprets in a more natural and true way [when he says] that the soul is source and cause of nourishing (τρέφεσθαι), augmenting and perceiving, which are in reality (τῷ ὄντι) activities of soul. But that he [Aristotle] does not say the soul is the source of that which nourishes and perceives (ἀρχὴν τοῦ θρεπτικοῦ καὶ αἰσθητικοῦ) he has made clear by his adding that it 'is defined by these, that which nourishes, that which perceives' and the rest – ['defined by these'] in place of 'the soul is given its boundaries in these, and has its being in these.' (Philoponus, *in de An.*, 237,11–23 transl. Charlton 2005)

This passage is a commentary on *de An.* II,2, 413b11–13: Νῦν δὲ τοσοῦτον εἰρήσθω μόνον, ὅτι ἐστὶν ἡ ψυχὴ πάντων τῶν εἰρημένων ἀρχὴ καὶ τούτοις ὥρισται, θρεπτικῷ, αἰσθητικῷ, διανοητικῷ, κινήσει (in Philoponus' text[71]). In Aristotle, this sentence is found just before the consideration whether the nutritive, sensitive and dianoetic powers are a soul of their own or a soul part and, if they are a part, whether they are separated in account or in place. According to Philoponus, two readings of the 413b11–13 passage have been offered. These two readings result from the ambiguity of two clauses: πάντων τῶν εἰρημένων, on the one hand, and the enumeration θρεπτικῷ, αἰσθητικῷ, διανοητικῷ, κινήσει, on the other hand.

In the first reading, the two clauses refer to the activities that are inseparable from the body, in opposition to the soul and its powers, which can be separated from the body. The obvious goal of this distinction is to preserve the immortality of the soul. As I. Kupreeva has shown, this position is the one of Numenius, whom Philoponus has already named in the proem of his commentary. For Numenius, the entire soul can be separated from the body, including the irrational soul and the nutritive soul. In the later tradition, this claim has been understood to endorse the immortality of the whole soul.[72]

---

[71] Philoponus omits the τούτων which our manuscripts have after εἰρημένων, but this does not alter the meaning. Charlton (2005) translates: "For the present let it suffice to say only that the soul is the source of all these things that have been said and is defined by them, that which nourishes, that which perceives, that which thinks, change."
[72] Philoponus, *in de An.* 9,35–38 (= Numenius, fr. 47 des Places); Kupreeva (2012), p. 125–126.

The second reading is Alexander's, for whom the soul is the principle of its activities. In talking about these activities, Alexander uses the infinitive: τρέφεσθαι, αὔξεσθαι, αἰσθάνεσθαι, which are "all the things that have been said." Accordingly, πάντων τῶν εἰρημένων does not refer to the same thing as the following enumeration (θρεπτικῷ, αἰσθητικῷ...), which is (correctly) interpreted by Alexander as a list of soul powers. The soul is not a "principle" of its parts or of its powers, because it is nothing else than these powers, since the soul "has its being in these". The soul is identical in definition with its powers.

Any soul, because it is necessarily a complex form, is then clearly a cluster of powers objectively distinct but unified by their common end: life for plants, sensation for non-human animals, intellect or reason for men. Hence, when Alexander says that the powers of the vegetative soul are "conjoined" (συνέζευκται, T.2), the verb is, so to speak, in the middle, rather than in the passive voice.

Bibliography

Accattino, Paolo (1995): "Generazione dell'anima in Alessandro di Afrodisia *De anima* 2.10–11.13?". In: *Phronesis*, 40,2, p. 182–201.
Accattino, Paolo/Donini, Pierluigi (ed.) (1996): *Alessandro di Afrodisia, L'anima*, Roma – Bari: Biblioteca universale Laterza.
Annas, Julia and Barnes, Jonathan (1985): *The Modes of Scepticism: Ancient Texts and Modern Interpretations*. Cambridge: Cambridge University Press.
Bergeron, Martin/Dufour, Richard (2008): *Alexandre d'Aphrodise, De l'âme*, Paris: Vrin.
Caston, Victor (transl.) (2012): *Alexander of Aphrodisias, On the Soul. Part 1, Soul as Form of the Body, Parts of the Soul, Nourishment, and Perception*, London: Bristol Classical Press.
Charlton, William (transl.) (2005): *Philoponus. On Aristotle On the Soul 2.1–6*, London – New-York: Duckworth – Cornell University Press.
Corcilius, Klaus/Gregorić, Pavel (2010): "Separability vs. difference: parts and capacities of the soul in Aristotle". In: *Oxford Studies in Ancient Philosophy 39*, p. 81–119.
Dooley, William E. (transl.) (1993): *Alexander of Aphrodisias, On Aristotle's Metaphysics 5*, London – New-York: Duckworth – Cornell University Press.
Gregorić, Pavel (2007): *Aristotle on the Common Sense*. Oxford: Oxford University Press.
Groisard, Jocelyn (ed./transl.) (2013): *Alexandre d'Aphrodise. Sur la mixtion et la croissance (De mixtione)*, Paris: Belles Lettres.
Guyomarc'h, Gweltaz (2015): *L'Unité de la métaphysique selon Alexandre d'Aphrodise*. Paris: Vrin.
Hangai, Attila (2017): "Alexander of Aphrodisias on *Phantasia:* An Aristotelian account of mental representation in 2nd-3rd centuries CE". Ph.D. Dissertation, Central European University, Hungary.
Hintikka, Jakob (1974): "Knowledge and its objects in Plato". In: Hintikka, J. (ed.), *Knowledge and the Known: Historical Perspectives in Epistemology*. Dordrecht: Reidel, p. 1–30.
Inwood, Brad (1985): *Ethics and Human Action in Early Stoicism*. Oxford: Clarendon Press.

Inwood, Brad (2014): "Walking and Talking: Reflections on Divisions of the Soul in Stoicism". In: Corcilius, Klaus/Perler, Dominik (eds.), *Partitioning the Soul. Debates from Plato to Leibniz*. Berlin: De Gruyter, p. 63–83.

Johansen, Thomas Kjeller (2012): *The Powers of Aristotle's Soul*. Oxford: Oxford University Press.

Kupreeva, Inna (2012): "Alexander of Aphrodisias and Aristotle's *De anima:* what's in a commentary?". In: *Bulletin of the Institute of Classical Studies* 55,1, p. 109–129.

Lefebvre, David (2003): "Comment bien définir une puissance ? Sur la notion de puissance des contraires en *Métaphysique* Thêta 2". In: *Philosophie Antique* 3, p. 117–139.

Menn, Stephen (2002): "Aristotle's Definition of Soul and the Programme of the *De anima*". In: *Oxford Studies in Ancient Philosophy* xxii, p. 83–139.

Morel, Pierre-Marie (2010): "Parties du corps et fonctions de l'âme en *Métaphysique* Z". In: Van Riel, Gerd/Destrée, Pierre (eds.). *Ancient Perspectives on Aristotle's "De anima"*, Leuven: Leuven University Press, p. 125–139.

Polansky, Ronald (2007): *Aristotle's De anima*. Cambridge: Cambridge University Press.

Rashed, Marwan (2007): *Essentialisme. Alexandre d'Aphrodise entre logique, physique et cosmologie*. Berlin: De Gruyter.

Sharples, Robert W. (transl.) (2004): *Alexander of Aphrodisias, Supplement to On the soul*, New-York: Cornell University Press.

Shields, Christopher (transl.) (2016): *Aristotle. De anima*, Oxford: Clarendon Press.

Tieleman, Teun (1996): *Galen and Chrysippus on the Soul: Argument and Refutation in the de Placitis Books ii – iii*. Leiden – Boston: Brill.

Todd, Robert B. (ed./comm.) (1976): *Alexander of Aphrodisias on Stoic Physics. A Study of the De Mixtione with preliminary essays, text, and commentary*, Leiden–Boston: Brill.

Tommaso Alpina
Is Nutrition a Sufficient Condition for Life?
Avicenna's Position Between Natural Philosophy and Medicine

Abstract: In the opening lines of the *Qānūn fī l-ṭibb* (*Canon of Medicine*) Avicenna outlines the epistemological status of medicine: it is a derivative natural science, whose epistemological underpinnings are given in natural philosophy – the theoretical science to which medicine is said to be subordinated –, and their investigation is declared off-limits to the physician.
In providing the theoretical setting of the medical investigation in the first book of the *Qānūn*, Avicenna lists the things that the physician must accept on authority because their existence has been already ascertained elsewhere (i.e. in natural philosophy). Among those things there are the psychic faculties, their existence, their number, and their location. Consequently, in dealing with the diseases related to and affecting the psychic faculties, Avicenna has to assume their ascertainment provided in natural philosophy and, notably, in psychology. Nutrition and the nutritive soul seem not to escape this paradigm: Avicenna provides a formal account of nutrition in the *Kitāb al-Nafs* (*Book of the Soul*), i.e. the psychological section of the *Kitāb al-Šifāʾ* (*Book of the Cure* or *the Healing*), and its physiological account in the first book of the *Qānūn*.
However, is it really indisputable that the physiology of nutrition provided in medicine supplements its formal account provided in psychology, and is subordinated to it? A close inspection of the texts devoted to nutrition and the nutritive soul reveals that, with respect to the psychic faculties, medicine is not entirely subordinated to the conclusions of natural philosophy, but it integrates them with another theoretical framework inherited by the previous medical tradition.

Note: This article has been written under the aegis of the project "Animals in Philosophy of the Islamic World", which has received funding from the European Research Council (ERC) under the European Union's Horizon 2020 research and innovation programme (grant agreement No. 786762). I would like to thank Peter Adamson for his valuable comments and suggestions on a first draft of this paper.

https://doi.org/10.1515/9783110690552-013

1 Introduction[1]

Asking whether according to Avicenna (Ibn Sīnā, d. 1037) nutrition is a sufficient condition for life is not as naïve a question as it might appear *prima facie*. For though in a philosophical, mainly Aristotelian, perspective nutrition, (together with growth and reproduction) represents one of the activities that are minimally constitutive of sublunary life, and in which every sublunary living being is engaged,[2] dealing with nutrition is not an exclusive prerogative of (natural) philosophy. Actually, the issue of nutrition pertains to a field that we can call biology (borrowing the term from modern science to refer to the study of living things and their vital processes), an area which stands at the crossroads between natural philosophy and medicine. Natural philosophy and medicine represent two distinct traditions, having their own authorities, Aristotle and Galen (together with his digest of Hippocrates' medical theories) respectively, which occasionally overlap, intersect, and conflict.

Avicenna is at the same time the heir and collector of these two traditions, to which he devoted his two masterpieces (Musallam 1987, p. 94), namely the *Kitāb al-Šifāʾ* (*Liber Sufficientiae* or *Sufficientia* in Latin, *Book of the Cure* or *the Healing* in English, henceforth *Šifāʾ*), and the *Kitāb al-Qānūn fī l-ṭibb* (*Liber Canonis* in Latin, *Canon of Medicine* in English, henceforth *Qānūn*). The former, composed

---

[1] All quotations from and the translations of Avicenna's *Nafs* are based on the edition by Rahman (1959, 1970²). Quotations from Avicenna's *Nafs* are usually followed by the reference to the page and the line number of the corresponding passage in the Latin translation in square brackets. See Avicenna Latinus, *Liber de anima seu sextus de naturalibus*, ed. by S. van Riet (1968 [*IV–V*] and 1972 [*I–II–III*]). The same quotation scheme is followed in the case of other sections of the *Šifāʾ* whose Latin translation is edited in the Avicenna Latinus series (*Samāʿ ṭabīʿī*, *Kawn wa-Fasād*, *Afʿāl wa-Infiʿālāt*, *Ilāhiyyāt*). All quotations from and the translations of Avicenna's *Qānūn* are based on the edition by the Institute of History of Medicine and Medical Research, Ǧāmiʿa Hamdard, New Delhi (1981–1996), and are followed by the reference to the page and the line number of the corresponding passage in the Latin translation in square brackets. See Avicenna, *Liber canonis*, reprography of the Venetian edition 1507 (1964). The English translation of Aristotle's works is that provided by Barnes (1984) slightly modified.

[2] See *de Anima* II, 1, 412a13–15: "Among natural bodies some have life (τὰ μὲν ἔχει ζωήν), while some other do not have it (τὰ δ' οὐκ ἔχει). We call life (ζωή) the capacity for self-nourishing, growing, and perishing"; II, 2, 413b1–2: "Living (τὸ ζῆν) belongs to living beings (τοῖς ζῶσι) in virtue of this principle (διὰ τὴν ἀρχὴν ταύτην, sc. nutrition)"; II, 4, 415a23–25: "The nutritive soul belongs also to the other [living beings: other than plants], and it is the first and most common faculty (πρώτη καὶ κοινοτάτη δύναμις) of the soul, in virtue of which living (τὸ ζῆν) belongs to them all"; II, 4, 416 b9–11: "Since nothing is nourished which does not share in life (μὴ μετέχον ζωῆς), that which is nourished is the ensouled body insofar as it is ensouled, so that nourishment is related to that which is ensouled, and not by accident (οὐ κατὰ συμβεβηκός)."

approximately between 1020 and 1027, encompasses all divisions of theoretical philosophy (natural philosophy, mathematics and metaphysics) with the addition of logic, and represents a *summa* of Aristotelian philosophy. The latter, whose composition started in Ǧurǧān in approximately 1013, is a formidable textbook of Galenic medicine. These two works have been conceived as complementary, one providing the theoretical background of the other. This complementarity is not only reflected in structural analogies[3] and cross-references,[4] but also emerges from Avicenna's teaching praxis. For, in his *Biography* it is reported that, during his stay in Hamadān (1015–1024) serving at the court of the Kākūyid emir Šams al-Dawla, Avicenna wrote the part on natural philosophy of the *Šifāʾ* (with the exception of botany and zoology, which were written approximately in 1027),[5] approximately ten years later the completion of the first book of the

---

[3] Avicenna's *Šifāʾ* consists of four parts (*ǧumal*, sg. *ǧumla*), internally subdivided into sections (*funūn*, sg. *fann*) with the exclusion of the metaphysical part, which has no internal subdivision into sections. Each section corresponds to a book (*kitāb*), and consists of treatises (*maqālāt*, sg. *maqāla*), which in turn consist of chapters (*fuṣūl*, sg. *faṣl*). In the structure of the *Qānūn* we find the same terminology with the addition of "teaching" (*taʿlīm*); however, this terminology is not used in the exact same manner as in the *Šifāʾ*. The *Qānūn* consists of five books (*kutub*, sg. *kitāb*). Taking the structure of the first book as a model, we can say that each book consists of sections (*funūn*). Each section is internally subdivided into teachings (*taʿālīm*, sg. *taʿlīm*), and each teaching into parts (*ǧumal*). Each part contains chapters. However, it can be the case that sections and teachings are immediately subdivided into chapters. Structural dissimilarities between books can be detected. For instance, in books II–V treatises (*maqālāt*) are introduced: in some cases they seem to replace sections (e.g. in book II), while in other cases they seem to replace teachings (e.g. in book III–IV). Thus, the main structural difference between the *Šifāʾ* and the *Qānūn* is that the former is primarily made of *ǧumal*, whereas the latter is primarily made of *kutub*. These considerations, however, are based on the current printings of the *Qānūn*, which still awaits a proper critical edition.

[4] Cross-references can be found in the natural philosophy of the *Šifāʾ*, for example in the *Kitāb al-Nafs* (*Liber de Anima* in Latin, *Book of the Soul* in English, henceforth *Nafs*), which is the psychological section of the *Šifāʾ*, and corresponds to Aristotle's *De anima*. There Avicenna twice mentions *medical books*, by which in all likelihood he refers to his *Qānūn*: 1) II, 4, 76.20 [146.21] (*fī l-kutub al-ṭibbiyya*) in connection with the classification of flavours (see *Qānūn*, III, 6, i, 2); and 2) III, 8, 156.15 [275.60] (*fī kutub al-ṭibb*) in connection with the causes of vertigo (see *Qānūn*, III, 1, v, 1). In addition to those mentions, in *Nafs*, IV, 4 Avicenna refers again to his *medical books* (*fī kutubinā l-ṭibbiyya*, 201.13 [67.70]); however, in this case he seems to refer to his *Maqāla fī l-adwiya al-qalbiyya* (*De Medicinis cordialibus*, or *De Viribus cordis* in Latin, *On Cardiac Remedies* in English). Moreover, in the *Qānūn* there are several passages in which Avicenna defers the settlement of controversies concerning philosophical issues to (natural) philosophy, since they go beyond the prerogatives of the physician. More on these passages in Alpina (2017), p. 376–77.

[5] See Avicenna, *Biography* 65.5–67.4. In another passage of Avicenna's biography it is said that he firstly composed the natural philosophy of the *Šifāʾ* with the exclusion of zoology, and the

Qānūn, and gave lessons to a certain number of students who every night gathered in his house, and read in parallel passages from the two works.[6]

On several occasions Avicenna claims that medicine is subordinated to natural philosophy.[7] In his *Maqāla fī Aqsām al-'ulūm al-'aqliyya* (*Treatise on the Divisions of the Intellectual Sciences*), a writing in which sciences are divided into fundamental (*aṣlī*) and derivative (*far'ī*), medicine is classified as a derivative natural science (*al-ḥikma al-ṭabī'iyya al-far'iyya*), which investigates the states of human body only in terms of health (*ṣiḥḥa*) and sickness (*maraḍ*), their causes, and their symptoms.[8] In a similar vein, in the opening lines of the *Qānūn*, in outlining the epistemological status of medicine, Avicenna maintains that its philosophical and epistemological underpinnings, mainly, but not exclusively, the theory and principles of humoral pathology, are given in natural philosophy – the theoretical science to which medicine is said to be subordinated –, and their investigation is declared off-limits to the physician.[9]

---

metaphysics, with no reference to botany; see Avicenna, *Biography* 58.7. The botanical and the zoological sections of the *Šifā'* can be considered to lie at the crossroads between natural philosophy and medicine. For the fact that Avicenna's zoology contains material transplanted from the first and the second book of the *Qānūn*, see Musallam (1987).

6 See Avicenna, *Biography* 54.4–56.1.

7 On Avicenna's hierarchical arrangement of sciences with metaphysics on top, see *Ilāhiyyāt*, I, 1–2. For a thorough analysis of these chapters, see Bertolacci (2006), p. 116–126. On the peculiar status of psychology in Avicenna's system of sciences, see Alpina (2021a) and Alpina (2018b).

8 Avicenna *Maqāla fī Aqsām al-'ulūm al-'aqliyya* (*Treatise on the Divisions of the Intellectual Sciences*, ed. in: *Rasā'il fī l-ḥikma wa-l-ṭabī'iyyāt*, 1980²) 110.7–10: "Divisions of the derivative natural science. Among these [divisions] there is medicine (*al-ṭibb*): its goal (*ġaraḍ*) is the knowledge of the principles of human body and its states in terms of health (*ṣiḥḥa*) and sickness (*maraḍ*), their (sc. of health and sickness) causes and their symptoms in order to repel sickness and preserve health." A similar passage can be found at beginning of the *Qānūn* (I, 1, i, 1, 33.8–9).

9 *Qānūn* I, 1, i, 2, 36.8–14 [fol. 1v, a41–55]: "Were a physician to begin discussing the proof [of the existence] of the elements, the temperament, etc. – all of these things being the subject (*mawḍū'* [*subiecta*]) of natural philosophy – he would commit an error insofar as he would introduce into the discipline of medicine something which does not belong to it, and he would commit another error insofar as he would be thinking that he is explaining something while [in reality] he will not have explained it at all. The things whose quiddity (*bi-l-māhiyya* [*quid sit*]) the physician has to conceive (*an yataṣawwarahū*), whereas he has to assume ([*an*] *yataqallada*) that they are (*bi-l-haliyya* [*utrum sit vel non sit*]), though their existence is not [immediately] clear, are the following: the elements, whether they are and how many they are; the temperaments, whether they are and how many they are; likewise, the humors, whether they are, how many they are and how they are; the faculties, whether they are, how many they are, and where they are; the pneumata, whether they are, how many they are, where they are; and if each [of them] has a transformation in [its] state or a stability due to some cause, and,

However, though required, the theoretical foundation of medical theories within a philosophical framework is far from being a straightforward operation: for sometimes natural philosophy and medicine are patently in conflict, and consequently it is hard to find in the former the appropriate theoretical premises of the latter. What is more, especially in the field of physiology and anatomy, the authority of Aristotle, to whom Avicenna refers as *al-muʿallim al-awwal* ("the First Teacher") and to whose works he refers as *al-taʿlīm al-awwal* ("the first teaching"),[10] is challenged by that of Galen, who is referred to by the epithet *fāḍil al-aṭibbāʾ* ("the excellent among the physicians"),[11] because Aristotle's knowledge of some issues, common to both natural philosophy and medicine, turns out to be too rudimentary to provide a solid, theoretical basis for medical understanding. In the *Qānūn*, Avicenna generally highlights the points of conflict between philosophers and physicians and explains them as the result of an illegitimate disciplinary trespassing on the part of physicians, who believe they are explaining something but, in reality, are transcending the boundaries of their discipline. Their settlement, however, is usually deferred to philosophy.[12] Conversely, in philosophy, notably in the *Šifāʾ*, when philosophical and medical perspectives on specific issues diverge, Avicenna grants primacy to the philosoph-

---

if there are more causes, how many they are" (the English translation is the one provided by Gutas 2003, p. 150, slightly modified). For the different levels of investigation belonging to physician and philosopher respectively, see *Qānūn* I, 1, vi, 1, 123.6–11 [fol. 23r, a40–53]. For the subordination of medicine to natural philosophy a crucial passage is *Burhān* II, 7, 163.14–20, where medicine is said to be subordinated to natural philosophy because it investigates the subject of the part of natural philosophy that deals with the human body, insofar as the latter is qualified by *health* and *sickness*, which are two *per se* accidents of the human body. For a discussion of this passage, see Strobino (2017), p. 111. See also *Ilāhiyyāt* I, 2, 14.18–15.3 [15.74–79].

10 Aristotle's being *first* refers not only to his chronological priority (al-Fārābī, for instance, is called *al-muʿallim al-ṯānī*, "the Second Teacher"), but also to his philosophical primacy. For Avicenna's use of these epithets, see Endress (1991a); Endress (1991b); Endress (1997); Gutas (2014), p. 325, n. 12–13.

11 An alternative formulation of this epithet is *al-ṭabīb al-fāḍil* ("the excellent physician"). The epithet by which Avicenna refers to Galen is similar to the epithets by which he refers to Alexander of Aphrodisias, namely *fāḍil al-qudamāʾ al-mufassirīna* ("the excellent one among the ancient commentators", *Nafs* III, 7, 149.5 [265.78]), and *fāḍil al-mutaqaddimīna* ("the excellent among the predecessors", *Ilāhiyyāt* IX, 3, 393.16–17 [464.92–93]). This can be considered as a sign of the high esteem in which Avicenna held Galen, who seems to be one of the most important and admired philosophers of the Peripatetic tradition.

12 See n. 4 above.

ical account of the issue, because (natural) philosophy provides the principles of medical investigation.[13]

Recently, two studies have tackled the question of the relationship between philosophy and medicine in Avicenna's thought by focusing on two topics with respect to which the two sciences overlap. In 2013, P. E. Pormann compared Avicenna's treatment of internal senses in the *Nafs* to that in the *Qānūn*. In the former, Avicenna lists five internal senses (*al-ḥawāss al-bāṭina*), namely *common sense* (*al-ḥiss al-muštarak* or *banṭāsiyā/fanṭāsiyā*), *imagery* or *form-bearing faculty* (*al-muṣawwira* or *al-ḫayāl*), *imaginative* or *cogitative faculty* (*al-mutaḫayyila* or *al-mufakkira*), *estimation* (*wahm/al-wahmiyya*), and *memory* or *recollective faculty* (*al-ḥāfiẓa* or *al-mutaḏakkira*).[14] In the latter, by contrast, as to the internal cognitive faculties (*quwà mudrika fī l-bāṭin*) falling under the more general category of psychic faculties (*quwà nafsāniyya*) and belonging to human beings,[15] Avicenna distinguishes the position of physicians, according to which they are three (*al-ḥiss al-muštarak* or *al-ḫayāl*, imagination; *al-mutaḫayyila* or *al-mufakkira*, cogitation; *al-ḥāfiẓa* or *al-mutaḏakkira*, memory), and that of philosophers, according to which they are five.[16] To settle the doctrinal conflict, Avicenna refers to the specific and limited goal of the physician: unlike the philosopher, the physician considers only the faculties that can be affected in their functions by the occurrence of damage, whereas with the other faculties he limits himself to acknowledging their existence.[17] On the basis of this comparison, P. E. Pormann concludes that "Avicenna's medical ideas were heavily influenced by his philosophy" (Pormann 2013, p. 102–107).

In 2014, A. Tawara argued – wrongly, as we shall see – that in the botanical section of the *Šifā'*[18] Avicenna denies that plants, though possessing a soul, possess also life.[19] The author explains Avicenna's unique and surprising denial of

---

[13] Examples of this philosophical praxis can be found, for instance, in *Nafs* V, 8, and *Ḥayawān* III, 1. For a thorough analysis of these chapters, see Alpina (2021b).
[14] See, for instance, the list of internal senses provided in *Nafs* I, 5, 44.3–45.16 [87.19–90.60], and in IV, 1, and their thorough analysis in *Nafs* IV, 2–3. For a study of Avicenna's theory of internal senses and its background, see Alpina (2020a).
[15] More on this terminology in the medical context, *infra*.
[16] *Qānūn* I, i, vi, 5, 128.17–129.30 [fol. 24v, b41–25r, b26].
[17] *Qānūn* I, i, vi, 5, 129.13–20 [fol. 25r, a42–b4]. For the fact that faculties are among those things whose existence the physician must assume, see the passage quoted in n. 9 above.
[18] More on this work, *infra*.
[19] Here I preliminarily say that A. Tawara's reconstruction is based on a sentence drawn from the first chapter of Avicenna's botany, which he wrongly considers as providing Avicenna's positive position on plant life (see p. 129: "Surprisingly, in his treatise *On Plants*, the seventh book of the section on Natural Science in his work *al-Šifā'*, Avicenna declares, 'It is impossible to attrib-

life in plants in botany by referring to a passage from the *Qānūn*, where life is associated with the *vital faculty* (*quwwa ḥayawāniyya*)[20] situated in the cardiac pneuma. According to Tawara's reconstruction, given that plants lack this faculty, and given that in the *Qānūn* the possession of natural/vegetative faculties is said not to be a sufficient condition for a thing to be considered living, Avicenna is forced by his medical beliefs to argue that plants are lifeless. A. Tawara therefore credits with a more decisive role "the influence of Avicenna's medical findings on his philosophical arguments when examining the development of his thought" (Tawara 2014, p. 127). This interpretation has been recently endorsed by Aileen Das.[21]

Though I do not entirely agree with all their conclusions, these studies have the merit of pointing out two crucial and complementary aspects of the relationship between (natural) philosophy and medicine in Avicenna: on the one hand, the perfect subordination of the latter to the former in the case of a specific doctrine and, on the other hand, the radical divergence of their respective theoretical frameworks. An example of the first aspect is the treatment of internal senses: the selective medical interest in and, consequently, the specific treatment of some of them perfectly integrate into the more comprehensive philosophical account of them. An example of the second aspect is, by contrast, the different taxonomies of the faculties of the soul provided both in medicine and in natural philosophy, to be precise in psychology, as the theoretical background against which every faculty is introduced: for instance, no reference to the *vital faculty* (*quwwa ḥayawāniyya*), which is mentioned in medical context, can be found in psychology. What is more, in medicine there seems to be no reference to a fully-developed concept of *soul* (*nafs*).

In this connection, nutrition offers an unparalleled case study for untangling Avicenna's thought: for it is an issue dealt with in both natural philosophy (psy-

---

ute life to plants in any way (*lā yağūzu an tajʻala li-al[sic]-nabāt ḥayāt bi-wağh min al-wuğūh*)'. He believes them to be lifeless, although he admits that they possess soul."). However, the sentence quoted by Tawara does not express Avicenna's own position (which Tawara qualifies as "dramatic shift", p. 131); rather, it is the apodosis of a conditional sentence (in his translation Tawara omits to translate *fa-* (*then*), which introduces the quoted sentence). What is more, this conditional sentence represents one of two alternative positions about life in plants, to none of which Avicenna seems to immediately commit himself (at least at this level of the argument). For a thorough analysis of this passage, see *infra*.
20 In his article A. Tawara translates *ḥayawāniyya* as *animal*, which is of course a possible translation. However, I think that in medical context the translation of *ḥayawāniyya* as *vital* is preferable since it marks the theoretical discontinuity between medicine and natural philosophy with respect to the classification of the faculties of the soul.
21 See Das (2017). More on the interpretation advanced in this article, *infra*.

chology and botany) and medicine and, what is more, it represents an intermediate case between proper subalternation and mere intersection. On the doctrinal level, the account of nutrition provided in medicine is perfectly integrated into that offered in natural philosophy: medicine can be said to supplement natural philosophy with the physiology of nutrition. Nonetheless, these two accounts seem to presuppose different theoretical backgrounds, which prevents the former, i.e. that provided in medicine, from being entirely grounded on the latter, i.e. that provided in philosophy.

2 Avicenna's Account of Nutrition

In the *Nafs* Avicenna deals *ex professo* with nutrition (*tagḏiya, nutrire*) in two chapters, i.e. I, 5 and II, 1. *Nafs* I, 5 provides a survey of all the psychic faculties (and their activities) that will be treated in this writing.[22] These faculties are grouped into three main divisions, corresponding to the three kinds of soul that are traditionally associated with them, namely (i) vegetative soul (*nafs nabātiyya, anima vegetabilis*); (ii) animal soul (*nafs ḥayawāniyya, anima vitalis vel sensibilis*);[23] (iii) human soul (*nafs insāniyya, anima humana*). Though abiding by the established tradition of enumerating and then presenting the psychic faculties instead of the souls themselves,[24] Avicenna also envisages the possibility of dealing with the souls themselves, not with the psychic faculties, by focusing on the arrangement of souls according to degree of perfection, and by making the lower soul a genus for the higher one.[25] This possibility, however, is not ac-

---

[22] *Nafs* I, 5, 39.13–14 [79.3–4]: "Let us now enumerate the faculties of the soul by way of convention ('alà sabīl al-waḍ' [quasi ponendo]), then we shall engage in the clarification of the state of every faculty (ṯumma li-naštagil bi-bayān ḥāl kull quwwatin [deinde procedemus ad declarandum unamquamque illarum (sc. virium)])." A similar survey of psychic faculties can be found in all other Avicennian *summae*. For a comparison of these surveys, see Alpina (2021a), Chapter 6.

[23] On the possible translations of the term *ḥayawāniyya*, see n. 20 above. For the discussion of this faculty see § 3.1 below.

[24] *Nafs* I, 5, 39.14 [79.4–5]: "We say, thus, that the psychic faculties are divided, according to the primary division, into three parts." The tripartition of the soul is common to both Platonic and Aristotelian tradition. The primacy granted to faculties and their activities over the soul itself in the psychological investigation echoes *de An.* I, 1, 402b9–16; and II, 4, 415a14–22, where Aristotle maintains that the investigation of the correlative objects must come first, and then the activities and the parts of the soul responsible for them must follow.

[25] *Nafs*, I, 5, 40.4–13 [80.17–81.28]: "If there were not [such a] custom, it would be best to make every [perfection coming] first a condition that is mentioned in the description of the following

tualized, since the purpose that Avicenna explicitly assigns to psychology *qua* part of natural philosophy is to treat the soul as an operational principle of activities observable in bodies by focusing – in the Aristotelian fashion – on its faculties.[26]

Following the established tradition, the vegetative soul is said to have three faculties: (i) the nutritive faculty (*quwwa ġāḏiya, vis/virtus nutritiva*); (ii) the faculty of growth (*quwwa munammiya, vis/virtus augmentativa*); (iii) the generative faculty (*quwwa muwallida, vis/virtus generativa*), whose respective activities are concisely outlined (40.14 – 41.3 [81.29 – 82.39]). Concerning the nutritive soul, Avicenna briefly says that it

> is a faculty that transforms a body different from the body in which it is (*sc.* nourishment) is into something similar to the body in which it is (*sc.* the nourished body), and attaches it to the body [in which it is] as a compensation for what dissolves from it (40.14 – 16 [81.29 – 82. 32]).[27]

[perfection], if we want to describe the soul, not the psychic faculty belonging to it in accordance with that activity. For 'perfection' is included in the definition of the soul, not in the definition of the faculty of the soul. You will learn the difference between the animal soul and the faculty of perception and of setting in motion, and the difference between the rational soul and the faculty concerning the aforementioned things with respect to discernment, etc. If you want a close scrutiny, the right thing [to do] would be to make the vegetative [soul] a genus for the animal [soul], and the animal [soul] a genus for the human [soul], and to include the more general in the definition of the more specific. However, if you take into consideration the souls insofar as they have specific faculties in their animality and in their humanity, then you may be satisfied with what we have mentioned (*sc.* the traditional tripartition mentioned in 39.14 above)." Here, however, the term *genus* needs not to be taken in the technical sense as to imply that for Avicenna souls can be arranged according to genera and species. In *de An.* II, 3, 414b28 – 33, Aristotle maintains that souls are ordered in series in the very same way in which rectilinear figures are, and that the lower soul/figure can be potentially found in the higher one. However, Aristotle does not say that the lower soul, e. g. the vegetative soul, can be considered – not even in a non-technical sense – a genus for the higher soul, e. g. the animal soul. On the presence of this Aristotelian theme in Avicenna, see Alpina (2018a), and Alpina (2021a), Chapter 3, n. 35.

26 More on Avicenna's global project in the *Nafs* in Alpina (2018a), (2018b), and (2021a), Chapter 3 and 4.

27 Cf. *Nafs* I, 5, 39.15 – 18 [79.5 – 80.10]: "One of them is the vegetative soul, which is the first perfection of a natural, organic body in virtue of the fact that it generates [the like], grows, and nourishes itself. The nourishment is a body such that it becomes like the nature of the body of which it is said to be the nourishment. It adds to the body the exact quantity of what has been dissolved, or more or less."

Nafs I, 5 is the last chapter of the first treatise, where Avicenna provides the theoretical basis of the psychological investigation he is embarking on. From the second treatise onwards Avicenna engages in "determining the state of the aforementioned faculties", as has been announced at the beginning of I, 5.[28] As happens in the Aristotelian treatise, the first faculty to be dealt with is nutrition: "the first of these [activities to be dealt with] are the activities of the vegetative soul, and the first among them is the state of nutrition (II, 1, 52.5 [103.5–7])", to which Avicenna devotes the first part of *Nafs* II, 1 (52.5–53.2 [103.5–105.30]).

The treatment of nutrition is immediately integrated into the more general account of increase (or growth) provided in *Kawn wa-Fasād*[29] 8, which is entitled *Chapter on the discourse about increase (Faṣl fī l-kalām fī l-numuww [⟨Capitulum⟩ de sermone augmenti])* and which corresponds to Aristotle's *de Generatione et Corruptione* I, 5. There, Avicenna says, the relation of nourishment to what is nourished, their definition and characteristics, as well as the process of transformation of nourishment into blood and humors in the case of sanguineous animals, have been already explained.[30] Therefore, the reader is somehow exhorted to look at that chapter in order to get more information about these topics.[31]

Avicenna then refers to the two major activities of this faculty: the replacement of what dissolves from the body as a consequence of the activities in which it is engaged, and the procurement of what is necessary for growth. The first ac-

---

28 *Nafs* II, 1, 52.4–5 [103.4–5]: "Let us begin determining the state of the aforementioned faculties one by one (*fal-nabdaʾ bi-taʿrīf ḥāl al-quwà l-maḏkūra quwwatan quwwatan* [*incipiemus nunc notificare sigillatim virtutes quas diximus*]), and let us determine them with respect to their activities (*wa-l-nuʿarrifhā min ǧihat afʿālihā* [*et demonstrabimus eas ex suis actionibus*])." See also the passage quoted in n. 22 above.
29 It is the third section of the part on natural philosophy of the *Šifāʾ*, and corresponds to Aristotle's *GC*.
30 See *Kawn wa-Fasād* 8, 144.11–146.9 [84.31–87.91].
31 See the following references that Avicenna makes in *Nafs* II, 1: "you have already learned [this] in what precedes" (*qad ʿalimta fīmā salafa*, 52.6 [*ex praemissis cognovisti*, 103.8]); "according to what we have shown elsewhere" (*ʿalà mā bayyannā fī mawāḍiʿ uḫrà*, 52.12–3 [*sicut iam ostendimus alias*, 104.17–8]). The reference, though not explicit, to *Kawn wa-Fasād*, fits with the conclusive remark in *de An.* II, 4, where Aristotle suggests to look at the appropriate works for a more exhaustive account of nourishment than the outline provided in *de An.*: τύπῳ μὲν οὖν ἡ τροφὴ τί ἐστιν εἴρηται· διασαφητέον δ' ἐστὶν ὕστερον περὶ αὐτῆς ἐν τοῖς οἰκείοις λόγοις (II, 4, 416b30–1). Assuming that Aristotle never wrote a writing on nutrition, or that he only planned to write it (see *De Somno et Vigilia* 3, 456b5–6: εἴρηται δὲ περὶ τούτων ἐν τοῖς Περὶ τροφῆς), or that, if he wrote it, it was lost, some commentators interpret Aristotle's comment as referring to *GC* (in all likelihood to I, 5), and/or to *de Generatione Animalium* (possibly to II, 4) (see Philoponus, *in de Anima*, ed. by Hayduck 1897, 289.2–7; Ps.-Simplicius, *in de Anima*, ed. by Hayduck 1882, 116.16–17).

tivity is referred to as the greatest benefit of this faculty (*akṯar manāfiʿihī*, 52.16 [*eius plures utilitates*, 104.22]), whereas the second one is limited in time, and lasts until growth is brought to completion. Lastly, Avicenna points out the peculiarity of the nutritive faculty: unlike the rest of the vegetative faculties it is the only one to perform its activity throughout the individual's existence and, as long as its activity is performed, the individual continues to exist.[32] This statement chimes with what Avicenna says afterwards, namely that the nutritive faculty is aimed at preserving the individual, the faculty of growth at completing its substance, and the generative faculty at guaranteeing the continuance of the species (54.18–55.5 [108.78–86]).[33]

Avicenna does not delve into the manner in which nutrition occurs: he concisely says that nourishment firstly undergoes a transformation from its own quality and, then, is disposed to be transformed into the substance of what is nourished, that the digestive faculty,[34] one of the servants of nutrition (*quwwa min ḥadam al-quwwa al-ġāḏiya wa-hiya l-hāḍima*, 52.9–10, [*una ex virtutibus servientibus virtuti nutritivae, quae est digestiva*, 104.13–14]), is responsible for the transformation of nourishment, and that every organ has its own nutritive faculty which is responsible for the transformation of nourishment into what is similar to the temperament of that organ. Moreover, once the faculty of growth and the generative faculty have been presented (53.2–54.18 [105.30–108.77]), and the wrong opinions of the predecessors on nutrition have been refuted (55.5–19 [108.87–110.4]),[35] Avicenna briefly adds that the nutritive faculty has four instruments (*āla*, sg. [*instrumentum*]): the active ones, i.e. (i) hotness (*ḥārr* [*calor*]), i.e. innate heat, to set matters in motion, and (ii) coldness (*bard* [*frigiditas*]) to quiet-

---

[32] Here Avicenna makes no reference to respiration and blood circulation, which seem to be the most basic vital activities. In the *Qānūn* they seem to be associated with the vital faculty. More on the vital faculty and its connection with life in § 3.1.

[33] For the Aristotelian background of Avicenna's position, see *de An.* II, 4, 415a26–b7.

[34] For the reference to digestion in Aristotle's account of the nutritive faculty, see *de An.* II, 4, 416b25–30.

[35] The opinions of the predecessors that Avicenna presents here are the same as those presented by Aristotle in *de An.* II, 4, 415b28–416a18, namely the opinion of Empedocles, who explained the growth of plants downwards (e.g. the roots) and upwards (e.g. the branches) by arguing that their nature is earthy and fiery; and the opinion of those who believe that the cause of nutrition is fire. However, two major differences from the Aristotelian text are detectable: (i) firstly, the opinions are presented in an inverted order (fire and Empedocles, who however is not mentioned by name); (ii) secondly, and this might be relevant to Avicenna's method in dealing with doxography, both opinions are duly analyzed: in each of them Avicenna singles out two specific mistakes (one already pointed out by Aristotle, the other added by Avicenna, but not fully developed), and refutes them.

en them; and the passive ones, i.e. (iii) moistness (*ruṭūba* [*humiditas*]) to shape matters, and (iv) dryness (*yubūsa* [*siccitas*]) to preserve their shape (55.19–56.4 [110.5–10]).

That the nutritive faculty has some subordinate faculties and four instruments, together with the internal distinction of the faculty responsible for reproduction into generative and formative that is introduced in the *Qānūn*, represents a medical, that is, Galenic, integration of Avicenna's fundamentally Aristotelian account of the vegetative soul. The integration of medical elements within an Aristotelian account of psychic faculties clearly, though cursorily, emerges at the end of *Nafs* I, 5. There Avicenna arranges the psychic faculties into a hierarchy, where each faculty rules (*ra'usa* [*imperare*]) the faculty that follows and serves (*ḫadama* [*deservire/famulari/servire/subesse*]) the one that precedes (50.13–51.16 [99.79–102.15]). In the case of the nutritive faculty, Avicenna distinguishes four subservient faculties, which are called 'natural faculties' (*al-quwà l-ṭabīʿiyya al-arbaʿ*, 51.10 [*quattuor virtutes naturales*, 101.6–7]) by borrowing Galen's terminology: digestive (*hāḍima*, *digestiva*), attractive (*ǧāḍiba*, *attractiva*), retentive (*māsika*, *retentiva*), expulsive (*dāfiʿa*, *expulsiva*). These faculties, in turn, are said to be served by the four elementary qualities (*al-kayfiyyāt al-arbaʿ*, 51.12 [*quattuor qualitates*, 101.10]): hot, cold, moist, and dry.[36]

Thus, the combination of information provided in *Nafs* I, 5 and II, 1 represents the most exhaustive (basically, though not exclusively, Aristotelian) account of nutrition that can be found in Avicenna's philosophical writings, because there are no specific writings on either vegetative soul or nutrition, and the account of nutrition provided in the psychological section of Avicenna's other *summae* is even more concise than the one provided in the *Šifāʾ*.[37]

However, a supplement to the account of nutrition of the *Šifāʾ* can be found in the *Qānūn*. For there Avicenna provides a survey of all psychic faculties, not because their general investigation pertains *per se* to the physician, but because the knowledge of their functioning and, in particular, of the organs in which they are located, supplies to the physician the necessary theoretical background to

---

[36] For the Galenic background of these four subservient faculties, see *De facultatibus naturalibus* III, 4–8, where Galen distinguishes four powers (δυνάμεις) belonging to plant and animal bodies, the actions of which account for biological functions such as nutrition. These powers are: attractive, retentive, alterative, expulsive. It is worth mentioning that Avicenna composed a brief treatise entitled *Refutation of Ibn al-Ṭayyib's Treatise* On the Natural Faculties (*Fī Naqḍ Risālat Ibn al-Ṭayyib fī l-quwà l-ṭabīʿiyya*), which is a refutation of Abū l-Faraǧ Ibn al-Ṭayyib's (d. 1043) essay on the four natural faculties, which are precisely the bodily functions of attraction, retention, transformation (or digestion), and expulsion.

[37] See Alpina (2021a), Chapter 6.

deal with the diseases that might affect them, of course in the case of human beings.[38]

Qānūn I, 1, vi is entirely devoted to psychic faculties. It consists of six chapters. In the first chapter, which is entitled *Chapter on the genera of faculties by means of a general discourse* (*Faṣl fī aǧnās al-quwà bi-qawl kullī* [*De generibus virtutum secundum sermonem universalem*, fol. 23r, b6–7]), Avicenna lists the psychic faculties, the activities ensuing from them, and their internal subdivisions, to which the subsequent five chapters are devoted. As in *Nafs* I, 5, the psychic faculties are arranged in a triadic scheme. However, this tripartition is not that of the philosophers, but rather that of the physicians (*'inda l-aṭibbā'*, 122.19 [*apud medicos*, fol. 23r, b13]). The distinction between the two is immediately detectable. Firstly, unlike the tripartition outlined in *Nafs* I, 5, where the soul is immediately subdivided into three parts (*aqsām ṯalāṯa*, 39.14 [*tres partes*, 79.5]) distinguished by their specific faculties, in the *Qānūn* there is no reference to the notion of soul (*nafs*, *anima*) as the principle from which faculties ensue. Rather, here faculties are divided into three genera (*aǧnās*, 122.18 [*genera*, fol. 23r, b12]), standing on their own with their own activities. Secondly, in naming the different genera of faculties, Aristotelian terminology is replaced by the Galenic one: instead of vegetative, animal, and human soul, there are the genus of psychic faculties (*al-quwà l-nafsāniyya*, *virtutes vitales*), the genus of natural faculties (*al-quwà l-ṭabīʿiyya*, *virtutes naturales*), and the genus of vital faculties (*al-quwà l-ḥayawāniyya*, *virtutes animales*). Lastly, these genera of faculties seem to be arranged in a random rather than a hierarchical order.

Although the activities for which these three genera of faculties are responsible are briefly sketched, in this introductory chapter Avicenna's interest is primarily medical, that is, focused on the bodily seats of these faculties in which diseases might occur. The majority of philosophers and the totality of physicians, among whom the figure of Galen stands out (*wa-kaṯīr min al-falāsifa wa-ʿāmma al-aṭibbā' wa-ḫuṣūṣan Ǧālīnūs*, 122.21 [*et multis quidem philosophorum et medicis omnibus*, fol. 23r, b15–6]), assign to each (genus of) faculty a chief organ (*ʿuḍw ra'īs*, 122.22 [*membrum principale*, fol. 23r, b17]), namely the brain to psychic faculties, the liver and testicles to, respectively, the nutritive and the generative faculty among natural faculties, and the heart to the vital faculties. Despite this apparent agreement, the major point of conflict between philosophers and physicians is the general anatomical model: for Aristotle's cardiocentrism grants primacy to the heart over any other organ, and to the connection of the soul with the cardiac pneuma at the moment of generation, whereas Galen's tripartite

---

38 See n. 17 above.

model maintains that there is no primary organ in the body; rather the heart, liver, and brain are all on equal footing.[39]

The underlying theoretical framework of Avicenna's treatment of psychic faculties in the *Qānūn* and its relation to that provided in the *Šifā'* will be expounded in detail in due course. Preliminarily, however, the account of the natural faculties, with a particular focus on nutrition, which is provided in the *Qānūn* will be analyzed.

First of all, the natural faculties, whose scope is broader than that of the Aristotelian vegetative faculties, are immediately divided into serving (*ḫādima* [*ministrantes*]) and served (*maḫdūma* [*ministratae*]) faculties. The former are treated in the third chapter, whereas the latter in the second. The served faculties, which correspond to Aristotle's vegetative faculties, are in turn divided into those aimed at preserving the individual's existence (*li-baqā' al-šaḫṣ* [*ad hoc ut individuum remaneat*]), and those aimed at preserving the species' existence (*li-baqā' al-naw'* [*ad hoc ut remaneat species*]).[40] To the first group belong the nutritive faculty (*al-ġāḏiya* [*v. nutritiva*], also named *taṣarruf fī l-ġiḏā'* [*ministrare in re nutrientis*])[41] and the faculty of growth (*al-nāmiya* [*v. crescitiva*]), whereas to the second group the generative (*al-muwallida* [*v. generativa*]), and the formative (*al-muṣawwira* [*v. formativa*])[42] faculty, namely the two faculties in which physicians subdivide the more general faculty of reproduction.

As in the *Nafs*, nutrition is described as the faculty that transforms nourishment into the likeness of what is nourished and replaces what dissolves from the body as a consequence of the activities in which it is engaged. Then the functioning of this faculty is dealt with in detail. The activity of nutrition is said to be accomplished by means of three specific activities (*bi-afʿāl ġuzʾiyya ṯalāṯa* [*tribus particularibus operationibus*]): (i) attainment of the substance of the body (*taḥṣīl ġawhar al-badan* [*permutatio substantiae corporis*]), that is, the nourishment being firstly transformed into blood and humor; (ii) adhesion (*ilzāq* [*adherentia*]), that is, the transformed nourishment becoming part of a specific organ; (iii) assimilation (*tašbīh* [*similitudo*]), that is, the nourishment becoming exactly

---

39 See *Nafs* V, 8, and *Ḥayawān* III, 1, where Avicenna discusses the positions of philosophers and physicians on this issue. For a thorough analysis of these chapters, see Alpina (2021b).
40 Here Avicenna's significant use of diairetic method is clearly detectable.
41 In *de An.* II, 4 the nutritive faculty is also called τροφῇ χρῆσθαι (415a26).
42 This passive faculty of reproduction is also called *al-ṭābiʿa*, a name that stresses its similarity with the disposition of matter or of an underlying nature (*ṭabīʿa*) to receive the form. It is noteworthy that the term *muṣawwira*, by which here Avicenna refers to the formative faculty, is the same term by which in the Nafs Avicenna refers to one of the internal senses, namely the form-bearing faculty, also called imagery (*ḫayāl*). For the same consideration, see Hall (2004), p. 82.

the same as the organ with which it is united. A defect in one of these sub-processes of nutrition causes a specific disease: the failure of *taḥṣīl* causes atrophy (*aṭrūfiyā*, which is also referred to as '*adam al-ġiḏā*', deficiency of nourishment), the failure of *ilzāq* causes fleshy dropsy (*istisqā' laḥmī*), i.e. dropsy of bodily tissues, and the failure of *tašbīh* causes illnesses like leprosy (*baraṣ*) and vitiligo (*bahaq*). The entire process of transformation and assimilation of nourishment into a certain organ is said to be ruled by the transformative faculty (*al-quwwa al-muġayyira*), a single faculty branching out from the liver to every organ, where the nourishment is transformed into the specific temperament of that organ.

In order for nutrition to perform its specific activities, it needs the assistance of four subordinate faculties: the attractive faculty (*ǧāḏiba*), which attracts nourishment to the body; the retentive faculty (*māsika*), which keeps the nourishment within the body while it awaits transformation; the digestive faculty (*hāḍima*), which is responsible for digestion; and, the expulsive faculty (*dāfi'a*), which expels the waste resulting from digestion. These subordinate faculties are in turn served by the four elementary qualities (*al-kayfiyyāt al-arba' al-ūlà*), i.e. heat (*ḥarāra*), coldness (*burūda*), moistness (*ruṭūba*), and dryness (*yubūsa*), which directly or indirectly contribute to the activities of one or more subordinate faculties.[43]

On the whole, the account of nutrition provided in the *Qānūn* seems to integrate perfectly with that provided in the *Nafs*. For, the *Qānūn* deals in detail with the functioning of nutrition, and the activities of the subordinate faculties and of the four elementary qualities, which are essential to fully understand the way in which the nutritive faculty works but have been just hinted at, or completely disregarded in the *Nafs*. That being said, however, nothing can be inferred about what the relation of nutrition to life is, if any, according to Avicenna. In the context of the *Nafs*, however, this might be unsurprising: for, in line with Aristotle's view,[44] Avicenna considers the soul to be the principle of the activities of life in organic bodies, nutrition included, and perhaps it would have been trivial to specify its connection with life.

---

[43] For the Galenic background of the doctrine of the four subordinate faculties and the four elementary qualities as their instruments, see n. 36 above.
[44] See n. 26–27 above.

3 Is Nutrition a Sufficient Condition for Life?

3.1 A Breath of Life: Animals and Humans

At the beginning of *Nafs* I, 1 Avicenna says that the soul is the principle for a set of activities observable in bodies, which traditionally distinguish living beings from lifeless ones:

> We thus say: we do sometimes see bodies that sense (*aḥassa*) and move at will (*taḥarraka bi-l-irāda*); indeed, we see bodies that nourish themselves (*iġtaḏà*), grow (*namà*) and generate their like (*wallada l-miṯl*). And this does not belong to them due to their corporeality; therefore, it remains that in these themselves there are principles for that other than their corporeality, that is, the thing from which these activities derive. *In general, whatever is a principle for the derivation of activities that are not in the same manner [as if they were] devoid of will, we call 'soul'.*" (*Nafs* I, 1, 4.5–10 [14.71–15.78], emphasis mine)[45]

Though among the activities of life for which the soul is said to be responsible, both the activities common to plants and animals (nutrition, growth, and reproduction) and those peculiarly distinguishing animals (sensation and voluntary motion) are mentioned,[46] the sentence with which the passage ends seems to restrict those activities to the voluntary ones (*kull mā yakūnu mabda' li-ṣudūr afāʿīl laysat ʿalà watīra wāḥida ʿādima li-l-irāda fa-innā nusammīhu nafsan*, 4.9–10 [*quicquid est principium emanandi a se affectiones quae non sunt unius modi et sunt voluntariae*, 15.77–78]). This would exclude the activities that do not involve volition, like nutrition, growth and reproduction, from the group of those belonging to what is ensouled and, consequently, living.

The distinction that seems to be implied here, namely that between voluntary and involuntary activities, has its theoretical background in *Samāʿ ṭabīʿī* I,

---

[45] It is noteworthy that in a branch (B) of the Latin tradition the translation of this passage conveys a different meaning from that of the Arabic text, namely it explicitly distinguishes living beings from lifeless ones on the basis of the activities common to both plants and animals: *Et dicemus quod nos videmus corpora quaedam quae* non *nutriuntur* nec *augmentantur* nec *generant; et videmus alia corpora quae nutriuntur et augmentantur et generant sibi similia,* [...] (14.71–4, and n. 71–74, emphasis mine).

[46] This list of activities echoes that provided in *de An.* II, 2, 413a20–25, where Aristotle says that all them are instances of life, which is precisely the notion distinguishing what is ensouled from what is not (διωρίσθαι τὸ ἔμψυχον τοῦ ἀψύχου τῷ ζῆν).

5.[47] There, before commenting upon Aristotle's definition of nature and a later reformulation of it, Avicenna distinguishes, among activities (and motions) occurring in bodies, (i) those resulting from an external cause from (ii) those proceeding from the bodies themselves, which however – explains Avicenna, in a way similar to the aforementioned passage of *Nafs* I, 1 – do not derive from the body *qua* body, but from a power in it (*bi-quwwa fīhi* [lacuna in the Latin translation]).

This power in the body is in turn subdivided into two types: (ii.i) uniform, i.e. capable of producing one single effect, and (ii.ii) non-uniform, i.e. capable of producing contrary effects.[48] Both uniform and non-uniform powers can perform their activities (and motions) through volition or without volition. The uniform power producing its effect without volition is nature (*ṭabīʿa*); the same power producing its effect through volition is celestial soul (*nafs falakiyya*); the non-uniform power producing its effects without volition is vegetative soul (*nafs nabātiyya*); and the same power producing its effects through volition is animal soul (*nafs ḥayawāniyya*). Sometimes – adds Avicenna – both vegetative and animal soul can be called nature, but nature in the proper sense is only the first one.[49] For, although both the notion of nature and that of soul can be included within that of power (*quwwa*) since they both are principles for activities (and motions) in bodies, Avicenna aims at distinguishing them: the former is found in inanimate beings, like minerals, whereas the latter in living beings, like plants and animals.

At the beginning of *Nafs* I, 1, however, the term *soul* seems to designate only the principle of animals (and, perhaps, of celestial bodies) with the exclusion of plants, since plants do not perform their activities through volition. This perspective jibes with the prologue to the *Nafs*. There, in pointing out what distinguishes animated bodies from inanimate ones, Avicenna refers exclusively to sensation and voluntary motion, which is problematic for the inclusion of plants into the roster of ensouled, living beings:[50]

---

47 *Samāʿ ṭabīʿī* is the first section of the second part, i.e. that on natural philosophy, of the *Šifāʾ*, and corresponds to Aristotle's *Physica*. For a thorough analysis of *Samāʿ ṭabīʿī* I, 5, see Lammer (2018), p. 213–306.
48 This echoes the Aristotelian distinction between irrational, one-way capacities and rational, two-way capacities in *Metaphysica* IX, 2, 1046a36–b7.
49 In *Samāʿ ṭabīʿī* IV, 9 Avicenna says that by 'natural power' (*quwwa ṭabīʿiyya*) he refers to every power belonging to a thing that produces motion without volition, being it either nature in an absolute sense (*ṭabīʿiyya ṣirfan*), or the soul of plants (*nafs al-nabāt*).
50 This distinction cannot be dismissed as inaccurate. For, in order to include in the group of animals those that do not have local motion, Avicenna extends the notion of 'voluntary motion'

It remained for us to deal with generated things (*wa-baqiya lanā an natakallama 'alà l-umūr al-kā'ina* [*remanserat autem ut post haec loqueremur de rebus generatis*]), whereas the inanimate bodies and what has neither sensation nor voluntary motion (*fa-kanāt al-ǧamādāt wa-mā lā ḥiss lahū wa-lā ḥaraka irādiyya* [*sed quia res congelatae et insensibiles et quae non habent motum voluntarium*]) are prior to them and closer than them to elementary generation, and we dealt with that [issue] in the fifth section (*sc.* in *Ma'ādin wa-Āṯār 'ulwiyya*). (*Nafs*, prologue, 1.9–11 [9.12–10.15])[51]

Therefore, although in the prologue to the *Nafs* plants are mentioned together with animals among the generated things that are going to be dealt with specifically in the botany and the zoology of the *Šifā'* respectively,[52] and in *Nafs* I, 1 soul is said to be the part of the subsistence of plants as well as of animals through which they are what they are in actuality,[53] from the combination of the aforementioned passages life seems to be primarily associated with sensation and locomotion, and soul seems to be primarily conceived as a principle for the activities peculiarly distinguishing animal life. By contrast, vegetative activities, though ensuing from a vegetative soul, as maintained in *Nafs* I, 5 and II, 1, are apparently below the threshold of what is sufficient to account for life, and to some extent outside of that for which a soul is said to be primarily responsible.[54] This scenario can be compared to that of Aristotle's *de An.*, which can be considered a treatise mainly on animal psychology.[55] There Aristotle seems to

---

which, together with sensation, properly distinguishes animals, to the motion of contraction and expansion. See *Nafs*, II, 3, 68.6–19 [132.7–133.24].

51 On this passage and, in general, on Avicenna's prologue to the *Nafs*, see Alpina (2018a).
52 *Nafs*, prologue, 3.9–11 [13.59–62].
53 *Nafs* I, 1, 5.3–6, 1 [16.87–18.10].
54 This could be confronted with a passage from the commentary on Plato's *Timaeus* where Galen raises the same question. More on this passage in Wilberding (2014).
55 In Pellegrin (1996), P. Pellegrin claims that in the *de An.* Aristotle reduces the general notion of soul to the animal soul: for instance, in *de An.* III, 3, 427a17–19 Aristotle says that the soul is primarily defined by locomotion and perception, being the latter the general capacity of discriminating ('Επεὶ δὲ δύο διαφοραῖς ὁρίζονται μάλιστα τὴν ψυχήν, κινήσει τε τῇ κατὰ τόπον καὶ τῷ νοεῖν καὶ φρονεῖν καὶ αἰσθάνεσθαι [...]). Locomotion and perception are the powers by which the animal is defined. Therefore, the *De Anima*, Pellegrin concludes, is neither a treatise of general biology which deals with all the instances of sublunary life progressively more articulated, nor a treatise whose ultimate subject is the human soul; rather, it is a treatise of general zoology or 'psychologie naturelle' (Pellegrin 1996, p. 470–71). For the central role played by sensation in Aristotle's *de An.*, see also Giardina (2009), p. 76, n. 7. A different position is argued by A. Falcon, who maintains that, even though the *zoological orientation* of the *de An.* cannot be disputed, that is to say the *de An.* does provide the explanatory resources and the conceptual framework for an optimal study of animal life, Aristotle cannot consider his investigation of the soul preliminary only to the study of animals. See Falcon (2010), p. 168–69.

deal with plants not because of a genuine interest in plant life,[56] but because of the functional analogy with animals that plants exhibit on the physiological level: in fact, both plants and animals share in nutrition, growth, and reproduction, namely in the activities that are minimally constitutive of sublunary life (Repici 2000, p. 17).

The *zoological orientation*[57] detectable at the outset of Avicenna's psychological investigation emerges also at the end of the work. In the last chapter of the *Nafs*, in showing what the instruments of the soul are,[58] Avicenna says that

> the primary vehicle of psychic faculties in the body (*al-quwà l-nafsāniyya l-badaniyya maṭiyyatuhā l-ūlà, virtutum animalium corporalium vehiculum*) is a subtle body (*ǧism laṭīf, corpus subtile*), which passes through the outlets (*nāfiḏ fī l-manāfiḏ, diffusum in concavitatibus*), and spiritual (*rūḥānī, spirituale*), and that this body is the pneuma (*rūḥ, spiritus*) (*Nafs* V, 8, 263.9–10 [175.49–51]),

that the soul unites primarily with the pneuma and then, through its mediation, with the body; and that the primary origin (*awwal ma'din, primus locus*) of the pneuma is the heart (*qalb, cor*) – the first organ to be generated – from which it spreads out and penetrates into all bodily organs (*Nafs* V, 8, 263.20–264.3 [176.64–70]).[59]

The fact that, according to Avicenna, the cardiac pneuma is the primary vehicle of the psychic faculties performing their activities in and through the body, and, ultimately, of the soul itself, poses some problems in the case of plants:

---

[56] In spite of what is announced in the prologue to *Meteorologica* (I, 1, 339a5–10: "When the inquiry into these matters is concluded let us consider what account we can give, in accordance with the method we have followed, of animals and plants, both generally and in detail. When that has been done, we may say that the whole of our original undertaking will have been carried out"), Aristotle did not write a specific treatise on plants. He refers to it in several places (see, for instance, *Historia Animalium* V, 1, 539a21; *de Generatione Animalium* I, 2, 716a1; 23, 731a29–30); however, in his commentary on *De Sensu et Sensilibus* (ed. by Wendland 1901), Alexander of Aphrodisias implies that, even if Aristotle had actually written a treatise on plants, at his time that work was not extant anymore (καὶ ἔστι Περὶ φυτῶν Θεοφράστῳ πραγματεία γεγραμμένη· Ἀριστοτέλους γὰρ οὐ φέρεται, 87.11–12). On the fact that, according to Aristotle, plants also have soul as their principle, see *de An*. I, 5, 411b27–28: ἔοικε δὲ καὶ ἡ ἐν τοῖς φυτοῖς ἀρχὴ ψυχή τις εἶναι.

[57] I am using here the same expression that G.E.R. Lloyd used to refer to the relation of Aristotle's *de An*. to his zoological writings; see Lloyd (1992), p. 148.

[58] *Nafs* V, 8 is entitled "*[Chapter] concerning the clarification of the organs belonging to the soul*" (*Fī bayān al-ālāt allatī li-l-nafs*, 262.18 [*Capitulum de ostensione instrumentorum animae*, 174.36]).

[59] Avicenna also expounds a cardiocentric position in the *Adwiya Qalbiyya*. For Avicenna's position, and the connections between psychology and medicine in this particular respect, see Alpina (2017).

though possessing the soul as a principle for their activities and acting in and through a body, they seem not to comply with the model just outlined because they do not possess the heart.⁶⁰ Furthermore, the fact that the pneuma is the primary vehicle of the psychic faculties in the body is said to be evident when the obstruction of bodily passages occurs, according to the opinion of those who undertook medical experiments (*'inda man ǧarraba l-taǧārib al-ṭibbiyya*, 263.13 [*secundum eum qui cognovit experimenta physica*, 175.54–5]), and the fact that the heart is the origin of generation of the pneuma is said to be verified by accurate dissection (*mā ḥaqqaqahū l-tašrīḥ al-mutqan*, 264.4 [*qui certificavit hoc chirurgia vera*, 176.70–1]).

Cardiocentrism (and its corollaries), which is argued for in psychology in the case of animals (both irrational and rational), seems therefore to be grounded – at least in the case of human beings – on medical practice (e. g. medical experiments, accurate dissection). The discussion of the same issue concerning irrational animals seems to be deferred to *Kitāb al-Ḥayawān* (*Liber de Animalibus* in Latin, *Book of Animals* in English, henceforth *Ḥayawān*):⁶¹ "We shall supply an explanation of what is meant [by that] in the section on animals (sc. *Ḥayawān* XIII, 3)⁶² (*wa-sanazīdu hāḏā l-maʿnà šarḥan fī l-fann allaḏī fī l-ḥayawān* [*postea autem hoc clarius faciemus in libro qui est De Animalibus*])" (*Nafs* V, 8, 264.4–5 [176.71–2]).

---

60 This might be also the reason why in *Nafs* II, 1 Avicenna presents nutrition, which is shared by all sublunary living beings, as the most fundamental vital activity without mentioning respiration and blood circulation: the vital faculty, which does not belong to plants, seems to be responsible for them (see n. 32 above). A similar difficulty arises also in Aristotle. For, having related nutrition to innate heat, which is a sort of organ for digestion (*de An.* II, 4, 416b28–29) and having assigned nutritive soul to plants, Aristotle is forced to assign innate heat also to plants. However, this attribution is not straightforward: in *de Juventute et Senectute, de Vita et Morte, de Respiratione* the source of innate heat is said to be situated in the middle of the body (2, 468a21), in the heart in sanguineous animals and in an intermediate point between stem and root in plants (23(17), 478b32–479a1). Things get complicated with the digestive process: Aristotle has to maintain that the organ of digestion in plants is placed outside, in the earth, which has some internal heat and, therefore, digests the nourishment so that plants can directly assimilate it (*de Partibus Animalium* II, 3, 650a2–23). On this tension within Aristotelian biology, see Freudenthal (1995), p. 70–73.
61 It is the eighth and last section of the part on natural philosophy of the *Šifāʾ*, and corresponds to Aristotle's *HA*, *PA*, and *GA*.
62 *Ḥayawān* XIII, 3 is entitled "*[Chapter] on the anatomy of the heart and of what originates from it in terms of arteries.*" In the Avicenna Latinus this is taken as a reference to *Ḥayawān* XII, 2, see Avicenna, *Liber de anima seu sextus de naturalibus*, p. 176, n. 71.

Nafs V, 8, together with the other aforementioned passages, seems therefore to connect psychology directly with zoology[63] and, indirectly, with medicine. By contrast, botany, especially *qua* study of plant life, can be hardly slotted into a framework where preeminence is given to animal life with a specific focus on human beings.[64] Though ensuing from a soul, and being necessary to keep something alive, vegetative activities, which properly distinguish plants, seem not to be sufficient for distinguishing what is living from what is lifeless.

In the *Qānūn* Avicenna explicitly tackles the question concerning what makes an organism living, of course from an explicitly human perspective. As has been already shown, there the notion of *soul* is absent, the adjective *psychic* (*nafsānī*) is used to refer exclusively to the faculties that are responsible for sensation and voluntary motion;[65] and the *vital* (*ḥayawāniyya*) faculty, an intermediate faculty between natural and psychic faculties, and equivalent to Galen's ζωτική δύναμις, is introduced to account for what imparts life to organs. This faculty is said to originate in the cardiac pneuma at the moment of generation, to prepare for the reception of the so-called psychic faculties and, in general, for the activities of life.[66] Grasping its essence, however, is not easy: for, on the one hand, it resembles the natural faculties because it acts without volition, whereas on the other hand, it resembles the psychic faculties because it is responsible for manifold activities.[67] What is more, it is not included in the taxonomy of psychic faculties traditionally provided by philosophers.

---

63 Apart from the aforementioned reference, in *Nafs* V, 8 Avicenna refers three more times to the *Ḥayawān*: (i) 265.1 [177.95] (*Ḥayawān* XII, 8); (ii) 266.4 [179.27] (*Ḥayawān* III, 1); (iii) 269.14–5 [185.26] (*Ḥayawān* XV, 1). These references seem to suggest that zoology will be treated immediately after psychology.

64 On Avicenna's *specific orientation* towards human soul in the *Nafs*, see Alpina (2018b).

65 For Avicenna's taxonomy of the faculties of the soul in the *Qānūn*, see § 2 above.

66 Avicenna, *Qānūn* I, i, vi, 4, 126.19–28 [fol. 24r, b30–54]: "As for the vital faculty, they (*sc.* physicians) intend by it the faculty which, when it comes to be in the organs, prepares them for the reception of the faculty of sensation and motion, and for the activities of life, and [they] add to them (*sc.* to the activities of life) the motions of fear and anger because they find in this extension and contraction occurring to the pneuma related to this faculty. Let us [now] set forth in detail this whole [amount of information]. [(…)] The psychic (*sc. nafsāniyya*) faculties do not come into being in the pneuma and in the organs, except after the coming into being of this faculty (*sc.* the vital faculty). Even if an organ is deprived of the psychic (*nafsāniyya*) faculties, but it is not deprived of this faculty (*sc.* the vital faculty), it [remains] living."

67 Avicenna, *Qānūn* I, i, vi, 4, 127.26–27 [fol. 24v, a51–55]: "This faculty (*sc. al-quwwa al-ḥayawāniyya*) resembles the natural faculties (*al-quwà l-ṭabīʿiyya*) because of its privation of will in what derives from it, and resembles the psychic faculties (*al-quwà l-nafsāniyya*) because of its manifold activities, since at the same time it contracts and extends, thus moving according to two motions opposed to each other ([…])."

In this context, however, the contribution of natural (or vegetative) faculties and, notably, of nutrition, to life, whether it is decisive or not, cannot be unequivocally established, since Avicenna wavers on this point. For, in the chapter in which the vital faculty is introduced, life is said to be either *primarily* or *exclusively* connected with it. Initially, he says that the faculty of nutrition alone cannot be what prepares to perception and locomotion so that, as long as this faculty is active, a thing is living while, as soon as it ceases to be active, that thing is dead:

> This preparing thing is not the faculty of nutrition alone (*wa-laysa hāḏā al-muʿidd huwa quwwat al-taġḏiya waḥdahā, et praeparans non est virtus nutriendi tantum*) such that, when its faculties of nutrition remain, it (*sc.* the thing sharing in nutrition) is living, and when they cease [to exist], it is dead. (*Qānūn* I, i, vi, 4, 127.4–5 [fol. 24v, a9–12])

Thus, the disposition to life seems to be the result of a combination of the nutritive and the vital faculties. However, immediately afterwards, Avicenna denies that the nutritive faculty performs any preparatory activity to life, since, in that case, plants also would be prepared to receive perception and locomotion, which is not the case:

> If it were the nutritive faculty inasmuch as nutritive faculty, to prepare to sensation and motion, then [also] plants would be prepared to receive sensation and motion. Therefore, it remains that what prepares [to receive sensation and motion] is something else (*amr āḫar, alia res*), following a specific temperament. This thing is called 'vital faculty', which is the first faculty coming into being in the pneuma, when the pneuma comes into being from the thinness of the gametes. (*Qānūn* I, i, vi, 4, 127.7–9 [fol. 24v, a16–22])

Therefore, it remains that the vital faculty *alone* imparts life to organs, and prepares for the reception of the *psychic* faculties.

In the *Qānūn* a narrower notion of *life*, modelled on animal life, is certainly presupposed (which is not in contrast with the purpose of the work): for, the activities of life, to which the vital activity prepares, are those proper to animals, i.e. sensation, locomotion, and even the capacity for feeling emotions, without which something cannot be said to be *animal*. Though in the *Nafs* as well as in Avicenna's philosophical psychology in general there is no reference to the vital faculty,[68] and the existence of plant life and vegetative soul is acknowl-

---

[68] In the *Qānūn* Avicenna defers the ascertainment of the nature of the vital faculty to (natural) philosophy, even though there no reference to this faculty can be found. This might have prompted al-Ǧūzǧānī, Avicenna's secretary and disciple, to add an excerpt (chaps. 2–9) from Avicenna's

edged,[69] here some echoes of the clearly narrower notion of *life* of the *Qānūn* are detectable.

The primacy – implied or explicitly claimed – of animal life can be considered not only as the effect of Avicenna's zoological and medical interest on his psychology, but also as an attempt to narrow the investigation of the principle of life, i.e. the soul, to what the human being shares with other sublunary living beings, as is recommended in the prologue to the *Nafs*.[70]

3.2 Staying Alive: Plants or the Necessity of a Broader Notion of Life

Despite the primacy of animal life in Avicenna's philosophical psychology, which ideally connects with the investigation of animal and human anatomy in the *Ḥayawān* and the *Qānūn* respectively,[71] Avicenna devotes the *Kitāb al-Nabāt* (*Liber De Vegetabilibus* in Latin,[72] *Book of Plants* in English, henceforth *Nabāt*), i.e. the seventh section of the part on natural philosophy of the *Šifāʾ*, to an autonomous treatment of plants. At the beginning of this writing Avicenna wonders whether plants have to be considered living or lifeless, by discussing the same alternatives introduced in the *Qānūn*. However, before analyzing Avi-

---

Adwiya Qalbiyya between the end of the fourth treatise and the beginning of the fifth treatise of Avicenna's *Nafs*. More on this in Alpina (2017), p. 373–381.

69 Plant life and vegetative soul are accounted for not only in *Nafs* I, 5 and II, 1, where they are *ex professo* dealt with. For instance, in *Nafs* I, 1, 13.8–14.8 [31.85–33.5], in presenting two possible objections to the Aristotelian standard definition of the soul, Avicenna contrasts the notion of life belonging to celestial substances with that belonging to all sublunary living beings considered as a whole (plants included), because the former are not engaged in any of the activities that are minimally constitutive of sublunary life shared by the latter. Furthermore, in *Nafs* I, 3, 30.5–31.11 [62.82–64.12], three possible meanings of 'vegetative soul' are discussed. For a thorough analysis of the first passage, see Alpina (2020b).

70 According to Avicenna, the investigation of the soul has to focus on what the soul of the human being shares with that of the other sublunary living beings since, on the one hand, grasping the specific differences of what is remote from us, e.g. of plants, is difficult because they fall outside our cognitive faculties due to their extreme specificity, and, on the other hand, we have a direct acquaintance with our own soul. For a thorough analysis of this passage of Avicenna's prologue to the *Nafs*, see Alpina (2018a).

71 For the dependence of the investigation of animal anatomy conducted in the *Ḥayawān* on the parts on human anatomy of the *Qānūn*, and Avicenna's copying and pasting portions of the anatomical treatment from the *Qānūn* in the *Ḥayawān*, see n. 5.

72 The Latin translation of this section of the *Šifāʾ* is only attested but no longer extant. See Avicenna (1987), p. 65, and n. 3, and Bertolacci/Alpina (2017), p. viii.

cenna's argument, some preliminary information about this writing and its position within the overall investigation of organic life has to be provided.

According to present knowledge, *Nabāt* is Avicenna's only work on botany. It consists of one single book subdivided into seven chapters.[73] In composing it Avicenna primarily relies on the pseudo-Aristotelian treatise *De Plantis*, in all likelihood composed by Nicolaus of Damascus (*fl.* I c. BC), which Isḥāq ibn Ḥunayn (d. 910) translated into Arabic from an intermediate Syriac version.[74] Nicolaus' *De Plantis* consists of two books, the first subdivided into seven chapters, and the second into ten; however, Avicenna's *Nabāt* reproduces the structure and to some extent the content of the first book only, and ignores the second book. The reason for this selective approach might have to do with the content of the second book: for, unlike the first book, the second book includes, together with botanical issues, materials exceeding the boundaries of botany and properly pertaining to other sciences, for instance to meteorology (e. g. the treatment of primary qualities, and the process of concoction).[75] Hence, Avicenna might have considered their inclusion in a discussion on botany a reduplication of what he has treated elsewhere in the *Šifā'*, for example in the meteorological section.[76] However, given that the second half of the second book of *De Plantis* does deal with botanical issues, Drossaart Lulofs and Poortman have formulated

---

[73] The table of contents of Avicenna's *Nabāt* is the following: (i) Chapter on the generation of plants, their taking nourishment, their male and female, and the fundamental principle of their temperament; (ii) Chapter on the organs of plants at the beginning of the development and after that; (iii) Chapter on the principles of nutrition, reproduction, and generation in plants; (iv) Chapter on the state of the generation of the parts of plants, and on the state of their difference, and the difference of plants depending on countries; (v) Chapter on determining in particular the states of stems, branches, and leaves; (vi) Chapter on what is generated from plants in terms of fruits, seeds, thorns, resins, and what is alike; (vii) Chapter in which a general discourse on the kinds of plants is followed by the discourse on the temperaments of things having a nutritive soul.

[74] Actually there are five translations of this writing: (i) a fragmentary Syriac translation, (ii) an Arabic translation, (iii) a Hebrew translation, (iv) an Arabic-Latin translation, and (v) a Latin-Greek translation. For more information about this writing and its transmission, as well as the edition of its five translations and the English translation of the Syriac fragments and of the Arabic and Hebrew versions, see the edition of Nicolaus Damascenus' *De Plantis* by Drossaart Lulofs/Poortman (1989).

[75] This is particularly evident in the first half of the second book, where the author explicitly refers to the meteorological investigation as the place in which the causes of the appearance of rivers and springs have been ascertained (*qad qaddamnā l-'illa li-ẓuhūr al-anhār wa-l-'uyūn fī l-kawn l-'ulwī*, II, 2). See Nicolaus Damascenus, *De Plantis* II.2 § 150.

[76] See *Ma'ādin wa-Āṯār 'ulwiyya*, which is the fifth section of the part on natural philosophy of the *Šifā'*, and corresponds to Aristotle's *Mete.* I–III.

the hypothesis that the first book enjoyed an independent circulation and, therefore, Avicenna's selective approach reflects the status of the source at his disposal.[77]

Avicenna outlines the botanical (and the zoological) investigation conducted in the *Šifāʾ* and its relation to psychology in his prologue to the *Nafs*. There, Avicenna seems to assign priority to form over matter in the investigation of organic life. The line of reasoning is as follows: before studying issues specifically concerning plants and animals, the constituents by means of which plants and animals are rendered substances must be studied, because they are composite substances. These constituents are form, i.e. the soul, and matter, i.e. body and limbs. However, since – Avicenna says – knowledge of something is preferably knowledge of that something with respect to its form, because a thing is what it is primarily in virtue of its form, the soul, insofar as it is the form of organic composite substances, has to be studied first.[78] Avicenna reiterates this position later on: once the investigation of the soul is accomplished, which pertains to one single science, and is contained in one single work (*fī kitāb wāḥid*, 2.18 [*in uno libro*, 12.44]), it might be supplemented by a specific discourse (*kalām muḥaṣṣaṣ*, 3.1 [*verba propria*, 12.45]) on plants and animals, the very same discourse which at the beginning of the prologue was deferred to after the general investigation of the soul. This discourse, however, will depend no longer on the soul, but rather on their bodies and the properties of their bodily activities (*bi-abdānihā wa-bi-ḫawāṣṣ min afʿālihā l-badaniyya*, 3.2 [*ex corporibus eorum et ex proprietatibus suarum affectionum corporalium*, 12.46–8]). The fact is that their formal principle, by which they are defined, is always that soul which is ascertained in general in psychology, regardless of the level of complexity of the activities issuing from it; indeed, their specificity, which is difficult to ascertain from inside, on the formal level may be sufficiently accounted for from outside on the basis of the external differences observable in their material substratum (Avicenna has already pointed out that grasping the specific differences of each

---

[77] See Drossaart Lulofs-Poortman's introduction to the Arabic translation of Nicolaus Damascenus, *De Plantis* (Drossaart Lulofs/Poortman (1989), p. 121).

[78] *Nafs*, prologue, 1.11–2.1 [10.16–21]: "What remained to us of [natural] science is to investigate the issues concerning plants and animals (*fī umūr al-nabātāt wa-l-ḥayawānāt* [*de rebus vegetabilibus et animalibus*]). Since plants and animals are rendered substance as to [their] essences through a form, that is the soul (*mutaǧawhirat al-ḏawāt ʿan ṣūra hiya l-nafs* [*ea quorum essentiae constituuntur ex forma quae est anima*]), and a matter, that is body and limbs, and [since] it is more appropriate (*awlà* [*melior*]) that what is science of something is [science] with respect to its form, it seemed to us [more convenient] to deal firstly with the soul." On the fact that this represents a break with the tradition with respect to the place of psychology, and the kind of investigation conducted therein, see Alpina (2018a), p. 449, n. 14.

instance of soul (and of its bearer) falls outside our cognitive faculties due to their extreme specificity; hence, we have to deal only with what is shared).[79] Ultimately, the specific discourse on plants and animals, supplementing the general one on the soul, is an investigation of their matter.

A further confirmation of the fact that the specific inquiries into plants and animals depend on the general investigation of their principle, i.e. the soul, which provides their theoretical background, seems to be provided by the summary of the topics of the conclusive sections of natural philosophy, i.e. botany and zoology, at the end of the prologue. There, Avicenna maintains that these disciplines inquire into the states (*aḥwāl*, 3.10 [*dispositiones*, 13.60 – 1]) of plants and animals respectively. Here the use of the term *ḥāl* (pl. *aḥwāl*) is noteworthy: it designates a *state* that presupposes a subject of inherence, and cannot exist without it (*per se* accidents or attributes). Botany and zoology should abide by this theoretical model. In this connection, the prologue to the *Nafs* contains a lucid, though concise, exposition of what can be labelled Avicenna's *essentialism*.

Recently, A. Tawara (2014) and A. Das (2017) have challenged this model in the case of botany by arguing that Avicenna's medical findings affected his philosophical arguments. The gist of Tawara's argument about Avicenna's alleged denial of life in plants as a direct consequence of his medical beliefs have been briefly presented in the introduction to the present paper,[80] and the following analysis of *Nabāt* 1 will corroborate our assessment of his interpretation. By contrast, though not explicitly arguing that for Avicenna plants are lifeless, Das maintains that the integration of medicine and philosophy in botany "indicates that the subject of plant life defies categorization" and, ultimately, "challenges the notion of a strict separation between medicine and philosophy" (Das 2017, p. 217). In particular, Das considers the identification of the principle of plant life with the same powers that regulate nutrition, i.e. heat and moistness, which are explicitly – though not exclusively[81] – connected with medicine, and the lack of any reference to the vegetative soul to explain vegetative functions (with the exclusion of a cursory reference at the end of *Nabāt* 3) a sign

---

[79] *Nafs*, prologue, 2.5 – 17 [11.27 – 12.43]. See also *Ilāhiyyāt* V, 4, 220.13 – 18 [255.70 – 256.78]. There, in dealing with the *differentia* that specifies the genus, Avicenna says that we cannot grasp what is proper to the specific difference of every genus with respect to every species, nor what is proper to the specific differences of the species of a single genus, because this knowledge escapes our cognitive capacities; rather, we can grasp the rule in virtue of which a *differentia* enters a genus and specifies it.
[80] See n. 19 above.
[81] See n. 98 below.

Is Nutrition a Sufficient Condition for Life? — 247

of the fact that Avicenna "was not entirely convinced that plants were endowed with a form of soul" (Das 2017, p. 216–17). In Das' interpretation, *Nabāt* 7 plays a crucial role: there, Avicenna discusses plant temperaments and their capacity for interacting with and affecting the temperament of the human body, a discussion that exceeds the boundaries of natural philosophy and protrudes into medical pharmacology (here Avicenna refers to the temperament of compound drugs [*mizāǧ al-adwiya al-murakkaba*, 34.11] as examples of secondary temperaments arising from individual, primary temperaments). For Das this chapter "stands in contrast to the other more speculative sections of the work, and seems to defy Avicenna's broader conception of the relationship between medicine and philosophy" (Das 2017, p. 214).

Das' reconstruction of Avicenna's position, however, completely disregards the outline of botanical (and zoological) investigation provided in the prologue to the *Nafs* and underestimates the reference to *nafs*/soul as the principle of plants in the *Nabāt*. In particular, in *Nabāt* 3, in dealing with the principles of nutrition and reproduction in plants, Avicenna clarifies that, though regulated by two distinct faculties, they ultimately derive from one single soul once the organs disposed to receive these faculties are created. Here Avicenna refers back to the *Nafs* for the theoretical background of this discussion: for, unlike the medical model, in which the faculties somehow stand on their own, in philosophical psychology they ensue from a unitary principle, i.e. the soul.[82] However, even if Das were right, the fact that Avicenna does not explicitly refer to the soul as the principle of plants would be perfectly in line with what he says in the prologue to the *Nafs*, namely that the general investigation of the soul *qua* principle of sublunary life (vegetative life included) has to be conducted in psychology before the specific investigation of plants (and animals).[83]

---

[82] *Nabāt* 3, 13.17–14.5: "This (*sc.* the fact that two distinct faculties are the principle of nutrition and reproduction) is well-known and evident, except that the truth is that the soul is one single [thing], and has [several] faculties proceeding from it in accordance with the existence of [their] receptor (*illà anna l-ḥaqq huwa anna l-nafs wāḥida wa-lahā quwà tanbaʿiṯu ʿanhā bi-ḥasab wuǧūd al-qābil*), and that these faces are like the part of the soul, which is in the fundamental principle from which the seed is engendered. [(...)] Thus, when it (*sc.* the soul) comes about in the seed, the seed is a receptacle for the nutritive faculty due to the nutritive faculty's suitability for its (*sc.* of the seed-receptacle) use. In order for the organ of reproduction to be created, the generative [faculty] not existing in actuality is generated. When the organ [of reproduction] is there, the generative [faculty] proceeds from this first (*sc.* nutritive) soul, which is actually nutritive and generative. We have already explained this in our discourse on the soul (*wa-qad šaraḥnā hāḏā fī kalāminā fī l-nafs*, *sc. Nafs* I, 3, 31.11–32.14, and II, 1)."
[83] Besides the passage quoted in n. 78 above, see the sentence that immediately follows it, that is, *Nafs*, prologue, 2.1–3 [10.21–11.24] "It did not seem [convenient] to us to sever the science of

Furthermore, in *Nabāt* 7, a chapter crucial for Das' interpretation, Avicenna seems to distinguish the botanical inquiries carried on by not further specified people (*ğamā'a min al-nās, a group of people*), who tried to account individually for every property of plants, like colors and fragrances, from his own botanical investigation. The main point of differentiation is that those people's inventories of botanical knowledge were based on the idea that properties of plants depend on their nature and matter, which renders their enterprise virtually impossible due to the extreme variety of matters, and ultimately useless, whereas Avicenna's botany is grounded on the fact that plant properties are necessitated by a psychic (*nafsānī*) principle, that is, the vegetative soul (*nafs nabātiyya*) diffused in the body, which operates through matter.[84] This consideration echoes Avicenna's essentialistic approach to the study of sublunary living beings, which was outlined in the prologue to the *Nafs:* the soul *qua* form is the overarching principle that accounts for what sublunary living beings are in themselves and how they must be.[85]

As for the fact that, according to Das' interpretation, *Nabāt* 7 displays that in the case of botany medicine and philosophy are intermingled, and the former is not subordinate to the latter, it is worth mentioning a passage of this chapter in which Avicenna explicitly addresses this issue:

> We must elucidate the discourse on the issue of the temperaments of plants with regard to [their] relation to our bodies (*al-kalām fī amr amziğat al-nabāt bi-ḥasab al-qiyās ilà abdā-*

---

the soul so as to deal firstly with vegetative soul and plants, then with animal soul and animals, [and] then with human soul and human being."

84 *Nabāt* 7, 33.5–12: "A group of people undertook the task of explaining the causes of plants. Some of them began to look for the cause of each property [of plants], until they endeavored to explain the cause of variegated colors and of different fragrances. However, in terms of endeavor this is impossible, for none of those things (*sc.* colors and fragrances) follows the need of natures and the necessity of matter; rather, they follow the managing of the vegetative soul and its distribution [in the plant] (*fa-innahū laysa šay' min tilka yatba'u muğib al-ṭabā'i' wa-ḍarūrat al-hayūlà bal yatba'u tadbīr al-nafs al-nabātiyya wa-tawzī'ahā*), even though they do not occur except through the mediation of these natures. [(...)] When we become thoroughly acquainted with the causes of that and its reasons, we know that this does not occur in plants and animals except from those causes, but those causes do not occur in their own places in plants due to a natural reason, but due to a psychic reason (*bal bi-sabab nafsānī*), every cause occurring in what is concealed. Thus, the occupation in which these men engaged is superfluous." This is not in contrast with the characterization of the specific discourse on plants (and animals) as a discourse on their bodies and the properties of their bodily activities: for this specific discourse is ultimately grounded on a unitary account of the soul *qua* principle of every instance of sublunary life.

85 See n. 31 above. More on this in Alpina (2021b).

ninā), so that it is a principle for medicine (*mabda' mā li-l-ṭibb*) and what is analogous to it. (*Nabāt* 7, 33.16 – 7)

Botanical knowledge does not mix natural philosophy and medicine, but is essential to medicine. For, being part of natural philosophy and, consequently, higher than medicine, botany provides the principles for a branch of medical science, i.e. pharmacology. What is more, the distinction between primary mixtures, in which the ingredients are simple (elemental) bodies, and secondary mixtures, in which the ingredients themselves are composite, to which Avicenna refers in *Nabāt* 7, 34.9 – 16, though thematically connected with the pharmacological part of the *Qānūn*, has been already theoretically founded in *Kawn wa-Fasād* 6 – 7 and *Afʿāl wa-Infiʿālāt* (*Activities and affections*)[86] II, 1 – 2.[87] In this connection, the conclusion of the *Nabāt* does not serve "as an apology for the text's failure to provide more details about the nature of plant life" (Das 2017, p. 214), but, rather, it clearly establishes the boundaries of the botanical investigation which belongs to natural philosophy: it has to provide the foundation of the inquiry into the properties of plants' concrete instances and their actions, which pertains to a particular discipline (*ṣināʿa ǧuzʾiyya*), that is, in all likelihood, to medical pharmacology.[88]

Once the content and the goal of the *Nabāt* have been outlined, we can address the issue of the kind of life enjoyed by plants, which Avicenna tackles at the beginning of the writing.

The botanical investigation in the *Nabāt* begins *in medias res* by accounting for what plants share with animals: "Plants share with animals activities and affections connected with nourishment (*ammā l-nabāt fa-qad yušāriku l-ḥayawān fī l-afʿāl wa-l-infiʿālāt al-mutaʿalliqa bi-l-ǧiḏāʾ*), in terms of bringing [it] to the body, distributing [it in the body], separating [from it] the surplus [aimed at reproduction], and generating the seed that is generated from it" (*Nabāt*, 1, 3.4 – 5).

Introducing a discourse about plants with a reference to what plants have in common with animals might be a way in which the specific treatment of something that is remote from us, like plants, can be connected with something that is much closer and, consequently, more evident to us, like animals (this seems to

---

[86] It is the fourth section of the part on natural philosophy of the *Šifāʾ*, and corresponds to Aristotle's *Mete.* IV.
[87] See Stone (2008) and Gannagé (2018).
[88] *Nabāt* 7, 38.4 – 5: "This extent (*sc.* of the botanical investigation conducted so far) is sufficient for us to give the foundation (*al-aṣl*). Let us conclude our discourse about plants, for if we were to devote ourselves to the properties of their particular instances and of their activities (*bi-ḥawāṣṣ ǧuzʾiyyātihī wa-afʿālihī*), we would descend into a particular discipline (*ilà ṣināʿa ǧuzʾiyya*)."

be also the strategy of Nicolaus' *De Plantis*.[89] Then, Avicenna immediately tackles the question of the kind of life that plants enjoy, since this seems to be considered fundamental, though potentially problematic, for any further inquiry into plant faculties and activities.

Having excluded the possibility that plants have sensation,[90] two alternatives are envisaged: "If administrating nourishment (*al-taṣarruf fī l-ġiḏāʾ*) is called *life* (*ḥayā*) such that, when the body is [in the condition] to endure due to [its capacity for] self-nourishing, it is living, whereas when it is unable to make its individual endure through nourishment, and the corrupting [factor] coming from outside imposes on it (*sc.* on the body) so as to change its temperament and dissolve its faculty, it is dead; then, plants must be said to have life. If, by contrast, it is part of the condition of life (*min šarṭ al-ḥayā*) that, together with that (*sc.* capacity for self-nutrition) there are also perception and voluntary motion, then life cannot be assigned to plants in any respect (*fa-lā yaǧūzu an tuǧʿala li-l-nabāt ḥayā bi-waǧh min al-wuǧūh*)" (*Nabāt* 1, 3.15–19).[91]

Two notions of *life* are thus contrasted: (i) a broader notion, essentially connected with the capacity for managing nourishment, i.e. the nutritive faculty; and (ii) a narrower notion, similar to the one hinted at in the aforementioned passages of the *Nafs*, and explicitly argued for in the *Qānūn*, according to which life results from the simultaneous possession of nutrition, and perception and locomotion. According to the former, plants can be said to be living, whereas, according to the latter, they must be considered lifeless.

Though not expressly endorsing either of these two alternatives, Avicenna points out that this controversy is mostly verbal: "The greater part of the dispute about this is verbal (*wa-akṯar al-ḫiṣām fī hāḏā lafẓī*)" (*Nabāt* 1, 3.19–20). Through this consideration, however, Avicenna does not mean to dismiss the issue as trivial. On the contrary, the problem is philosophically relevant, but requires a conceptual distinction not supported by ordinary language. As Avicenna puts it:

> The term *animal* seems to be coined for what has sensation and voluntary motion. Therefore, it seems that plants cannot be called *animal* at all. A group of people (*qawm*) has

---

89 Nicolaus Damascenus, *De Plantis*, I.1 § 1.
90 *Nabāt* 1, 3.13–5: "One ought not to grant sensation to plants. If it (*sc.* sensation) were granted [to them], it would be inactive, since they do not have the capacity to flee what is harmful and to seek for what is useful. The men most distant from the truth were those who assigned to plants intellect (*ʿaql*) and understanding (*fahm*) together with sensation, like Anaxagoras, Empedocles, and Democritus." In this passage dependence on Nicolaus' work is evident. See Nicolaus Damascenus, *De Plantis*, I.1 § 10.
91 This passage shows that Tawara's interpretation is untenable, because it is based on the last sentence of the passage, extrapolated from its context. See n. 19.

made this kind of distinction between *living* (*al-ḥayy*) and *animal* (*al-ḥayawān*). This distinction between the signification of the expression *having life* (sc. living) and [the signification of] the term *animal* (*bayna mafhūm lafẓa ḏī l-ḥayā wa-lafẓat al-ḥayawān*) is a difference that linguists (*aṣḥāb al-luġāt*)[92] are not aware of. (*Nabāt* 1, 4.1–4)

In order to include plants into the roster of living beings and, at the same time, to distinguish them from animals, we must keep the meaning of the term *ḥayy* distinct from that of the term *ḥayawān*: although they can be used interchangeably in ordinary language (they share the same stem, and linguists do not distinguish between the two), the latter must be used to refer exclusively to animals, that is, to living beings capable of perception and locomotion, whereas the former, being broader than *ḥayawān*, must be used to refer to living beings lower than animals in the *scala naturae*, i.e. to plants.[93]

Avicenna seems to consider the issue concerning plant life settled, because at the end of *Nabāt* 1 he explicitly ascribes to plants a specific form of life, i.e. vegetative, nutritive life:[94]

> Hence, vegetative *life*, and in general nutritive [life] (*fa-iḏan al-ḥayā al-nabātiyya wa-bi-l-ğumla al-ġiḏā'iyya*), depends on moistness and heat. The temperament of every plant is then in itself moist and hot. [(…)] Since this [kind of] *life* is by means of moistness and heat, the death, which is the opposite, occurs only in the case of the annihilation of the matter of moistness and the extinction of heat. For this *life* belongs to a moist and hot body. [(…)] Then, when the matter of moistness ceases [to exist] and the heat connected with it is extinguished through the process of nutrition (*'alà sabīl al-taġaḏḏī*),[95] as was said in other places (*wa-'alà naḥw mā qīla fī mawāḍi' uḫrà*, sc. in *Nafs* I, 5 and II, 1), and as we have explained thoroughly in our big book on the discipline of medicine (*wa-'alà mā basaṭnāhu kull al-basṭ fī kitābinā l-kabīr fī ṣinā'at al-ṭibb*,[96] sc. *Qānūn*), the substance having this [kind of] *life* necessarily corrupts. Then, the transformation of a temperament similar to this into coldness and dryness is annihilation. (*Nabāt* 1, 7.12–20)

---

92 For Avicenna's attitude towards philology (*al-luġa*), see Avicenna, *Biography* 68.6–72.8.
93 The need to distinguish between *having life/living* and *animal* seems to emerge, though cursorily, also in *Nafs* I, 1, 12.15–6 [30.73–5]: "Moreover, if the process of nutrition (*taġaḏḏī* [*nutrimentum*]) constitutes life, why do you not call the plants animals (*ḥayawān* [*viva sive animalia*])?" The Latin rendering of *ḥayawān* is noteworthy.
94 It is noteworthy that in a few lines the term *ḥayā* (*life*) in connection with plants occurs four times: 7.12, 14, 16, 19.
95 Because nutrition is no longer able to provide a replacement for the moist that dissolved.
96 The reference is in all likelihood to *Qānūn* I, i, vi, 3, 126.6–8, where Avicenna says that the process of digestion needs heat and moistness. This is, to my knowledge, the only occurrence of this circumlocution to refer to the *Qānūn* in Avicenna's œuvre. However, it might make us wonder about the way in which Avicenna used to refer to this work of his.

The kind of life ascribed to plants is primarily connected with food processing (in this connection, see the activities that Avicenna ascribes to plants at the beginning of the chapter). For this reason, vegetative life can be altogether referred to as nutritive. The reference to moistness and heat as the factors that preserve this form of life is not surprising and does not seem to imply a purely physical account of nutrition.[97] Rather, it fits with what has been already said in *Nafs* I, 5 and II, 1, and in *Qānūn* I, 1, vi, 3, as Avicenna himself seems to suggest: there, heat and moistness are presented as the two main instruments, active and passive respectively, for nutrition or, to be precise, for the digestive process (*ṭabḫ*, concoction, πέψις in Greek).[98] The implicit reference to the *Nafs*, and the explicit reference to the *Qānūn*, where a narrower notion of life is explicitly argued for, might therefore be Avicenna's attempt to integrate the specific treatment of plant life into a broader and unitary theoretical account of (sublunary) life.

4 Conclusion

Three notions of *life* seem to emerge from the passages drawn from the *Nafs* and the *Nabāt* of the *Šifāʾ*, and from the *Qānūn*:
(i) life as exclusively connected with sensation and locomotion, which is the narrowest notion;
(ii) life as related to the concomitant presence of nutrition, sensation and locomotion;
(iii) life as essentially related to nutrition, which is the broadest notion.

Despite some hesitations concerning the first and the second notion of *life*, in the *Qānūn* Avicenna seems to incline toward the narrowest one: life is connected primarily with the vital faculty (*quwwa ḥayawāniyya*), which prepares the organic body to receive the activities of life, that is, sensation, locomotion, and the capacity for feeling emotions. Nutrition, by contrast, being identified with one of

---

[97] *Pace* Das, who considers the reference to heat and moistness in connection with the explanation of plant life a sign of the fact that Avicenna conceives nutrition as a purely physical process, with no reference to the soul as its principle. In this connection, Das assigns great importance to the explicit reference to the *Qānūn*, "rather than to any philosophical work," which "indicates that he (*sc.* Avicenna) does not see this subject (*sc.* nutrition) as pertaining to theoretical philosophy" (216). However, she overlooks that the reference to the *Qānūn* is preceded by a reference to unspecified *other places*, which in all likelihood are *Nafs* I, 5 and II, 1, where nutrition and its instruments are dealt with.
[98] For the Aristotelian background of this doctrine, see n. 60.

the Galenic natural faculties, seems to be placed below the threshold of what can impart life to the body. Hence, life can be ascribed only to animals, human beings included, but not to plants. This might depend on the goal of the work, namely the knowledge of the states of human body in terms of health and sickness.

Though some passages of the *Nafs* (e.g. the prologue, and I, 1) seem to imply the narrowest notion of life, plants as well as animals are ultimately considered living for the very fact of being ensouled. In the *Nabāt*, by contrast, two competing notions of life are presented, i.e. the second and the third in the above-mentioned list. According to the third and broadest notion, life is essentially connected with nutrition and, consequently, can be also ascribed to plants because they have the capacity for administrating nourishment aimed at the continuance of the individual (through nutrition and growth) and of the species (through reproduction).

Avicenna's position on life is therefore at an impasse: for, in spite of the conciliatory approach displayed at the end of *Nabāt* 1 with the ascription of a basic form of life to plants, he has outlined at least two mutually exclusive notions of life (the narrowest and the broadest) and, consequently, two mutually exclusive ways in which nutrition and life relate to each other, which can be hardly reconciled into a unitary framework.

Responsible for this impasse is apparently the theoretical divergence between (natural) philosophy and medicine, specifically the lack of a coherent doctrine of the soul as the basis of the medical discourse. As has been shown, in the *Qānūn* faculties are presented as three independent genera, with no reference to the soul as the unitary principle from which they ensue. What is more, life is associated only with one of them, i.e. with the genus of the vital faculties; consequently, life turns out to be a univocal notion, since it is spoken of everything belonging to that genus. In philosophical psychology, by contrast, the arrangement of faculties is different: they are all called *psychic* (*nafsāniyya*) because they ensue from a *soul* (*nafs*) and are not divided into genera. Furthermore, faculties themselves are modified: the label *natural faculties* refers only to the Galenic subordinate faculties, whereas the served natural faculties of the *Qānūn* are called – in the Aristotelian fashion – vegetative; psychic faculties are split into animal (*ḥayawāniyya*, term used in a narrower sense) and human (*insāniyya*) faculties; and the vital faculty (*ḥayawāniyya*) is obliterated, and its etymological meaning, which is connected with life, is distributed among all the faculties of the soul at various degrees. The notion of life becomes, therefore, equivocal.

The equivocity of *life* is a fundamental outcome of Avicenna's psychological investigation,[99] which directly stems from the equivocity of its principle, i.e. the soul.[100] Firstly, Avicenna refers to the commonly acknowledged and more manifest notion of life, i.e. animal life, which is defined by perception and locomotion. Then, by extending the scope of the notion of life through its connection to nutrition, Avicenna manages to stretch it downwards so as to include also plants in the roster of living beings. In a similar manner, but in the opposite direction, that is, by stretching the notion of life upwards, in metaphysics Avicenna manages to account also for the celestial and divine form of life.[101] However, by extending the notion of life downwards or upwards, something inevitably escapes: in the first case, celestial life, in the second case, vegetative life.

That being said, with respect to the treatment of nutrition, a form of 'disjunctive subalternation' is detectable between (natural) philosophy and medicine: for, on the lower level of the specific doctrine, the medical account of nutrition can be said to be subordinated to that of natural philosophy, since the former contributes to refining and supplementing the latter. However, on the higher level of its theoretical underpinnings, there is a gap between natural philosophy and medicine which is difficult to bridge, except at the cost of a radical reorganisation of the theoretical framework of medicine, which consequently results in an unavoidable discontinuity between the two disciplines.

Bibliography

Alpina, Tommaso (2017): "Al-Ǧūzǧānī's Insertion of *On Cardiac Remedies* in Avicenna's *Book of the Soul:* the Latin Translation as a Clue to his Editorial Activity on the *Book of the Cure?*". In: *Documenti e studi sulla tradizione filosofica medievale* 28, p. 365–400.
Alpina, Tommaso (2018a): "Knowing the Soul from Knowing Oneself. A Reading of the Prologue to Avicenna's *Kitāb al-Nafs* (*Book of the Soul*)". In: *Atti e Memorie dell'Accademia Toscana di Scienze e Lettere 'La Colombaria'* 82 (68), p. 461–476.
Alpina, Tommaso (2018b): "The soul of, the soul in itself, and the *Flying Man* Experiment". In: *Arabic Sciences and Philosophy* 28.2, p. 187–224.

---

[99] In the *De Anima* Aristotle says that the notion of life is equivocal: "living is spoken of in many ways (πλεοναχῶς δὲ τοῦ ζῆν λεγομένου)" (*de An.* II, 2, 413a22); a position also held in *Topica* VI, 10, 148a29–31 (ed. Brunschwig) where, though in a dialectical context, the Stagirite refers to life as an example of equivocal notion: "life seems to be not one kind of thing only, but one thing in animals and another thing in plants (ἡ δὲ ζωὴ οὐ καθ' ἓν εἶδος δοκεῖ λέγεσθαι, ἀλλ' ἑτέρα μὲν τοῖς ζῴοις ἑτέρα δὲ τοῖς φυτοῖς ὑπάρχειν)."
[100] More on this in Alpina (2018b) and (2020b).
[101] See *Ilāhiyyāt* VIII, 7, 368.4–6 [430.36–7]. More on this in Alpina (2020b).

Alpina, Tommaso (2020a): "Retaining, Remembering, Recollecting. Avicenna's Account of Memory and Its Sources". In: Decaix, Véronique/Thomsen Thörnqvist, Christina (eds.), *Aristotle's* De memoria et reminiscentia *and Its Reception*. Brepols Publishers, Studia Artistarum, forthcoming.

Alpina, Tommaso (2020b): "Is the Heaven an Animal? Avicenna's Celestial Psychology at the Intersection between Cosmology and Biology". In: Salles, Ricardo (ed.): *Biology and Cosmology in Ancient Philosophy: from Thales to Avicenna*. Cambridge University Press, forthcoming.

Alpina, Tommaso (2021a): *Subject, Definition, Activity. Framing Avicenna's Science of the Soul*, De Gruyter – Scientia Graeco-Arabica series, forthcoming.

Alpina, Tommaso (2021b): "Exercising Impartiality to Favor Aristotle: Avicenna and 'the accomplished anatomists' (aṣḥāb al-tašrīḥ al-muḥaṣṣilūna) in Ḥayawān, III, 1". In: *Arabic Sciences and Philosophy. A Historical Journal*, forthcoming.

Avicenna (1964): Avicenna, *Liber canonis*, Reprographischer Nachdruck der Ausgabe Venedig 1507, Hildesheim: G. Olm.

Avicenna (1980[2]): *Ibn Sīnā, Maqāla fī Aqsām al-ʿulūm al-ʿaqliyya (Treatise on the Divisions of the Intellectual Sciences)*, in *Rasāʾil fī l-ḥikma wa-l-ṭabīʿiyyāt*. 2 vols. Cairo: Dār al-ʿarab.

Balme, D.M./Gotthelf, Allan (ed.) (2002): *Aristotle, Historia animalium*. Cambridge: Cambridge University Press.

Barnes, Jonathan (ed.) (1984): *The complete works of Aristotle. The revised Oxford translation*. Princeton: Princeton University Press.

Bertolacci, Amos (2006): *The Reception of Aristotle's Metaphysics in Avicenna's* Kitāb al-Šifāʾ. *A Milestone of Western Metaphysical Thought*. Leiden/Boston: Brill.

Bertolacci, Amos/Alpina, Tommaso (2017): "Introduction". In: *Documenti e studi sulla tradizione filosofica medievale* 28, p. vii–xviii.

Brock, A.J. (transl.) (1916): *Galen, On the Natural Faculties*. [Loeb Classical Library 71.] Cambridge, MA: Harvard University Press.

Brunschwig, Jacques (ed./transl.) (1967–2007): *Aristoteles, Topiques*. Paris: Les Belles Lettres.

Das, Aileen (2017): "Beyond the Disciplines of Medicine and Philosophy: Greek and Arabic Thinkers on the Nature of Plant Life". In: Adamson, Peter/Pormann, Peter E. (Eds.): *Philosophy and Medicine in the Formative Period of Islam* (Warburg Institute Colloquia 31). London: Warburg Institute, p. 206–217.

Drossaart Lulofs, H.J. (ed./comm.) (1965, 1972): *Aristotelis de generatione animalium*. Oxford: Clarendon Press.

Drossaart Lulofs, H.J./Poortman, E.L.J. (ed.) (1989): *Nicolaus Damascenus, De Plantis*. Amsterdam/Oxford/New York.

Endress, Gerhard (1991a): "La 'Concordance entre Platon et Aristote'. L'Aristote arabe et l'émancipation de la philosophie en Islam médiéval". In: Mojsisch, Burkhard/Pluta, Olaf (eds.): *Historia Philosophiae Medii Aevi. Studien zur Geschichte der Philosophie des Mittelalters* (Festschrift K. Flasch). Amsterdam/Philadelphia 1991, p. 237–257.

Endress, Gerhard (1991b): "'Der erste Lehrer'. Der arabische Aristoteles und das Konzept der Philosophie im Islam". In: Tworuschka, Udo (ed.): *Gottes ist der Orient. Gottes ist der Okzident. Festschrift für Abdoldjavad Falaturi zum 65. Geburtstag*. Köln/Wien: Böhlau, p. 151–181.

Endress, Gerhard (1997): "L'Aristote arabe. Réception, autorité et transformation du Premier Maître". In: *Medioevo* 23, p. 1–42.

Falcon, Andrea (2010): "Aristotle on the Scope and Unity of the *De anima*". In: Van Riel, Gerd/Destrée, Pierre (Eds.): *Ancient Perspectives on Aristotle's* De anima. Leuven University Press, p. 167–181.

Fobes, F.H. (ed.) (1919, 1967): *Aristotelis meteorologicorum libri quattuor*. Cambridge, Mass.: Harvard University Press.

Freudenthal, Gad (1995): *Aristotle's theory of material substance. Heat and pneuma, form and soul*. Oxford: Clarendon Press.

Gannagé, Emma (2018): "Between Medicine and Natural Philosophy Avicenna on Properties (*khawāṣṣ*) and Qualities (*kayfiyyāt*)". In: El-Bizri, Nader/Orthmann, Eva (eds.): *The Occult Sciences in Pre-modern Islamic Cultures*. Würzburg in Kommission, Beirut: Ergon Verlag, p. 41–66.

Giardina, Giovanna R. (2009): "'Se l'anima sia entelechia del corpo alla maniera di un nocchiero rispetto alla nave'. Plotino IV 3, 21 su Aristotele *De anima* II. 1, 413 a8–9". In: Di Pasquale Barbanti, Maria/Iozzia, Daniele (eds.): *Anima e libertà in Plotino. Atti del Convegno Nazionale (Catania, 29–30 gennaio 2009)*. Catania: CUEM, p. 70–112.

Gohlman, W.E. (ed./transl.) (1974): *Avicenna, The Life of Ibn Sina. A Critical Edition and Annotated Translation*. Albany/New York: State University of New York Press.

Gutas, Dimitri (2003): "Medical Theory and Scientific Method in the Age of Avicenna". In: David C. Reisman, David C., with the assistence of Ahmed H. Al-Rahim (eds.): *Before and After Avicenna. Proceedings of the First Conference of the Avicenna Study Group*. Leiden/Boston: Brill, p. 145–162.

Gutas, Dimitri (2014): *Avicenna and the Aristotelian Tradition. Introduction to Reading Avicenna's Philosophical Works*. Leiden: Brill 1988 (second, revised and enlarged edition 2014).

Hall, Robert E. (2004): "Intellect, Soul and Body in Ibn Sina: Systematic Synthesis and Development of the Aristotelian, Neoplatonic and Galenic Theories". In: McGinnis, Jon, with the assistance of David C. Reisman (eds.): *Interpreting Avicenna: Science and Philosophy in Medieval Islam. Proceedings of the Second Conference of the Avicenna Study Group*. Leiden/Boston: Brill, p. 62–86.

Hayduck, M. (ed.) (1882): *Simplicii in libros Aristotelis de anima commentaria* [Sp.?] *(fort. auctore Prisciano Lydo)*. [Commentaria in Aristotelem Graeca vol. 11.] Berlin: G. Reimer.

Hayduck, M. (ed.) (1897): *Ioannis Philoponi in Aristotelis de anima libros commentaria*. [Commentaria in Aristotelem Graeca vol. 15.] Berlin: G. Reimer.

Helmreich, G. (ed.) (1893): *Galen, De naturalibus facultatibus*. In: id. (ed.), *Galen. Scripta minora*. Leipzig, p. 101–257.

Institute of History of Medicine and Medical Research (ed.) (1981–96): *Avicenna, Qānūn fī l-ṭibb*. New Delhi, India: Ǧāmiʿa Hamdard.

Lammer, Andreas (2018): *The Elements of Avicenna's Physics. Greek Sources and Arabic Innovations*. Scientia Graeco-Arabica 20, Berlin/Boston: De Gruyter.

Lloyd, Geoffrey Ernest Richard (1992): "Aspects of the Relationship between Aristotle's Psychology and his Zoology". In: Nussbaum, Martha C./ Amélie O. Rorty, Amélie O. (Eds.): *Essays on Aristotle's* De anima, Oxford: Clarendon Press, p. 147–167.

Louis, P. (ed.) (1956): *Aristote, Les parties des animaux*. Paris, Les Belles Lettres.

Musallam, Basim F. (1987): "Avicenna. x. Biology and medicine". In: *Encyclopaedia Iranica*, vol. III, fasc. 1. 1987, 2011², p. 94–99.
Mūsā, M.Y./Dunyā, S./Zāyid, S. (ed.) (1960): *Avicenna, Al-Šifāʾ, al-Ilāhiyyāt* (2). Cairo, al-Hayʾa al-ʿāmma li-šuʾūn al-maṭābiʿ al-amīriyya.
Pellegrin, Pierre (1996): "Le *De anima* et la vie animale. Trois remarques". In: Romeyer Dherbey, Gilbert/Viano, Cristina (Eds.): *Corps et âme. Sur le* De anima *d'Aristote*. Paris: Vrin, p. 465–492.
Pormann, Peter E. (2013): "Medical Practice, Epistemology, and the Physiology of the Inner Senses". In: Adamson, Peter (Ed.): *Interpreting Avicenna. Critical Essays*. Cambridge: Cambridge University Press, p. 91–108.
Qanawatī, G. Š./Zāyid, S. (ed.) (1960): *Avicenna, Al-Šifāʾ, al-Ilāhiyyāt* (1). Cairo, al-Hayʾa al-ʿāmma li-šuʾūn al-maṭābiʿ al-amīriyya.
Qāsim, M. (ed.) (1969): *Avicenna, Al-Šifāʾ, al-Ṭabīʿiyyāt, al-Samāʾ wa-l-ʿĀlam, al-Kawn wa-l-Fasād, al-Afʿāl wa-l- Infiʿālāt*. Cairo, Dār al-kitāb al-ʿarabī li-l-ṭibāʿa wa-l-našr.
Rahman, F. (ed.) (1959, 1970²): *Avicenna's De Anima [Arabic Text], being the Psychological Part of Kitāb al-Shifāʾ*. London/New York/Toronto: Oxford University Press.
Rashed, Marwan (ed./transl.) (2005): *Aristote, De la generation et la corruption*. Paris: Les Belles Lettres.
Repici, Lucina (2000): *Uomini capovolti. Le piante nel pensiero dei Greci*. Roma/Bari: Editori Laterza.
Ross, W.D. (ed./comm.) (1924): *Aristotle's Metaphysics*. Oxford: Clarendon Press.
Ross, W.D. (ed.) (1955, 1970): *Aristotle, Parva naturalia*. Oxford: Clarendon Press.
Ross, W.D. (ed./comm.) (1961): *Aristotle's De Anima*. Oxford: Clarendon Press.
Stone, Abraham D. (2008): "Avicenna's Theory of Primary Mixture". In: *Arabic Sciences and Philosophy* 18, p. 99–119.
Strobino, Riccardo (2017): "Avicenna's *Kitāb al-Burhān* II.7 and its Latin Translation by Gundissalinus: Content and Text". In: *Documenti e studi sulla tradizione filosofica medievale* 28, p. 105–147.
Tawara, Akihiro (2014): "Avicenna's denial of life in plants". In: *Arabic Sciences and Philosophy* 24, p. 127–138.
Van Riet, S. (ed.) (1968): *Avicenna Latinus, Liber de anima seu sextus de naturalibus IV–V*, édition critique de la traduction latine médiévale. Louvain/Leiden: E. Peters/E. J. Brill.
Van Riet, S. (ed.) (1972): *Avicenna Latinus, Liber de anima seu sextus de naturalibus I–II–III*, édition critique de la traduction latine médiévale. Louvain/Leiden: E. Peters/E. J. Brill.
Van Riet, S. (1977): *Avicenna Latinus, Liber de Philosophia prima sive Scientia divina, I–IV*. Louvain/Leiden: E. Peters/E. J. Brill.
Van Riet, S. (ed.) (1980): Avicenna Latinus, *Liber de Philosophia prima sive Scientia divina, V–X*. Louvain/Leiden: E. Peters/E. J. Brill.
Van Riet, S. (ed.) (1983): *Avicenna Latinus, Liber de Philosophia prima sive Scientia divina, I–X*. Louvain/Leiden: E. Peters/E. J. Brill. (6*–14*: *addenda et corrigenda* regarding the previous two volumes.)
Van Riet, S. (ed.) (1987): *Avicenna Latinus, Liber tertius naturalium De generatione et corruptione*. Louvain/Leiden: E. Peters/E. J. Brill.
Wendland, Paul (ed.) (1901): *Alexandri in librum de sensu commentarium*. [Commentaria in Aristotelem Graeca vol. 3.1.] Berlin: G. Reimer.

Wilberding, James (2014): "The Secret of Sentient Vegetative Life in Galen". In: Adamson, Peter/Hansberger, Rotraud E./Wilberding, James (Eds.): *Philosophical Themes in Galen*. Institute of Classical Studies, School of Advanced Study – University of London, p. 249–268.

Zāyid, S. (ed.) (1983): *Avicenna, Al-Šifāʾ, al-Ṭabīʿiyyāt, al-Samāʿ al-ṭabīʿī*. Cairo, al-Hayʾa al-miṣriyya al-ʿāmma li-l-kitāb.

Martin Klein
Digestive Problems
John Buridan on Human Nutrition

Abstract: The problem of human nutrition in medieval natural philosophy was closely connected with metaphysical claims about the human soul. The human soul was considered to be ungenerable and incorruptible, since it is created by God and not naturally derived from the potency of matter. This raises a question about human nutrition: How can an immaterial soul be engaged in obviously material processes such as nutrition? This problem is particularly pressing for John Buridan (ca. 1300–1358/60), who identifies nutritive powers with the soul; and since the human soul is immaterial, the human nutritive powers are immaterial as well. Though Buridan subscribes to the view that the process of nutrition involves the corruption of food and the partial substantial generation of the soul, he nevertheless believes that general features of nutrition can be explained for human beings. I argue that Buridan conceives of nutrition as a merely material change, a view which is in line with his broader conception of substantial generation and the relation between a substantial form and its coming to existence in suitably disposed matter. Ultimately, the way in which Buridan accounts for nutrition turns out to be another example of a rising dualism between body and soul, pointing to developments some centuries later which will render substantial forms superfluous.

1 Aperitif: Longing for Food

John Buridan spent his entire academic career as a master of arts (or as we would say today, a professor of philosophy) in Paris. This led to him being considered "the most important philosopher at the most important university in the world for three decades in the mid-fourteenth century" (Pasnau 2017, p. 59), and among those who commented the most extensively on Aristotle's corpus. His in-

---

Note: I am grateful to Chiara Beneduce and Christoph Sander for helpful comments on earlier versions of this paper and for pointing out important literature on the topic. I have also profited from excellent suggestions by Ian Drummond concerning my English. Special thanks to Gyula Klima for his generous permission to use the forthcoming new edition and translation of Buridan's *QDA(3)*. All other translations in this paper are mine.

fluence on European universities in the 14th and 15th centuries is not to be underestimated either.[1]

Apparently, Buridan had problems digesting food. He surely noticed the empirical fact that he ate and grew, as we all do, and that we would die if we stopped eating for a longer period. He was also familiar with Aristotle's dictum that nutrition, alongside reproduction, is that "in virtue of which all are said to have life" (de Anima II,4, 415a25; cf. Buridan, *Quaestiones super De Anima Aristotelis* [QDA(3)] II, q. 2, n. 2). Furthermore, Buridan was well aware of the danger of excessive desire for food (see Grellard 2017), which was held to be a sin, and a deadly one at that. However, in order to explain how a human being like him could be nourished in the first place, Buridan is forced to develop a rather involved account of how humans are capable of nutrition at all. The reason for this is the special metaphysical nature which he attributes to the human soul, namely, that it is immaterial. All animals are material substances, whose essential components are matter on the one hand a substantial form on the other, which is the soul. However, unlike what Buridan calls the material soul of non-human living beings – say, Brunellus the horse – the human soul neither derives from matter as its generative principle nor is it dependent for its existence on matter. Therefore, it does not share features that are due to matter, such as extension and quantification; furthermore, it is neither generated naturally, for it is created by God, nor does it suffer from corruption, since it is immortal (see Buridan, QDA(3) III, q. 6; Klein 2019, ch. 3).

The problem that Buridan has to face regarding human nutrition is how an immaterial soul like his own can be engaged in obviously material processes such as nutrition. In particular, if nutrition is defined as a partial substantial change in which food gets corrupted, and an animal, as a composite of body and soul, is maintained in being by being partly regenerated, how can the immaterial human soul be involved in nutrition at all? This problem has immediate consequences for explanations of the natural phenomena that accompany nutritive processes in human beings, for instance, the fact that we grow and increase when we eat (a.k.a. augmentation).

Granting to the human soul its immaterial nature, if only as a matter of religious belief (see Klein 2019, ch. 4), Buridan defends his conception of an immaterial human soul by admitting a crucial difference between humans and other animals in how to conceive of the partial substantial change that takes place in nutrition: unlike in other animals, the human soul is not partially generated. But he also claims that despite this metaphysical difference, the general features of

---

[1] On the life and work of Buridan, see Michael (1985) and Zupko (2003).

nutrition can be explained for humans as well. In his commentary on Aristotle's *de An.* (*QDA(3)* III, q. 17) Buridan discusses several objections that arise for his conception of the human soul in relation to the body, including problems regarding human nutrition. Since in that text he does not further substantiate his claims in his reply to these objections, it will be my purpose here to investigate how his suggestion can be accommodated to his general conception of nutritive processes as we find them in his other works, especially in his commentaries on the *Physica* and *De Generatione et Corruptione*.[2]

2 Hors-d'Oeuvre: the Immaterial Soul and its Powers

With Buridan we can examine how the problem of human nutrition in medieval natural philosophy was closely connected with metaphysical claims about the human soul. It is precisely in the context of the late medieval debate about how the soul is related to its powers and whether there is a plurality of souls or powers in a human being, in which Buridan is faced with the objection that it would follow from his conception of the human soul that a human being cannot be nourished.[3]

How is the soul related to its powers and is there a plurality of souls and powers in a human being? According to Buridan, every natural substance is the composite of matter and a single substantial form, which in the case of living beings is the soul. Thus, all living beings have just one soul, even if this soul is considered to be immaterial, as in the case of human beings. All powers of the soul, such as understanding or nutrition, are neither distinct from one another nor from the soul. But since the powers are identical to the soul, they also share its nature. Thus, the powers of the human soul are also immaterial (*Quaestiones super libros De generatione et corruptione* [*QGC*] I, q. 8; *QDA(3)* II, q. 4 and III, q. 17; *Quaestiones super decem libros Ethicorum Aristotelis ad Nicomachum* VI, q. 4).

---

[2] My focus will be on Buridan's metaphysics of the soul and specific issues arising from it for nutrition in relation to general principles in natural philosophy. For more strictly biological details of Buridan's natural philosophy, including nutrition, see Beneduce (2016, 2017, 2019, and 2020). Important works on medieval conceptions of nutrition in general are Cadden (1971 and 1980).

[3] On medieval conceptions of the relation between soul and its powers, see King (2008), Wood (2011), Perler (2015), and Bakker (2019); on medieval debates about the plurality of substantial forms, see Callus (1961), Pasnau (2007), and Duba (2012).

This seems to be a bold claim for Buridan to make in order to reconcile the various duties of the soul: in the case of human beings all powers of the soul are immaterial, including the vegetative powers! To get a better understanding of Buridan's position it is useful to compare it to two earlier accounts, namely, those of Thomas Aquinas (1225–1274) and William of Ockham (ca. 1288–1347). Aquinas holds that the human soul is the principle of its intellectual, sensory, and vegetative functions. However, it seems to follow from the immateriality of the intellect that, as Aquinas faces one objection, the intellect on the one hand and the vegetative and sensory soul on the other should differ essentially, since what is incorruptible (and hence immaterial) cannot belong to the same essence as what is corruptible (and hence material). Therefore, there must be more than one soul in a human being: the rational soul on the one hand and a vegetative and sensory soul on the other (*Summa Theologiae* [*ST*] I, q. 76, art. 3, arg. 1). Against this objection, Aquinas forcefully defends the unity of the soul in human beings. Regarding the impossibility of the soul being both immaterial/incorruptible and material/corruptible, he claims that the human sensory and vegetative soul is incorruptible only on account of the intellect, which "gives incorruptibility" to it. A vegetative or sensory soul in itself is corruptible but if it is connected with an immaterial intellect it cannot be withdrawn from incorruptibility (*ST* I, q. 76, art. 3, ad 1).

Hence, when we speak of the vegetative soul, the sensory soul, and the intellectual soul of a human being, what we really mean are different *powers* of one and the same soul. However, between these powers and the soul there has to be a difference, Aquinas claims. Powers are said to "flow" from the essence of the soul, and it is not through the essence of the soul that a human being understands, senses, and nourishes. That is, the soul does not perform these functions immediately; rather they are mediated by intellectual, sensory, and vegetative powers, which are said to be necessary qualities (*propria*) of the soul. As such, they are not really distinct from the soul, but are also not identical with it (*Quaestiones disputatae de anima*, art. 12, co.; *ST* I, q. 77 art. 6, co.; see also Pasnau 2002, p. 151–64). However, Aquinas's claim that powers of the soul are really distinct from one another is made precisely against the background of a human being's intellectual and biological functions – they are why we have to assume that there are both material and immaterial powers (*ST* I, q. 77, art. 2, co.). But whereas the intellectual operations have only the immaterial soul as their subject, operations which are performed through bodily organs have to be traced back to powers which are in the animated body, and not in the soul alone (*ST* I, q. 77, art. 6, co.).

Ockham agrees with Aquinas that powers are in some sense distinct from the soul. But by those powers Ockham means the bodily and material dispositions for the soul to operate. Hence, powers in this sense are reduced to the body as a partial cause of material operations. Ockham could be seen here as interpreting Aquinas's claim that material powers have the animated body as its subject. However, when it comes to the soul itself – that is, the soul not considered in relation to the body – Ockham fundamentally disagrees with Aquinas in stating that powers of the soul are not distinct from each other nor from the soul (*Quaestiones in tertium librum Sententiarum* [*Rep. III*], q. 4, p. 135). This identification of the powers of the soul with the soul itself forces Ockham to introduce a real distinction within the human soul, between the vegetative and sensory soul, which has material operations, and the intellectual soul, which has immaterial operations. For one and the same thing cannot be, for instance, both indivisible, like an immaterial soul, and divisible, like a material soul. Properly speaking, therefore, human beings have two souls, one that is responsible for purely intellectual operations, the other for bodily functions (*Quodlibeta septem* [*Quodl.*] II, q. 10–11; cf. Perler 2010).

Buridan holds a middle position between Aquinas and Ockham. Buridan agrees with Ockham in distinguishing between powers as dispositions for the soul to act, which he calls instrumental powers, and powers of the soul itself, which he calls principal powers. Instrumental powers are identified with the body, and are distinct from each other and from the soul; we call them powers of the soul only because the soul uses them as instruments (*QDA(3)* II, q. 5, n. 24).[4] But they do not seem to be more than bodily dispositions for the soul to act, although it is not entirely clear whether Buridan's instrumental powers are purely material requisites for the soul or rather belong to the body-soul composite (see De Boer 2013, p. 241–51 and Bakker 2019, p. 71–72). If the latter, Buridan's position will turn out to be not very different from Aquinas's position (*pace* Wood 2017). Principal powers, on the other hand, are not distinct from the soul or from each other; rather, they are identical with the soul insofar as it performs an act. For instance, insofar the soul understands we call it intellect, and insofar as it nourishes we call it the vegetative soul (*QDA(3)* II, q. 5, n. 20).[5]

---

4 [...] *licet anima sit principale activum nutritionis, tamen calor naturalis et plures dispositiones animae vel corporis coagunt ad nutritionem tamquam agentia instrumentalia quibus anima utitur ad agendum nutritionem, sicut faber igne et malleo* [...] *manifestum est quod, loquendo de potentiis instrumentalibus quae vocantur potentiae animae quia sunt instrumenta animae, illae different ab anima et ab invicem.*

5 *Et si non sit in uno supposito nisi unica anima, tunc in homine illa anima est potentia intellectiva, potentia sensitiva, et potentia vegetativa, potentia secundum locum motiva, et potentia appetitiva, secundum praedicta. Est enim principium activum et passivum nutritionis, intellectionis et sensationis.*

At the same time, however, Buridan vehemently defends Aquinas's claim that a human being has only one soul. The human soul is immaterial on account of the intellect (though according to Buridan we cannot philosophically prove this but can only believe it). Since the whole soul is immaterial, the intellectual, sensory, and vegetative powers (in the sense of principal powers) are immaterial as well. Once separated from the body, the soul will keep all these powers although it cannot further exercise those which require material dispositions for their actualization.[6]

What consequences does all this entail for human nutrition? Ockham, for his part, did not pay much attention to the vegetative soul and nutritive powers. He mentions nutrition only a few times, and claims that a vegetative soul together with heat in the body is required for nutrition (*Scriptum in librum primum Sententiarum – Ordinatio*, dist. II, q. 10, p. 348; *Quaestiones in secundum librum Sententiarum*, q. 19, p. 413–14, 420–21; *Rep.* III, q. 6, p. 163). He probably thought that his conception makes it easy to explain all the biological details, on account of the metaphysical fragmentation of the soul. Although his account faces serious metaphysical problems regarding the unity of a human being, it makes nutrition perfectly possible in all living beings, since in all living beings it is a material form that enables the material process of nutrition.[7]

Unlike Ockham, Buridan thinks that a plurality of souls should not be posited, since "it is vain to posit more in nature if with fewer everything can be explained; but everything can be explained by positing just one single soul or substantial form in a *suppositum*" (QGC I, q. 8, p. 85).[8] Ockham's Razor, of course, does not mean having a parsimonious ontology at any price. It is at play when Ockham argues against a distinction between the sensory soul and the vegetative soul (*Quodl.* II, q. 11), but he deems it necessary to posit a plurality of souls because the unicity thesis cannot explain every natural event, contrary to what Buridan claims. But can Buridan in fact explain all the functions of

---

[6] Buridan exemplifies this for human sensitive powers: [...] *eadem anima quae est sensitiva et intellectiva in omni actu suo sentiendi utitur organo corporeo, sed non in actu suo intelligendi. Et negatur quod anima sensitiva in homine sit extensa. Sed bene informat materiam corpoream et extensam, et habet actum sentiendi coextensum organo corporeo* [...] *Et negatur quod anima sensitiva hominis corrumpatur in morte. Sed bene corrumpuntur corporales dispositiones requisitae ad naturaliter sentiendum* (QDA(3) III, q. 17, n. 16).

[7] For Aquinas on nutrition, see Cadden (1971) and Brower (2014), p. 241–46.

[8] [...] *frustra ponuntur plura in natura, si per pauciora possunt omnia salvari; sed ponendo in uno supposito unam unicam animam seu formam substantialem, tam in animatis quam in aliis, omnia possunt salvari; igitur non sunt ponendae plures. Et minor huius apparet, si possent solvi rationes ad oppositum.*

the soul without giving up the unicity thesis? Here nutrition turns out to be a particularly pressing case.

3 Grosse Pièce: Nutrition and Substantial Change

Buridan believes that "regarding humans and horses and other animals which require a lot of heat and blood, an animal is nourished as long as it lives, because there is remarkable natural heat, especially in the heart, and conjoined with this heat is nourishment, such as blood and humors" (*QGC* I, q. 17, p. 133).[9] However, for blood to be produced, one needs to digest food (*QGC* I, q. 16, p. 126). As to this, Buridan addresses an objection in his commentary on *De anima*, which focuses on how we can account also for humans with an immaterial soul, which must be involved in nutrition if this immaterial form is the only substantial form in the substantial composite:

> It would follow that a human being is not be nourished, which is false. The consequence is clear, because nutrition does not occur without some partial substantial generation, for the nourishment must be converted into the substance of what is nourished. And yet nothing is substantially generated in a human being: the matter is not generated, since it is ungenerable and incorruptible; nor is the intellective soul or any part of it generated in nutrition, since it is indivisible; nor is some other substantial form generated, because those who do not posit that the intellective soul is distinct from the sensitive soul also do not posit a substantial form in a human being other than the intellective soul itself. (*QDA(3)* III, q. 17, n. 3)[10]

The objection aims essentially at the conception of nutrition as some sort of substantial change (cf. *QGC* I, q. 17, p. 132). This conception of nutrition as substan-

---

9 [...] *bene credo de hominibus et equis et aliis animalibus quae ad vitam requirunt multum calorem et multum sanguinem, quod animal quamdiu vivit nutritur, quia quamdiu vivit est calor naturalis notabilis (saltem circa cor) et illi calori coniunctum est nutrimentum, ut sanguis vel humores.*
10 [...] *sequeretur quod homo non nutriretur, quod est falsum. Consequentia patet, quia nutritio non est sine aliqua partiali generatione substantiali, oportet enim nutrimentum converti in substantiam nutriti. Et tamen nihil in homine generatur substantialiter, quia nec materia, cum sit ingenerabilis et incorruptibilis; nec anima intellectiva generaretur in nutritione nec aliqua pars eius, cum sit indivisibilis; nec alia forma aliqua substantialis, quia non ponentes animam intellectivam distingui a sensitiva non ponunt in homine aliam formam substantialem quam ipsam animam intellectivam* (transl. slightly revised). See also *QDA(3)* III, q. 17, n. 4.

tial change is not unusual in late medieval philosophy.[11] Buridan takes it to be obvious and known by his students as much as the conception's underlying general principles of Aristotelian natural philosophy; to find a more detailed explanation we need to refer to his commentaries on the *Ph.* and *GC*. The principles are: (1) substantial change is the generation of one thing and implies the corruption of some other thing; (2) this substantial change is only partial. For food to be digested means for it to be substantially corrupted by losing its substantial form and to be changed into the animated body, which is a substance composed of matter and form (*QGC* I, q. 10, p. 98).[12]

Take Brunellus the horse again. If we suppose that he eats some grass, what happens is that the substance of the grass – a composite of matter and substantial form – is destroyed and changed into the substance of Brunellus. Brunellus's substance is also a composite of matter and form, which together make up his ensouled body. Matter and form as the essential parts of a substance do not exist separately in nature. Hence, food being destroyed means that its matter loses its form and gets informed by another form. What makes this substantial change partial is the fact that in nutrition, although food gets completely corrupted, it is not an entirely new substance that comes into being. For Brunellus has already been generated himself and does not come into existence when he is eating; rather, only some quantitative parts of him are generated. In other words, Brunellus is augmented. What we would commonly describe as Brunellus gaining weight, is, technically speaking, the quantity of nourishment being applied to and conjoined with the pre-existing quantity of his body (*QGC* I, q. 10, p. 98; I, q. 17, p. 134; *QDA(3)* II, q. 8, n. 8).

But then why isn't food entirely indigestible for humans? Substantial change as the basic principle of nutrition seems to be violated once we suppose that a human being is composed of matter and a single soul which is immaterial, for the simple reason that there does not seem to be any possible term-to-which of nutrition, neither the matter, nor the form, nor the composite of form and mat-

---

11 See, e.g., Aquinas, *In librum primum Aristotelis De generatione et corruptione*, cap. V, lect. 14, p. 313a. However, the idea that nutrition does not occur without some partial substantial *generation* is, as a far as I can see, absent in the *Auctoritates Aristotelis*, a collection of authoritative passages of Aristotle which was compiled around 1300 and was widely circulated; see Hamesse/Meirinhos (eds.) 2017.

12 *Et etiam ita est in nutritione, quia in nutritione est vera corruptio cibi et per consequens etiam est ibi vera generatio substantialis, non tamen alicuius totius totalis, quia non generatur totale animal* [...]. See also *QDA(3)* II, q. 8, n. 19. For the general principle '*omnis generatio unius sit corruptio alterius*' see *QGC* I, q. 7, p. 74–79. See also *Auctoritates Aristotelis* [AA] 4,7, p. 167 (ed. Hamesse 1974).

ter. It seems not to be the matter that food gets transformed into, since Aristotelian physics generally hold that matter is the underlying principle of substantial change and itself remains unchanged.[13] What does change is the substantial whole, however, on account of some form coming to be and ceasing to be, and thereby leading to the existence of something new. But the human intellective soul does not seem to be suitable for what is generated since it is indivisible. It does not seem to make sense that something indivisible has parts into which it could be divided; rather, it seems it should have no parts at all. Therefore, an immaterial intellect seems to be incompatible with nutrition as partial substantial change because it requires that a part of the intellect be generated by food being corrupted. However, for a form to have parts is a feature which it owes to matter insofar as matter is quantified. Since the intellect is immaterial it lacks parts into which it could be divided (*Quaestiones super octo libros Physicorum Aristotelis (secundum ultimam lectionem)* [QP(U)] I, q. 20, p. 204).[14]

This brings us to the final point in the objection. The indivisibility of an immaterial form, as we have seen above, is one of Ockham's arguments for a real distinction between the intellective soul as an immaterial form and the sensitive and vegetative soul as a material form. Hence, the argument from indivisibility does not concern Ockham, since he can point to an *additional soul* which is divisible and can thus account for nutritive powers. If neither matter nor an immaterial soul can be that into which food is transformed, only an additional material form could do the job.

Other than what the passage quoted insinuates at the end, there actually have been late medieval authors who denied plurality of souls and still maintained the plurality of substantial forms. For instance, Buridan's contemporary Gerald of Odo (ca. 1285–1348) posits just one immaterial soul, but presupposes the additional substantial form of corporeity, which, together with matter, constitutes the human body before it gets informed by the soul (*In secundum librum Sententiarum, d. 18, q. unica*). A more prominent defender of this view is John Duns Scotus (ca. 1266–1308; see Duba 2012). Since the form of corporeity is clearly material in nature and hence divisible, it can count as what is partially

---

13 *Quaestio super octo libros Physicorum Aristotelis (secundum ultimam lectionem)* [QP(U)] I, q. 17, p. 171 [...] *hoc nomen 'generatio substantialis' non supponit pro materia et sic generatio substantialis non est ipsa materia* [...] *materia prima, quam* [Aristoteles] *vocat hyle* [...] *est maxima proprie subiectum susceptibile generationis et corruptionis; et ipsa non est subiectum susceptibile sui ipsius.* See also Aristotle, *Ph.* I,9, 192a31–33; *GC* I,5, 320a2–3, and *AA* 4,9, p. 168.
14 *licet materia esset sine aliqua forma sive substantiali sive accidentali sibi inhaerente, tamen ipsa esset actu quanta, quia hoc nomen 'quantum' vel 'quantitas' extendimus ad numerum; et ipsa sicut nunc est suae partes duae vel tres, ita adhuc esset ipsa illae partes* [...].

generated through nutrition, as is claimed by Francis of Marchia (1285/90–ca. 1344), another contemporary of Buridan. According to Francis, it is not the soul that is nourished, since it is immaterial, but the body in being operated on by the soul (*Quaestiones in secundum librum Sententiarum Reportatio*, q. 38, p. 121–22). But since any additional substantial form is precluded by Buridan's unicity claim, Buridan himself seems to be unable to tell us which essential part of a human being could be the terminus of partial substantial generation in nutrition.

4 Entrée: Buridan's Reply

And yet, to say that a human being is not nourished is obviously false. How then can Buridan account for nutrition without giving up the principles of the unity of form and the identity of the soul and its powers? In his reply to the objection, Buridan points first to the fact that the human soul is not the product of natural generation but of divine creation. Thus, since it is not derived from a material potency, which would entail that is has material characteristics, the mode of nutrition in human beings also has to be different from that of other living beings. However, this does not imply simply that human beings cannot be nourished, nor that their nutritive process is *entirely* different from that of other animals:

> There is, to be sure, similarity in one respect, and difference in another. There is similarity, because in both cases the substantial form of the food is corrupted, i.e., in the matter of the food there begins to be the substantial form or part of the substantial form of the living thing, which is said to be nourished. And so in both cases, something comes to be not absolutely speaking, but in a certain respect, namely, in a three-part predication, because the matter is informed by a form by which it was not previously informed, namely, by the soul or by a part of the soul. But there is certainly a difference, because in brute animals something of the substantial form is generated which did not exist before; however, nothing is generated in a human being. But the substantial form, namely, the soul, comes to be in that matter and begins to be in that matter in which it was not before. And this suffices for nutrition because in this way all the bodily quantities are preserved: both the shapes of the limbs and other dispositions suited to every activity of the soul. (*QDA(3)* III, q. 17, n. 18)[15]

---

15 *Est enim hic et illic convenientia et differentia. Convenientia enim est quia utrobique, forma substantialis alimenti corrumpitur, id est quod ibi in materia alimenti, incipit esse forma substantialis vel pars formae substantialis viventis, quae dicitur nutriri. Et sic utrobique fit aliquid non simpliciter sed loquendo secundum quid, scilicet cum praedicatione de tertio adiacente, quia materia sit formata forma qua non ante erat formata, scilicet anima vel parte animae. Sed bene est differentia, quia in brutis aliquid generatur de forma substantiali sic quod illud non erat ante, sed in homine nihil. Sed forma substantialis, scilicet anima, fit in materia et incipit esse in materia in*

Buridan refutes the original objection that in human beings no substantial change can occur. Nutrition in humans is not different from nutrition in other animals, in the sense that in both cases the substantial form of the food is corrupted, and the matter of the food is informed by a new form, namely, the soul. This suffices to account for nutrition as substantial change in living beings which have a material soul, as well as those with an immaterial soul, though in a qualified sense.

Since Buridan does not care to argue much for his point, his reply can be best described as suggestive. But what he is pointing at must be gathered again from his commentaries on the *Ph.* and *GC*. There, however, he does not address the particular problem of human nutrition, but is concerned with substantial change, nutrition, and augmentation in general. In what follows, I will focus on the following questions: First, how can we speak of partial substantial change with respect to the matter of food which gets newly informed by the soul? Second, how can it come about that the immaterial soul newly informs matter, given that it is ungenerable? Third, how is this conception related to explanations of bodily quantity and shape? And finally, how can we explain those features given the rather special sort of nutrition in bodies which are informed by an immaterial soul?

5 Sorbet: Matter

As to the first question, Buridan makes clear in his *Ph.* commentary that the terms 'substantial generation' and 'to be generated' must be put to a nuanced semantic analysis. What we mean by the term 'substantial generation' is the generation of the substantial form which newly appears in matter and acts on it. Neither matter nor the substantial composite of matter and form is what 'generation' refers to, because matter as the subject of generation is rather what can receive generation, and the composite cannot be that which matter receives. Hence, Buridan concludes, it is form that 'substantial generation' stands for (*QP(U)* I, q. 17, p. 170–72). Nevertheless, it is not just form, but also matter and the substantial composite that can be said to be 'what is generated', but in different ways: form, because it begins to exist after it did not exist; and matter, not because it begins to exist when it previously did not exist but precisely

---

qua ante non erat. Et hoc sufficit ad nutritionem, quia sic salvantur quantitates corporis, et figurae membrorum et aliae dispositiones convenientes omnibus operationibus animae (transl. slightly revised).

because it is the underlying subject of generation and gets a new form which it did not have before. Finally, the substantial composite can be said to be that which is generated because, as the name says, it is composed of generation (i.e., form) and the subject of generation (i.e., matter) (*QP(U)* I, q. 19, p. 199).[16]

These different senses in which matter, form, and the substantial composite can be understood as what 'generation' and 'corruption' refer to lurk in the background of Buridan's reply. He uses very technical vocabulary from the domain of logic in claiming that in nutrition something is generated not simply speaking (*simpliciter*) but in a certain respect (*secundum quid*), namely, in a three-part predication (*cum praedicatione de tertio adiacente*). These technical terms are generally meant to specify how we are to understand the use of 'is' in a sentence, whether as a copula or as an absolute attribute.

This is crucial for conceiving of substantial change and generation since generation is a change from non-being to being, that is, a change by means of which something is made which did not exist before. Accordingly, Buridan distinguishes two senses in which we can say that something *is* generated. Either the thing itself is brought into existence after it did not exist, or something which already exists undergoes some change in getting an added feature which it did not have before. In the first case, something is said to be generated in an unqualified sense (*simpliciter*), for example, when we say that something exists which did not previously exist. In the second case, something is said to be generated with some qualification (*secundum quid*), for example, in the sentence 'Something is white, and it was not white before'. In such a sentence 'is' is taken as a "third adjacent" in serving as the copula which is followed by a predicate. Hence, to say that something is made white does not mean that the thing itself has been made, but that it has been made in a certain way. In short, generation takes place *secundum quid* when an already existing subject takes on a new feature which it did not have before. This type of generation is appropriate for things which are permanent or have a permanent nature. Generation (as well as corruption) is attributed to them in respect of a form which they gain (or lose).

---

16 *Forma enim dicitur generari, quia est ipsa generatio et quia secundum se totam est, cum ante non esset. Materia autem dicitur generari, quia subicitur illi generationi et quia est formata tali forma qua non erat prius formata, et non qui sit, cum ante non esset [...] Compositum autem dicitur generari, quia componitur ex ipsa generatione et subiecto generationis, et quia atiam ipsum est, cum ante non esset; sed hoc, scilicet esse, cum ante non esset, non convenit sibi primo, sed ratione partis, scilicet formae, quae est generatio.*

The paradigmatic example of permanent things is of course matter, which is the persistent substrate for the production of a new form (*QGC* I, q. 7, p. 75).[17]

It is precisely with respect to matter that Buridan defends nutrition in both human and non-human animals as substantial change. Nutrition is a substantial change, but not one in which something is made entirely new; instead, an already existing subject takes on a new feature. Buridan relates this to the matter of food being digested. The substantial form of the food is corrupted, and the remaining matter is newly informed by the substantial form of the soul. Thus, in reply to the objection from the impossibility of human nutrition, Buridan argues as follows: according to the objection, a human being could not be nourished because for none of its substantial parts could it be true that the nourishment is converted into that part. Matter could not be substantially generated, since it is the ungenerable and incorruptible substrate in which a form is generated. Buridan replies that there is a sense in which matter can be said to be what is generated: not because it begins to exist when it previously did not exist, but because it is the underlying subject of generation and because it gets a form which it did not have before. We can say correctly that matter is generated, since we take the existential predicate not unqualifiedly but in a certain respect. It is precisely in this sense that we can say that nutrition is substantial generation in human beings as well.

This claim is compatible with the other sense in which we can say that something is generated, namely, regarding a form. Remember that the generation of a form implies the corruption of a previous form. Matter cannot take on a new form before it is deprived of the form that previously informed it. Hence, regarding both of the essential parts of the food, we can properly say that in both human and non-human animals nutrition takes place: the form of the food is corrupted, and its matter is informed by a new form.

---

[17] *Nota quod generatio est mutatio de non esse ad esse, id est generatio est mutatio qua aliquid est, cum ante non erat; et hoc vel simpliciter loquendo vel secundum quid. Et dico simpliciter loquendo, si hoc verbum 'est' capitur secundum adiacens, scilicet sine additione alterius praedicati; nam quod sine additamento dico, simpliciter dico. Sicut si ego dico 'hoc est et ipsum ante non erat', sequitur simpliciter loquendo quod ipsum est generatum seu factum. Sed dico secundum quid, si hoc verbum 'est' capiatur tertium adiacens, scilicet cum additione tertii praedicati, sicut si dico 'hoc est album et ipsum ante non erat album'. Tunc enim non sequitur quod ipsum est factum, sed sequitur cum additione quod ipsum est factum album.* See also *Summulae de Dialectica*, tract. 1, cap. 3,2; tract. 3, cap. 6,1; tract. 7, cap. 4,2.

6 Salade: Immaterial Substantial Form

After rejecting the first horn of the original objection, according to which nutrition in humans cannot be generation since matter cannot be generated, Buridan concentrates on the second horn, according to which the human soul cannot be generated. Whereas the objection defended this claim with the argument from the indivisibility of immaterial forms, Buridan concentrates first on generation in terms of the matter of the corrupted food: for the matter of the food to count as what is generated by nutrition it must now be informed by the human soul. However, with respect to the human soul we cannot say that it is generated in matter; rather, it begins to exist in matter.

Immaterial souls are not generated, but are created or produced by God in a supernatural way. For if they were naturally generated they would necessarily be corruptible (*QDA(3)* III, q. 6, n. 10),[18] but according to church doctrine, the human soul is incorruptible and survives the death of the body. This has implications for the way in which the soul is related to matter. As an immaterial form, the human soul cannot come into existence from the potency of matter; again, however, "according to the catholic faith", as Buridan likes to put it, the human soul informs matter and inheres in matter and the human body (*QDA(3)* III, q. 6, n. 18–19).[19]

Hence, there is crucial difference between human and non-human souls and how they become related to the matter of digested food. Take Brunellus the horse again. Nutrition as a partial substantial change implies that for Brunellus's soul to inform the matter of corrupted food it has to be brought forth from that matter's potency, just as is true for Brunellus's soul coming into existence in the first place by being brought forth from the potency of matter. A human soul on the other hand merely begins to exist non-generatively in the matter of food after the form of food is corrupted as a result of the human soul informing what was the matter of the food. The question, however, is whether this new existence

---

18 [...] *si intellectus non est perpetuus a parte ante et a parte post, ipse est genitus et corruptibilis et eductus de potentia materiae et extensus extensione materiae et multiplicatus multiplicatione hominum. Nam si intellectus non sit perpetuus, ipse est factus, et ratio naturalis non dictaret, sine fide vel supernaturali revelatione, quod aliquid esset factum scilicet per modum creationis, sed quod omne factum esset de novo factum per modum naturalis generationis ex subiecto praesupposito, de cuius potentia educitur forma ab agente.*

19 [...] *conclusiones vel propositiones quae in hac materia secundum fidem catholicam sunt tenendae, quarum prima est quod intellectus humanus non est perpetuus a parte ante, sed a parte post. Secunda est quod intellectus iste non est proprie genitus generatione naturali, sed creatus, nec est proprie corruptibilis corruptione naturali, sed annihilabilis. Tamen non annihilabitur.*

comes about supernaturally, just as the human soul coming to exist in the body at birth is supernatural. If so, Buridan will have a hard time defending nutrition in human beings as something which still is the concern of natural philosophy. In what follows, I will interpret the corollary of Buridan's reply as making a claim for his conception of human nutrition as a natural process, despite some appearances of, as it were, artificial nourishment.

7 Entremet de Douceur: Augmentation and Quantification

Although human nutrition can be conceived of as a partial substantial generation only in a qualified sense, Buridan thinks that several crucial features of nutrition are sufficiently explained also for human beings, as he makes clear at the end of his reply. Of those features, he mentions first the quantity of the body and the shape of the limbs. This is a surprising claim to make without further arguing for it, given that nutrition is accompanied by augmentation of the body, which Buridan seems to explain in terms of quantitative parts being added to the animal. It is questionable in what sense this change in quantity of the body in nutrition can be explained without also positing that the human soul is partially generated in the process, as the material soul of other animals is. In what sense can we explain the quantity of the human body and the shapes of its limbs? That is, how can we explain augmentation of the human body, if the human soul begins to non-generatively inform the matter of the corrupted food?

Here we have to notice first that, according to Buridan, the corruption of food and the generation of flesh and bones is not augmentation properly speaking, since it is not through this generation alone that an animal becomes bigger and occupies more space. Rather, the veins and pores are filled with the generated flesh and nerves, which in turn are stretched apart from one another so that the animal gets bigger. This leads to an extension of limbs, which itself does not come about by any "apposition of quantity" (*QGC* I, q. 16, p. 126–27).[20] Augmentation thus is apparently not what would properly allow for *quantitative* increase.

---

20 [...] *concurrit ibi quaedam extensio membrorum, quae non est per rarefactionem nec etiam per appositionem quantitas, immo solum per elongationem partium ab invicem* [...]. *Et ideo videtur quod illa extensio non est nisi motus localis partium* [...]. On the relation between rarefaction/condensation, augmentation/diminution, locomotion and quantitative change see also *QP(U)* I, q. 8, p. 87–90 and *QP(U)* IV, q. 11, p. 300–3. See also Pasnau (2011), p. 304–8; Sylla (2015), p. c–cvii and Sylla (2016), p. cxc–cxcii.

However, in another passage of his commentary on *GC*, Buridan makes what appears to be the opposite claim when he argues that the substantial change in nutrition implies also a quantitative change since "the *quantity of nourishment* is applied to (*apponitur*) and conjoined with (*coniungitur*) the pre-existing quantity of the body" (*QGC* I, q. 17, p. 134).[21] This apposition of matter is then related to the extension of the body: because pores and veins would remain filled up or blocked with generated flesh and nerves, it must be that they are opened, and for this to happen parts of the body are stretched apart from one another during or after the conversion of food (*QGC* I, q. 17, p. 134; *QP(U)* III, q. 9, p. 97).

But the contradiction between the two claims might be only apparent. If anything, quantitative change belongs to the corruption of food and the generation of flesh and bones and the like. Because pores and veins are filled up, they need to be widened, which is the reason for limbs to extend. Buridan thinks that this extension of the limbs is the main factor of bodily expansion, whereas by the substantial change of food into flesh and bones alone an animal would not appear bigger. Hence, along with the substantial change there occurs some apposition of quantity of corrupted food to the generated parts of the body. But what makes a body get bigger is not this conjoined quantity but rather the extension of its limbs.

There are therefore two distinct questions that have to be asked regarding augmentation within the process of an immaterial soul's informing the matter of corrupted food. First, how is the extension of limbs explained? And second, how can we account for partial substantial generation as the apposition of quantitative parts to the body? As to the former, Buridan compares the extension of the limbs with the violent pulling when the parts of a snare or skin are stretched apart, but notes the crucial difference that the pulling of parts of flesh and nerves during the consumption of food is not just forced. Rather, their motion is caused by the soul as an instance of locomotion (*QGC* I, q. 16, p. 127).[22]

Obviously, then, for the extension of the limbs to be possible in the case of the human body we must assume that an immaterial soul can account for the

---

21 [...] *per 'augmentationem' intelligitur aliquis motus simplex de motibus requisitis ad hoc quod animal fiat maius* [...] *Primo enim alimentum sumitur et alteratur et digeritur et ad membra localiter movetur per venas et poros et substantialiter convertitur in corpus animatum, et ita quantitas alimenti quantitati corporis praeexistenti apponitur et coniungitur.* See also *QGC* I, q. 14, p. 118 and *QP(U)* III, q. 5, p. 49–50.

22 *Et ideo videtur quod illa extensio non est nisi motus localis partium, quas natura elongat ab invicem, sicut si partes retis aut pellis elongarentur ab invicem per tractionem, licet non sit totaliter simile, quia illa tractio esset violenta et illa extensio in animato est per naturam, scilicet per ipsam animam, quae est natura viventis.* See also *QDA(3)* III, q. 20, n. 13.

locomotion of the body and its parts. Indeed, Buridan takes this to be generally possible. He speaks of such a causal influence of the soul on the body in other domains too. For instance, when a person feels anger or fear, the size of her limbs is changed due to calefaction or cooling down, which Buridan traces to the appetitive and cognitive powers of sense and intellect (*Expositio in librum de motibus animalium* 5; *QDA(3)* III, q. 20, n. 18). Once we accept that an immaterial soul has the power of locomotion of parts of the body, then such motion should not be a peculiar problem for the case of the limbs' increase or decrease being accompanied by nutrition in human beings.

As to the other question, that is, how an immaterial soul informing the matter of corrupted food without being itself generated explains or is in line with the acquisition of quantity, it is crucial to take a closer look at precisely how Buridan describes the relationship between partial substantial change and quantitative apposition in augmentation. The substantial change in question is partial because it is not the whole animal that is generated but only a part of it, which is to say, some *quantitative parts of the animal* (*QGC* I, q. 10, p. 98).[23] Nourishment is substantially converted into the animated body and thus the quantity of nourishment is added to the *quantity of the animated body* (*QGC* I, q. 17, p. 134).[24] But there is an ambiguity here, since in the case of souls which are material substantial forms, the animated body, in terms of both of its components (i.e., matter and soul), is what is generated and hence is quantitatively changed. In the case of a human being, however, since the soul is ungenerable, it can only be the body informed by a soul that is quantitatively changed, and not the soul. Therefore, Buridan needs to assume that what is quantitatively added to the animated body in the partial substantial change belongs exclusively to the matter of the corrupted food. This matter does not quantify the immaterial soul, nor is its quantity due to the soul. Rather, when food is corrupted, its matter preserves its quantity, which is then added to the human body by the matter being informed by the soul.

Obviously, this implies a robust conception of matter. Even if matter does not need to be quantified by itself – that is, without any substantial or accidental form – it is presupposed that matter can undergo substantial change while preserving accidental features such as quantity. This means ascribing to matter

---

23 [...] *in nutritione est vera corruptio cibi et per consequens etiam est ibi vera generatio substantialis, non tamen alicuius totius totalis, quia non generatur totale animal, etsi aliquae partes quantitativae eius generantur* [...].

24 [...] *alimentum sumitur et alteratur et digeritur et ad membra localiter movetur per venas et poros et substantialiter convertitur in corpus animatum, et ita quantitas alimenti quantitati corporis praeexistenti apponitur et coniungitur.*

some actuality which is not due to any particular form informing it. Indeed, according to Buridan, although matter never exists in nature without being informed by substantial and accidental forms, it is not dependent on these forms for its own existence. Regarding quantity, Buridan claims that matter in and by itself is quantified, in the sense that it has parts. However, it is not magnified by this: for the parts to be spread out, and hence for matter to be extended, the matter needs to be actualized by real quantity (*QP(U)* I, q. 20, p. 202–5; IV, q. 2, p. 214).[25] Although Buridan does not say so explicitly, quantity does belong to those accidental features which directly inhere in matter without being mediated by a substantial form. This not only allows him to explain the transfer of the quantity of the corrupted food to the augmented body, but is also in line with what he claims in others contexts. For instance, it was hotly debated in his time how a corpse can share striking similarities with a formerly ensouled body even though they are strictly speaking two different substances, since death entails the corruption of the soul and hence of all other accidental features. Buridan's solution to this problem is that what appears to be similar between the living and the dead body is grounded in accidents which are preserved during substantial change since they inhere directly in matter; that is, they are not mediated by a substantial form (*QGC* II, q. 7, p. 226).[26]

Thus, there is a difference in augmentation and quantitative change between human and non-human animal nutrition. The apposition of quantity in non-human animals is due to a partial generation of the soul and therefore has to do with the animated body as the composite of matter and material soul. Human bodies on the other hand are quantitatively changed only at the material level as a result of parts of the quantified matter of food being apposed to the matter of the human body. But it is precisely this difference which makes it possible to account for the quantitative change of the human body in nutrition even though the human soul is not partly generated but only starts to inform the previous matter of food. Since the matter of the human body is informed by the human soul, new parts of already quantified matter will also be informed by the human soul once they are conjoined or mixed with the matter of the human body.

---

[25] See also *AA* 4,10, p. 168. See Pasnau (2011), chap. 4 and 14 on matter, quantity, and extension in late medieval philosophy, including Buridan's views.

[26] [...] *subiectum de cuius potentia egreditur caliditas vel frigiditas passive et receptive est prima materia. Igitur si ipsa manet, quamvis forma substantialis non maneat, nihil est inconveniens tales qualitates manere.* Cf. *QDA(3)* II, q. 2, n. 24 and II, q. 9, n. 31. For this debate and Buridan's position see De Boer (2017).

8 Entremet de Fromage: Dispositions

One might expect that the soul has a greater role in nutrition and augmentation than just informing new parcels of the matter of digested food. Limbs and organs are shaped not just by quantitative parts being added to the body; they are also shaped and thus arranged in a certain way in order to exhibit the complexity which is needed for the various operations of the soul. One could argue that this is exactly what the generation of a material soul by nutrition contributes to: in order to become an essential part of the nourished body, the matter of food is informed by the soul also in the sense that the soul informs matter in a certain way so that it can become part of an *organized* body. That is, parcels of the matter of food become functionally conjoined with the matter of the animated body. It seems questionable how this could come about in human beings, however, since here it seems that matter is merely added to the body by becoming newly informed by an immaterial soul that happens to begin to exist in matter.

The case appears to be different if we allow that this coalescence of quantitative parts as a material process is not mediated by the information of the soul but *precedes* it. In this sense, digestion will be a qualitative change of the matter of corrupted food in becoming rightly disposed so as to be newly informed by the soul. As I understand Buridan in the final remark of his reply in *QDA(3)* III, q. 17, n. 18, he is claiming a picture like this when he says that also the "dispositions suited to all the operations of the soul are explained" by his conception of human nutrition. Moreover, he seems to suggest that this is not merely an adjustment for humans with immaterial souls but is somehow not very different from ordinary nutritive processes.

In fact, Buridan thinks of nutrition considered as a partial substantial change as analogous to complete substantial change. It is therefore necessary to look at how he treats the more general problem of how the soul as a substantial form comes to be related to matter. This more fundamental question is about how the soul makes a living body by informing matter. Here, the same problem arises of how substantial generation can come about if the soul of the animal to be generated is both the single substantial form of the substantial composite and is immaterial. According to another objection from *QDA(3)* III, q. 17 to Buridan's claim that a human being is composed only of matter and a single soul that is immaterial, it seems that a father would not generate any offspring, since he generates neither the matter of the animate substance nor its substantial form, given that the human soul is supposed to be created directly by God (*QDA(3)* III, q. 17,

n. 9).²⁷ How then can human beings be considered to be the result of their parents bringing about a substantial change?

Buridan replies that if we understand generation in terms of the production of the soul, in human as well as non-human animals, neither the male nor the semen is strictly speaking the agent in bringing about offspring. The male is hardly more than a semen donor, as it were, since he could die immediately after insemination and hence would contribute nothing during the time of substantial generation. Nor is the semen the proper agent in substantial generation, since it has a lower state of being than a living animal, which precludes it from being the cause of a higher state of being, that is, a substance. Thus, neither the father nor the semen substantially contributes to the production of a new offspring (*QDA(3)* III, q. 17, n. 23).²⁸ Generally, no corporeal agents involved in substantial generation, from the elements up to the celestial spheres, is sufficient to generate a new animated substance, for ultimately, every substantial form needs to be induced by an immaterial form-giver. All causal agency below this Avicennian *dator formarum* contributes to substantial change in producing what Avicenna called the "aptitude" of matter for a form suitable to it. Thus, the soul does not make up the body by organizing matter, but rather, as it were, meets matter which is already somehow organized (*QP(U)* II,5, p. 274–76; cf. Avicenna Latinus, *Liber de philosophia prima sive scientia divina* I, tract. 9, cap. 5).

Of course, this does not mean that the parents do not contribute in any sense to the production of new offspring. On the contrary, they produce the right material dispositions for the infusion of the soul. In this respect, Buridan ascribes generative powers to both the male and the female (*Quaestiones de secretis mulierum*, q. 3–7; see Beneduce 2017, p. 98–124; Beneduce 2020). Strictly speaking, however, they only prepare the right material conditions for the animation of the embryo, and this is true for the generation of both material and immaterial souls. While material souls are generated by this process, immaterial souls are created

---

27 [...] *inconveniens est quia sequeretur quod pater tuus non genuit te nec aliquid de substantia tua, quia non genuit materiam tuam nec animam intellectivam tuam; immo illa creatur a deo. Et tamen nihil plus est de substantia tua. Et sequitur ulterius quod non esset in homine potentia nutritiva, nec sui similis generativa, quod est falsum quia dicit Aristoteles 'homo generat hominem et sol'.*

28 [...] *non solum in generatione hominis, sed etiam in generatione equi, nec equus nec sperma equi generat proprie et principaliter equum, non enim equus, quia forte mortuus est; nec sperma equi, quia minoris est entitatis et non potest dare plus quam habet. Sed generans principale est dator formarum, qui est deus benedictus. Pater autem et sperma sunt tamquam agentia instrumentalia, et instrumentaliter dispositiva materiae ad illas animas recipiendas. Unde ad hanc intentionem et non ad aliam videtur mihi esse quod homo generat hominem et equus equum.* See also *QP(U)* II, q. 13, p. 339–42 and *QGC* II, q. 12, p. 251–53.

by God. To be sure, a material soul is said to be brought forth from the potency of matter and is thus more closely connected with matter; but according to Buridan, it is also true of such a soul that it does not only evolve from matter but is ultimately infused from outside.

In this respect, human procreation can be compared with human nutrition. In fact, the new information of matter by an immaterial soul, once the matter has been deprived of the substantial form of food, is not too different from how an immaterial soul comes into existence by being supernaturally infused. This admittedly miraculous event shares a crucial feature with the natural processes involved in the substantial generation of living beings. For the substantial form in any living being does not evolve directly from matter, but is infused by an immaterial principle from high above (which the theologians like to call God) once the right material conditions are provided. In both cases, before a new substantial form enters, there is a complex causal agency at work by which matter is made suitable for being informed by the soul. If this is the case for complete substantial generation in the procreation of living beings, whether they have a material or an immaterial soul, it should also be possible for partial substantial changes in which an immaterial soul newly informs matter. Thus, although the immaterial soul itself does not undergo the partial substantial generation at play in nutrition, it is sufficient, according to Buridan, that it informs matter after parts of food have been transformed in such a way that they became conjoined with the human body.

9 Dessert: Nutritive Mechanisms

The problem of human nutrition faced by medieval thinkers was how the nutritive soul which Aristotle speaks of as the principle of nutrition (*de An.* II,4, 415a23–25) could be reconciled with the immaterial nature which church doctrine attributed to the human soul. The 14th-century thinker John Buridan is an excellent example here. Buridan denies that material processes such as nutrition make it necessary to assume a material substantial form – e.g., a vegetative soul – distinct from the human immaterial soul, which is identical with its powers, among them the nutritive powers. He claims that the partial substantial change of the nourished animated body can be explained even if the human immaterial soul is not that which is partially (re)generated. Since Buridan is so parsimonious here, it is questionable whether it can be granted that the nutrition that he got from eating a meal was anything more than some *ad hoc* artificial nourishment.

How can Buridan account for nutrition as a natural process in humans too? Analysing the terms and concepts of substantial change in a commendably nominalist manner, he tries to safeguard nutrition in human beings as a partial substantial change by making the matter of food that which is generated. By the corruption of food, the matter of the food is deprived of its form and newly informed by the immaterial soul without the latter being generated, unlike in animals which have a material soul. Despite this metaphysical difference between human and non-human nutrition, a human being is equally augmented and quantitatively increased by the digestion of food, because the immaterial soul acts on parts of the body and because the quantity of the parcels of food is transferred to the human body. Crucially, this merely material change is in line with Buridan's broader conception of substantial generation and the relation between a substantial form and its coming into existence in suitably disposed matter.

This conception of human nutrition is also in line with how Buridan thinks the soul is related to its powers. On the one hand, the soul is identical with the nutritive powers insofar as it carries out the nourishment of the body; on the other hand, it is distinct from those material or bodily dispositions which are required for the incorporation of food. However, if we were to ask precisely how food is digested and thus loses its form, and how its matter is transformed to become rightly disposed in order to be conjoined with the human body, Buridan would have a hard time allowing the soul an important explanatory role in this process. The biological details, which I have not gone into here, seem to be easily explained without it. To be sure, the human soul would use the various bodily parts to perform the various motions required for digestive processes. But it seems that in order to understand those we would need to investigate purely material processes.

Ultimately therefore, the way in which Buridan accounts for nutrition turns out to be another example of a stronger sort of dualism between body and soul, which points to the development some centuries later of mechanistic principles that rule the internal structure of a body regardless of any substantial form, which would make substantial forms superfluous. But Buridan is surely not already some sort of mind-body dualist (see De Boer 2019; *pace* Lagerlund 2004), and might claim that substantial forms are needed in order to account for the difference between substantial change and mere alteration. But his flexibility regarding how to account for substantial change in the case of ungenerable and incorruptible immaterial soul forms runs the risk of raising suspicions about what substantial change in terms of pure change of matter would actually be.

Be that as it may, Buridan makes it clear that human beings, despite the supernatural status of their souls, can function well enough within the order of natural processes that they do not have to starve, as much as, by the by, "Buridan's ass" will eventually be satiated, for different reasons, of course.[29]

Bibliography

Aquinas, Thomas (1886): *In librum primum Aristotelis De generatione et corruptione expositio*. Ed. Commissio Leonina. [Opera Omnia (Editio Leonina) vol. 3.] Rome: Ex Typographia Polyglotta S. C. de Propaganda Fide, p. 259–322.

Aquinas, Thomas (1889): *Pars prima Summae theologiae a quaestione L ad quaestionem CXIX*. Ed. Commissio Leonina. [Opera omnia (Editio Leonina) vol. 5.] Rome: Ex Typographia Polyglotta S. C. de Propaganda Fide. [*ST*]

Aquinas, Thomas (1996): *Quaestiones disputatae de anima*. Ed. Bernardo B. Bázan. [Opera omnia (Editio Leonina) vol. 24.1.] Rome: Commissio Leonina.

Avicenna Latinus (1980): *Liber de philosophia prima sive scientia divina V–X*. Ed. Simone van Riet. Louvain: Peeters.

Bakker, Paul J.J.M. (2019): "The Soul and Its Parts: Debates About the Powers of the Soul". In: Stephan Schmid (ed.): *Philosophy of Mind in the Late Middle Ages and Renaissance*. London: Routledge, p. 63–82.

Barnes, Jonathan (ed.) (1984): *The Complete Works of Aristotle: The Revised Oxford Translation*. Princeton: Princeton University Press.

Beneduce, Chiara (2016): "Conoscenza sensibile e nutrizione. Il cardiocentrismo di Giovanni Buridano tra filosofia naturale e medicina". In: Giancarlo Garfagnini/Anna Rodolfi (eds.): *'Scientia humana' e 'scientia divina'. Conoscenza del mondo e conoscenza di Dio*. Pisa: ETS, p. 133–146.

Beneduce, Chiara (2017): *Natural Philosophy and Medicine in John Buridan: With an Edition of Buridan's Quaestiones de secretis mulierum*. Ph.D. dissertation. Radboud Universiteit, Nijmegen and Università di Pisa.

Beneduce, Chiara (2019): "La teoria buridaniana dell'umido radicale tra filosofia naturale e medicina". In: *Rivista di Filosofia Neo-Scolastica* 111.3, p. 597–605. DOI: 10.26350/001050_000129, visited on 30 November 2019.

Beneduce, Chiara (2020): "Filosofia naturale e medicina nella teoria buridaniana della generazione". In: *Rivista di Filosofia Neo-Scolastica* 112.1, p. 165–186. DOI: 10.26350/001050_000175, visited on 12 June 2020.

Boer, Sander W. de (2013): *The Science of the Soul: The Commentary Tradition on Aristotle's De anima, c. 1260–1360*. Leuven: Leuven University Press.

Boer, Sander W.de (2017): "Where Should We Discuss the Soul? On the Relation between the Doctrines of *De anima* and *De generatione et corruptione*". In: Gyula Klima (ed.): *Questions on the Soul by John Buridan and Others: A Companion to John Buridan's Philosophy of Mind*. Cham: Springer, p. 21–43.

---

[29] What has been mockingly ascribed to Buridan later (see Rescher 1960), is what Buridan himself actually considers foolish (*Expositio in Aristotelis De caelo* II, tract. 3, cap. 2, p. 148–51).

Boer, Sander W. de (2019): "Dualism and the Mind-Body Problem". In: Stephan Schmid (ed.): *Philosophy of Mind in the Late Middle Ages and Renaissance*. London: Routledge, p. 207–228.

Brower, Jeffrey (2014): *Aquinas's Ontology of the Material World: Change, Hylomorphism, and Material Objects*. Oxford: Oxford University Press.

Buridan, John (1967): "Jean Buridans De motibus animalium". Eds. Frederick Scott/Herman Shapiro. In: *Isis* 58.4, p. 533–552.

Buridan, John (1968 [1489]): *Quaestiones super decem libros Ethicorum Aristotelis ad Nicomachum*. Paris: P. de Preux, Repr. Frankfurt a. M.: Minerva.

Buridan, John (1996): *Expositio et Quaestiones in Aristotelis De caelo*. Ed. Benoît Patar, Louvain-la Neuve: Éditions de l'Institut supérieur de philosophie.

Buridan, John (2001): *Summulae de Dialectica: An annotated translation with a philosophical introduction*. Ed./transl. Gyula Klima, New Haven: Yale University Press.

Buridan, John (2010): *Quaestiones super libros De generatione et corruptione Aristotelis: A Critical Edition with an Introduction*. Eds. Michiel Streijger/Paul J.J.M. Bakker/Johannes M.M.H. Thijssen, Leiden: Brill. [*QGC*]

Buridan, John (2015): *Quaestiones super octo libros Physicorum Aristotelis (secundum ultimam lecturam). Libri I–II*. Eds. Michiel Streijger/Paul J.J.M. Bakker. Leiden: Brill. [*QP(U) I–II*]

Buridan, John (2016): *Quaestiones super octo libros Physicorum Aristotelis (secundum ultimam lecturam). Libri III–IV*. Eds. Michiel Streijger/Paul J.J.M. Bakker Leiden: Brill. [*QP(U) III–IV*]

Buridan, John (2019): "John Buridan's *Quaestiones de secretis mulierum:* Edition and Introduction". Eds. Chiara Beneduce/Paul J.J.M. Bakker. In: *Vivarium* 57.1–2, p. 127–181. DOI: 10.1163/15685349–12341364, visited on 30 November 2019.

Buridan, John (forthcoming): *Questions on Aristotle's 'On the Soul' by John Buridan: Latin Edition with an Annotated English Translation*. Eds./transl. Gyula Klima/John Peter Hartman/Peter G. Sobol/Jack Zupko. Springer: Cham. [*QDA(3)*]

Cadden, Joan (1971): *The Medieval Philosophy and Biology of Growth: Albertus Magnus, Thomas Aquinas, Albert of Saxony and Marsilius of Inghen on Book I, Chapter V of Aristotle's De generatione et corruptione*. Ph.D. dissertation. Indiana University, Bloomington.

Cadden, Joan (1980): "Albertus Magnus' Universal Physiology: The Example of Nutrition". In: James A. Weisheipl (ed.): *Albertus Magnus and the Sciences*. Toronto: Pontifical Institute of Mediaeval Studies, p. 321–339.

Callus, Daniel (1961): "The Origins of the Problem of the Unity of Form". In: *The Thomist* 24.2, p. 257–285. DOI: 10.1353/tho.1961.0018, visited on 30 November 2019.

Duba, William O. (2012): "The Souls After Vienne: Franciscans' Views on the Plurality of Forms and the Plurality of Souls, ca. 1315–1330". In: Paul. J.J.M. Bakker/Sander W. de Boer/Cees Leijenhorst (eds.): *Psychology and the Other Disciplines: A Case of Cross-Disciplinary Interaction (1250–1750)*. Leiden: Brill, p. 171–249.

Grellard, Christophe (2017): "Le philosophe et le glouton. Le plaisir de boire et de manger dans les commentaires de l'Éthique à Nicomaque de Jean Buridan et Nicole Oresme". In: Christophe Grellard (ed.): *Miroir de l'amitié. Mélanges offerts à Joël Biard à l'occasion de ses 65 ans*. Paris: Vrin, p. 371–385.

Hamesse, Jaqueline (ed.) (1974): *Les Auctoritates Aristotelis. Un florilège medieval. Étude historique et édition critique*. Louvain: Publications Universitaires. [AA]

Hamesse, Jaqueline/Meirinhos, José F. (eds.) (2017): *Les Auctoritates Aristotelis, leur utilisation et leur influence chez les auteurs médiévaux. État de la question 40 ans après la publication*. Turnhout: Brepols.

King, Peter (2007): "Why Isn't the Mind-Body Problem Mediaeval?" In: Henrik Lagerlund (ed.): *Forming the Mind: Essays on the Internal Senses and the Mind/Body Problem from Avicenna to the Medical Enlightenment*. Dordrecht: Springer, p. 187–205.

King, Peter (2008): "The Inner Cathedral: Mental Architecture in High Scholasticism". In: *Vivarium* 46.3, p. 253–274. DOI: 10.1163/156853408X360911, visited on 30 November 2019.

King, Peter (2012): "Body and Soul". In: John Marenbon (ed.): *The Oxford Handbook to Medieval Philosophy*. Oxford: Oxford University Press, p. 505–524.

Klein, Martin (2019): *Philosophie des Geistes im Spätmittelalter. Intellekt, Materie und Intentionalität bei Johannes Buridan*. Leiden: Brill.

Lagerlund, Henrik (2004): "John Buridan and the Problems of Dualism in the Early Fourteenth Century". In: *Journal of the History of Philosophy* 42.4, p. 369–387. DOI: 10.1353/hph.2004.0071, visited on 30 November 2019.

Marchia, Francis of (2012): *Reportatio IIA (Quaestiones in secundum librum Sententiarum). Quaestiones 28–49*. Eds. Tiziana Suarez-Nani/ William O. Duba/Delphine Carron/Girard J. Etzkorn. Leuven: Leuven University Press.

Michael, Bernd (1985): *Johannes Buridan. Studien zu seinem Leben, seinen Werken und zu Rezeption seiner Theorien im Europa des späten Mittelalters*. Ph.D. dissertation. Freie Universität, Berlin.

Ockham, William of (1970): *Scriptum in librum primum Sententiarum (Ordinatio). Distinctiones II–III*. Eds. Stephen F. Brown/Gedeon Gál. [Opera theologica vol. 2.] St. Bonaventure, NY: Franciscan Institute.

Ockham, William of (1980): *Quodlibeta septem*. Ed. Joseph C. Wey. [Opera theologica vol. 9.] St. Bonaventure, NY: Franciscan Institute. [*Quodl.*]

Ockham, William of (1981): *Quaestiones in librum secundum Sententiarum (Reportatio)*. Eds. Stephen F. Brown/Gedeon Gál. [Opera theologica vol. 5.] St. Bonaventure, NY: Franciscan Institute.

Ockham, William of (1982): *Quaestiones in librum tertium Sententiarum (Reportatio)*. Eds. Francis E. Kelley/Gerald E. Etzkorn. [Opera theologica. Vol. 6.] St. Bonaventure, NY: Franciscan Institute. [*Rep. III*]

Odo, Girald of (2012): "Geraldi Odonis *In secundum librum Sententiarum*, distinctio 18, quaestio unica: *Utrum in homine sint duae formae substantiales*". Eds. Russell L. Friedman/ Christopher D. Schabel. In: Paul J.J.M. Bakker/Sander W. de Boer/ Cees Leijenhorst (eds.), *Psychology and the Other Disciplines: A Case of Cross-Disciplinary Interaction (1250–1750)*. Leiden: Brill, p. 250–272.

Pasnau, Robert (2002): *Thomas Aquinas on Human Nature: A Philosophical Study of Summa theologiae 1a 75–89*. Cambridge: Cambridge University Press.

Pasnau, Robert (2007): "The Mind-Soul-Problem". In: Paul J.J.M. Bakker/Johannes M.M.H. Thijssen (eds.): *Mind, Perception, and Cognition. The Tradition of Commentaries on Aristotle's De anima*. Aldershot: Ashgate, p. 3–19.

Pasnau, Robert (2011): *Metaphysical Themes 1274–1671*. Oxford: Oxford University Press.

Pasnau, Robert (2017): *After Certainty: A History of Our Epistemic Ideals and Illusions*. Oxford: Oxford University Press.

Perler, Dominik (2010): "Ockham über die Seele und ihre Teile". In: *Recherches de théologie et philosophie médiévales* 77.2, p. 329–366. DOI: 10.2143/RTPM.77.2.2062481, visited on 30 November 2019.

Perler, Dominik (2015): "Faculties in Medieval Philosophy". In: Dominik Perler (ed.): *The Faculties: A History*. Oxford: Oxford University Press, p. 97–139.

Rescher, Nicolaus (1960): "Choice without Preference: A Study of the Logic of the Problem of Buridan's Ass". In: *Kant-Studien* 51.1–4, p. 142–175. DOI: 10.1515/kant.1960.51.1–4.142, visited on 30 November 2019.

Sylla, Edith D. (2015): "Guide to the Text". In: John Buridan, *Quaestiones super octo libros Physicorum Aristotelis (secundum ultimam lecturam). Libri I–II*. Eds. Michiel Streijger/Paul J.J.M. Bakker. Leiden: Brill, p. xliii – clxxxvi.

Sylla, Edith D. (2016): "Guide to the Text". In: John Buridan, *Quaestiones super octo libros Physicorum Aristotelis (secundum ultimam lecturam). Libri III–IV*. Eds. Michiel Streijger/Paul J.J.M. Bakker. Leiden: Brill, p. xx – ccx.

Wood, Adam (2011): "The Faculties of the Soul and Some Medieval Mind-Body Problems". In: *The Thomist* 75.4, p. 285–636. DOI: 10.1353/tho.2011.0041, visited on 30 November 2019.

Wood, Adam (2017): "Aquinas vs. Buridan on the Substance and Powers of the Soul". In: Gyula Klima (ed.): *Questions on the Soul by John Buridan and Others: A Companion to John Buridan's Philosophy of Mind*. Cham: Springer, p. 77–93.

Zupko, Jack (2003): *John Buridan: Portrait of a Fourteenth-Century Arts Master*. Notre Dame, Ind.: University of Notre Dame Press.

Christoph Sander
Magnetism and Nutrition
An Ancient Idea Fleshed out in Early Modern Natural Philosophy, Medicine, and Alchemy

Abstract: Already in Antiquity, Galen linked magnetic attraction, by way of an analogy, to the idea that animal parts are able to attract their own 'specific quality'. For example, a kidney attracts urine just like the magnet attracts iron. In the Middle Ages, Averroes argued that foodstuff and iron possess a specific disposition which allows them to move themselves towards the body and a magnet respectively. Thus, the concepts of 'specific attraction' and 'dispositional self-movement' were regarded as crucial to understanding the powers of both a magnet and a living body. Particularly in the early modern period, these concepts were spelled out differently by Aristotelians, Galenists and Paracelsians. During this period, the magnetism-nutrition analogy was also transformed into a vitalist principle in order to explain magnetic attraction itself. Natural philosophers such as Gerolamo Cardano suggested that a magnet, being alive in some way, seeks out iron as its foodstuff – a popular idea among alchemists as well.
This paper aims to trace the complicated history of two intertwined concepts, 'nutrition' and 'magnetism', which were closely related in pre-modern times but appear to be unrelated from a modern perspective. By uncovering the historical origin(s) of this relation, its rationale, its subsequent transformation and its dissolution, the historical concept of 'nutrition' will come into sharper view from the perspective of the history of ideas. At the same time, from the perspective of the philosophy of science, this historical study presents a test case scenario for discussing the importance of metaphors and analogies in the formation of scientific theory.

In 1629, the Dutch philosopher and trained physician Isaac Beeckman jotted down some thoughts in his scientific diary after reading Galen's *De marcore*.[1]

Note: I would like to thank Dorothea Keller for her patient editing and refinement of the many references in this paper, and Michael Infantine for his linguistic revision.

[1] See Isaac Beeckman, *Journal tenu par Isaac Beeckman de 1604 à 1634*, ed. by de Waard (1939), vol. III, p. 124: *Hinc mihi in mentem venit quaerere an membra corporis nostri nutrimentum trahant per virtutem quandam magneticam, an potiùs eo modo quo cor sanguinem attrahit, videlicet*

In his short comment, Beeckman wonders whether the parts of our bodies attract nourishment by a certain magnetic power. At the margin of this page, he also delivered the answer: "Nutrition in our bodies does not work by magnetism." (*Nutrimentum in nobis non fit per magnetismos.*) In *De marcore* and other works, Galen presented his account of attractive and repulsive faculties or powers within a living body in order to explain physiological processes such as nutrition.[2] Moreover, since antiquity, scholars considered magnetic attraction one of the most exciting and miraculous instances of attraction within the natural world (Radl 1988; Sánchez Muñoz/Pajón Leyra 2015; Sander 2019). The magnet obviously was able to pull iron towards itself by some unknown power. Hence, it may be no wonder that two different natural instances of attraction, namely nutrition and magnetism, were compared to each other and stimulated the analogical reasoning of physicians and philosophers.

This article aims at investigating this intertwining of two processes which seem to be completely unrelated nowadays. On the one hand, there is the physiological process of nutrition and growth. Eating a piece of cake, now and then, somehow adds up to an increase of our body. As it seems, bodily parts do indeed attract what they need of this piece of cake, for example certain nutrients like carbs and fats, and expel the rest of it. On the other hand, there is the magnet attracting a piece of iron. In contrast to a living body, a magnet is today considered a dead mineral. This type of attraction, we might say with a layman, is something completely different than any physiological process, as it invokes concepts such as polarity and magnetic fields, and probably does not extend to organic substances.

However, several pre-modern authors stressed certain similarities. Obviously, both a magnet and a living body are natural bodies, made up of the same basic constituents of nature, for example the four elements. Moreover, both nutrition and magnetism seem to be selective in some sense, as they appear to be confined to their specific object of attraction. And among pre-modern authors,

---

per motum diastoles. Cf. also French (1994), p. 170. Cf. also Juan de Pineda, *In Salomonem commentarios Salomon praevius: sive De rebus Salomonis regis, libri octo* (Venice, Thomas Ballionus: 1611), p. 37, who refers to Galen's *De sanitate tuenda: Est vero inter ea mirabilis quaedam proportio, et sympathie, ut sicut magnes ferrum, ita semen attrahat sanguinem, propria sibi et innata facultate.* On Beeckman's understanding of Galen's account of attraction, see also Moreau (2012), p. 142–146. On Beeckman in general, see Berkel (2013).

2 On the issue of attraction in Galen's *De marcore*, see 3 K. VII,674; 5 K. VII,683; 9 K. VII,700. For a translation, see Theoharides (1971), p. 376, 380, 388. For more details, cf. section 1 of this article.

even the categorical difference of life sometimes was given up, as some claimed that a magnet was indeed endowed with a soul and therefore was alive.

In the course of this paper, the focus will be on two aspects or stages of the magnetism-nutrition analogy, with a focus on the early modern period. The first stage concerns the use of the analogy of magnetic attraction in order to illustrate ideas about nutrition, or to elucidate the philosophical concept of attraction itself. As will be explained, authors tried to catch two different aspects of nutrition when comparing it to magnetic attraction. On the one hand, they looked at the physiological process, that is, the way some organs attract nutriment inside the body. On the other hand, they considered the attraction by which a hungry animal is attracted by its nourishment due to its desire to eat. Both the physiological process inside animals and the motivational behaviour of animals were compared to magnetic attraction. And yet, some authors pointed out that such processes and such behaviour do not involve attraction at all, but must be conceived of as self-motion of the *agens* towards its object. This understanding did not disregard the similarity to magnetism, as magnetic attraction could be explained as a sort of self-motion of iron towards the magnet.

The second stage of the magnetism-nutrition analogy concerns the way in which the powers of the magnet itself were understood. As will be shown, the concept of nutrition informed the way in which the magnet's power and its ontological status were conceived. According to this more or less vitalist idea, the magnet itself became a living being and as such a hungry creature (Fletcher 2005; Henry 2001; Weill-Parot 2012).

1 Galen's Account of Attractive Faculties

It is an ancient idea to conceive of nutrition in terms of attraction. Already Empedocles had tried to account for nutrition by the implicit assumption of an attraction of likes by likes.[3] And Aristotle, in his *de Anima*, briefly compared the functions of plant roots with animal mouths; both organs "attract" (ἕλκω) their foodstuff, he says.[4] Thus the idea of describing nutrition as some kind of attraction is quite old. But the actual story of magnetism and nutrition begins

---

[3] Empedocles relies on the principle of *similia similibus*. See Kamtekar (2009).
[4] See Aristotle, *de An.* II,1 (412b1–4): αἱ δὲ ῥίζαι τῷ στόματι ἀνάλογον· ἄμφω γὰρ ἕλκει τὴν τροφήν (*radices vero ori similes sunt: utraque enim trahunt alimentum*). On Aristotle's notion of attraction in biology, cf. Furley (1983), p. 90: "Aristotle recognizes the existence of attraction, but does not elevate it to a principle of importance against his opponents, as Galen does in his polemic against Erasistratus, Asclepiades and others in De nat. fac." Cf. also Boylan (1982), p. 98.

with Galen and, after that, was fleshed out both in the Galenic and Aristotelian tradition.

Galen's very complex theory includes a specific idea of physiological attraction inside a living body, which is crucial for this article (Boylan 1982, p. 102; Brunn 1967, p. 108–109 n. 2; Hall 1975, p. 327–331; Holmes 2012, p. 67; McVaugh 2012; Meyer-Steineg 1913, p. 443–444; Temkin 1972, p. 61; 1977, p. 162–166). Owsei Temkin (1977, p. 162) gives the following short and helpful summary of it:

> Among the fundamental faculties (*dynameis*) of nature Galen counts the faculty of the bodily parts to attract, for nutritional purposes, what is appropriate for them. For his belief in this attractive faculty he claims the authority of Hippocrates. In particular, the formation of urine by the kidneys presupposes that the kidneys attract the thin, serous part of the venous blood. According to Galen, any theory of urine formation which dispenses with the attractive faculty can be proved to be impossible.

Galen's most elaborate account is presented in his *De facultatibus naturalibus* (I,4–5). In order to disprove his adversaries, who deny any attractive faculties in nature, Galen extensively and harshly criticizes the atomistic account of Epicurus. The Atomist, according to Galen, had denied that there is attraction in the natural world, properly speaking. Instead, the propulsion of atoms pushes objects together. Nutrition, accordingly, amounts to a mere and seemingly random collision of atoms in the body, any aspect of an orderly and selective process of attraction being removed. According to Galen, Epicurus also tried to explain magnetic attraction by hook-shaped atoms which drag magnet and iron towards each other.[5] Galen, however, accounts for both magnetic attraction and nutrition by what he calls an "attractive faculty".[6] Such a faculty is not governed by a soul, but by nature as Galen puts it.[7] Both the magnet and the kidney, for example, have their specific object of attraction which they drag towards them. This hap-

---

[5] For a comprehensive discussion, see Radl (1988), p. 172–190. On Epicurus, see also Hankinson (1998), p. 395–400.

[6] Cf. Galen, *De fac. nat.* I,15 K. II,60: "For if it be granted that there is any attractive faculty at all in those things which are governed by Nature, a person who attempted to say anything else about the absorption of nutriment would be considered a fool." (transl. here and in note 7 are taken from Brock 1916).

[7] Cf. Galen, *De fac. nat.* I,1 K. II,2: "we say that animals are governed at once by their soul and by their nature, and plants by their nature alone, and that growth and nutrition are the effects of nature, not of soul." (οὕτως ὀνομάζοντες ὑπὸ μὲν ψυχῆς θ' ἅμα καὶ φύσεως τὰ ζῷα διοικεῖσθαί φαμεν, ὑπὸ δὲ φύσεως μόνης τὰ φυτὰ καὶ τό γ' αὐξάνεσθαί τε καὶ τρέφεσθαι φύσεως ἔργα φαμέν, οὐ ψυχῆς.) On Galen's account of φύσις, see Kovačić (2001).

pens because *agens* and *patiens* of the attraction share similar qualities, i. e., they are mixed in a similar fashion.⁸

Galen's theory is much more complex, but here especially one aspect is crucial: Galen does not claim that nutrition works in any way magnetically, but he links both domains for the first time. He does so by attacking Epicurus's theory of attraction in general by means of his account of magnetic attraction in particular. Moreover, he does so by implying that both forms of attraction instead come about by what he calls a "natural faculty". Hence, to some degree at least, it could be argued that both powers, nutrition and magnetic attraction, are analogical for Galen.⁹

2 Alexander of Aphrodisias and Averroes against Galen

Galen's ideas about this link between magnetism and physiological attraction seem to have bothered a contemporary Peripatetic, Alexander of Aphrodisias.[10] Although Alexander does not name Galen in his *Quaestiones*, he argues against an assumption which seems to be very close to what Galen had argued for. In fact, Alexander points out that the way in which a part of the body attracts nourishment is not the way in which the magnet attracts iron.[11] Thus, Alexander criti-

---

8 In his pharmacological theory, Galen also uses his account of a natural faculty of attraction to explain the specific attraction of a certain type of drugs. Here, he uses magnetic attraction explicitly as an analogy to illustrate how the drugs attract specific fluids in the living body. See Galen, *De simplicium medicamentorum temperamentis et facultatibus* III,15 K. XI,612; Radl (1988), p. 73.
9 This claim is disputable, yet, it is precisely the way in which several of his readers understood his theory. Cf. Alexander's view in the following section and in note 21.
10 Cf. Alexander of Aphrodisias, *Quaest.* II,23, p. 128 in the edition by Gentian Hervet (Basileae: Ioannes Oporinus, 1548) = p. 73–74 in the edition by Bruns (1892). For an English translation, see Sharples (2014), p. 31, for a German one Radl (1988), p. 84. Cf. also Rashed (2011), p. 131, 143. On Alexander's criticism on Galen, see Al-Hasan ibn Musa al-Nawbakhti, *Commentary on Aristotle* De Generatione et Corruptione, ed. by Rashed (2015), p. 168–173; Pines (1961).
11 See Alexander of Aphrodisias, *Quaest.* II,23 (transl. Sharples 2014): "The womb, too, seems to attract the seed, and the veins and limbs [seem to attract] the nourishment. But the magnet does not attract the iron in this way. For of these things too those that attract by first drawing in the intervening air and moisture do this accidentally. Nourishment, too, and everything that is an object of desire and appetite attracts a living creature not by making what is between itself and the object of appetite like itself (for what is between does not become nourishment, nor is this attracted), but rather, what intervenes is set in motion by the object of appetite and trans-

cizes the magnetism-nutrition analogy which seemingly was implied in Galen's account. While the veins attract not only the nourishment but everything spatially between the vein and the nourishment as well, this is not to be observed in magnetic attraction. Yet, Alexander not only points to this disanalogy, but himself introduces another analogy instead when he compares the way in which an animal is attracted by its object of desire, e. g., its nourishment, with magnetic attraction.[12] Here, he does not think about a physiological process inside the body. He rather has in mind that animals get hungry, perceive their food by their senses and are moved towards it. Likewise, according to him, a magnet does not properly and "forcibly" attract iron, but instead the iron is moved towards the magnet by its own desire. The magnet has an elementary nature which the iron lacks, but strives to have, and therefore the iron is moved towards the magnet. In this view, the magnet thus does not change the iron. In the same way, animals are carried or moved to their foodstuff by their desire or hunger. According to Alexander, this natural inclination is not limited to animals endowed with sense perception but pervades all instances of nature, such as minerals like a magnet.[13]

It is roughly this understanding of magnetic attraction which was taken up by most medieval Aristotelians, although they did not know it was Alexander's theory.[14] Alexander's views on attraction entered the medieval stage through the

---

mits the form to what is set in motion, as occurs in the case of seeing. It is in this way that the iron, too, is carried towards the magnet; [the magnet] does not attract [the iron] to itself forcibly, but rather by desire for that which it lacks itself but the magnet possesses. For the magnet too seems to be iron-like, but to have had its moisture dried out either by time or for some other reason. For it is not only things that possess sensation and soul that have a desire for what is natural to them; this is so with many things that are without soul, too."

12 Radl (1988), p. 86, argues that this passage is corrupt and needs to be emended.
13 Alexander attributes this sort of "striving" (ἐφίεται) also to substances without soul (ἄψυχοι). See Alexander of Aphrodisias, *Quaest.* II,23 (Bruns p. 74): οὐ μόνον γὰρ τὰ αἴσθησιν ἔχοντα καὶ τὰ ἔμψυχα ἐφίεται τοῦ κατὰ φύσιν ἑαυτοῖς, ἀλλ' οὕτω πολλὰ καὶ τῶν ἀψύχων ἔχει.
14 Cf. esp. Weill-Parot (2012). On William of Ockham's account, see Goddu (1984), p. 195; Weill-Parot (2012), p. 99. Albertus Magnus instead claims that the magnet is moved to the iron, cf. *Physica* lib. 7 tract. 1, cap. 1, p. 523 in the edition by Hossfeld (1993): *Est etiam adhuc advertendum, quod licet quaedam trahant, tamen non omnia quae aliquo modo trahuntur, dicuntur moveri motu tractionis. Sed aliquando moventur plus motu naturali eius quod trahitur, sicut nutrimentum movetur ad membra non motu membri, quod trahit ipsum, sicut locus trahit locatum, sed potius proprio motu, quia cum assimilatum est secundum aliquid membro, movetur ad ipsum sicut ad suum locum salvantem se in forma, quam recipit. Et hoc etiam modo grave movetur deorsum et leve movetur sursum. Et hoc etiam modo magnes movetur ad ferrum propter similitudinem formae, quam cum ferro habet, et ideo ferrum est locus eius. Et ideo cum impeditur vis illius similitudinis, ferrum non movetur ad magnetem nec e converso. Et ideo cum fricatur suco allii vel lapis vel fer-*

Long and the *Middle Commentary on the Physics of Aristotle*, written by the Arabic philosopher Averroes in the twelfth century.[15] Aristotle had taught that nothing can act on something else unless it is in contact with it (Boas 1947; Buchheim 2007; Wardy 2007: p. 121–151; Weill-Parot 2013: p. 140–144). Thus, there is no action at a distance. According to Aristotle's account in the *Physica*, attraction – in Latin it is simply called *tractio* – is actually a sort of pulling by contact, as if a car is towing another car by a tow-rope.[16] Instances of apparent attraction at a dis-

---

rum, neutrum movetur ad alterum. A passage more closely resembling Alexander's views can be found, e.g., in Thomas Aquinas, *In Physic.* lib. 7, lect. 3, n. 7, p. 461 in the edition by Maggiòlo (1954): *Tertio quia ad hoc quod magnes attrahat ferrum, oportet prius ferrum liniri cum magnete, maxime si magnes sit parvus; quasi ex magnete aliquam virtutem ferrum accipiat ut ad eum moveatur. Sic igitur magnes attrahit ferrum non solum sicut finis, sed etiam sicut movens et alterans. Tertio modo dicitur aliquid attrahere, quia movet ad seipsum motu locali tantum. Et sic definitur hic tractio, prout unum corpus trahit alterum, ita quod trahens simul moveatur cum eo quod trahitur.* Cf. Jean Buridan, *Quaestiones super libros De generatione et corruptione Aristotelis* lib. 1, q. 18, ad 4, p. 141 in the edition by Thijssen/Streijger/Bakker (2010): *De magnete dicendum est quod agit in ferrum, sed prius in medium aerem, ut dicit Commentator octavo Physicorum, ita quod aliquam qualitatem imprimit in aerem et multiplicatur impressio eius usque ad ferrum; et tunc ferrum perillam qualitatem sibi impressam est natum moveri ad magnetem propter aliquam convenientiam. Ita etiam conceditur quod sol agit in illa inferiora, sed prius naturaliter, licet non tempore, agit et multiplicat lumen in sphaeras caelestes sibi coniunctas; igitur tangit suum primum passum.* Cf. Nicole Oresme, *Questiones super physicam* lib. 7, q. 3, ad 4, p. 733 in the edition by Caroti et al. (2010): *Respondetur sicut dicit Commentator, quod movetur ab intrinseco, videlicet a qualitate inducta in eo per ipsum magnetem, ita quod magnes alterat aerem usque ad ferrum, deinde ferrum.* Cf. also in Rommevaux (2010), p. 624. More examples are: Robert Grosseteste (Ps.), *Summa philosophiae* tract. 17, cap. 13; vol. II, p. 613–614 in the edition by Baur (2010); Walter Burley, *Super Aristotelis libros De physica auscultatione lucidissima commentaria* ad VII,10, p. 865–867; ad VIII,35, p. 1006 (Venice: Michele Bernia, 1589). Similar thoughts also can be found in Nikolaus of Cusa, *Idiota de sapientia* I,16, p. 34 in the edition by Baur and Steiger (1983): *Nisi enim in ferro esset quaedam praegustatio naturalis ipsius magnetis, non moveretur plus ad magnetem quam ad alium lapidem.* Cf. however also Nikolaus of Cusa, *Sermo* CLVIII,7, p. 175–176 in the edition by Pauli (2001): *Species seu forma magnetis trahit ad se ferrum, sed non nisi species a forma et virtute procedens mittatur ad ferrum. Postquam spiritus ille missus est ad ferrum ita, quod ibi maneat, tunc movetur ferrum in illo spiritu ad virtutem et essentiam magnetis, et prout movetur magnes sic et ferrum. Nam spiritus magnetis iunctus est spiritui ferri ita,* 15 *quod nexus est indissolubilis. Et non movetur magnes ad motum ferri, sed e converso.*
15 On Averroes's knowledge of Alexander's text, cf. Glasner (2009), p. 159, n. 110.
16 Cf. Aristotle, *Ph.* VII,2, 243a. On this issue, see Wardy (2007), p. 127. On different types of attraction in Aristotle, see Furley (1983), p. 90. Cf. also Maier (1943), p. 174: "Aristoteles und Averroes kennen keine attractio, sondern nur einen tractus, d. h. einen Zug im eigentlichen mechanischen Sinn des Worts, bei dem sich sowohl der Ziehende wie das Gezogene bewegen, also etwa der von einem Pferd gezogene Wagen. Hier ist natürlich der erforderliche Kontakt zwischen Beweger und Bewegtem vorhanden. Aber ein Analogon eines derartigen Vorgangs kommt für die Anziehung durch den natürlichen Ort nicht in Frage." Cf. also Weill-Parot (2012), p. 91–94.

tance naturally posed a challenge to this view.[17] Aristotle does not mention magnetic attraction here or anywhere else in his works as an obstacle to his theory, but Alexander had already tried to cope with this puzzle of magnetism. Centuries later, Averroes also deals with magnetism in the context of Aristotle's understanding of attraction.

In his *Middle Commentary*, Averroes summarized Alexander's account of the intrinsic movement of the iron, probably based on Alexander's now lost *Commentary on the Physics of Aristotle*.[18] Averroes's *Middle Commentary* was translated into Latin and Hebrew a few times but was used rather rarely by Latins.[19] In his widely used *Long Commentary*, Averroes did not mention Alexander's name in this instance but instead presented Alexander's position as his own.[20] He elucidates that in attraction, properly speaking, the substance that attracts cannot be at rest. Thus, the magnet does not actually attract the iron, nor does the part of the body attract its nutriment. In both cases, it is the iron and the nutriment which are moved *ex se* towards the magnet and the bodily part, as they have a

---

[17] On the concept of 'action at a distance' in the Middle Ages, cf. Decaen (2007); Delaurenti (2006); Hesse (2005); Jammer (1999); Kovach (1979), (1986), (1987); Weisheipl (1965).
[18] On the passage of the *Middle Commentary*, see Wolfson (1929), p. 562. On translations of Averroes's works, see Hasse (2010).
[19] The *Middle Commentary* has been translated twice into Hebrew (by Zerahia ben Isaac ben Shealtiel Hen and Qalonymos ben Qalonymos) and three times into Latin (by Vitalis Dactylomelos, Abraham de Balmes, and Jacob Mantino). The seventh book has not been edited until today.
[20] See Averroes, *Commentaria in libros de Physico auditu* lib. 7, summa tertia, fol. 315r in *Aristotelis Opera cum Averrois commentariis* 4 (Venice, Giunta 1562–74, repr. 1962): *Attractio autem, in qua attrahens est quiescens et attractum motum, non est attractio in rei veritate, sed attractum movetur ex se ad attrahens, ut perficiat se, ut lapis movetur ad inferius et ignis ad superius. Et similiter oportet hoc intelligere de motu ferri ad magnetem et nutrimenti ad membra [...]. Nutrimenta vero non moventur ad nutriendum, nisi cum fuerint in quadam dispositione de nutrito, et similiter ferrum non movetur ad magnetem, nisi cum fuerit in aliqua qualitate de magnete. Et ideo quando magnes fricatur cum alleo, amittit virtutem. Nam ferrum non acquirit de lapide in illa dispositione qualitatem, per quam innatum est moveri per se ad lapidem. Et hoc manifestum est in ambra, quod attrahit paleam, quando calefit.* Cf. also lib. 8, cap. 3 (fol. 374v): *[...] et similis ferrum est quoquo modo de numero eorum, quae naturali moventur, cum non moventur a magnete, nisi per alterationem, quam acquirit, mediante aere a magnete. Et non quum complexio magnetis transmutatur, non attrahit; sicut accidit ei, quando confricatur cum alliis, et ut dicitur.* For a philological issue with the first passage, cf. Rommevaux (2010), p. 623, n. 23. According to Ioannes Baptista Montanus, *Medicina universa*, pars 2, p. 375 (Frankfurt: Wechel, 1587), Averroes is contradicting himself in these two passages. Cf. also Avicenna, *The Physics of The Healing* II,8, transl. by McGinnis (2009), p. 188: "Unless, that is, that body emits to [the nature] a certain influence or power, and that influence and power are a certain principle that triggers the moved body to move naturally toward [the body], as in the case of the magnet and iron, in which case the motion is forced, not natural."

certain disposition to do so. He compares this to the elements and their movement to their natural place.

Averroes's theory of magnetic attraction, and thereby his analogue of nutrimental attraction, became widely accepted in Latin medieval Scholasticism.[21]

3 Early Modern Receptions of Galen, Alexander, and Averroes

Averroes, more or less incidentally, had cemented the link between magnetism and nutrition, both being improper forms of attraction. In the early modern period, Galen's account of a proper bodily attraction and Averroes's account co-existed. One of Galen's sixteenth-century commentators, for example, wrote:

> Galen believed that the action [of the magnet] is natural, insofar as everything that strives does so for its own perfection. This is why a part of the body attracts in order to be nourished. [...] There is a certain natural attractive power in every thing [to attract] what is suitable for it, just like the iron for the magnet. This faculty thus is discovered in seeds, plants, animals, stones, and metals.[22]

According to this reading of Galen, living and non-living entities are endowed with the same faculty of attraction, which is the cause both of magnetism and nutrition.

---

[21] Some philosophers also elaborated on or even argued against Averroes's account, for example William of Ockham and Walter Burley. Cf. note 14. A few Arabic and Hebrew philosophers dealt with Averroes's ideas on this issue as well. Cf. Langermann (2011), p. 86–89; Wolfson (1929), p. 90–92, 253–257, 562–568.

[22] Luis de Lemos, *In tres libros Galeni de Naturalibus facultatibus commentarii* in I,14, p. 94 (Salamanca: G. Foquel, 1591): *Galenus autem actionem naturalem esse credidit, quoniam omne appetens propter sui perfectionem appetit: propterea enim membra attrahunt ut nutriantur [...]. Naturalis enim quaedam potentia inest unicuique enti, eorum, quae sibi conveniunt, attractiva, sicut lapidi Herculeo ferri. Quare latissime haec attractrix facultas patet in seminibus, in plantis, in animalibus, in lapidibus, in metallis.* Cf. also Antonio Luiz, *De occultis proprietatibus* II, praef., fol. 16r (Lisboa: Luis Rodrigues, 1540): *Latissime autem haec attractix facultas patet in seminibus, in plantis, in metallis, in animalibus. Et denique ausim affirmare attracticem quandam facultatem, per omnem naturam diffusam esse, quae singula nexu indissolubili devinciat. Nec enim aliquam rem reperire quis facile possit, quae non ad aliam quampiam: vel amicam familiaritatem habeat, vel naturae communione non dissideat, ex qua convenientia, vel disconvenientia attractiones fieri docebimus. Per hanc virtutem mundus ipse connectitur, et mundi partes invisibilibus nodis: quamvis longissime distantes, ne diffluant continentur. Hac facit ut similia similibus coniugantur.*

Averroes's theory of ascribing a form of accidental self-motion to iron was, for example, elaborated in at least three different works by Gabriele Falloppio.[23] The Italian university-trained physician also pointed out that the movement of nutriment and iron is caused by their inherent nature and cannot be inverted. Thus, he says, this form of attraction must not be imagined like the following scene: friend A pulls friend B to a dinner meal, but B pulls stronger in another direction and therefore they do not end up at the dinner table but somewhere else.[24] In magnetism and nutrition however the causal relation is pre-established. There is no stronger party in any quantitative sense. Even a large piece of iron cannot pull a small piece of magnet – it will always be the iron that is moved towards the magnet.

But Averroes's concept was also criticized by some physicians, e.g., by Bartolomeo Eustachi and Ioannes Baptista Montanus.[25] In natural philosophy, an outgrown controversy emerged about Averroes's question of causal order within magnetic attraction, reaching far into the seventeenth century and including famous scholars such as Julius Caesar Scaliger and Fortunio Liceti.[26] In 1641, Athanasius Kircher even claimed that several physicians hold the view that nutrition is properly magnetic.[27] This cannot be, the Jesuit explained, as different parts of

---

23 This issue is discussed in Falloppio's *Observationes anatomicae* (1561), p. 442 in the 1584 edition (*Opera, quae adhuc extant, omnia*, Frankfurt: Wechel); *De tumoribus praeter naturam* (1563) 3, p. 703, *De simplicibus medicamentis purgantibus* (1565) 3, p. 27. On this issue, cf. also briefly in Gadebusch Bondio (2005), p. 150–151.

24 See Gabriele Falloppio, *De simplicibus medicamentis purgantibus*, p. 27. Cf. also *Observationes anatomicae*, p. 442 (both cited after the 1584 edition, cf. note 23): *Sed quoniam se ipsum movet ad aliquem determinatum motum, aut scopum: ideo dicitur trahi ab illo, veluti ferrum trahi a magnete, et alimentum a membris ipsis dicitur, cum tamen re vera non attrahantur, neque vera sit haec attractio, sed analoga, quia motus iste localis non fit ab alio, sed a virtute intima, qua ferrum vel alimentum movent se ipsa ad determinatum magnetem, aut membrum.*

25 Cf. Bartolomeo Eustachi, *De rerum structura, officio & administratione* 32, p. 106–109, in the edition by Pietro Matteo Pini (in: *Opuscula anatomica*, Venice: Vicenzo Luchino, 1564); Ioannes Baptista Montanus, *Medicina universa*, pars 2, p. 375–376, 383, 387 (Frankfurt: Wechel, 1587); Fabius Pacius, *Commentarius in Galeni libros methodi medendi* ad I,6, p. 170–171 (Vicenza: Grecus, 1597). Cf. also Hall (1975), p. 340–341.

26 Cf. Julius Caesar Scaliger, *Exotericarum exercitationum liber quintus decimus, De subtilitate, ad Hieronymum Cardanum*, ex. 102,6, fol. 156v–157r (Paris: Michael Vascosanus, 1557). Cf. also Fortunio Liceti, *De spontaneo viventium ortu libb. quatuor, in quibus de generatione animantium, quae vulgo ex putri exoriri dicuntur, accurate aliorum opiniones omnes [...] examinantur* I,116, p. 115 (Vicenza: Franciscus Bolzeta, 1618); *Litheosphorus, siue, De lapide Bononiensi lucem in se conceptam ab ambiente claro mox in tenebris mire conseruante* 42, p. 184–187 (Udine: Nicolaus Schirattus, 1640).

27 Cf. Athanasius Kircher, *Magnes; sive, De arte magnetica opus tripartitum* lib. III, pars 5, cap. 7, p. 721 (Rome: Hermann Scheus, 1641): *Plerique Medicorum existimant, attractionem, qua animal*

a plant, for example, attract different substances out of the ground. A magnet does not exhibit such distinction with regard to its object of attraction. However, Kircher is fully wont to accept magnetism as an analogy to describe physiological instances of attraction and repulsion.[28]

4 Paracelsus's Idea of Magnetic Nutrition

The magnet-nutrition analogy reached far beyond the Latin, more or less academic realms of early modern medicine. If we look for example in a Carthusian manuscript of the late fifteenth century, which deals with practical medical topics in the vernacular, one finds that the liver's attraction of foodstuff is straightforwardly compared to magnetic attraction.[29] Especially Paracelsus seemed to be fascinated by this type of analogy (Kerner 1973; Rutschow 1965; Schott 2002). Although awarded with a bachelor in medicine at Vienna, Paracelsus distinguished himself sharply from the academic establishment of his time, altogether with Galenic medical theory and Aristotelian natural philosophy (Dopsch et al. 1993; Grell 1998; Letter 2000; Pagel 1958).[30] Yet, the nutrition-magnetism analogy remained. He even pushed it one step further.

Already in his earliest writings of the 1520s, he explained that each part of the body attracts its suitable nourishment just like a magnet attracts iron.[31] Whilst only the stomach gets nourished in a visible way, all other parts are nourished invisibly, just like the magnet works by an invisible power. In Paracelsus's

---

aut planta conveniens sibi nutrimentum attrahit, esse vere Magneticam, perperam; magnes enim ob multos alios fines cum consimili sibi corpore coit, quam planta cum nutrimento, ordinantur enim illi coitus, in ordine ad situm naturae convenientem cum toto obtinendum, sive ad individuum sibi subiectum conservandum, quae admodum diversae sunt. Cf. also William Gilbert, *De magnete, magneticisque corporibus, et de magno magnete tellure; physiologia noua, plurimis et argumentis et experimentis demonstrata* II,39, p. 112 (London: Short, 1600); Andreas Libavius, *Singularium, Pars tertia: continens octo libros bituminum et affinium, historicè, physicè, chymicè, cum controuersiis difficilimis, expositorum indicatorumq[ue]* I,10, p. 86 (Frankfurt: Kopff, 1601).

28 Cf. Athanasius Kircher, *Magnes*, lib. III, pars 7, cap. 5, p. 817: *Pari ratione attrahimus insecus alimentum humidum et solidum, superfluum vero noxium expellimus, ita ut nullum membrum sit, quod non attractiva vi, aut expulsiva sit imbutum.* Cf. also Čermáková (2018).
29 Cf. Sailer (2012), p. 138, transcribing UB München, 2° Cod. ms 578, fol. 26r: "Und als das tranck und das essen in den darm kumbt, so zeucht die leber das tranck an sich mit ainem swais, recht als ain magnet das eysen an sich reuchet zuhant, als das tranck in die lebern kumet."
30 Cf. also Dopsch (1993); Jaeckle (1993); Pagel (1982b), p. 203–310; Schipperges (1993).
31 Cf. Paracelsus, *Elf Traktat von Ursprung, Ursachen, Zeichen und Kur einzelner Krankheiten. I. Von der Wassersucht [Andere Redaktion]*, ed. by Sudhoff (1929a).

astrological medicine it is the "heavenly kitchen" from which the body attracts its spiritual nutriment by some magnetic power.[32] Especially in his *Philosophia sagax* Paracelsus develops this idea of an outright magnetic nutrition.[33] As known, his ideas are not always easy to understand but a sketch of his basic idea might be this: "Man", Paracelsus claims, "is a double magnet." On the one hand, he attracts visible food, being an "elemental magnet". On the other hand, by means of his "heavenly magnet", he attracts thoughts, for example by the power of imagination. A magnet sucks the juice out of a piece of iron, in the same way the elemental magnet of the body absorbs food and drink. Thoughts are attracted from the stars directly by the heavenly magnet.

This idea of magnetic nutrition even informs Paracelsus's religious ideas. In a work on piously motivated abstinence from food, the Swiss Paracelsus refers to his fellow countryman, the fifteenth-century ascetic and saint Nikolaus von Flüe, who allegedly survived nineteen years without food, except for the Eucharist.[34]

---

32 Cf. Paracelsus, *Von der Wassersucht [Andere Redaktion]*, p. 12–13: "es ist die kuchin des himels, aus der do ziehen unser glider durch magnetische kraft ir solche narung; die ist ein ursach der krankheiten, so sie also gekocht wird."

33 Cf. Paracelsus, *Astronomia Magna oder die ganze Philosophia sagax der großen und kleinen Welt*, ed. 1929 by Sudhoff (1929b), p. 163–164: "zu gleicher weis, wie der magnet das eisen an sich zeucht und saugt im den saft aus und leßt den rost fallen, also ist der mensch ein zweifacher magnet, des leibs halben, darumb er sein speis an sich zeucht, der weisheit halben, darumb er das gestirn an sich zeucht. in elementen fint er die narung seines bluts und fleischs, im gestirn fint er die weisheit seiner sinn und gedanken durch die anziehende kraft, so ein ieglicher mensch zwifach an im hat wie gemelt ist also zeucht der mensch auch an sich durch den elementischen magneten die speis und trank, und was nichts sol, das leßt er fallen durch die excrementa und behalt den saft bei ime." Cf. also Paracelsus, *Die Bücher von den unsichtbaren Krankheiten*, 1531/32, ed. by Sudhoff (1925), p. 311, 316, 363; *Astronomia Magna oder Philosophia sagax*, p. 39. Cf. also Daniel (2006), p. 137; Athanasius Kircher, *Magnes*, lib. III, pars 7, cap. 5, p. 818.

34 Cf. Paracelsus, *Von Fasten und Casteyen*, ed. by Goldammer (1965), p. 435–436: "Iedoch aber soll sich niemandt verwundern, daß der mensch also ohne gessen und trunken sein leben verhalten kan und auswendig gespeist werden mag dann zugleicherweis wie der magnet sein speis vom eisen an sich zeucht, allein so er darein gelegt wird, also tut auch der magen im mensche; der auch wie ein magnet ist, muß auch sein tägliche speis haben, es sei dann inwendig oder auswendig. wie ein magnet liebe, lust und begird zum eisen als zu seiner speis hat, also auch der magen im menschen ein magnetische begirigkeit zur speis hat, also daß ers auswendig hinein kan ziehen. aber nit den corpus, sonder den spiritus und quinta essentia der speis, welliche kein unlust oder kat im magen macht, als so sie mit dem mund corporalisch gessen wird. Darumben sag ich hie, daß der magen ein magnet, und die speis, so soll gessen werden, das eisen ist. und wie der magnet sein speis muß vom eisen haben, also der magen auch sein nahrung von kreutern empfacht wo er aber sein speis, das ist das eisen, nimmer haben kan, hebt er sich selbst an zu fressen an seinen kreften, also lang und vil, bis er abstürbt. Deßgleichen tut

This "hunger artist" did it like the magnet, Paracelsus explained with severe admiration. He only attracted the spirit and the essence of the food, not its physical body – just like a magnet does not increase in weight when attracting iron. If the stomach and the magnet are kept from their nourishment they begin to consume themselves, leading to starvation eventually.

This alleged magnetic diet leads to the second part of this paper. Paracelsus transformed the Galenic analogy into a full-fledged concept of magnetic nutrition. Linguistically, Paracelsus does not only compare magnetism and nutrition, but he spells out nutrition in terms of magnetism. In his words, man really is a magnet. As a side effect, the magnet attracts its food as well. Iron then is not just the object of attraction but the magnet's nutriment, thereby making the magnet somehow alive. Paracelsus has no problems with these vitalist implications and, for example, he calls the two sides or poles of a magnet its "stomach" and its "back".[35]

5 Cardano's Vitalist Ideas about Magnetic Attraction

The idea of a living magnet is ancient as well. According to Aristotle, Thales held the magnet to be ensouled, as the soul was considered the principle of motion and the magnet moved iron.[36] The magnet was called "living stone" in several

---

auch der magen im menschen, wann er nimmer gespeiset wird, es sei auswendig oder innwendig, das ist corporalisch oder spiritualisch: so hebt er sich auch an zu fressen selbst und verleurt seine kräft der däuung und greift sich selbs an, zugleicherweis wie ein mensch, der hunger sterben soll, der auch sein eigen fleisch frißt." Cf. also Eis (1965), p. 175. Cf. also Paracelsus's *De sensu et instrumentis* (1530s), ed. by Goldammer (1965), p. 91: "item machst dir dein aug zu einem wolf und magneten und agstein, daß du fressest, was du sihest und an dich reißest; das alles nichts soll. drumb sein deine augen nit da. dann der magen isset für all unsere glieder; derselbig ist an einem kleinenll gespeiset."
35 Cf., e.g., Paracelsus, *"Herbarius" von den Heilwirkungen der Niswurz der Persicaria, des Salzes, der Engel-Distel, der Korallen und des Magneten*, ed. by Sudhoff (1930), p. 52; 53; 56. Cf. also Johann Baptist van Helmont, *De magnetica vulnerum curatione* (Amsterdam: Ludovicus Elzevirius, 1652), p. 597. On this, see Schott (2011), p. 107. Cf. also note 46.
36 Cf. Aristotle, *de An.* I,2, 405a19–21: "Ἔοικε δὲ καὶ Θαλῆς ἐξ ὧν ἀπομνημονεύουσι κινητικόν τι τὴν ψυχὴν ὑπολαβεῖν, εἴπερ τὴν λίθον ἔφη ψυχὴν ἔχειν, ὅτι τὸν σίδηρον κινεῖ. For other mentions in the presocratics cf. Wöhrle/Strohmaier (2009), p. 52–53, as well as p. 36; 52; 138; 180; 198; 268; 330; 346; 460; 477. Cf. also Hankinson (1998), p. 12. Porphyry agrees with Thales in his *de Abstinentia.* Cf. Porphyrius, *Abst.*, ed. by Nauck (1886), IV,20: ὁ δὲ μάγνης λίθος σιδήρῳ

magical recipes of ancient papyri and later medical and magical writings.[37] The Latin poet Claudianus and the Christian philosopher Nemesius already referred to iron as the magnet's food, yet only by way of metaphorical or as-if parlance.[38] In the Middle Ages, Thales's account was known through Aristotle, but virtually all Scholastic philosophers disagreed with this idea. A piece of iron may move itself towards the magnet but not in the way an animal is capable of locomotion. Peter of Spain also discussed the apparent ability of the magnet to select its object of attraction, as it does not attract gold, for example.[39] This may seem like

---

ψυχὴν δίδωσι πλησίον γενομένῳ, καὶ ὁ βαρύτατος ἀνακουφίζεται σίδηρος πνεύματι προσανατρέχων λίθου. Cf. also the translation given by Clark (2014), p. 117; Radl (1988), p. 90.

[37] Cf., e.g., Aëtius of Amida, *De re medica* VII,39, p. 308 (Basel: Froben, 1535 [*Libri XVI: In Tres Tomos divisi* [...] *In quo opere cuncta quae ad curandi artem pertinent congesta sunt* 1]); Janus Cornarius, *Pedanii Dioscoridis De materia medica* V,111, p. 477 (Basel 1557); *Kyranides*, ed. by Delatte (1942), p. 48, 86, 87, 97; Hans Friedrich Herwart von Hohenburg, *Admiranda Ethnicae Theologiae mysteria propalata: Ubi Lapidem Magnetem Antiqvissimis Passim Nationibus Pro Deo Cultum* [...] *commonstratur; Accessit Exacta temporum ratio* 54, p. 189–190 (Ingolstadt: Eder, 1623); *Moderante Auxilio Redemptoris Supremi, Kirani Kiranides, Et ad eas Rhyakini Koronides: Quorum ille In Quaternario tam Librorum, quam Elementari, e totidem Linguis, Primo de Gemmis XXIV, Herbis XXIV, Avibus XXIV ac Piscibus XXIV quadrifariam semper et fere mixtim ad Tetrapharmacum constituendum agit*, ed. 1638 by Andreas Rivinus (Leipzig), p. 30; 61; 69; 82; 123; *Kyranides*, ed. by Kaimakēs (1976), I,7,63–67; I,21,90–103; I,24,97; II,20,21; *Orphei Lithica. Accedit Damigeron De lupidibus*, ed. by Abel (1881), p. 24–25; *Papyri graecae magicae*, ed. by Preisendanz/Henrichs (1973) IV,2875; IV,2627; IV,1721; II,18; Röhr (1923), p. 92–93.

[38] See Nemesius of Emesa, *De natura hominis* 1, p. 3 (ed. by Morani, 1987): διαλλάττει μὲν γὰρ καὶ λίθος λίθου δυνάμει τινί, ἀλλ' ἡ μαγνῆτις λίθος ἐξεληλυθέναι δοκεῖ τὴν τῶν ἄλλων λίθων φύσιν τε καὶ δύναμιν ἐν τῷ προφανῶς ἕλκειν πρὸς ἑαυτὴν καὶ κατέχειν τὸν σίδηρον, ὥσπερ τροφὴν αὐτὸν ποιήσασθαι βουλομένη, καὶ μὴ μόνον ἐφ' ἑνὸς σιδήρου τοῦτο ποιεῖν, ἀλλὰ καὶ ἄλλον δι' ἄλλου κατέχειν τῷ μεταδιδόναι τοῖς ἐχομένοις πᾶσιν τῆς δυνάμεως ἑαυτῆς· κατέχει γοῦν καὶ ὁ σίδηρος σίδηρον ὅταν ὑπὸ τῆς μαγνήτιδος ἔχηται. Cf. Claudian 29(48), vol. II, p. 236 (ed. by Platnauer 1922): *nam ferro meruit vitam ferrique rigore / vescitur; hoc dulces epulas, hoc pabula novit; / hinc proprias renovat vires; hinc fusa per artus / aspera secretum servant alimenta vigorem; / hoc absente perit: tristi morientia torpent / membra fame, venasque sitis consumit apertas.* Cf. also Radl (1988), p. 98, 105. On physiological metaphors for describing dead objects, cf., e.g., Aristotle, *de An.* II,4, 416a1–30. Cf. Falcon (2005), p. 119–120; Taub (2012). Pagel (1982a), p. 139, assumes that Nemesius was known to Paracelsus. For further ancient examples, cf., e.g., Achilles Tatius, *Leucippe and Clitophon* III,2; p. 50 (ed. by Vilborg 1955); Theophylaktos Simokates, *Epistulae*, ed. by Zanetto (1985), p. 15; Johannes Tzetzes, *Biblion historikēs tēs dia stichōn politikōn alpha de kalumenēs: Graece = Joannis Tzetzae historiarum variarum chiliades*, Chiliada 4, Epistula, p. 145 (ed. by Kiessling 1963); Letoublon (2014).

[39] Cf. e.g. Lawn (1993), p. 151; Petrus Hispanus, *Comentario al "De anima" de Aristóteles* lib. I, lect. 13, p. 442–444 (ed. 1944 by Alonso). Cf. also the anonymous *Tractatus utriusque philosophie* (14th century), edited in Lohrmann (2008), p. 240: *Illi enim qui per motum animam investigarunt dixerunt omne corpus esse animatum quod per se movetur, et convenienter opinati sunt quod*

being alive or making a choice, but, as Peter points out, it is not. Albert the Great, the most important authority in the field of medieval mineralogy, generally discredited the view that minerals or gemstones are alive.[40]

In the early modern period, the idea of a living magnet was reanimated. A very prominent example is Gerolamo Cardano who extensively argued that all mixed bodies are alive in some basic sense.[41] His main arguments came from mineralogy. As he put it, minerals or mixed things,

> are nourished, and nourishing does not take place except by the soul, and what possesses a soul is alive. But if you deny that they are nourished, you will anyway accept that they are generated; yet nothing is generated except by a soul, because, as I said, it alone does mixing.[42]

Moreover, metals are born in the mountains, nourished as they show growth (*De subtilitate* V, p. 345). Yet, distinct from animals, metals lack respiration, which is why they can be generated under the ground like a mole or worms. Likewise, with a more empirical approach, Cardano points out that metals often have veins inside their material structure, which must be considered their organs for nutrition, just as we find veins in plants and animal bodies.[43]

The very idea of the generation of metals by seeds and a form of nutrition also dates back to ancient metallurgy.[44] Minerals often were also divided into

---

anima nihil erat nisi quedam virtus motiva. Huius opinionis fuit Tales Milesius, unus de septem sapientibus Atheniensium qui etiam credidit lapidem adamantinum esse animatum eo quod motum per se causabat, quia ferri naturaliter et de se dictus lapis est attractivus. Cf. also Oelze (2018), p. 102.

40 Cf. e. g. Georgius Agricola, *De natura fossilium* I, p. 171 (Basel: Froben, 1546); Marsilio Ficino, *De vita* III,3, p. 257 (ed. 1989 by Kaske/Clark); Weill-Parot (2012), p. 96. On Albert's mineralogy, cf. Riddle/Mulholland (1980); Strunz (1951). For ancient theories, cf. Halleux (1974).

41 On Cardano's natural philosophy, cf. Fierz (1983); Schütze (2000).

42 Cf. Gerolamo Cardano, *De subtilitate* V, p. 356 (Basel: Henric Petrina, 1560): *itaque omnia mixta vel vivere, vel vixisse necessarium est. Sumatur autem ratio hoc modo; quia nutriuntur, et nutrimentum non fit nisi ab anima, et quod animam habet, vivit. Quod si nutriri neges, concedes saltem generari; at nihil generatur nisi ab anima, quia, ut dixi, sola illa miscet*. Translations are taken from Forrester (2013).

43 Cf. Cardano, *De subtilitate* V, p. 360 (1560): *Dixeramus metallica, ac metalla, tum lapides vivere. Quorum enim est maturitas, et acerbitas, ac senium, eorum etiam est vita. Nam lapides quidam immaturi, et colore diluto, et substantia haud concocta inveniuntur; pars quoque velut et in eiusdem arboris fructibus purior, alia impurior cernitur. Adsunt praeterea venae et instrumenta nutritionis, et meatus non laxiores, sed molliores, ut in lapidi bus cernere licet, quibus facile possumus persuaderi, non aliter quam plantas et ossa in animalibus nutriri; nam si per accessum et additionem augerentur, his haud quaquam indigerent.*

44 Cf. note 40.

male and female specimens, leaving room for the idea of a sexual reproduction (Halleux 1970). Moreover, Neo-Platonists such as Marsilio Ficino had suggested that metals and gems have a generative faculty.[45] And, of course, Paracelsus was sympathetic to view of living minerals as well.[46] What is important for this paper is that Cardano for the first time applied some of these mineralogical ideas directly and extensively to the magnet. He even used the magnet as a paradigmatic example to demonstrate his vitalist mineralogy. In addition to those indications for metal life that have been mentioned, he invoked the magnet as an example to demonstrate that some minerals also show a life-like activity, such as magnetic attraction.

Cardano's mineralogical ideas were first presented in his encyclopaedic work, *De subtilitate*, of 1550, but he had related to the magnetic power of nutrition before that in the second edition of one of his chief medical works of 1548.[47] It is precisely in the context of a discussion of the ideas of Galen, Alexander, and Averroes regarding the attractive faculty of nutrition, that Cardano also suggests that the magnet seeks iron as its food. The magnet, Cardano says, may not be alive, but has a certain kind of perception as it is able to detect iron as his food.[48] Moreover, probably following an allusive idea in Pseudo-Alexander's *Problemata*, when a magnet is covered in iron fillings, its attractive powers are preserved as the iron fillings serve as a kind of nutrimental refreshment.[49]

---

45 Cf. note 40.
46 Cf. also e.g. Paracelsus, *De natura rerum neun Bücher* 11, p. 331 (ed. by Sudhoff 1928): "Das leben des magnets ist ein spiritus ferri" (*Vita magnetis est spiritus ferri*). On Paracelsus's mineralogy, cf. Hiller (1952a/b); Schröter (1941). On metallogenesis, cf. Hirai (2008).
47 See Cardano, *Contradicentium medicorum* lib. II, tract. 1, contr. 3, p. 18–21 (Lyon: Gryphius, 1548). Not included in the edition of 1545 (Venice: Hieronymus Scotus).
48 Cf. Cardano, *Contradicentium medicorum* lib. II, tract. 1, contr. 3, p. 20 (1548) = *Opera Omnia* (Lyon: Ioannes Antonius Huguetan/Marcus Antonius Ravaud, 1663), vol. VI, p. 442. Cf. also *Contradicentium medicorum* lib. II, tract. 4, tract. 6, p. 434 (1548) = (1663), vol. VI, p. 641.
49 Cf. Cardano, *Contradicentium medicorum* lib. II, tract. 1, contr. 3, p. 21 (1548). Cf. also lib. II, tract. 5, contr. 9, p. 277 = (1663) vol. VI, p. 442, 566. Cf. already in Agricola, *De natura fossilium* V, p. 251 (1546): *Quinetiam magnes si diu ferro aut eius vena careat, aliquam virium iacturam facit; quod ne fiat, squama ferri est obruendus*. Pseudo-Alexander already alluded to the vivifying power of iron fillings. Cf. Alexander of Aphrodisias (Ps.), *Problemata* 1,19–21, p. 4 (ed. by Ideler 1841): ἢ διὰ τί λίθος ἡ μαγνῆτις ἕλκει μόνον τὸν σίδηρον, ὑπό τε τῶν τούτου ῥινημάτων ζωοποιεῖται ἡ λίθος. This however was only included in the Latin translation by Angelo Poliziano but left out in Giorgio Valla's translation. Cf. Angelo Poliziano, *Omnia opera Angeli Politiani: et alia quaedam lectu digna, quorum nomina in sequenti indice uidere licet* (Venice: Aldus Romanus, 1498), fol. T4v: *cur magnetis lapis ferrum tantum attrahit deque eius scobe vivificatur*. The 1501 edition of the *Pr.* (Venice: A. Vercellensis), fol. 3r, lacks this passage. Theodorus Gaza translates *iuvari* (*Aristotelis de natura animalium. lib. ix.: Eiusdem de partibus animalium. lib. iiii. Eius-*

However, two years later in his *De subtilitate,* Cardano changed his mind and no longer hesitated to call the magnet alive, just like all mixed things. But where there is life, there is death, he infers. Thus Cardano points out: "There is death in [metals] too, hence life as well. In my hands, the [magnet] has quite often actually become powerless in a few years; although it used to attract iron with vigour, later as time passed it stopped doing so."[50] With regard to the magnet's alleged nutrition he claims that the magnet "longs for iron as food; since the magnet cannot attract the iron to itself, it is carried towards the iron."[51] This statement

---

dem de generatione animalium. lib. v. Theophrasti de historia plantarum. lib. ix. Et decimi principium duntaxat. [sic] Eiusdem de causis plantarum. lib. vi. Aristotelis problemata [...] Alexa[n]dri Aphrodisiensis problemata [...] [Venice: Aldus, 1504), fol. 256v]), while a later editor, Jean Davion, adds a margin to correct that the Greek term (ζῳοποιεῖται) should be translated with *nutriri* (Paris: E. Tusana, vidua C. Neobarii, 1541, fol. 5r). In the Greek edition of Alexander of Aphrodisias (Ps.) *(Iatrika aporēmata, kai physika problēmata.* Paris: E. Tusana, vidua C. Neobarii, 1540), this passage is included. Also Luis de Lemos, *Galeni de Naturalibus facultatibus commentarii* ad I,14, p. 92; Giambattista della Porta, *Magiae naturalis* VII,50 (Naples: Salvian, 1589), p. 146, ascribe this idea to Alexander and rely on Poliziano's translation. On the complicated textual history of the *Pr.*, cf. Goynes/Leemans (2007). Cf. also *Kyranides* I,21,103 (Kaimakēs' 1976 text): καὶ ὀλίγου ῥινίσματος σιδήρου, διὰ τὸ εἶναι ζῶντα τὸν μαγνήτην. In the Latin translation the causal διά is translated as final *ut.* Cf. Delatte's 1942 text, p. 87: *limaturae ferri, ut magnes vivat.* A collection of passages mentioning the vivifying power of iron fillings can be found in a manuscript by the Vatican librarian Leone Allacci. See his *De magnete* preserved in Rome, Biblioteca Vallicelliana, Allacci LXXVII, here fol. 18r–v. Among the passages quoted there, Allacci also refers to Jean de Renou, *Institutionum pharmaceuticarum* IV,12, p. 128, *De materia medica* II,9, p. 205 (both Paris: De la Nouë, 1608); Kaspar Schwenckfeld, *Stirpium & fossilium Silesiae catalogus. In quo praeter etymon, natales, tempus ; natura & vires cum varijs experimentis assignantur* III, p. 384 (Leipzig: David Albertus, 1600); Marin Mersenne, *Quaestiones celeberrimae in Genesim: cum accurata textus explicatione* ad Gen 1,13 art. 3, p. 948 (Paris: Cramoisy, 1623); Anselmus de Boodt, *Gemmarum et lapidum historia qua non solum ortus, natura, vis & precium, sed etiam modus quo ex iis olea, salia, tincturae, essentiae, arcana & magisteria arte chymica confici possint, ostenditur: opus principibus, medicis, chymicis, physicis, ac liberalioribus ingeniis utilissimum: cum variis figuris, indiceq. duplici & copioso* II,244 (Hanau: Wechel 1609). On Allacci's study, see Sander (2020). See also Johannes Eck, *Aristotelis Stagyritae Acroases Physicae,* fol. 91r (Augsburg: Grimm/Wirsung, 1518): *magnes etiam consumit ferrum et utitur eo quasi nutrimento, ut omnes norunt qui habent magnetes.*
50 Cf. Cardano, *De subtilitate* V, p. 361 (1560): This observation of a maget loosing its power already appears in *Contradicentium Medicorum* lib. II, tract. 2, contr. 7, p. 101. Cf. also *De sublimitate* VII, p. 441 (1560): *Vivere superius omnia quae miscentur demonstravimus hoc autem maxime lapidibus convenit. Neque solum vivunt, sed morbos, et senectutem, et post etiam mortem patiuntur. Nam senio confectus Herculeus lapis, ferrum non trahit; pedore etiam ac squalore debilitatur, ut etiam animal. Neque enim qualitate id agunt, sed vita.*
51 Cf. Cardano, *De subtilitate* VII, p. 494 (1560 = fol. 158v in the Parisian edition [M. Fezandat/R. Granion, 1550]): *Hoc ideo contingit, quoniam ferrum magnes ut pabulum* [1550: *alimentum*] *desid-*

is particularly interesting as it openly contradicts Averroes's and Alexander's claims, namely that the iron is moved towards the magnet. It is the magnet, Cardano emphasizes, which is moved to the iron due to its appetite.

6 Alchemical Theories of Magnetic Attraction and Nutrition

Cardano's vitalist ideas about the generation and nutrition of minerals in general and about the magnet in particular were widely criticised. Amongst his most severe adversaries was Julius Caesar Scaliger.[52] He ridiculed, for example, that Cardano revived Thales's ideas and that the magnet's alleged soul would be even nobler than the human soul, as the human soul was only able to attract an object which it touches.[53] Cardano's ideas about mineral nutrition were also attacked by Scaliger on the grounds of an Aristotelian account of the soul and its powers. For example, it is mistaken, he points out, that all generation implies nutrition.[54]

But the idea of the magnet's nutrition did not die with Cardano and Paracelsus. Particularly later alchemists followed their leads.[55] The pseudographic al-

---

erat; qui cum illud ad se trahere nequeat, ad ipsum mutata vice fertur. Others critics of this vitalist idea were Garzoni (ed. Ugaglia 2005), p. 119, 221–222; Mersenne, *Quaestiones celeberrimae in Genesim*, p. 948. Cf. Allacci, *De magnete*, in Rome, Biblioteca Vallicelliana, Allacci LXXVII, fol. 16r–20r; Giulio Cesare LaGalla, *Disputatio de sympathia et antipathia*, in Rome, Biblioteca Vallicelliana, Allacci XXX.4, fol. 61r.

[52] Cf., e.g., Scaliger, *De subtilitate, ad Hieronymum Cardanum* 101,15–18, fol. 149v–152v. On this dispute in general, cf. Giglioni (2015); Leinsle (2009); Maclean (1984). On Scaliger, cf. Sakamoto (2016). Cardano replied in a later work (*In calumniatorem librorum de subtilitate*, 1559; text available in *Opera omnia* [Paris: Ioannes Antonius Huguetan/Marcus Antonius Ravaud], vol. III, p. 693).

[53] Cf. Scaliger, *De subtilitate, ad Hieronymum Cardanum* 102, fol. 155v.

[54] Cf. Scaliger, *De subtilitate, ad Hieronymum Cardanum* 101,18, fol. 151v. Scaliger, e.g., refers to the generation of water out of air.

[55] Cf., e.g., Tommaso Campanella, *De sensu rerum et magia, libri quatuor, pars mirabilis occultae philosophiae, ubi demonstratur, mundum esse Dei vivam statuam, beneque cognoscentem* I,8, p. 27 (Frankfurt: Egenolphus Emmelius/Godefridus Tampachius, 1620); Pietro Maria Canepari, *De atramentis cuiuscunque generis opus sanè novum hactenus à nemine promulgatum in sex descriptiones digestum*, descr. 1, cap. 6 (Venice: Euangelista Deuchinus, 1619), p. 26; Petrus Magirus, *Antilogia inutilis futilisque: discursus duorum cultrivoracum*, p. 151 (Linz: 1639); Heinrich Nolle, *Naturae Sanctuarium: Quod Est, Physica Hermetica: In Studiosorum Sincerioris Philosophiae gratiam, ad promovendam naturalium rerum veritatem, methodo perspicua et admirandorum Secretorum in Naturae abysso latentium philosophica explicatione decenter in undecim libris*

chemist Basilius Valentinus, for example, in one of his later writings (1626) points out, just like Cardano, that a magnet recovers and is nourished when put into iron fillings looking like a hedgehog.[56] Among the many further examples, only three shall be given in what follows.

In a German manuscript of the well-known alchemist Joachim Tancke, the magnet is considered a hybrid creature, possessing a nature between metal and stone.[57] The spirit of iron included inside the magnet seeks the body of iron, e. g., a piece of iron. This piece of iron is attracted by the magnet as it serves as a spiritual nutriment of the magnet's living spirit. Tancke's manuscript work was read by a couple of other alchemists and thereby his theories about the magnet and its nutrition spread, too.

---

tractata XI,2, p. 653 (Frankfurt: Rosa, 1619); Jean de Renou, *Institutiones pharmaceuticarum* IV,12, p. 95 (in: *Dispensatorium Galeno chymicum: continens primo Ioannis Renodaei Institutionum pharmoceuticarum* [sic] *lib. V. De materia medica lib. III et Antidotarium varium et absolutissimum: secundo Iosephi Quercetani Pharmacopoeam dogmaticorum restitutam*, Hanau: David Aubrius, 1631). Cf. also Campanella, *Disputationes in quatuor partes suae Philosophiae realis: Physiologica, Ethica, Politica et Oeconomica*, q. 29, art. 7, p. 273 (Paris: D. Houssaye, 1637): *Magnes quoque ferri scobe nutriri probatur Portae et Cardano, licet Gilberto non probetur*. Étienne de Clave argues against Cardano. See Clave, *Paradoxes; ou, Traittez philosophiques des pierres et pierreries, contre l'opinion vulgaire. Ausquels sont demonstrez la matiere, la cause efficiente externe, la semence, la generation, la definition et la nutrition d'icelles* II,27, p. 474–475 (Paris: La veufue Pierre Chevalier, 1635); Hirai (2001), p. 67. Andreas Libavius also critizizes Cardano and Paracelsus for their vitalist ideas. See Libavius, *Antigramania*, p. 606–607 (in: *Neoparacelsica: In Quibus Vetus Medicina Defenditur aduersus Teretismata*, Frankfurt: Kopff, 1594), *Singularium, Pars tertia* I,10, p. 99, 101; Libavius, Andreas/Riolan, Ioannes (1607): *Alchymia triumphans. De Iniusta in se Collegii Galenici spurii in Academia Parisiensi censura* 63, p. 646–654 (Frankfurt: Ioannes Saurius/Petrus Kopff, 1607). On Paracelsus's followers, cf. esp. Clericuzio (2000); Debus (1966, 1977, 1991).

56 Cf. Basilius Valentinus, *Fratris Basilii Valentini Benedicter Orden Letztes Testament und Offenbahrung der himmlischen und irdischen Geheimnüß* (Jena: Eyring, 1626), p. 225, id., *Geheime Bücher oder letztes Testament: vom grossen Stein der uralten Weisen und andern verborgenen Geheinussen der Natur; auß dem Original, so in dem hohen Altar zu Erffurt unter einem marmorsteinernen Täfflein gefunden, nachgeschrieben* (Straßburg: Dietzel, 1645), p. 132; id., *Basilius Innovatus, das ist: Fr. B. Valentini O.S.B. Chymische Schrifften* (Hamburg: Heyl, 1717), p. 678: "[...] daß er [der Magnetstein] mit dem Eisen muß Gemeinschaft halten / mit seinen Feil-Spänen erfrischt und ernähret werden." On this author, cf., e. g., Sudhoff (1933).

57 Cf. the manuscript treatise *Schatzkammer der Natur* (1609) in Kassel, Universitätsbibliothek, MS chem. 99, here fol. 20r–21v. Cf. ibid., fol. 21v: "auff daß sein Geist lebendig und ihme den Magneten / als ein Nutrimentum, mitgetheilet werde." Shortly sketched in Moran (1991), p. 138–139. On Tancke, cf. Benzenhöfer (1987).

Robert Fludd, as another example, simply plagiarized this idea of magnetic nutrition.[58] The famous mid-seventeenth century Spagyric, Pierre Jean Fabre, thought about the magnet's nutrition in a slightly different way.[59] He claimed that the magnet and iron were generated from a similar substance within the earth. Iron and magnet emit material spirits which then are converted into the substance of the recipient, just like food. They exchange, so to speak, their emit-

---

[58] Cf. Robert Fludd, *Philosophia moysaica* sect. 2, lib. III, membr. 1, cap. 1, fol. 126v (Gouda: Petrus Rammazenius, 1638): *Sed, quoniam ferreum corpus non tam cito relinquet Spiritum suum internum, ideo sequitur, simul cum suo Spiritu attahi ad Magnetem: (experientia enim docemur, quod magnes a ferri Spiritibus nutriatur et in suo vigore augeatur) atque iterum, ipsum ferrum Spiritus sibi similes in Magnete reperiens, appetit pariter aequali coitione, ut ipsorum fieret particeps atque in conjunctio sive unio haud aliter inter eos facta, quam inter marem et foeminam.* Cf. also sect. 2, lib. II, membr. 2, cap. 2 (fol. 111r); lib. III, membr. 1, cap. 1 (fol. 126v): *Quae quidem ejus nutritio seu spiritus nutriens praecipue in farre* [sic] *invenitur. Accidit ob hanc causa, spiritum internum Martialem in Magnete attrahere ferrum ad se, et occulto quodammodo videri nutrimentum ex eo ad se fugere et allicere.* [...] *Sed, quoniam ferreum corpus non tam cito relinquet Spiritum suum internum, ideo sequitur, simul cum suo Spiritu attrahi ad Magnetem: (experientia enim docemur, quod magnes a ferri Spiritibus nutriatur et in suo vigore augeatur) atque iterum, ipsum ferrum Spiritus sibi similes in Magnete reperiens, appetit pariter aequali coitione, ut ipsorum fieret particeps atque in conjunctio sive unio haud aliter inter eos facta, quam inter marem et foeminam.* Another alchemist, Andreas Tentzel, copied most of Tancke's ideas, yet magnetic nutrition was not among them. Cf. Tentzel, *Medicina diastatica; hoc est, Singularis illa et admirabilis ad distans et beneficio mumialis transplantationis operationem et efficaciam habens, quae ipsa loco commentarii in tractatum tertium De tempore seu philosop. D. Theoph. Paracelsi, multa, eaque selectissima abstrusioris philosophiae et medicinae arcana continent* 4, p. 58–67 (Jena: Ioannes Bircknerius, 1629); English version: *Medicina Diastatica, or, Sympatheticall Mumie Containing Many Mysterious and Hidden Secrets in Philosophy and Physick, by the Construction, Extraction, Transplantation and Application of Microcosmical and Spiritual Mumie: Teaching the Magneticall Cure of Diseases at Distance, &c.* (London: T. Newcomb: 1653), p. 50–57. Tancke's work was reprinted in Jacob Lupius, *Schatzkammer der Natur: Gründliche Erklärung Dreyer grossen Geheimnüssen* (Leipzig, 1651), here p. 21–22.

[59] Cf. also Pierre Jean Fabre, *Panchymici seu Anatomiae totius universi opus* (Tolosae: Petrus Bosc, 1646), IV,26, p. 242: *hinc magnes trahit ferrum, et ferrum magnetem, quod habeant invicem eandem et similem substantiam primordialem et seminalem, ex qua fiunt et componuntur in visceribus terrae, quae substantia similis, et eadem similis et eosdem de se mittit et eijicit spiritus subtiles et tenues, qui ab attrahente substantia, in se ipsam convertuntur, tanquam in alimentum sui ipsius, hinc fit ut attractis hisce spiritibus, attrahatur et ipsa substantia, ex qua oriuntur hi spiritus, ut fonte potiatur ipsorum spirituum.* Cf. also Savinien de Cyrano de Bergerac, *Les Estats et Empires de la Lune et du soleil: (avec le Fragment de physique)*, ed. by Alcover (2004), p. 295: "Or le fer se nourrit d'aimant, et l'aimant se nourrit de fer si visiblement, que celui-là s'enrouille et celui-ci perd sa force, à moins qu'on les produise l'un à l'autre pour réparer ce qui se perd de leur substance." In Bergerac's view, iron and magnet are generated from two trees.

ted spirits. Attraction between magnet and iron then occurs because they whish to drink from the source which the received spirits originated from.

It is noteworthy that Fabre clearly has overcome a strictly asymmetrical account of magnetic attraction. It is neither, as in Averroes's theory, the iron, which is moved towards the magnet, nor, as in Cardano's account, the magnet that is moved to the iron, but it is a bidirectional movement. In the course of the seventeenth century, most natural philosophers gave up this one-directional causal scheme and instead emphasized the mutual attraction between magnet and iron due to their similar nature.[60] And Fabre's account proves that this conception could be spelled out in terms of nutrition as well. The pre-established causal relation between that which is nourished and that by what it is nourished is blurred if not overcome in his idea. Magnet and iron feed and consume each other, so to speak. In Fabre, this did not lead to a novel theory of nutrition, but at least it seems to imply a rather original understanding of a reciprocal "eat and be eaten".

7 Summary

Since antiquity, several philosophers conceived of nutrition as a form of attraction. This basic understanding can also be found in Aristotle. However, at first, especially within the Aristotelian tradition, e.g., in Alexander and Averroes, this idea was spelled out in detail in two different regards. The first regard is the one particularly promoted by Galen, and later by Averroes, Paracelsus, and others. According to them, certain parts of the body or organs attract specific fluids with-

---

[60] Gilbert (*De magnete*) coined the notion of *coitio* as a reciprocal approaching of two magnetic bodies. Cf. King (1959); Roller (1959), p. 141–144; Wang (2016), p. 712. Fludd ascribed this idea already to Lucretius and identified Giovanni Costeo as a follower of this idea. Cf. Fludd, *Philosophia moysaica*, sect. 2, lib. II, membr. 2, cap. 2, fol. 97v: *Lucretius Carus, quidam sectae Epicureicae Poeta, videtur somniare, ferri attractionem procedere ab atomorum ex subiecto effluxione: Nam, quemadmodum (inquit ille) iuxta Epicureorum opinionem, atomi subtiles ex re qualibet emanant, ita pariter atomi, quasi semina Magnetica a ferro per quandam coitionem ipsius cum Magnete in locum sive spatium interpositum, quod est inter eos, emittuntur, et per unionem aut complicationem corporis utriusque, ferrum attrahitur etc.* Cf. already in Gilbert, *De magnete*, lib. I, cap. 1, p. 3, who, on the contrary, does not use the word *coitio* with regard to the Atomist theory. Costeo seems to have agreed to this idea of a mutual approach of iron and magnet (*mutua ergo utrique est opera et mutuus fructus*) in 1589, while in 1578 he had followed Averroes's unidirectional account. Cf. Giovanni Costeo, *De universali stirpium natura, libri duo* (Turin: Nicolai Beuilaqua, 1578), p. 268; id., *Disquisitiones physiologicae in primam primi Canonis Avic.* sect. VI,3, p. 515 (Bologna: Ioannes Roscius, 1589).

in the animal body or elements of the food which has been consumed. This attraction happens inside the body. Another regard is the one that has been invoked by Alexander, Cardano and Fabre. According to them, animals seek their foodstuff by sense perception and move themselves towards their food. This movement is due to an internal desire of the self-moving substance. This form of attraction happens at a much wider distance and not inside the body, but between the animal and its prey, so to speak.

Both forms of biological attraction were compared to magnetic attraction. However, one should note that the apparent similarity between these forms of movement, that is the very thing called "attraction," itself is a metaphor, at least in the understanding of many. Those philosophers, as we have seen, were keen on emphasizing that both are no proper forms of attraction, but only apparent ones. They pointed out that the seemingly attracted object in reality is moved towards the apparent attractor due to a specific disposition. In this regard, Galen's account is much closer to Paracelsus's than to Averroes's or most medievalists'. Galen and Paracelsus really considered nutrition a form of attraction while Alexander and Averroes probably did not.

As has been shown, not only was nutrition linked to magnetic attraction, or pseudo-attraction, by way of similarity, but it also happened the other way around. A is similar to B, thus B is similar to A. Some ancient authors compared magnetic attraction to nutrition insofar as they described iron as the magnet's food. Paracelsus followed this line of analogy but only Cardano came up with a straightforward account of it. According to him, magnets are living minerals, endowed with a nutritive faculty and an appetite for iron. Alchemists, especially those who were interested in explaining chemical effects in terms of vital motions, applauded this idea.

The variety of authors contributing to the debate of how magnetism and nutrition relate to each other also demonstrates that this debate was not limited to the Aristotelian or Galenic tradition, which, to be sure, might have laid the debate's foundation. Enigmatic figures like Cardano and Paracelsus, and an entire tradition in alchemy, fleshed out the magnetism-nutrition analogy. Categories such as "Aristotelians" and "Anti-Aristotelians" are not even very helpful in order to sort their positions and claims, as all these camps overlap to some degree and thereby shed light on surprising and often over-looked dependencies and similarities among them.

8 Concluding Remarks

How do we assess this story of magnetism and nutrition in more general, philosophical terms? First of all, the magnetism-nutrition analogy was by no means an exception. Magnetism was used as an analogy in many aspects of medical theory, for example, in ideas about blood circulation, procreation, the efficacy of drugs, the powers of human imagination, the medical influence of the heavens and many, many more.[61]

For the present case study, it cannot be proved that the magnetism analogy changed or influenced the medical understanding of nutrition to a considerable degree. Of course, the concept of nutrimental attraction could be contested vis-à-vis magnetic attraction. Moreover, particularly in the case of Paracelsus, his readers encountered a completely novel and much wider understanding of nutrition – an understanding of nutrition that was not just compared to magnetic attraction but was even conceived to be magnetic itself. Yet, the analogy or metaphor of magnetism did not lead to analogical reasoning, strictly speaking. This is to say that the authors covered in this paper did not use magnetism as a heuristic tool to find out more about the very nature of nutrition.[62] Instead, they used magnetism to illustrate their medical theory and to render it more understandable. At best, they invoked magnetism in order to undergird certain philosophical assumptions about physiological processes. When Galen, for example, refuted the atomist account of magnetism, he did so in order to demonstrate that the assumption had to be accepted that there really is a faculty of attraction in nature, not just colliding atoms or pulling hooks.

There seems, however, to be a more intriguing perspective on the story, if one conceives of the medical magnetism-nutrition analogy from the point of view of cognitive linguistics. This approach promoted by George Lakoff and Mark Johnson has often been used quite fruitfully to analyze scientific or philosophical ideas and theories (Johnson 2010; Johnson/Lakoff 2003). According to

---

[61] For a general overview, cf. Sander (2018). I am inclined to say, as a result of my research, that there is almost no domain of early modern natural philosophy or medicine that could not be compared to or explained by magnetism. Cf. also Sander (2019).

[62] Cf. e.g. Provijn (2014), p. 219: "The application of analogies can serve different purposes. First of all, they can be useful in communication, both for pedagogical reasons in view of clarifying a concept by means of a well-accepted analogue or for rhetorical reasons aiming at persuasion, in most situations combined with an informative component. Secondly, they may allow for predications, for example, in the specific case of extrapolation. Finally, in the domain of problem solving and discovery processes they may enable the change of existing concepts and methods and they may even facilitate the creation of completely new concepts."

them, as all language depends on a metaphorical use of concepts, thus also philosophical ideas about reality are often expressed in metaphors. The deliberately or conventionally employed metaphor, however, has consequences for the way the philosophical ideas are conceived of. In their theory, a so-called target domain is understood in terms of a so-called source domain. For example philosophers or scientists frequently describe causal relations in spatial metaphors like "paths" or "directions", like the phrase "heat leads to pressure". Yet, causality or causal relationality itself is not a spatial object, of course. A thing called "heat" does not move or "lead" towards a thing called "pressure" in any spatial sense. Still, the way we talk about causality has a considerable effect on how we think about causality. And there is no entirely clean or pure way to speak about it, not even by logical formula.

In the example of medical theory the source domain is "magnetism" and the target domain is "nutrition", i.e., certain features of nutrition are understood in terms of magnetic attraction. Magnetism and nutrition may be comparable to each other because both are instances of "attraction". Averroes on the contrary thought that neither the organs nor the magnet show attraction at all, but instead referred to the notion of "motion *ex se*". Probably both concepts, "attraction" and "self-motion," are metaphorical to some degree when applied to the stomach or to our nourishment. And this certainly informs the way we think of our bodily functions.

If one thinks of nutrition, however, as an entirely magnetic process, as Paracelsus did, for example, certain semantic features of physical magnetism are mapped onto anthropology, most importantly the idea of attraction itself. This was summed up by his contemporaries writing in Latin: *Homo est magnes*.[63] In the case of Paracelsus, his readers encounter a web of magnetic properties and effects within human and animal nature, reaching far beyond nutrition and far beyond the scope of this paper. Yet, the idea of magnetic nutrition was an important asset to this anthropology. By relying on his magnetic metaphors and concepts, he could draw on an established tradition, and still push it much further. While some of his predecessors were inclined to say "nutrition

---

[63] Paracelsus, *Philosophia sagax* I,1, p. 99 (in: *E chymicis secundus, Continens, vires, efficacias et proprietates Rerum Naturalium et earum quoad Medicinam, praeparationes: Cum multis Alchimicam scientiam secretis spectantibus* [Operum medico-chimocorum siue paradoxorum 7] ed. 1605 by Zacharias Palthenius, Frankfurt am Main: A Collegio Musarum Palthenianarum in nobili Francofurto); I,7; vol. XII, p. 163–164 (Sudhoff). Libavius argues against Paracelsus's *magnes humanus*, cf. Andreas Libavius, *Res Chymicae Epistolica Forma Ad Philosophos Et Medicos Quosdam In Germania excellentes descriptae*, lib. III, ep. 13, p. 121 (Frankfurt: Kopff, 1599).

is like magnetism", Paracelsus, on a linguistic level, simply crossed out the word "like": "nutrition is magnetism."

As a consequence, both the concepts of magnetism and nutrition were reshaped vis-à-vis, leading to much wider and more abstract ideas about them: "nutrition" was no longer confined to physiological processes, and "magnetism" reached far beyond the physical powers of a specific mineral. Following the approach of cognitive linguistics means to accept that such a metaphorical mapping gives rise to a novel understanding of a certain aspect of reality, just like, for example, taking the idea "time is money" seriously. Following Paracelsus's ideas about biological magnetism probably leads straightforwardly to Mesmerism and "animal magnetism" in the eighteenth century, which certainly involved a very unique and idiosyncratic view of reality.[64]

What is quite fascinating about the idea of equating magnetism and nutrition is its potential to be inverted. Just like a hungry animal could become a magnet, the magnet itself could become a hungry animal as well. Cardano and several later alchemists testify to this consequence. For them, this idea did not just fit in some vaguely defined vitalist concept of the world but really had explanatory power to account for magnetic phenomena. They did explain magnetic attraction by tying it to processes which were, according to them, basic powers of nature, such as self-maintenance or nutrition. Even new ideas about magnets themselves could be explained in terms of nutrition and strive, for example, for the finding that magnet and iron mutually attract each other and unite due to their similar nature.

The argument of the present case study, i.e, the conceptual connection between magnetism and nutrition, is *mutatis mutandis* important for the history of philosophy and science. Two realms of phenomena or two concepts are linked to each other in the mindset or ontology of some author of the past, but these supposedly interconnected domains appear to be completely unrelated from a modern perspective. Such hermeneutical challenges are the historians' stumbling blocks, as modern readers necessarily have to question or at least reflect on their own concepts about the world.

---

[64] Cf. e.g. Joseph Ennemoser, *Der Magnetismus, nach der allseitigen Beziehung seines Wesens, seiner Erscheinungen, Anwendung und Enträthselung* (Leipzig: Brockhaus, 1819), p. 20–23; Gerabek (2009); Lessing (1839), p. 197–202; Johann Bernhard Wilbrand, *Darstellung des thierischen Magnetismus als einer in den Gesetzen der Natur vollkommen gegründeten Erscheinung* (Frankfurt a. M.: Sauerländer, 1824), p. 36.

Bibliography

Abel, Eugen (ed.) (1881): *Orphei Lithica. Accedit Damigeron De lapidibus*. Berlin: Calvary.
Alcover, Madeleine (ed.) (2004): *Savinien Cyrano de Bergerac. Les Estats et Empires de la Lune et du soleil: (avec le Fragment de physique)*. [Champion classiques.] Paris: H. Champion.
Alonso, Manuel (ed.)(1944): [*Petrus Hispanus.*] *Comentario al "De anima" de Aristóteles*. Madrid: Instituto de Filosofía Luis Vives.
Averroes (repr. 1962): *De physico auditu libri octo*. [Aristotelis Opera cum Averrois commentariis 4.] Frankfurt a. M.: Minerva [repr. of the edition 1562–74, Venice: Giunta].
Baur, Heinrich/Steiger, Renate (ed.) (1983): *Nicolai de Cusa: Idiota de sapientia, de mente, de staticis experimentis*. [Opera omnia V.] Hamburg: Felix Meiner.
Baur, Ludwig (ed.) (2010): *Die philosophischen Werke des Robert Grosseteste, Bischofs von Lincoln*. [Beiträge zur Geschichte und Theologie des Mittelalters 9.] Münster i. W.: Aschendorff.
Benzenhöfer, Udo (1987): "Joachim Tancke (1557–1609). Leben und Werk eines Leipziger Paracelsisten." In: Sepp Domandl (Ed.), *Paracelsus und Paracelsisten: Vorträge 1984/85*. Salzburger Beiträge zur Paracelsusforschung 25. Wien: Verband der Wissenschaftlichen Gesellschaften Österreichs Verlag, p. 9–81.
Berkel, Klaas van (2013): *Isaac Beeckman on Matter and Motion: Mechanical Philosophy in the Making*. Baltimore: Johns Hopkins University Press.
Boas, George (1947): "Aristotle's Presuppositions about Change." In: *The American Journal of Philology* 68. No. 4, p. 404–413.
Boylan, Michael (1982): "The Digestive and 'Circulatory' Systems in Aristotle's Biology." In: *Journal of the History of Biology* 15. No. 1, p. 89–118.
Brock, Arthur John (ed./transl.) (1916): *Galen. On the Natural Faculties*. [Loeb Classical Library 71.] London/ New York: W. Heinemann/G.P. Putnam's Sons.
Brunn, Walter A.L. von (1967): *Kreislauffunktion in William Harvey's Schriften*. Berlin; New York: Springer.
Bruns, Ivo (ed.) (1892): *Alexander Aphrodisiensis. Scripta minora. Quaestiones. De Fato. De Mixtione*. [Commentaria in aristotelem graeca, Supplementum aristotelicum 2.2.] Berlin: G. Reimer.
Buchheim, Thomas (2007): "Effective Primary Causes: The Notion of Contact and the Possibility of Acting without Being Affected in Aristotle's *De Generatione et Corruptione*." In: S. Stern-Gillet/K. Corrigan (Eds.), *Reading Ancient Texts. Volume II: Aristotle and Neoplatonism*. Brill's Studies in Intellectual History 162. Leiden: Brill, p. 65–96.
Caroti, Stefano/Celeyrette, J./Kirschner, Stefan/Mazet, Edmond (ed.) (2010): *Nicole Oresme, Questiones super Physicam (Books I–VII)*. [Studien und Texte zur Geistesgeschichte des Mittelalters 112.] Leiden/Boston: Brill.
Čermáková, Lucie (2018): "Athanasius Kircher and Vegetal Magnetism: Analogy as a Method." In: *Early Science and Medicine* 23. No. 5–6, p. 487–508.
Clark, Gillan (transl.) (2014): *Porphyry. On Abstinence from Killing Animals*. London: Bloomsbury.
Clave, Étienne de (1635): *Paradoxes; ou, Traittez philosophiques des pierres et pierreries, contre l'opinion vulgaire. Ausquels sont demonstrez la matiere, la cause efficiente*

externe, la semence, la generation, la definition et la nutrition d'icelles. Paris: La veufue Pierre Chevalier.

Clericuzio, Antonio (2000): *Elements, Principles, and Corpuscles: A Study of Atomism and Chemistry in the Seventeenth Century.* Dordrecht: Kluwer Academic.

Daniel, Dane T. (2006): "Invisible Wombs: Rethinking Paracelsus's Concept of Body and Matter." In: *Ambix* 53. No. 2, p. 129–142.

De Waard, Cornelis (ed.) (1939): *Journal tenu par Isaac Beeckman de 1604 à 1634.* La Haye: M. Nijhoff.

Debus, Allen G. (1966): *The English Paracelsians.* Watts History of Science Library. New York: F. Watts.

Debus, Allen G. (1977): *The Chemical Philosophy: Paracelsian Science and Medicine in the Sixteenth and Seventeenth Centuries.* New York: Science History Publications.

Debus, Allen G. (1991): *The French Paracelsians: The Chemical Challenge to Medical and Scientific Tradition in Early Modern France.* Cambridge/New York: Cambridge University Press.

Decaen, Christopher A. (2007): "The Impossibility of Action at a Distance." In: Peter A. Kwasniewski (Ed.), *Wisdom's Apprentice: Thomistic Essays in Honor of Lawrence Dewan, O.P.* Washington D.C.: Catholic University of America Press, p. 173–200.

Delatte, Louis (ed.) (1942): *Textes latins et vieux français relatifs aux Cyranides.* Bibliothèque de la Faculté de philosophie et lettres de l'Université de Liège 93. Liège/Paris: Faculté de philosophie et lettres/E. Droz.

Delaurenti, Béatrice (2006): "La fascination et l'action a distance: questions medievales (1230–1370)." In: *Médiévales* 50, p. 137–154.

Dopsch, Heinz (1993): "Humanismus, Renaissance und Reformation Paracelsus und die geistigen Bewegungen seiner Zeit." In: Dopsch, Heinz/Goldammer, Kurt/Kramml, Peter F. (eds.), *Paracelsus (1493–1541): keines andern Knecht.* Salzburg: A. Pustet, p. 249–258.

Dopsch, Heinz/Goldammer, Kurt et al. (eds.) (1993): *Paracelsus (1493–1541): keines andern Knecht.* Salzburg: A. Pustet.

Eis, Gerhard (1965): *Vor und nach Paracelsus: Untersuchungen über Hohenheims Traditionsverbundenheit und Nachrichten über seine Anhänger.* Stuttgart: Fischer.

Falcon, Andrea (2005): *Aristotle and the Science of Nature: Unity without Uniformity.* Cambridge: Cambridge University Press.

Fierz, Markus (1983): *Girolamo Cardano, 1501–1576: Physician, Natural Philosopher, Mathematician, Astrologer, and Interpreter of Dreams.* Boston: Birkhäuser.

Fletcher, Angus (2005): "Living Magnets, Paracelsian Corpses, and the Psychology of Grace in Donne's Religious Verse." In: *ELH* 72. No. 1, p. 1–22.

Forrester, John M. (2013): *The De subtilitate of Girolamo Cardano.* Tempe, Ariz.: Arizona Center for Medieval and Renaissance Studies.

French, Roger K. (1994): *William Harvey's Natural Philosophy.* Cambridge: Cambridge University Press.

Furley, David J. (1983): "The Mechanics of Meteorologica IV: a Prolegomenon to Biology." In: Paul Moraux/Jürgen Wiesner (Eds.), *Zweifelhaftes im Corpus Aristotelicum: Studien zu einigen Dubia: Akten des 9. Symposium Aristotelicum, Berlin, 7.–16. September 1981.* Peripatoi 14. Berlin/New York: W. de Gruyter, p. 3–93.

Gadebusch Bondio, Mariacarla (2005): "'Officinae sanguinis' – Theorien zur Hämopoese in der Renaissance." In: Mariacarla Gadebusch Bondio (Ed.), *Blood in History and Blood Histories*. [Micrologus' Library 13.] Firenze: SISMEL/Edizioni del Galluzzo, p. 137–167.

Gerabek, Werner E. (2009): "Romantische Medizin und Religiosität." In: Dinzelbacher, Peter (ed.), *Mystik und Natur zur Geschichte ihres Verhältnisses vom Altertum bis zur Gegenwart*. [Theophrastus Paracelsus Studien 1.] Berlin; New York: Walter De Gruyter, p. 141–154.

Giglioni, Guido (2015): "Scaliger versus Cardano versus Scaliger." In: Lines, David A. / Kraye, Jill/Laureys, Marc (eds.), *Forms of Conflict and Rivalries in Renaissance Europe*. Göttingen: V&R unipress, Bonn University Press, p. 109–130.

Glasner, Ruth (2009): *Averroes' Physics: A Turning Point in Medieval Natural Philosophy*. Oxford/New York: Oxford University Press.

Goddu, André (1984): *The Physics of William of Ockham*. [Studien und Texte zur Geistesgeschichte des Mittelalters 16.] Leiden: Brill.

Goldammer, Kurt (ed.) (1965): *Ethische, soziale und politische Schriften; Schriften über Ehe, Taufe, Busse und Beichte*. [Sämtliche Werke [von] Theophrast von Hohenheim gen. Paracelsus Abt. 2, Bd. 2.] Wiesbaden: F. Steiner.

Goynes, Michele/de Leemans, Pieter (eds.) (2007): *Aristotle's* Problemata *in Different Times and Tongues*. [Mediaevalia Lovaniensia 39.] Leuven: Leuven University Press.

Grell, Ole Peter (ed.) (1998): *Paracelsus: The Man and His Reputation, His Ideas and Their Transformation*. [Studies in the History of Christian Thought 85.] Leiden/Boston: Brill.

Hall, Thomas S. (1975): "Euripus; Or, the Ebb and Flow of the Blood." In: *Journal of the History of Biology* 8. No. 2, p. 321–350.

Halleux, Robert (1970): "Fécondité des mines et sexualité des pierres dans l'Antiquité gréco-romaine." In: *Revue belge de philologie et d'histoire* 48. No. 1, p. 16–25.

Halleux, Robert (1974): *Le problème des metaux dans la science antique*. [Bibliothèque de la Faculté de philosophie et lettres de l'Université de Liege 209.] Paris: Les belles lettres.

Hankinson, Robert James (1998): *Cause and Explanation in Ancient Greek Thought*. Oxford/New York: Clarendon Press/Oxford University Press.

Hasse, Dag Nikolaus (2010): *Latin Averroes Translations of the First Half of the Thirteenth Century*. Hildesheim/New York: Olms.

Henry, John (2001): "Animism and Empiricism: Copernican Physics and the Origins of William Gilbert's Experimental Method." In: *Journal of the History of Ideas* 62. No. 1, p. 99–119.

Hesse, Mary B. (2005): *Forces and Fields: The Concept of Action at a Distance in the History of Physics*. Mineola, N.Y: Dover Publications.

Hiller, Johannes Erich (1952a): "Die Mineralogie des Paracelsus. Teil I: Die naturphilosophischen Gedanken zur Genese und Heilwirkung der Mineralien." In: *Philosophia naturalis* 2, p. 293–331.

Hiller, Johannes Erich (1952b): "Die Mineralogie des Paracelsus. Teil II: Mineralien und Bergbau bei Paracelsus." In: *Philosophia naturalis* 2, p. 435–478.

Hirai, Hiro (2001): "Les Paradoxes d'Etienne de Clave et le concept de semence dans sa minéralogie." In: *Corpus: revue de philosophie* 39, p. 45–71.

Hirai, Hiro (2008): "Les logoi spermatikoi et le concept de semence dans la minéralogie et la cosmogonie de Paracelse." In: *Revue d'histoire des sciences* 61. No. 2, p. 245.

Holmes, Brooke (2012): "Sympathy between Hippocrates and Galen: The Case of Galen's Commentary on Epidemics II." In: Pormann, Peter E. (ed.), *Epidemics in Context: Greek*

Commentaries on Hippocrates in the Arabic Tradition. [Scientia Graeco-Arabica 9.] Berlin/Boston: De Gruyter, p. 49–70.

Hossfeld, Paul (ed.) (1993): *Albertus Magnus. Physica, libri 5–8*. [Opera omnia 4.2.] Münster i. W: Aschendorff.

Ideler, Jules Louis (ed.) (1841): "Alexandri Aphrodisiensis Problemata". In: id. (ed.), *Physici et medici graeci minores*. Berlin: Reimer, p. 3–80.

Jaeckle, Erwin (1993): "Der Naturphilosoph Paracelsus." In: Dopsch, Heinz/Goldammer, Kurt/Kramml, Peter F. (eds.), *Paracelsus (1493–1541): keines andern Knecht*. Salzburg: A. Pustet, p. 173–180.

Jammer, Max (1999): *Concepts of Force: A Study in the Foundations of Dynamics*. Mineola, N.Y: Dover Publications.

Johnson, Mark (2010): "Philosophy's Debt to Metaphor." In: Raymond W. Gibbs (Ed.), *The Cambridge Handbook of Metaphor and Thought*. New York: Cambridge University Press, p. 39–52.

Kaimakēs, Dēmētrios V. (ed.) (1976): *Die Kyraniden*. [Beiträge zur klassischen Philologie 76.]. Meisenheim am Glan: Hain.

Kamtekar, Rachana (2009): "Knowing by Likeness in Empedocles." In: *Phronesis: A Journal for Ancient Philosophy* 54. No. 3, p. 215–238.

Kaske, Carol V./Clark, John R. (ed.) (1998): *Three Books on Life*. [Medieval and Renaissance Texts and Studies 57.] Binghamton, NY: Center for Medieval and Early Renaissance Studies.

Kerner, D. (1973): "Paracelsus und die 'magnetische Kraft'." In: *Münchener Medizinische Wochenschrift* 115. No. 11, p. 466–470.

Kiessling, Gottlieb (ed.) (1963): *Ioannu tu Tzetzu Biblion historikēs tēs dia stichōn politikōn alpha de kalumenēs: Graece = Joannis Tzetzae historiarum variarum chiliades*. Hildesheim: Olms.

King, W. James (1959): "The Natural Philosophy of William Gilbert and His Predecessors." In: *Contributions from the Museum of History and Technology Series Bulletin* 218, p. 121–139.

Kovach, Francis J. (1979): "The Enduring Question of Action at a Distance in Saint Albert the Great." In: *The Southwestern Journal of Philosophy* 10. No. 3, p. 161–235.

Kovach, Francis J. (1986): "Action at a Distance in St. Thomas Aquinas." In: Kennedy, Leonard/Marler, Jack C. (eds.), *Thomistic Papers II*. Houston: Center for Thomistic Studies, University of St. Thomas, p. 85–132.

Kovach, Francis J. (1987): "Aquinas Theory of Action at a Distance: A Critical Analysis." In: idem, *Scholastic Challenges to Some Mediaeval and Modern Ideas*. Stillwater Okla.: Western Publications, p. 149–177.

Kovačić, Franjo (2001): *Der Begriff der Physis bei Galen vor dem Hintergrund seiner Vorgänger*. Philosophie der Antike 12. Stuttgart: Steiner.

Lakoff, George/Johnson, Mark (2003): *Metaphors We Live By*. Chicago: University of Chicago Press.

Langermann, Y. Tzvi (2011): "Different Hue to Medieval Jewish Philosophy: Four Investigations into an Unstudied Philosophical Text." In: Fontaine, Resianne/Glasner, Ruth/Leicht, Reimund/Veltri, Giuseppe (eds.), *Studies in the History of Culture and Science a Tribute to Gad Freudenthal*. [Studies in Jewish History and Culture 3.] Leiden/Boston: Brill, p. 71–90.

Lawn, Brian (1993): *The Rise and Decline of the Scholastic* Quaestio Disputata: *With Special Emphasis on Its Use in the Teaching of Medicine and Science*. [Education and Society in the Middle Ages and Renaissance 2.] Leiden: Brill.

Leinsle, Ulrich Gottfried (2009): "Wie treibt man Cardano mit Scaliger aus? Die (Nicht) Rezeption Cardanos an der Jesuitenuniversität Dillingen." In: Muslow, Martin (ed.), *Spätrenaissance-Philosophie in Deutschland 1570–1650: Entwürfe zwischen Humanismus und Konfessionalisierung, okkulten Traditionen und Schulmetaphysik*. [Frühe Neuzeit 124.] Tübingen: Niemeyer, p. 253–277.

Lessing, Michael Benedict (1839): *Paracelsus, sein Leben und Denken: Drei Bücher*. Berlin: Reimer.

Letoublon, Françoise (2014): "The Magnetic Stone of Love. Greek Novel and Poetry." In: Cueva, Edmund P./Byrne, Shannon P. (eds.), *Companion to the Ancient Novel*. [Blackwell Companions to the Ancient World.] Chichester: Wiley-Blackwell, p. 330–351.

Letter, Paul (2000): *Paracelsus: Leben und Werk*. Krummwisch: Königsfurt.

Lohrmann, Dietrich (2008): "Motus continuus und motus perpetuus in der mittelalterlichen Technik und Physik." In: Andreas Speer/David Wirmer (Eds.), *Das Sein der Dauer*. Miscellanea mediaevalia 34. Berlin; New York: De Gruyter.

Maclean, Ian (1984): "The Interpretation of Natural Signs: Cardano's *De Subtilitate* versus Scaliger's *Exercitationes*." In: Vickers, Brian (ed.), *Occult and Scientific Mentalities in the Renaissance*. Cambridge/New York: Cambridge University Press, p. 231–252.

Maggiòlo, Mariani (ed.) (1954): *Thomas Aquinas. In octo libros Physicorum Aristotelis*. Turin/Rome: Marietti.

Maier, Anneliese (1943): *An der Grenze von Scholastik und Naturwissenschaft: Studien zur Naturphilosophie des 14. Jahrhunderts*. Essen: Essener Verlagsanstalt.

McGinnis, Jon (transl.) (2009): *Avicenna, The Physics of the Healing*. Provo, Utah: Brigham Young University.

McVaugh, Michael Rogers (2012): "Losing Ground. The Disappearance of Attraction from the Kidneys." In: Zittel, Claus/Horstmanshoff, Manfred/King, Helen (eds.), *Blood, Sweat and Tears: The Changing Concepts of Physiology from Antiquity into Early Modern Europe*. [Intersections 25.] Leiden/Boston: Brill, p. 103–137.

Meyer-Steineg, Theodor (1913): "Studien zur Physiologie des Galenos." In: *Archiv für Geschichte der Medizin* 6. No. 6, p. 417–448.

Moran, Bruce T. (1991): *The Alchemical World of the German Court: Occult Philosophy and Chemical Medicine in the Circle of Moritz of Hessen, 1572–1632*. [Sudhoffs Archiv, Beihefte 29.] Stuttgart: F. Steiner.

Morani, Moreno (ed.) (1987): *De natura hominis*. [Bibliotheca scriptorum Graecorum et Romanorum Teubneriana.] Berlin: De Gruyter.

Moreau, Elisabeth (2012): "Le substrat galénique des idées médicales d'Isaac Beeckman (1616–1627)." In: *Studium* 4. No. 3, p. 137–151.

Nauck, August (ed.) (1886): *Porphyrii Platonici Opuscula Selecta*. Leipzig: Teubner.

Oelze, Anselm (2018): *Animal Rationality. Later Medieval Theories: 1250–1350*. Leiden; Boston: Brill.

Pagel, Walter (1958): *Paracelsus: An Introduction to Philosophical Medicine in the Era of the Renaissance*. Basel: Karger.

Pagel, Walter (1982a): *Joan Baptista van Helmont: Reformer of Science and Medicine*. Cambridge; New York: Cambridge University Press.

Pagel, Walter (1982b): *Paracelsus: An Introduction to Philosophical Medicine in the Era of Renaissance*. Basel: Karger.
Pauli, Heinrich (ed.) (2001): *Nicolai de Cusa: Sermones III*. [Opera omnia XVIII/2.] Hamburg: Felix Meiner.
Pines, S. (1961): "*Omne quod movetur necesse est ab aliquo moveri:* A Refutation of Galen by Alexander of Aphrodisias and the Theory of Motion." In: *Isis* 52. No. 1, 21–54.
Platnauer, M. (ed.) (1922): *Claudian*. London; Cambridge, Mass.: Heinemann; Harvard University Press.
Preisendanz, Karl/Henrichs, Albert (eds.) (1973): *Papyri graecae magicae: die griechischen Zauberpapyri*. Stuttgart: Teubner.
Provijn, Dagmar (2014): "Bloody Analogical Reasoning." In: Weber, Erik/Wouters, Dietlinde/Meheus, Joke (eds.), *Logic, Reasoning, and Rationality*. [Logic, Argumentation and Reasoning 5.] Dordrecht: Springer, p. 217–232.
Radl, Albert (1988): *Der Magnetstein in der Antike: Quellen und Zusammenhänge*. [Boethius 19.] Wiesbaden/Stuttgart: F. Steiner Verlag.
Rashed, Marwan (2011): "Introduction." In: id. (ed.), *Alexandre d'Aphrodise, commentaire perdu à la "Physique" d'Aristote (livres IV-VIII) les scholies byzantines: édition, traduction et commentaire*. [Commentaria in Aristotelem Graeca et Byzantina 1.] Berlin; Boston: De Gruyter, p. 33–165.
Rashed, Marwan (ed.) (2015): *Al-Hasan ibn Musa al-Nawbakhti. Commentary on Aristotle* De Generatione et Corruptione. [Scientia Graeco-Arabica 19.] Berlin: De Gruyter.
Riddle, John M./ Mulholland, James A. (1980): "Albert on Stones and Minerals." In: Weisheipl, James A. (ed.), *Albertus Magnus and the Sciences: Commemorative Essays*. [Studies and Texts, Pontifical Institute of Mediaeval Studies 49.] Toronto: Pontifical Institute of Mediaeval Studies, p. 203–234.
Röhr, Julius (1923): *Der okkulte Kraftbegriff im Altertum*. [Philologus, Supplementband 17.1.] Leipzig: Dieterich.
Roller, Duane H.D. (1959): *The* De Magnete *of William Gilbert*. Amsterdam: Hertzberger.
Rommevaux, Sabine (2010): "Magnetism and Bradwardine's Rule of Motion in Fourteenth- and Fifteenth-Century Treatises." In: *Early Science and Medicine* 15. No. 6, p. 618–647.
Rutschow, Hans (1965): "Über den Magnetismus bei Paracelsus." Inaugural-Diss. Universität Köln.
Sailer, Vera-Sophia (2012): "Das Buch als Medium der Bildung für die Lebenspraxis bei den Kartäusern." Diss. Universität Wien.
Sakamoto, Kuni (2016): *Julius Caesar Scaliger, Renaissance Reformer of Aristotelianism: A Study of His Exotericae Exercitationes*. History of Science and Medicine Library 54. Leiden; Boston: Brill.
Sánchez Muñoz, Luis/Irene Pajón Leyra (2015): "The Magnetic Stone of Posidippus' Poem Nr. 17: The Earliest Description of Magnetic Polarity in Hellenistic Egypt." In: *Zeitschrift für Papyrologie und Epigraphik* 195, p. 30–37.
Sander, Christoph (2018): "Magnetism in Renaissance Science." In: Sgarbi, Marco (ed.), *Encyclopedia of Renaissance Philosophy*. Dordrecht: Springer International Publishing, p. 1–6 (https://doi.org/10.1007/978-3-319-02848-4_944–1).
Sander, Christoph (2019): "Magnets and Garlic: An Enduring Antipathy in Early-Modern Science." In: *Intellectual History Review* (forthcoming), p. 1–38 (https://doi.org/10.1080/17496977.2019.1648924).

Sander, Christoph (2020): "Magnetism for Librarians. Leone Allacci's *De magnete* (1625) and Its Relation to Giulio Cesare LaGalla's *Disputatio de sympathia et antipathia* (1623)." In: *Erudition and the Republic of Letters* (forthcoming).

Schipperges, Heinrich (1993): "Das Menschenbild des Paracelsus." In: Dopsch, Heinz/Goldammer, Kurt/Krammel, Peter F. (eds.), *Paracelsus (1493–1541): keines andern Knecht*. Salzburg: A. Pustet, p. 181–186.

Schott, Heinz (2002): "Paracelsus and van Helmont on Imagination: Magnetism and Medicine before Mesmer." In: Scholz Williams, Gerhild/Gunnoe, Charles D. (eds.), *Paracelsian Moments: Science, Medicine and Astrology in Early Modern Europe*. [Sixteenth Century Essays and Studies 64.] Kirksville, Mo.: Truman State University Press, p. 135–147.

Schott, Heinz (2011): "Paracelsus und die Magie der Natur." In: Albrecht Classen (Ed.), *Religion und Gesundheit: Der heilkundliche Diskurs im 16. Jahrhundert*. Berlin: Walter de Gruyter, p. 99–112.

Schröter, Joachim (1941): "Die Stellung des Paracelsus in der Mineralogie des 16. Jahrhunderts." In: *Schweizerische mineralogische und petrographische Mitteilungen* 21. No. 2, p. 313–331.

Schütze, Ingo (2000): *Die Naturphilosophie in Girolamo Cardanos De subtilitate*. [Humanistische Bibliothek 49.] München: Fink.

Sharples, R.W. (transl.) (2014): *Alexander of Aphrodisias. Quaestiones 1.1–2.15*. London et al.: Bloomsbury.

Strunz, H. (1951): "Die Mineralogie bei Albertus Magnus." In: *Acta Albertina Regensburg* 20, p. 13–19.

Sudhoff, Karl (1933): "Die Schriften des sogenannten Basilius Valentinus." In: *Philobiblon* 6, p. 163–170.

Sudhoff, Karl (ed.) (1925): *Medizinische, naturwissenschaftliche und philosophische Schriften*. [Sämtliche Werke [von] Theophrast von Hohenheim gen. Paracelsus Abt. 1, Bd. 9.] Hildesheim: Olms.

Sudhoff, Karl (ed.) (1928): *Medizinische, naturwissenschaftliche und philosophische Schriften*. [Sämtliche Werke [von] Theophrast von Hohenheim gen. Paracelsus Abt. 1, Bd. 2.] Hildesheim: Olms.

Sudhoff, Karl (ed.) (1929a): *Medizinische, naturwissenschaftliche und philosophische Schriften*. [Sämtliche Werke [von] Theophrast von Hohenheim gen. Paracelsus, Abt. 1, Bd. 1.] Hildesheim: Olms.

Sudhoff, Karl (ed.) (1929b): *Medizinische, naturwissenschaftliche und philosophische Schriften*. [Sämtliche Werke [von] Theophrast von Hohenheim gen. Paracelsus, Abt. 1, Bd. 12.] Hildesheim: Olms.

Sudhoff, Karl (ed.) (1930): *Medizinische, naturwissenschaftliche und philosophische Schriften*. [Sämtliche Werke [von] Theophrast von Hohenheim gen. Paracelsus, Abt. 1, Bd. 2.] Hildesheim: Olms.

Taub, Liba Chaia (2012): "Physiological Analogies and Metaphors in Explanations of the Earth and the Cosmos." In: Zittel, Claus/Horstmanshoff, Manfred/King, Helen (eds.), *Blood, Sweat and Tears: The Changing Concepts of Physiology from Antiquity into Early Modern Europe*. [Intersections 25.] Leiden/Boston: Brill, p. 41–63.

Temkin, Owsei (1972): "Fernel, Joubert, and Erastus on the Specificity of Cathartic Drugs." In: Debus, Allen G. (ed.), *Science, Medicine and Society in the Renaissance: Essays to Honor Walter Page*. New York: Science History Publ., p. 61–68.

Temkin, Owsei (1977): *The Double Face of Janus and Other Essays in the History of Medicine.* Baltimore: Johns Hopkins University Press.
Theoharides, Theoharis C. (ed./transl.) (1971): "Galen on Marasmus". In: *Journal of the History of Medicine and Allied Sciences* 26, p. 369–390.
Thijssen, Johannes M.M. Hans/Streijger, Michiel/Bakker, Paul J.J.M. (ed.) (2010): *John Buridan. Quaestiones super libros De generatione et corruption Aristotelis.* [History of Science and Medicine Library 17.] Leiden/Boston: Brill.
Ugaglia, Monica (ed.) (2005): *Leonardo Garzoni. Trattati della calamita.* [Filosofia e scienza nell'età moderna 3.] Milano: Franco Angeli.
Vilborg, Ebbe (ed.) (1955): *Achilles Tatius. Leucippe and Clitophon.* [Studia Graeca et Latina Gothoburgensia 1.] Stockholm: Almqvist & Wiksell.
Wang, Xiaona (2016): "Francis Bacon and Magnetical Cosmology." In: *Isis* 107. No. 4, p. 707–721.
Wardy, Robert (2007): *The Chain of Change: A Study of Aristotle's Physics VII.* [Cambridge Classical Studies.] Cambridge: Cambridge University Press.
Weill-Parot, Nicolas (2012): "Magnetic Attraction as a Challenge to the Inanimate Realm. The Example of Walter Burley." In: Jacquart, Danielle/Weill-Parot, Nicolas (eds.), *Substances minérales et corps animés: de la philosophie de la matière aux pratiques médicales, 1100–1500.* Histoire de savoirs Montreuil: Omniscience, p. 87–110.
Weill-Parot, Nicolas (2013): *Points aveugles de la nature: la rationalité scientifique médiévale face à l'occulte, l'attraction magnétique et l'horreur du vide (XIIIe-milieu du XVe siècle).* [Histoire 20.] Paris: Les Belles Lettres.
Weisheipl, James A. (1965): "The Principle *Omne quod movetur ab alio movetur* in Medieval Physics." In: *Isis* 56. No. 1, p. 26–45.
Wöhrle, Georg/ Strohmaier, Gotthard (eds.) (2009): *Die Milesier: Thales.* [Traditio praesocratica 1.] Berlin: Walter De Gruyter.
Wolfson, Harry Austryn (1929): *Crescas' Critique of Aristotle; Problems of Aristotle's Physics in Jewish and Arabic Philosophy.* [Harvard Semitic Series 6.] Cambridge: Harvard University Press.
Zanetto, Joseph (ed.) (1985): *Theophylacti Simocatae epistulae.* Leipzig: B.G. Teubner.

Elisabeth Moreau
From Food to Elements and Humors
Digestion in Late Renaissance Galenism

Abstract: In late Renaissance medicine, the example of digestion was frequently invoked to prove the elemental composition of the human body. Food was considered as being decomposed in its first elements by the stomach, and digested into a thick juice which is assimilated by the liver and the body parts. Such a process points to the structure of the human body into four elements that are transformed into different types of humors during several stages of "concoction". This chapter examines the Galenic interpretation of digestion expounded by the French physician Jean Fernel (1497–1558) in his *Physiologia* (1567). In this treatise, Fernel states the body composition into elemental portions, while stressing the role of the "innate heat" as the physiological counterpart of the body's essence or "substantial form". He applies this view in his account of digestion, where he states that the conversion of food follows the rule of "mixture". This chapter aims to explore how Fernel applies his interpretation of elements and innate heat to the process of digestion, as well as his sources in Galen's *De facultatibus naturalibus,* Avicenna's *Canon*, and Aristotle's *Meteorologica*. It first examines the role of the natural soul and its "nourishing" faculties in nutrition as a physiological function. It then considers the role of elements, humors and innate heat during the "concoction" of food in the stomach, liver and veins.

1 Introduction

In the early modern period, the theme of nutrition pervaded all theoretical and practical branches of medicine. The functioning of the digestive system was studied in physiology, while its malfunctions and diseases were examined in pathology. Therapeutics prescribed numerous alimentary remedies to facilitate digestion and cure sicknesses, while dietetics advised the most adapted regimen to preserve health. Semiotics as the study of medical "signs" and symptoms considered urine as a major indication of a healthy or sick temperament. However, despite the wide scope of nutrition in the early modern medical disciplines, most of the historical surveys on this theme have been concentrated on dietetics and therapeutics, by exploring the various regimens and treatments based on herbals, spices and other food-based drugs (Gentilcore 2015; Albala 2002; Margolin and Sauzet 1982; Giacomotto-Charra and Vons 2017). It is mostly from the viewpoint of alchemy and Paracelsianism that the recent research has examined nutrition as a process of food digestion and assimilation (Clericuzio 2012; McKee 1998; Temkin 2002, p. 180–194). In this perspective, early modern alchemists have been considered as putting forward the body's chemistry, in particular the processes of coagulation and fermentation as well as sensory qualities like colors and flavors. Although these themes emerged in early medical and natural philosophy, they have received little attention from the historians, except for the medieval period (Jacquart 2006; Lyndon Reynold 1999; Cadden 1980). This chapter aims to fill this gap by exploring the early modern reception of the Galenic theory of digestion in a major treatise on theoretical medicine: the *Physiologia* of the French physician Jean Fernel (c. 1497–1558). Centered on the structure and functioning of the healthy body, this treatise is part of a broader work, the *Universa medicina* (1567), which is divided in physiology, pathology, therapeutics, and the study of hidden or "occult" causes.[1] As a concise and systematic account of Galenic medicine in a humanist framework, it was widely read by early modern physicians and saw multiple re-editions throughout the seventeenth century (Kany-Turpin 2002; Henry and Forrester 2003 and 2005).

In his works, Fernel aims to combine Galenic medicine and Aristotelian physics with Platonic philosophy in order to enhance the divine origin of life and the soul. Such a framework has been explored in the history of medicine concerning Fernel's theory of seeds, "occult" qualities and diseases of the "total substance" in his treatise *On the Hidden Causes of Things* [*De abditis rerum causis*] (1546) (Hirai 2011, p. 46–79; Deer Richardson 1982; Blank 2010). Interestingly, Fernel's interpretation also applies to his account of physiological functions, above all nutrition. Along the lines of the Galenic and Aristotelian tradition, he considers nutrition as a vital function which is operated by the soul. The latter accomplishes the physiological functions by means of a vital principle that has a celestial origin: the body's "innate heat". This principle thus points to the celestial facet of nutrition as a vital process connected to the soul.

While nutrition is a physiological function related to the soul and innate heat, it is also connected to the body composition in elements and humors. Fernel follows the medical tradition by stating that food is composed of the four elements (air, water, earth, fire) and that the digestion of food produces the four humors. During this process, the elements and their primary qualities within food are decomposed and assimilated into blood (hot–moist), yellow bile (hot–dry), phlegm (cold–moist), and black bile or "melancholia" (cold–dry).

---

[1] Fernel, *Physiologia*. I am referring to the edition and translation by Forrester (2003).

As the balance or imbalance of these qualities and humors determine the state of health, the digestion of food is a key process in the body's physiology.

In this perspective, this chapter examines Fernel's explanation of digestion from two main angles. First, it looks at his application of the Aristotelian theory of elements to digestion as a transformation of nutrimental matter. Second, it appraises the influence of Platonic philosophy on his interpretation of nutrition as a vital function directed by the soul, in particular its relation to "occult" qualities and the body's "total substance". Third, this chapter explores the "concoction" of food as a process of coagulation and fermentation. As will be argued, Fernel develops these three themes by synthetizing the philosophy of Galen, Aristotle, and Avicenna. For his synthesis of medieval Latin-Arabic texts in the light of ancient sources, he proves to be emblematic of the current of Renaissance medical humanism (McVaugh and Siraisi 1990). His account of nutrition indeed aims to articulate Greek texts with medieval Arabic textbooks in order to develop medicine as a theoretical field of knowledge based on the Aristotelian natural philosophy.

In the following section, I examine Fernel's presentation of nutrition as a physiological function related to the faculties of the "natural" soul. The three next sections investigate the successive "concoctions" of food in the stomach, liver and veins, in particular the formation of the four humors. To this purpose, I consider the fifth and sixth book of the *Physiologia* dedicated to the faculties of the soul and to the functions and humors, respectively.

2 The "Nourishing" Faculties of the Natural Soul

In his *Physiologia*, Fernel presents a clear synthesis on the structure and functioning of the healthy body, from its anatomical body parts and first components – elements and humors – to its main physiological functions. In order to explain the latter, he adopts a Galenic framework, and first recalls the "tripartition" of the soul within the body (Galen, *De placitis Hippocratis et Platonis* VI,3, K. V,519–532; Arist. *de An.* II,2, 413b1–10). The "natural" (*naturalis*), "sentient" (*sentiens*) and "intelligent" (*intelligens*) souls each predominate in the three realms of living beings: vegetal, animal, and human, respectively.[2]

In accordance with Galen's *De facultatibus naturalibus*, Fernel explains that the natural soul is common to all living beings (Fernel, *Physiologia*, p. 310–321).

---

2 Fernel, *Physiologia*, p. 304 : [...] *tres quoque animae species iisdem nominibus insignitas, quae sunt naturalis, sentiens, et intelligens: quibus haec respondent viventium genera,* φυτόν, *id est terra editum sive stirps, brutum, homo.*

Its faculties govern three main vital functions: reproduction, growth, and nutrition (Galen, *De fac. nat.*). During the embryonic development, the "procreative" (*procreatrix*) faculty overrides, followed by the "increasing" (*auctrix*) faculty until the mature age, while the "nourishing" (*altrix*) faculty predominates for the rest of life.[3] The "nourishing" faculty converts food into the body substance, which needs a constant repair of what has been lost. This occurs within the digestive system, which includes the actual digestive tract and the venous system. In addition, Fernel points out that nutrition needs to be distinguished from another important vital function, reproduction. Though both faculties relate to the natural soul, nutrition aims to the conservation of individuals, while reproduction is about the conservation of the whole species.

Fernel continues his synthesis of the medical tradition by applying Galen's account of the natural faculties to his explanation of nutrition. The "nourishing" faculty of the natural soul works by means of four "auxiliary" faculties. These are the "attracting" (*attrahens*), "expelling" (*expellens*), "retaining" (*continens*), and "concocting" (*concoquens*) faculties, which ensure the circulation of the ingested food within the digestive organs.[4] The auxiliary faculties operate during three successive "concoctions" in the stomach, liver, and veins, which each produce a specific humor as well as a residue (*excrementum*).[5] To do so, the attracting faculty in the stomach and liver draws the useful part of food. It works in concert with the retaining and concocting faculties to hold and "cook" the ingested food with the assistance of the body's "innate heat".

Along the lines of Galen's theory of nutrition, Fernel explains that the three phases of digestion as "concoctions" aim to sort the nutrimental part of food from its residual waste, which is evacuated afterwards. At the beginning of this process, the gastric concoction transforms the ingested food into a creamy white substance: chyle. As a humor and "first nutriment", chyle is sorted from its residual waste, the fecal matter that is produced in the intestines. Then, the hepatic concoction runs the conversion of chyle into blood (αἱμάτωσις). Chyle

---

3 Fernel, *Physiologia*, p. 312: [...] *opera autem sunt stirps surgens è semine, aut qui utero geritur foetus ac fingitur: augescens stirps aut animal: hoc salvum et vita conseruatum. Tres praesunt illis effectrices facultates, procreatrix, altrix et quae auctrix appellatur.*
4 Fernel, *Physiologia*, p. 320: *Facultatum naturalium numerus ut impleatur, ad altricem reversione facta, quae illi tanquam auxiliariae vel administrae parent, huc referendae videntur. Sunt autem hae attrahens, expellens, continens, et concoquens, quarum necessitate corpus alitur et sustinetur.*
5 Fernel, *Physiologia*, p. 438–441: *Hoc autem tribus tantummodo locis concessum videmus, è quibus etiam triplex existit concoctio, prima in ventriculo, secunda in iocinere, tertia in singulis particulis. Sua quanque antecedit praeparatio, suáque sequitur absolutio, suum cuiusque alimentum tanquam subiecta materia, suus genitus humor, suúmque excrementum.*

takes the form of a red humor or "juice" (χυμός) which contains the four humors, and whose remainder is the urine that is secreted in the kidneys. Finally, the venous concoction in the body parts operates the assimilation of the "secondary humors" in the veins, whose residues are sweat and the "vapors" that are evacuated by the skin pores during perspiration. Throughout this process, the auxiliary faculties achieve the "juxtaposition", "agglutination" and "assimilation" of the digested food into the body part following Galen's exposition in *De fac. nat.* – see section 4.

Within his synthesis of the Galenic account of nutrition, Fernel further describes the treatment of foodstuff at the level of its substance, elements and qualities in each phase of digestion. The following sections move on to explore his explanation of the three "concoctions", in particular the decomposition of food in elements and its transformation into humors in the stomach, liver and veins.

3 The Making of Chyle: Food Elements and the "Total Substance" of the Stomach

In the medical tradition, the first phase of digestion as a "concoction" in the stomach points to two important aspects for the understanding of nutrition. On the one hand, it raises the problem of the "attraction" of the ingested food by the stomach. Though Galen describes this process with the notion of "attracting faculty", the phenomenon of attraction in natural philosophy is difficult to explain (Jacquart 2006). For this reason, physicians often relate it to the equally challenging notion of hidden or "occult" quality related to the body's essence or "substantial form". On the other hand, the gastric digestion puts forward the decomposition of foodstuff into its basic ingredients, the four elements. In the early medical philosophy, this aspect is mostly treated in the context of the Aristotelian natural philosophy in relation to the notion of elemental "mixture", that is the union of elements through the moderation of their qualities.

Fernel, in turn, synthesizes the traditional account of the gastric digestion as a process of attraction and decomposition of food. He relates it as much to Galen's explanation of concoction as to his own account of the "substantial form" and "mixture" of elements, which is expounded at the beginning of his *Physiologia* and in his treatise *On the Hidden Causes of Things* (Hirai 2011, p. 46–79; Moreau 2018). As Fernel explains, the "highest and greatest mission" of nutrition is the purgation and transformation of foodstuff before its diffusion through all the body parts.⁶ This process starts with the preparation of the nutriment in the stomach. The masticated food is first treated by the heat and saliva of the mouth. Pushed by the gullet fibers, it moves towards the esophagus before being "drawn" by the stomach, which proceeds with the first concoction:

> In the interval the concocting faculty changes and prepares all the food, without assistance from the fibers, by its own innate heat and spirit, and by the heat that, like a fire set around a great caldron, is contributed by the adjacent parts, the liver, spleen, heart, diaphragm, vena cava and intestines [...] Moreover, the particular substance of the stomach brings this concoction to completion, and not only by those generally known qualities, but also by a secret and hidden property [...] It comes about not just by the force and ardor of heat, but also by the total substance of the stomach, and the "inserted" property which we mentioned. (Fernel, *Physiologia*, p. 404–407; translation slightly modified.)⁷

In this excerpt, Fernel states that the concocting faculty of the stomach transforms food thanks to the heat of the surrounding organs, which is coupled to the body's "innate heat". The latter is a vital principle related to the body's "total" or "whole" substance (*tota substantia*), which is related to its essence, and endowed with specific powers (Deer Richardson 1982; Bianchi 1982; Temkin 1972). Fernel further describes the attraction exerted by the innate heat as a phenomenon acting by "similitude of substance" so that the effects of the gastric attraction are comparable to those of the magnet and purgative drugs.⁸ By way of illustration, he takes the example of the ostrich's ability to digest iron thanks to the total substance of its stomach (Fernel, *Physiologia*, p. 406–407).⁹ The same

---

6 Fernel, *Physiologia*, p. 402: *Summum supremumque naturae munus est nutritio, quam ex alimento accepto illa perficit: hoc enim illius est tota materia, in quam penitus incumbit, quam variè tractans expurgat, convertit, propriisque ductibus in omnes extremasque corporis particulas diffundit atque confert.*

7 *Hoc interim spatio facultas concoquens, cibum omnem mutat et conficit, nullis quidem fibrarum adminiculis, sed tum proprio ingenitoque calore et spiritu, tum eo quem ut lebeti magno circumiectus ignis, ita et proximae partes illi inferunt, iecur, lienis, cor, diaphragma, vena caua, et omentum [...]. Caeterùm propria ventriculi substantia concoctionis illius absolutionem perficit, idque non modo vulgatis et notis illis qualitatibus, verùm etiam tacita et recondita proprietate. [...] Fit autem haec non modo vi et ardore caloris, sed etiam tota ventriculi substantia, et insita quam diximus proprietate.*

8 Fernel, *Physiologia*, p. 320–321: *Restat igitur ut sit sua cuiusque particulae vis attrahendi, quae quod illi familiare est et conveniens, ex sanguinis mole privata benignitate prolectet. [...] Sic stirpes e terra amoenum succum, sic lapis heraclius ferrum, sic et purgantia medicamenta unumquempiam è corpore humorem, naturarum substantiarumque similitudine trahunt.*

9 The wonderful ability of the ostrich to digest anything was stated in Pliny's *Naturalis Historia* X,1, and further discussed in the medieval and early modern period about the digestion of stones, iron, and gold (Buquet 2013). In medieval medicine, the relation between the ostrich's

idea is presented in Fernel's *On the Hidden Causes of Things*,¹⁰ where he explains that the concoction is operated by hidden or "occult" causes associated to the stomach's "total substance" (Fernel, *On the hidden causes*, p. 498–503).

Such a statement relies on Galen's analogy between the attracting faculty of the stomach and the force of the magnet, which are both related to the "total substance" (Galen, *De temperamentis* III,1, K. I,654–655; *De fac. nat.* II,7, K. II,106–107; *De elementis ex Hippocratis sententia* II,5, K. I,507–508). In turn, the "total substance" is a difficult concept in Galen's philosophy, which has been much theorized by medieval scholastic physicians (Gibbs 2013). Throughout his works, Galen sparsely mentions the body's total substance as the cause of a range of physiological, pathological and pharmacological phenomena (Galen, *De theriaca ad Pisonem* 3, K. XIV,224–225). They have the common feature of a great force of attraction which is attested by experience, but impossible to explain theoretically.

To Galen's account of the total substance, Fernel associates his own interpretation of the innate heat which is presented in the *Physiologia* and *On the Hidden Causes of Things*. As an instrument of the soul, the innate heat animates the body and presides over physiological functions such as reproduction, growth and nutrition. Because of its vital nature, it is not composed of the element fire but of aether. Aristotle establishes aether as a celestial entity which is additional to the four elements, and states that the stellar element enters in the composition of vital heat (Arist. *de Caelo* I,3, 270b20–25; *de Generatione Animalium* II,3, 736b34–737a2). According to Fernel, the celestial nature of the innate heat is due to its nature of instrument of the soul and its close connection to the body's essence or "substantial form" (Fernel, *Physiologia*, p. 256–263; *On the Hidden Causes*, p. 478–497). Recent studies have shown that his interpretation relies on the Platonic philosophy of Marsilio Ficino, which emphasizes the celestial origin of the form (Hirai 2011, p. 46–79; Walker 1958; Zanier 1957). In his medical philosophy, Fernel particularly insists that the living body has a substantial form of divine origin which comes from the world-soul as a cosmic "giver of forms".

Fernel's view on the attracting faculty of the stomach is also indebted to a major textbook of Galenic medicine in the late medieval and Renaissance period, the *Canon of Medicine* by the Persian physician Avicenna (Gibbs 2013; Copenhaver 1984). In his explanation of food concoction, Avicenna compares the force of

---

stomach and the total substance can be traced at least in Averroes about the virtues of food and drugs (Averroes, *Colliget libri VII* [Venice: Giunta, 1574], fol. 95vb).

10 Page numbers refer to the edition and translation of John M. Forrester (2005).

the substantial or "specific" form, which he associates to the Galenic notion of total substance, to the "attracting virtue" of the magnet (Gerard of Cremona, *Avicennae Arabum medicorum principis [Canon medicinae]*. Vol. I. [Venice: Giunta, 1595], p. 73a and 111b). Moreover, Avicenna establishes that such a specific form has a divine origin due to a celestial "giver of forms", a notion rooted in his emanationist philosophy (Hasse 2012). In this perspective, Fernel's approach to food concoction in the stomach is close to Avicenna's interpretation of the same theme. His description of the gastric concoction in the *Physiologia* is to be understood upon a Galenic and Platonic interpretation of innate heat, total substance and the superior form. For Fernel, it is the substantial form of the stomach, which is related to the Galenic notion of total substance, that supervises the preparation of the ingested food by means of the innate heat and the attracting faculty.

Having expounded the attraction of food by the stomach, Fernel further describes the decomposition of food and its transformation into chyle:

> And so, with the aid of all these, so to speak, assistants, [the stomach] starts by gathering the foods together, and mingles one with another, the dry with the moist. At the same time, it fragments and crushes up everything, so that all the portions can be seen to have taken on some uniformity of substance. While these [portions] are being thoroughly mixed in this way, they are inevitably softened, both by the rule of mixture and by the nature of the disposition. Having ceased fighting, all the outstanding qualities are brought into some middle range. The substance becomes of one and the same nature, reaching some likeness to cream. (Fernel, *Physiologia*, p. 404–407; translation slightly modified.)[11]

According to Fernel, the decomposition of food and its transformation into chyle follow the "rule of mixture". With these terms, Fernel refers to the Aristotelian theory of mixture which he develops in the *Physiologia* for his exposition of "temperament" (Fernel, *Physiologia*, p. 208–213; Moreau 2018). The latter is a fundamental concept in early medicine. It defines the state of health coming from the balance of the primary qualities (hot, cold, dry, and moist) that are related to the body's elements. From the perspective of Aristotelian natural philosophy, the temperament comes from the union of elements as a "mixture" (*mistio*) (Arist. *de Generatione et Corruptione* I,10, 327a30–328b24). As Fernel explains,

---

11 *Itaque his omnibus tanquam ministris adiutus ventriculus primùm quidem alimenta in unum cogit, et alia cum aliis, arida cum humidis confundit, simul verò omnia frangit et exterendo comminuit, ut iam partes omnes quandam substantiae aequabilitatem adeptae videri possint. Quae dum ita exquisitè permiscentur, necessariò ex mistionis lege, tum ex affectionis natura mitigantur, et deposita omnia qualitatum exuperantia in quendam mediocritatis ordinem adducuntur et recidunt, fitque uniusmodi eiusdémque naturae substantia, speciem quandam cremoris consequuta.*

the mixture of elements consists in the mingling of their qualities as a "battle" of opposites (hot/cold, dry/moist), which results in their common moderation. As the qualities reach a moderate state, the body's elements arrange in minute parts or "portions". Such a process is followed by the immediate introduction of the substantial form, which achieves the formation of a uniform and tempered substance.

Applied to the gastric concoction, Fernel's explanation of mixture implies that foodstuff is broken down into its mere elements as minute parts or "fragmented portions". At the same time, the elemental qualities "fight" and weaken to some "middle range" as a balanced disposition. The concocted foodstuff is then transformed into a new substance in the form of a "thick juice": chyle. For his explanation of food concoction in the stomach, Fernel thus offers a remarkably consistent explanation with his theory of mixture.

4 The Hepatic and Venous Concoctions: From Minima to Coagulated Blood

According to the Galenic tradition, the digestion of chyle into blood in the liver and veins consists in successive steps of filtration (Jacquart 2006). In this case, the ability of digestive organs to process the concoction of humors is not explained by their substantial properties, but by the fineness of their conducts, in particular these of the veins. To develop this reasoning, Fernel goes on using the general framework provided by Galen along with later contributions to this subject by medieval Latin-Arabic physicians. The latter draw particular attention to the transformation and texture of the resulting juices.

As Fernel specifies, the making of chyle in the stomach is only a preparation to the second concoction in the liver, for which chyle is transformed into blood. After the gastric concoction, the pylorus opens up to evacuate chyle and the waste materials towards the intestines, with the assistance of the "expelling" faculty. Chyle is then prepared in the "minute channels" of the mesenteric veins. With the aid of the attracting faculty, it is sent to the hepatic vena cava and its minuscule branches, where it acquires the texture of white wine.[12] There, it

---

[12] Fernel, *Physiologia*, p. 414–415: *Ad hunc igitur modum alimentum in iecur perlabitur, trahentibus quidem tum mesenterii venis, tum vero iecore per venas. [...] Ex qualitatis quidem cognatione, vinum quàm aqua ocyus in corpus assumitur: at substantiae tenuitas et vis quaedam penetrandi, causa est ut vinum album promptius quam rubrum et austerum in iecur pervadat, id enim expeditius trahentis vim consequatur.*

undergoes a "complete" concoction before obtaining the form of blood in the liver:

> The veins scattered throughout the liver are so very thin that all the juice closely approaches its flesh and particular substance, being all in contact with it on every side. Because of this prolonged stay, and the intimate contact in the liver, with the chyle dispersed by each of its *minima*, a more swift and unimpaired form of blood is acquired than if it were contained in a wide cavity, such as the heart or stomach. (Fernel, *Physiologia*, p. 420–421; translation slightly modified.)[13]

For Fernel, it is the fineness of the hepatic veins which achieves the concoction of chyle by contact with the liver wall as these veins process each of the smallest components (*minima*) of chyle. Such a "distribution" *per minima* denotes the elemental structure of chyle along the lines of the theory of elements and mixture which is presented in the *Physiologia* (Fernel, *Physiologia*, p. 210–212). Accordingly, elements are defined as minute, if not "minimal" parts in the compound. These terms mainly refer to Galen's conception of the element as the smallest or "minimal" part in bodies in *De elem. ex Hipp. sent.* In this treatise, Galen also states that mixture is facilitated by the decomposition of the compound into *minima* in the same manner as the mixing of wine and water (Galen, *De elem. ex Hipp. sent.* I,1 and 9, K. I,413 and 489). Moreover, medieval physicians such as Gentile da Foligno in his commentary to Avicenna's *Canon* raise the issue of the "quasi-minimal" diffusion of concocted blood by the tiny veins of the body parts during the third concoction (Jacquart 1998, p. 345–346; Gentile da Foligno, *Primus Avicennae Canonis* [...] *Liber* [Venice: Scotus, 1520] fol. 37v).

Fernel adds that after the hepatic concoction, chyle thickens, turns red, and reaches a certain moderation in order to become a mass of "natural blood" which contains the four humors – blood, yellow bile, phlegm and black bile or "melancholia". By describing blood as a tempered and finely structured substance, Fernel's again relates his theory of mixture to the way of producing humors and juices during digestion. The chyle is processed through its smallest elemental components, and transformed into the blood mass which has a balanced disposition.

After the hepatic concoction, the digested blood is subject to a third concoction in the veins of the body parts. In this regard, Fernel insists on the fineness of

---

[13] *Quum igitur tanta sit venarum per iecur disiectarum exilitas, succus totus ad eius carnem propriamque substantiam proximè accedit, à qua universus usquequaque attingitur. Ob hanc vero moram diutinam, et ob exquisitam in iecore contagionem, chylo per minima quaeque distributo, tum celerior, tum integrior sanguinis asciscitur species, quàm si ampla capacitate, qualis est aut cordis aut ventriculi, conciperetur.*

venous conducts which diffuse blood "vapor", in reference to Aristotle's *GA* (Fernel, *Physiologia*, p. 434–435; Arist. *GA* II,4, 738a35–38). He adds that chyle circulates not only through the orifices of the veins, but also through their pores, where it acquires the "highest degree of perfection". During this process, blood as an "alimentary humor" undergoes several steps of transformation, which end in the assimilation of blood into the body part.[14] Fernel here follows Galen's description of the last concoction as a triple process of juxtaposition, agglutination and assimilation that is applied to the digested blood in *De fac. nat.* (I,11, K. II,24; Hp. *De alimento* 8).

These phases of transformation are identified as "secondary humors" in medieval Latin-Arabic medicine (McVaugh 1974; Reynolds 1999, p. 105–119; Jacquart 2006). In this respect, Avicenna develops the notion of fourfold secondary "moistures" (*humiditates*) (Gerard of Cremona, *Avicennae Canon*, p. 20b–21a). As he explains in the *Canon*, the first moisture is contained in the orifices of the small veins that are close to the members, while the second moisture passes like "dew" through these veins to nourish and moisten the members. The third moisture is a coagulated "nutriment" which has acquired the temperament of the body part, though it is not converted yet into its complete essence. Received from the sperm at birth, the fourth moisture is responsible for the replenishment of the body parts. Fernel shares the same explanation of four nameless "secondary humors", although the scholastic tradition rather adopted the terminology of the *Pantegni*, one of the most important medieval textbooks of medicine along with Avicenna's *Canon*. In this treatise, the last three humors are named "dew" (*ros*), "glue" (*gluten*) and "change" (*cambium*), in allusion to their respective texture (Isaac Israeli, fol. 88rb).[15]

Having examined the successive concoctions of food in the stomach, liver and body parts, I shall now further explore the composition and formation of the four humors.

---

14 Fernel, *Physiologia*, p. 442–443: *Tertiae concoctionis opus et utilis succus est universum eorum humorum genus, quos aliquando diximus secundarios appellari, in quorum ordine eum primum collocavimus, qui in exilibus est partium venis, secundum qui tanquam ros in partes influit eisque apponitur, tertium qui affigitur, quartum qui iam in partis substantiam concedit eique assimilatur.*
15 Isaac Israeli (1515): *Liber Pantegni*. In: *Omnia opera Ysaac*. Lyon: Bartholomeus Trot, fol. 1r–144r.

5 Digesting the Four Humors: Fermentation, Combustion and Putrefaction

Upon his approach to concoction as a mixture, Fernel links the properties of the four humors to their structure in four elements. Their elemental content comes from the balance of chyle's primary qualities which produces the blood "mass" including the four humors. In the medical tradition, the body's humoral composition is fundamental for the understanding of health and disease. The excess of one of the humors is considered as an affection which requires to be purged by pharmacological means through the ingestion of drugs or by surgical means such as bloodletting. In this perspective, Fernel's exposition of the humors attempts to provide a theoretical counterpart to the main activity of physicians: identifying the qualitative disposition and possibly the humoral affection of their patients. Moreover, the formation of humors points to the inner "kitchen" of the body which operates the "cooking" of food. In this context, Fernel's description is not only centered on their diverse textures and properties, either in their natural and pathological states, but also on the nature of their transformation. He explains their successive alterations in metaphorical and chemical terms coming from Aristotelian philosophy as well as medieval Latin-Arabic medicine.

Resulting from the concoction of chyle in the liver, natural blood is a "mass" composed of the four humors. As Fernel explains, these fluids are each characterized by two of the primary qualities (hot, cold, dry, moist) as well as "secondary" qualities such as texture, color, and flavor (Fernel, *Physiologia*, p. 444–447; Hp. *De natura hominis* 5, L. VI,40–44; Galen, *De elem. ex Hipp. sent.* II). These qualities come from the mixture of the elements and their primary qualities within chyle. Phlegm (cold–moist) comes from the cold and "raw" portion of chyle, bile (hot–dry) from its hot and fine portion, blood (hot–moist) from its moderate portion, and melancholia (cold–dry) from its cold and "earthy" portion.[16] As Fernel underlines, the quantity of each humor depends on the constitution of the liver and on the composition of the ingested food. For this reason, each humor within blood has a certain "latitude" of temperament.[17] Established by Galen, this concept, also known as "latitude of health", is a gradual range of

---

[16] Fernel, *Physiologia*, p. 428–429: *Pituita enim ex frigida et cruda chyli fit portione, bilis ex calida atque tenui, sanguis ex mediocri, melancholicus humor ex frigida atque terrena.*

[17] Fernel, *Physiologia*, p. 426–428: [...] *nam unusquisque humor certa temperamenti latitudine circumscribitur, per quam divagari potest.* [...] *neque suus cuique ad unguem definitus est gradus. Ex quo intelligitur* [...] *varietatis plurimum afferri, ex efficientis caloris vi et alimenti materia, quarum iusta comparatio gignendi humoris speciem definiet.*

temperaments for each humor, from health to sickness, with a medium point as a "neutral" state (Ottosson 1984, p. 166–194; French 2001, p. 106–107; Galen, *Ars medica*, 4, K. I,316–317). This means that the proportion of qualities for each humor is variable and depends on the individual constitution. As a result, it can bring about a healthy ("natural") or morbid ("preternatural") variation of the humor.

Fernel adds that the humors are in a pure state only in their dedicated organ: phlegm in the brain, yellow bile in the gallbladder, and black bile or "melancholia" in the spleen. Consequently, they are not pure in the blood mass, otherwise the body would not be healthy.[18] For instance, the presence of pure yellow bile within blood is symptomatic of jaundice. Fernel anchors this reasoning in the Hippocratic treatise *De nat. hom.*, which states that the four humors are mingled within the blood, and that purgative drugs should be used when a humor is in excess (Hp. *De nat. hom.* 6, L. VI,44–46).

Like many Renaissance physicians, Fernel describes the process of digestion along the lines of Aristotle's *Meteorologica*. In this treatise, the notion of concoction (πέψις) is considered as a broad category which designates either ripening in the same way as fruits, boiling like the digestion of milk and food, or roasting (Arist. *Mete.* IV,3, 380a11–381b22). These models of cooking were used in a medical context to describe physiological processes such as the formation of blood and the seed (Martin 2002 and 2008). Following this framework, Fernel defines the gastric concoction as the boiling (ἕψησις or *elixatio*) of the moisture of ingested food by the innate heat (Fernel, *Physiologia*, p. 406–407). Concerning the digestion of chyle into the four humors, he refers to the Aristotelian definition of concoction as ripening and roasting. However, he extends it to an analogy with wine fermentation:

> We see that the innate heat of fresh must, collected into vats, makes it effervesce, be changed, and be digested. From this surplus items are forthcoming that were previously indistinctly present, and then get extracted and isolated for the first time by the force of concoction: one quite heavy and more earthy, which falls to the bottom and is called the lees; the other lighter and more airy, which floats on the top and is usually called the flower. It is assuredly in a similar manner that chyle traversing the liver, and being a moist liquid, boils after a fashion and is concocted, and in the end something thick appears, corresponding to

---

[18] Fernel, *Physiologia*, p. 454–455: *Ex quibus intelligitur quatuor humores, qui in sanorum sanguine continentur, synceros non esse, nec eorum naturam exprimere, qui in bilis folliculum atque in lienem coniecti sunt.*

the lees, and also something thin and light resembling the flower. (Fernel, *Physiologia*, p. 420–423; translation slightly modified.)[19]

As Fernel describes the chemical process of fermentation, he does not allude to the alchemical notions of *fermentum* or *fermentatio* (Clericuzio 2012; McKee 1998). In fact, the comparison of digestion to fermentation originates in ancient philosophy (Siraisi 1987, p. 36–40; Pagel 1982, p. 79–80). For instance, Hippocrates explains that the digestive organs are in a state of fermentation when they are processing food (Hp. *De prisca medicina* 11, L. I,594). Plato describes the decomposition of food within the small veins as a process of fermentation due to the movement of airy and earthy particles (Plat. *Timaeus* 66b). Aristotle includes the boiling of must (fresh grape juice) in the broader category of concoction (Arist. *Mete.* IV,3, 380b34). As for Fernel, he closely follows Galen's metaphor of vinification to describe the formation of the four humors (Galen, *De usu partium* IV,3, K. III,269–270). As parts of the concocted blood, the thin yellow bile and the thick melancholia (black bile) are analogous to the phases of fermenting grape juice, namely the scum or "flower" and the sediment or "lees". Phlegm (*pituita*) is presented as a nearly mature portion of blood that is dry and astringent in taste because of its composition in "raw" parts.[20] Pure blood, in turn, is the achieved result of concoction just like wine after a couple of years: "ripe, tasty, and full of heat and strength."

Finally, Fernel characterizes the four humors with particular sensory qualities such as color, texture and taste. Following the medical tradition, he describes blood as red and tempered, yellow bile as yellow and fine, melancholia as black and thick, and phlegm as white and liquid. As Fernel has previously established, each humor has a wide range of healthy and morbid variations. The latter include "superfluous" (residual) and "preternatural" (pathological) humors, in accordance with Galen's *De fac. nat.* (Fernel, *Physiologia*, p. 458–459;

---

19 *Mustum recens ab uvis expressum coniectumque in dolia, cernimus ab innato sibi calore effervescere, mutari, et concoqui: hinc supervacanea comparere quae prius quidem confusa inerant, ac tum primùm concoctionis vi secernuntur atque secedunt, alterum quidem grauius magisque terrenum, quod in fundo subsidens fecem appellant: alterum levius magisque aëreum, quod per summa innatat et flos vini dici solet. Simili profectò ratione chylus in iecur transfus, cùm humidus quidam liquor sit, quodammodo fervet et concoquitur, tandemque nonnihil crissum existit, feci respondens, nonnihil etiam tenue et leve flori consimile [...].*

20 Fernel, *Physiologia*, p. 424–425: *Quod bimestre adhuc est, quamvis à fecibus et spuma expurgatum, gustatu tamen subcrudum, austerum et astringens deprehenditur, quòd crudis partibus visum effugientibus perfusum est. Has dierum numero vini calor evincit et concoquit, indicio quod vinum annuum aut bimum qui degustabit, iam maturatum id, suave, plenum caloris et roboris percipient.*

Galen, *De fac. nat.* II,9, K. II,135–140). To describe their formation, Fernel does not refer to vinification, but to other chemical phenomena such as combustion and putrefaction. For instance, the preternatural bile is subject to a progressive burning into yolk-colored or "vitelline" (λεκιθοειδής) bile, leek-green (πρασοειδής) bile, verdigris-colored (ἰώδης) bile, woad–like or "cerulean" (ἰσατώδης) bile, and "black bile" (Galen, *De atra bile* I,1, K. V,104–148). The latter is distinct from the natural black bile or "melancholia". Fernel compares the preternatural "black bile" to a ferment of acidic nature, which effervesces and corrodes the body parts.[21] Such an "ashen" or "burnt" humor may also result from the putrefaction of the superfluous counterpart of black bile. It condenses into small lumps which decay afterwards.[22]

In addition to the preternatural sorts of bile, Fernel recounts four kinds of preternatural and superfluous phlegm. These types of phlegm are developed by Avicenna on the basis of Galen's distinction between acidic, salty and sweet phlegms, with liquid, thick, viscous and spreadable properties (Galen, *De alimentorum facultatibus* I,1, K. VI,463; Gerard of Cremona, *Avicennae Canon*, p. 21b). Among the preternatural phlegms, the "acidic" one, which is similar to fruit juice, results from an incomplete concoction in the liver, so that it effervesces but becomes sour due to the lack of innate heat. The "salty" phlegm arises from natural "sweet" phlegm, which is partly roasted by putrefaction.[23] In addition, the superfluous type of phlegm includes four possible varieties due to a gradual thickening: thin and aqueous, condensed and "mucilaginous", "vitreous", and "plastery".[24] The latter, which stems from the solidification of "vitreous" phlegm, has the consistency of stones or gypsum.

---

21 Fernel, *Physiologia*, p. 462–463: *Altera species ex ea fit bile quam vitellinam dixerunt. Exustione enim haec primùm in porraceam, deinde in aeruginosam, post in ceruleam, novissimè in atram omnium perniciosissimam commigrat* [...]. *Ea enim est quae in terram impacta, fermenti more et quodam quasi aestu efferuescens, hanc iactat et excutit* [...].
22 Fernel, *Physiologia*, p. 462–463: *Ex sanguine nulla proximè bilis atra profertur. Nam si sanguinem statuamus vel incendio vel putredine in venis torreri et conflagrare* [...]. *Quum autem sanguis à venis illapsus fuerit in ventrem, in intestina, aut in aliam interiorem capacitatem, illic quidem primùm concrescit in grumos, tandémque putrescit.*
23 Fernel, *Physiologia*, p. 464–465: *Una est quae acida vocatur, summè quidem cruda, et quae praeter primam ventriculi confectionem, vix ullam aut minimam in iecore atque venis accepit.* [...] *Altera pituitae species salsa appellatur, ex dulci haec putrescente nata* [...].
24 Fernel, *Physiologia*, p. 464–467: *Una est pituita tenuis et aquea* [...]. *Altera censetur muco similis* [...]. *Haec si caloris vi et appulsu tantam adipiscatur crassitudinem, ut iam partibus conclusa possit adhaerescere, vitrea tum appellabitur, quae tertium genus est. Ad extremum vero cum ex concretione sic iam durescet, ut à lapidis aut gypsi duritia non procul absit, quarta tum emerget quam multi gypseam appellant* [...].

6 Conclusion

It is argued that Fernel applies the framework of the Aristotelian physics of elements to the Galenic account of digestion. He indeed considers the process of "concoction" as a process of "mixture" which involves the elements and the primary qualities of the digested food. Such a scheme, in turn, involves two entities constitutive of the body: its essence or "substantial form" and its matter.

As a vital function, digestion implies the body's substantial form, which Fernel defines as a superior entity of celestial and divine origin. While this aspect of his philosophy relies on a Platonic cosmology, it also offers a physiological application. The body's substantial form is associated to its total substance and innate heat. The latter plays the role of instrument of the soul to exercise the vital functions. During the concoction of food in the digestive organs, the innate heat operates through the attracting faculty by unfolding hidden properties connected to the total substance.

On the other hand, the digestive organs process the matter of ingested food, which is composed of the four elements. The nutrimental matter is broken up in elemental portions in the stomach, and then "concocted" upon each of its *minima* in the mesenteric and hepatic veins. Furthermore, Fernel's view on the moderate constitution of chyle and blood, which consist in minute portions, follows his own interpretation of mixture that is expounded in the *Physiologia*. Accordingly, the four humors are described as composed of elemental portions endowed with a certain qualitative temperament.

Fernel's articulation of the Galenic account of nutrition with Aristotelian physics is close to that of Avicenna. In his *Canon*, the latter stresses the role of the specific form and the total substance in operating digestion, as well as the mixture of elemental parts which produce the humors. In his turn, Fernel synthesizes Avicenna's explanation in a fashionable Renaissance account that is based on Galenic medicine, Aristotelian physics and Platonic philosophy. This allows him to explain how the body's matter and substantial form, through the elements and innate heat, work in the physiological operations.

Moreover, Fernel's theory of nutrition gives an insight into the chemical nature of food digestion. Following the medical tradition, he adopts the Galenic metaphor of wine fermentation concerning the formation of the four humors. Also, Fernel expands on the different definitions of concoction as boiling, ripening and roasting which were developed in Aristotle's *Mete*. The bodily fluids are subject to diverse processes of coagulation, combustion, and putrefaction in the digestive system. While characterized by the four primary qualities, they feature other sensory qualities such as taste, color and texture.

Interestingly, the Aristotelian and Galenic frameworks which Fernel uses are further developed by early modern physicians in the context of an alchemical understanding of digestion. This mainly occurs in light of the Paracelsian interpretation of digestion as the transmutation par excellence in the "stomach" as an athanor. On the one hand, the sorting of the humor from its residual waste is considered as an alchemical extraction or "separation" of the pure from the impure. On the other hand, the processes of fermentation, coagulation and putrefaction are revised through the prism of three alchemical principles (Salt, Sulfur, Mercury) in replacement of the four elements and their primary qualities. Although the three principles aim to debunk the four humors and their morbid variations, Paracelsian physicians at times maintain traditional topics such as the chromatic variations of humors as a sign of different phases of transmutation, the comparison between the formation of blood and vinification, and the production of corrosive substances due to a bad digestion. From the perspective of seventeenth-century alchemical reinterpretation of digestion, Fernel's theory thus proves to be particularly valuable for tracing the ruptures and continuities of early modern innovations with the medical tradition.

Bibliography

Albala, Ken (2002): *Eating Right in the Renaissance*. Berkeley, London: University of California Press.

Bianchi, Massimo Luigi (1982): "Occulto e manifesto nella medicina del Rinascimento: Jean Fernel e Pietro Severino". In: *Atti e memorie dell'Accademia Toscana delle Scienze e Lettere La Colombaria* 47, p. 183–248.

Blank, Andreas (2010): "Jean Fernel on Divine Immanence and the Origin of Simple Forms". In: Vlad Alexandrescu and Robert Theis (Ed.): *Nature et surnaturel : philosophies de la nature et métaphysique aux XVIe-XVIIe siècles*. Hildesheim, Zürich, New York: Georg Olms Verlag, p. 9–21.

Buquet, Thierry (2013): "Fact Checking: Can Ostriches Digest Iron?". *Medieval Animal Data Network*. http://www.mad.hypotheses.org/131, visited on 1 December 2019.

Cadden, Joan (1980): "Albertus Magnus' Universal Physiology: the Example of Nutrition". In: James A. Weisheipl (ed.): *Albertus Magnus and the Sciences: Commemorative Essays*. Toronto: Pontifical Institute of Mediaeval Studies, p. 321–340.

Clericuzio, Antonio (2012): "Chemical and Mechanical Theories of Digestion in Early Modern Medicine". In: *Studies in History and Philosophy of Biological and Biomedical Sciences* 43, p. 329–337.

Copenhaver, Brian P. (1984): "Scholastic Philosophy and Renaissance Magic in the *De vita* of Marsilio Ficino". In: *Renaissance Quarterly* 37, p. 523–554.

Deer Richardson, Linda Deer (1982): "The Generation of Disease: Occult Causes and Diseases of the Total Substance". In: Andrew Wear, Roger K. French and Ian M. Lonie (Eds.): *The*

Medical Renaissance of the Sixteenth Century. Cambridge, Cambridge University Press, p. 175–194.

French, Roger K. (2001): Canonical Medicine: Gentile Da Foligno and Scholasticism. Leiden: Brill.

Forrester, John M. (transl.) (2003): "The Physiologia of Jean Fernel". In: Transactions of the American Philosophical Society 93 (2003), p. i–636.

Forrester, John M. (transl.) (2005): Jean Fernel's On the Hidden Causes of Things: Forms, Souls and Occult Diseases in Renaissance Medicine. Leiden–Boston: Brill, p. 3–66.

Gentilcore, David (2015): Food and Health in Early Modern Europe, 1450–1800. London–New York: Bloomsbury Publishing.

Giacomotto-Charra, Violaine, and Vons, Jacqueline (Ed.) (2017): Formes du savoir médical à la Renaissance. Pessac: Maison des Sciences de l'Homme d'Aquitaine.

Gibbs, Frederick W. (2013): "Specific Form and Poisonous Properties: Understanding Poison in the Fifteenth Century". In: Preternature 2, p. 19–46.

Hasse, Dag Nikolaus (2012): "Avicenna's 'Giver of Forms' in Latin philosophy, Especially in the Works of Albertus Magnus". In: Hasse, Dag Nikolaus, and Bertolacci, Amos (Ed.): The Arabic, Hebrew and Latin Reception of Avicenna's Metaphysics. Berlin-Boston: De Gruyter, p. 227–249.

Henry, John and Forrester, John M. (2003): "Introduction. Tradition and Reform: Jean Fernel's Physiologia (1567)". In: Transactions of the American Philosophical Society 93, p. 1–13.

Henry, John and Forrester, John M. (2005), "Jean Fernel and the Importance of the De abditis rerum causis, in Jean Fernel". In: Forrester, John M. (2005) (transl.): p. 3–66.

Hirai, Hiro (2011): Medical Humanism and Natural Philosophy: Renaissance Debates on Matter, Life and the Soul. Leiden–Boston: Brill.

Jacquart, Danielle (1998): La médecine médiévale dans le cadre parisien (XIVe-XVe siècles). Paris: Fayard.

Jacquart, Danielle (2006): "La nourriture et le corps au Moyen Age". In: Cahiers de Recherches Médiévales et Humanistes 13, p. 259–266.

Kany-Turpin, José (Ed.) (2002): Jean Fernel. Corpus, revue de Philosophie 41, p. 5–197.

Lyndon Reynolds, Philip (1999): Food and the Body: Some Peculiar Questions in High Medieval Theology. Leiden: Brill.

McKee, Francis (1998): "The Paracelsian Kitchen". In: Ole Peter Grell (Ed.): Paracelsus, the Man and His Reputation, His Ideas and Their Transformation. Leiden–Boston: Brill, p. 293–308.

McVaugh, Michael (1974): "The Humidum Radicale in Thirteenth-Century Medicine". Traditio 30, p. 259–283.

McVaugh, Michael and Siraisi, Nancy G. (1990): "Introduction". In: Osiris. Renaissance Medical Learning: Evolution of a Tradition 6, p. 6–15.

Margolin, Jean-Claude, and Sauzet, Robert (Ed.) (1982): Pratiques et discours alimentaires à la Renaissance. Paris: Maisonneuve et Larose.

Martin, Craig (2002): "Francisco Vallés and the Renaissance Reinterpretation of Aristotle's Meteorologica IV as a Medical Text". In: Early Science and Medicine 7, p. 1–30.

Martin, Craig (2008): "Scientific Terminology and the Effects of Humanism: Renaissance Translations of Meteorologica IV and the Commentary Tradition". In: Peter De Leemans, An Smets and Michèle Goyens (Ed.). Science Translated: Latin and Vernacular

Translations of Scientific Treatises in Medieval Europe. Leuven, Leuven University Press, p. 155–180.

Moreau, Elisabeth (2018): "Elements, Mixture and Temperament: The Body's Composition in Renaissance Physiology". In: Chiara Beneduce and Denise Vincenti (Eds.). *Oeconomia corporis: The Body's Normal and Pathological Constitution at the Intersection of Medicine and Philosophy*. Pisa: Edizioni ETS, p. 51–58.

Ottosson, Per-Gunnar (1984): *Scholastic Medicine and Philosophy: A Study on Commentaries on Galen's Tegni (ca. 1300–1450)*. Naples: Bibliopolis.

Pagel, Walter (1982): *Joan Baptista Van Helmont, Reformer of Science and Medicine*. Cambridge–New York–Melbourne: Cambridge University Press.

Siraisi, Nancy G. (1987): *Avicenna in Renaissance Italy*. Princeton: Princeton University Press.

Temkin, Owsei (2002): *"On Second Thought" and Other Essays in the History of Medicine and Science*. Baltimore–London: Johns Hopkins University Press.

Temkin, Owsei (1972): "Fernel, Joubert, and Erastus on the Specificity of Cathartic Drugs". In: Debus, Allen G. (Ed.): *Science, Medicine and Society in the Renaissance*. London: Heinemann, p. 61–68.

Walker, Daniel P. (1958): "The Astral Body in Renaissance Medicine". In: *Journal of the Warburg and Courtauld Institutes* 21, p. 119–133.

Zanier, Giancarlo (1987): "Platonic Trends in Renaissance Medicine". In: *Journal of the History of Ideas* 48, p. 509–519.

Bernd Roling
Standstill or Death
Early modern Debates on the Hibernation of Animals

Abstract: How the hibernation of animals could be explained physically? The question troubled natural scientists already in the Early Modern period, as this study will demonstrate. After a chain of examples, taken from 16th century zoological literature, the paper presents different Early Modern explanations of animal hibernation. How the power of the *vis nutritiva* could be satisfied and neutralized simultaneously? A first solution was offered by the Italian medicine professor Fortunio Liceti in the early 17th century, who made especially use of Albertus Magnus. A second more complex model was developed by the Danish scholar Ole Borch a few years later. Both models agreed in the idea that the Aristotelian key qualities in the process of digestion, heat and cold, had to neutralize themselves reciprocally. As a third case the almost encyclopedical survey of theories about hibernation will be summarized, given by the German physician Karl von Bergen in 1752.

1 Introduction

One of the occurrences in the cycle of the seasons that seems to have made an early impression on the human observer is the phenomenon of hibernation, the winter sleep. Unlike humans, who in the cold season at most retreat inside their homes to compose poetry or tax returns, large sections of the natural order seem to come to a total rest during winter; they halt all their activities and they sleep. Hibernation had its poetic potential, as can be seen from a glance at the most famous of all poems on the seasonal cycle, Thomson's *Seasons* (Thomson 1790, p. 166–167, 184–185); but it must also have presented a challenge to pre-modern scholars. A human slept seven hours and was already hungry on waking. A bear slept for months but evidently did not starve during this period, even though its food requirements were substantial in its waking state. How could this paradox be solved? What was taking place during winter in the physical system of the sleeping animal? These questions vexed natural scientists already in the early modern period. In what follows I shall offer an overview of the early modern discussion of animal hibernation, a discussion that was necessarily tied directly to the role of the *vis nutritiva* in the animal body; it will become clear that the proposed solutions remained astonishingly homogeneous over a long period. Two scholars from the seventeenth and eighteenth centuries will be given centre stage: the Danish natural scientist Ole Borch and the German medic and zoologist Carl von Bergen.

That animals hibernate had already been adequately demonstrated by Aristotle in his *Historia Animalium*. The bear, both male and female, as Aristotle knew, fattened itself up by feeding before the start of winter, then it slept for forty days, of which it was for two weeks totally motionless, the rest of the time sleeping with interruptions but without consuming any food. A similar behaviour could be observed in the hedgehog, according to Aristotle, but also in a large number of rodents, the *glires* of the Latin tradition, including the dormice. Aristotle had significantly expanded the catalogue of animals known to sleep through the winter. Birds of prey, too, could be found on the list, but also, as Aristotle continues in the *HA*, a range of fish and also insects such as mosquitos that required a period of rest in winter.[1] Other ancient zoologists had repeated Aristotle's information, including Pliny and Aelian, both focusing primarily on the sleeping bears and hedgehogs.[2] With the translations of Michael Scott and of William of Moerbeke, Aristotle's reflections were made available to the Latin Middle Ages.

In the early modern era, the period from 1500 onwards, it was the great zoologists, and above all Conrad Gesner, who sorted the ancient material and extended it with their own observations.[3] A rich collection of new illustrative material on sleeping animals was included in the *History of the Nordic Peoples* by Olaus Magnus; it discusses sleeping badgers, marmots and polar foxes, and was able to move hibernation beyond Aristotle, and beyond the horizons of central Europe, and raise it to the status of empirically demonstrated fact.[4] But as to the question of why so many animals spent the winter in lethargic rest, neither Gesner nor the Bishop of Uppsala had an answer. Edward Wotton, who after Gesner wrote what was perhaps the most important reference work on mammals in the sixteenth century, did not have a convincing theory of what could be prompt-

---

1 Aristoteles, *HA*, ed. and translated by A. L. Peck – D. M. Balme (Cambridge Mass.: 1991) (3 vol.), VII,13–17, 599a–600b, vol. III, Greek and English, p. 146–161.
2 As examples on bears, mice and hedgehogs see Pliny, *Naturalis historiae libri XXXVII*, with german translation by Roderich König (38 vol.) (Darmstadt: 1976–2004), VIII,126–135, vol. VIII, p. 96–103.
3 Conrad Gesner, *Historiae animalium Libri V* (Zürich: Christian Froschauer, 1551–1587), Liber de quadrupedibus, De urso, p. 1074–1075.
4 Olaus Magnus, *Historia de gentibus septentrionalibus* (Rom: 1555) (Reprint Kopenhagen: Rosenkilde, 1972), there on snowfoxes, bears and badges e.g. XVIII,17–23, p. 614–620; c. 38–40, p. 635–637.

ing the animals to halt their activities in such a thorough fashion.[5] Above all it was the altered physical constitution of the animals that left him baffled. Evidently, so it had been observed, the bear awoke from the long sleep anaemic and emaciated. But what had the animal subsisted on during this time? How could the nutritional processes during the long period of rest be plausibly explained?

2 Albertus Magnus and Fortunio Liceti: Total Abstinence and Hibernation

The first, and probably pioneering, answer to the question of hibernation was given by Albertus Magnus, who in his *Parva Naturalia* had addressed the *vis nutritiva* in detail. How did the bear manage to save itself from hunger for months? The *vis nutritiva* was a power of the soul, Albert made clear. The natural warmth of the digestive organs ensured that the *succositas*, which was the actual nutritional fluid, was distilled out of the food consumed; at the same time, the materials that the body was not able to utilise were separated from the *succositas* with the help of the liver and excreted through the intestines. During this process, so Albert continued, there was an opposition between a *virtus calidi ignei*, the fiery force of digestion, and the *humidum complexionale*, the moist and rather cold mixture of the humours. The transformation of the latter was guaranteed by the soul, as the shaping motor principle, which drove the nutritional fluid, and with it the *spiritus* that maintained life, into the various regions of the body. When an animal began its hibernation, Albert believed, the moist food solidified into a fat-like viscose phlegm, a *pituita*, as it would later be called, a *phlegma viscosum*. At the same time, the drop in the animal's body temperature prevented any further dissolution of this fluid. In addition there was a second factor. The outer envelope of the creature's body thickened, with the consequence that the remaining warmth in the body could no longer escape. The sleeping bear thus formed, as it were, a vacuum-insulated vessel, in which the two driving forces of digestion were minimised in their effect, while the thick brew of digestive aids sloshed around to no effect.[6]

---

5 Edward Wotton, *De differentiis animalium libri X* (Paris: Vascosanus, 1552), fol. 66rf.
6 Albertus Magnus, *De nutrimento et nutribili* (ed. 1890–99 by Auguste Borgnet, Paris, in: *Opera omnia* [38 vol.], vol. IX, p. 323–343). Basic was Galen, *De alimentorum facultatibus*, printed later on with latin translation as by Martin Gregorius as Galen, *De alimentorum facultatibus libri tres*

In the early modern period Albert's theory would develop an idiosyncratic influence that was linked to another, somewhat parallel debate. In the early seventeenth century a controversy had broken out among Italian and Iberian scholars over the physiological explanation of a phenomenon treated repeatedly in the scholarly literature, namely cases of total abstinence, *asitia*, that apparently lasted years – we would today call it extreme anorexia. The most famous case, cited repeatedly, was a young woman from Confolent in southern France, Jeanne Balant, who, after a severe fever, lived on for two years without, so it was claimed, consuming food of any kind. The French medic François Citois had devoted a study to this unfortunate woman in the early seventeenth century; a poisoned apple had apparently been the start of the peculiar episode.[7] A series of hypotheses had been developed on this case; in particular it moved two medics, the hypo-theses well-known major scholar Fortunio Liceti and the Portugese Estevão Rodrigo di Castro, to pursue a no-holds-barred dispute over it[8]. The Frenchman Laurent Joubert had been the first to draw parallels between animal hibernation and total abstinence among humans, as was the apparent case of the young Jeanne Balant from Confolent, and wanted to explain the two phenomena as analogous. Many animals, so Joubert, ate large quantities before the onset of the cold season, and then in their bodies a mass of phlegm formed, *pituita*, which could not be digested due to the low body temperature. Further, hibernators closed off the pores of their skin, so that the vapours, *vapores*, produced from the residual digestion within the body were no longer able to escape outside it, but could themselves be used as nutrition and to maintain the remaining heat. Further digestion, *concoctio*, did not take place, for the internal heat, the *calor nativus*, of the animal body was too low to process the phlegm or mucus. And it worked in just the same way, so Joubert's thesis, in the case of abstinents. Their body had halted its digestion and entered a state of rest.[9]

The explanation of the French medic from Montpellier did not meet much favour among his colleagues. The body of an anorexic young woman, so the

---

(Lyon: Rovillius, 1549), or once again in Leiden 1633. As new edition see Galen, *Sur les facultés des aliments*, ed. by John Wilkins (Paris: 2013).

[7] François Citois, *Abstinens confolentanea, cui obiter annexa est pro Iouberto Apologia* (Montpellier: Chouët, 1602), there p. 1–12, and afterwards in defence id., *Abstinentia puellae confolentaneae ab Israeli Harveti vindicata, ad Nicolaum Rapinum* (Aix-en-Provence: Blancét, 1602), passim.

[8] On the debates between Rodrigo di Castro and Liceti, which touched the idea of animal hibernation only very marginally, Roling (2016), p. 191–211.

[9] Laurent Joubert, *Paradoxa*, in: *Opera latina* (Lyon: Senneton, 1582, 2 vol.), vol. 1, Decas I, Paradoxum II, p. 11–19.

tenor of the argument, could not behave like the body of a rodent. Further, the abstinent had evidently continued to excrete bodily secretions such as sweat and tears. The property of being wholly insulated thus could not apply to her either. Yet the solution to the question of the hunger artist offered by the great scholar Fortunio Liceti, active in Bologna and Padua and one of the most versatile figures of the era, was in large part the same as Joubert's.[10] In the abstinent too, so Liceti's hypothesis, which he developed in an extensive treatise, the *vis nutritiva* had for a certain period halted its activity. If digestion consisted of the reciprocal interpenetration of heat and moisture in the human body, then it must be possible, so Liceti, for the two forces to achieve a perfect balance and cease their dynamism; a comparable example was quicksilver, which possessed internal heat and watery substance, without that combination forcing it to break up. This event must often occur together with a reduction in body temperature. The equilibrium in the belly came to an end when a fever, as with most patients, or some other new factor raised the body temperature and so gave heat the upper hand again in the digestive process. Thus animal hibernation and anorexia were indeed parallel.[11]

Liceti's great critic, the Portuguese doctor Rodrigo di Castro, would address serious criticisms to this model in a number of works directed against Liceti, which Liceti in his turn answered with a battery of works in response.[12] Although human hunger artists were the main focus, the question of the organic preconditions of hibernation also came into view. If *humor* and *calor*, so Rodrigo di Castro, could come into balance in the human body, why did this equilibrium at some point collapse? Would the aging process not come to a complete halt? The two forces were in a permanently dynamic state and behaved like act and potency; a state of rest was for that reason entirely ruled out. Hence the supposed stasis in the body of the sleeping animal, too, needed a different explanation. The remaining mucus, which Liceti, too, had left in the sleeper's digestive tract, would necessarily result in excrement, but if the zoologists were to be believed this was evidently not the case.[13] Rodrigo di Castro had another theory

---

[10] On Liceti's life and writings see e.g. Ongaro (1964), p. 235–244, on other aspects of his work see e.g. Hirai (2011a), p. 273–300, and Hirai (2011b), p. 123–150.

[11] Fortunio Liceti, *De his, qui diu vivunt sine alimento libri quatuor, in quibus diuturnae Inediae observationes, opiniones, et caussae summa cum diligentia explicantur* (Padua: Petrus Bertellius, 1612), there esp. III,42–48, p. 30–33; 70–71, p. 45; 78–79, p. 48–49.

[12] On the person of Rodrigo de Castro in general see Manuppella (1967), there on his biography p. 1–84.

[13] Estêvão Rodrigues de Castro, *De asitia tractatus* (Firenze: Pignonius, 1630), there esp. 13–14, p. 41–49; 17–18, p. 55–58.

about the lack of nourishment of the animal during its winter rest which would – probably with justice – move his opponents to sarcasm. Wandering insects, di Castro claimed, would provide the resting animal with nourishment during the winter. To place by every bear's bedside a column of ant-attendants that push honey into its mouth was just too absurd.[14]

3 Blocked Circulation: Ole Borch on Hibernation

The long-term influence of the ideas of Joubert and Liceti, and through them of Albertus, is revealed by the example that I wish to focus on, which is probably the most important treatment of the topic in the late seventeenth century. Although some of the scientific paradigms had in the meantime decisively changed and, above all, Aristotelian natural philosophy had long been in retreat, the discussion of animal hibernation could still be pursued on largely similar foundations. In 1680 the Danish medic Ole Borch published his study of animal hibernation.[15] Borch was one of the key figures at the University of Copenhagen and, as a chemist and doctor, like the majority of his colleagues in Denmark he was not without sympathies for Paracelsism.[16] His dissertations were reprinted far into the eighteenth century and were read well beyond Scandinavia. Although observation-based natural science had significantly expanded the empirical material, the core question was still open. How was the sleeping rodent, bear or hedgehog able to go without food for months, and what compensated this obvious deficit? As was usual at that time, Borch begins by locating the problem in theological terms. Sleep was common to both man and beast, in both of them the organic force needed to be regularly revived in periods of rest. Only in the postmortal state did all organic needs lapse in man. A tour of the animal kingdom showed that the need for sleep was even stronger there than among the upright creatures.[17] It was not only the now well-known badgers or marmots, so

---

14 As a reply see Fortunio Liceti, *De feriis altricis animae Nemeseticae Disputationes, in quibus encyclopedia medicinae, philosophiae celsiorisque sapientiae praesidio propulsantur ab olim culto mirabili mortalium ieiunio vulgatae recens oppugnationes Asitiastis de Castro* (Padova: Variscianus, 1631), disputatio IX, p. 82–86.
15 Ole Borch, *Dissertatio de animalibus hyeme sopitis*, in: id., *Dissertationes seu Orationes academicae selectioris argumenti* (Copenhagen: Johannes Sebastian Martini, 1715, 2 vol.) (first Copenhagen 1680), vol. I, there No. 7, p. 297–351.
16 On the life and writings of Ole Borch in general see e.g. Fink Jensen (2000), p. 35–68, and Fink Jensen (2006), p. 11–20.
17 Borch, *Dissertatio de animalibus hyeme sopitis*, p. 297–299.

Borch, that needed longer phases of rest; polar researchers like Johan van Linschoten had shown that the polar bears stopped all activity from 4 November to 24 January, in order to give the white polar foxes, their rivals, a chance to rest.[18]

But how was the phenomenon of hibernation to be explained physiologically? In the *machina* of the animal, so Borch informs his readers, the correct proportion of the temperaments ensured the circulation of the humours; a precondition of this circulation was the appropriate body temperature, which needed to be harmonised with the external temperature. Through the circulation of the blood, the energy-bearing *spiritus* was conducted through the body. Too fast circulation produced fever and exhaustion, too slow, as experience showed, produced general *stupor*. In order not to overtire the foundational system of the body, so Borch continued, the transfer of *spiritus* was blocked from time to time and for certain periods. The brain, which had the greatest need of energy, therefore paused, and with it large parts of the rest of the body, while the *discursus* of the energy-conductor came to a halt. The *spiritus* was not, so Borch, left frozen, so to speak, but achieved a state of rest in the form of vapour. The circulating blood was not able to provide the body with such a quantity of energy that its functioning would be able to continue without interruption.[19]

Nature revealed that these periods of rest could be so extensive that in many creatures one must speak of the phenomenon of hibernation. This was true even among bloodless insects, which transported the *spiritus* solely by means of the internal *liquor*, a gelatinous liquid, as Borch noted. Mosquitos wintered in hollow trees and cracks in stones, and many other insects did the same. When the warmth of spring arrived, they got their vitality back. Indeed, as Borch explained, the outside temperature must play the decisive role, for every farmer could observe that a swarm of mosquitos resting in winter could be brought back to life with a burning torch. The cold had made the creatures' *liquor vitalis* thicken and halt its circulation. The heat must reactivate the fluctuation of the *liquor*. The heat was thus like the finger of a musician that knew how to play on the instrument of the body's own *pneumata* and at the right moment struck the right string; the machinery of the insect's body was geared towards symmetry with the outside world, and was able to start up its activity once again.[20]

---

18 Borch, *Dissertatio de animalibus hyeme sopitis*, p. 299–300. Borch made use of the famous dutch travel report of Jan van Linschoten, *Reizen van Jan Huyghen van Linschoten naar het noorden (1594–1595)*, ed. by S. P. L'Honoré Naber ('s Gravenhage: 1914), but I was not able to identify the quotation.
19 Borch, *Dissertatio de animalibus hyeme sopitis*, p. 300–304.
20 Borch, *Dissertatio de animalibus hyeme sopitis*, p. 304–307.

Unlike insects, the crabs and fish, which held the next highest rung in the hierarchy of creatures, did not seem to need periods of rest. Whether fish slept at all was in this period discussed repeatedly and often simply denied, even though Aristotle had alluded to the opposite conclusion. However, already among the reptiles hibernation was necessary. In the autumn snakes reduced their activity levels; with the start of winter cohorts of them gathered and retreated together into caves and holes in the rock. Athanasius Kircher had entered a snake den like this, a cave near Bracciano in Italy; in December it had been possible to wake the snakes with a burning torch that the Jesuit had held up to them. The higher one went up the scale of the animal kingdom, the more often one found creatures, so Borch, that needed to hibernate.[21] The sleep of many rodents, the *glires*, seemed to be so deep that another naturalist, Johann Jonstonius, had even, so he claimed, been able to harry a sleeping dormouse with a knife without it exhibiting any reaction; only when this Polish scholar dropped the animal into warm water did it begin to wake up.[22] Borch cited the now familiar sleeping patterns of hedgehogs and bears. It was unclear what the situation was with birds: the reports could be interpreted now one way, now the other. Were swallows and storks migrants or were they among the hibernating animals? Some scholars argued in favour of hibernation by storks; others, including Jacob Golius, had regarded Africa as the winter habitat of many storks. The question of the winter behaviour of storks and swallows would still be pursued by more than half a dozen authors even after Borch. Many observations, so Borch, made it at least plausible that swallows hibernated; they nested in caves, sheds and rafters and only returned in spring; that the birds lost feathers in the process was an indication that swallows behaved little differently from other animals that needed to hibernate. The remaining body heat was concentrated inside the trunk of the animal, and superfluous extremities such as antlers or a snake's skin were cast off in order to reduce the circulation of blood as far as possible. And, further, the body thoroughly insulated itself: its outer *tunica*, so Borch, seemed to harden, in order to keep the heat loss as low as possible for the *aura vitalis*, the still virulent remaining life.[23]

---

[21] Borch, *Dissertatio de animalibus hyeme sopitis*, p. 307–308, and see Athanasius Kircher, *Magnes sive de arte magnetica opus tripartitum* (Rom: Zanobius, 1654), pars 7, lib. III, § 2, p. 548–552, and Borch's collegue Thomas Bartholin, *Historiarum anatomicarum Centuria I-II* (Copenhagen: Petrus Hauboldt, 1654), centuria II, historia 47, p. 245–248.

[22] Borch, *Dissertatio de animalibus hyeme sopitis*, p. 308–309, and see Johannes Jonstonius, *Historiae naturalis de quadrupedibus liber I* (Amsterdam: Johann Jacob Schipper, 1657), 16, p. 114–115.

[23] Borch, *Dissertatio de animalibus hyeme sopitis*, p. 309–311.

The bird kingdom directed attention to a phenomenon that changed views on hibernation yet again, namely birds that aestivate, i.e. go to sleep in summer. The naturalists of South America, including José Nierembergius and Francesco Ximenez, had described a bird in Mexico called the *hvitzitzil*, which succumbed to lethargy not in winter but in summer and which exhibited its greatest activity when the trees began to bloom in spring. For Borch the behaviour of this creature confirmed a basic assumption, namely that the *syncrasia* of an animal's humours, which in turn was a precondition for blood circulation and the fluctuation of *spiritus*, was directly related to the external temperature. The little Mexican bird, whatever it was, must therefore possess an extremely dry physical constitution, such that it depended on external moisture to compensate and allow its blood to circulate. Once the heat of summer set in, the bird needed to retreat to a place of rest.[24]

This prompted Borch to move on to a question that could not be separated from hibernation and which is of special interest for us. How did the animals go without food for so long and how did they manage to survive the periods of hunger unharmed? As already hinted, Borch adapted the theory of Joubert and Liceti, though without mentioning their names, but he also extended it through his own reflections. The process of digestion must be understood, just as Borch's predecessors had explained it, through the reciprocal interrelation of moisture and heat in the body. The *humidum*, that is the moisture, and the food were transformed by the work of the stomach and liver, the process of *concoctio*, into blood and then into *spiritus*, which was then conducted through the veins as a bearer of energy. Similar ideas could be found in the works of thinkers such as Thomas Willis too. At the same time, the liver dealt with the removal and excretion of materials for which the body detected no need. If body heat was wound down to a minimum, so Borch, its antagonist in the digestive process, namely the moist-cold food, would remain in a largely untransformed state. The two values, heat and food, would then be neutralised in their effect and would be stuck in a state of rest. Precisely this state occurred, a kind of digestive stasis, when a creature began to hibernate. The internal *viscosum*, as Borch calls it, the cold and moisture, was no longer absorbed by the body heat; the internal mucus, the residual product of digestion, remained in the digestive tract of the animal and was kept in a viscose consistency, without being processed. This con-

---

[24] Borch, *Dissertatio de animalibus hyeme sopitis*, p. 311–312, and see Johannes Eusebius Nierembergius, *Historia naturae maxime peregrinae libris XVI distincta* (Antwerpen: Balthasar Moret, 1635), X,1, p. 206, where the bird is named 'hoitzitziltototl'.

text also permitted an explanation of why many animals, as Borch believed, suffered no significant weight loss during hibernation.[25]

Borch did not want to commit himself fully to the logic of a self-contained bodily system that was no longer dependent on receiving nourishment. The Jesuit La Roquette, as Borch reports, had shut a lizard in a glass flask for a period of three months. The opening of the bottle was closed with perforated canvas. The animal, lethargic and evidently fallen into a state of motionlessness, did not die, for during its period of rest it was evidently able, so Borch, to maintain the *succus*, the body's mucus, at a minimal level and perhaps even to allow blood to circulate. The humours were no longer transformed into *vapores* or *spiritus* and the beast no longer produced excrement.[26] In the light of this experiment, and perhaps because he had a certain proximity to Paracelsus, Borch also allowed another nutritional factor to be considered, namely air. Boyle's experiments had shown that no creature could survive more than perhaps twenty minutes without an air supply. Liceti and other medics had cast doubt on the elementary nutritional value of air, as Borch probably knew, but the Dane did not share their view.[27] Mere breathing in itself guaranteed an animal a first basic provision of nutrition. Salt, oil and water were ingredients of air, which could in this way be supplied to every living creature. Their presence in air had been confirmed by the simplest experiments. Oil could be confirmed by the smudges that air left on clean linen; salt precipitated in the form of crystals on iron chains that were exposed to fresh air for too long; and water could be condensed from the atmosphere by the pressure of a simple air pump. A sleeping animal thus received from the external effect of air alone a basic supply of nutrition, which was able to support the otherwise self-contained digestive cycle of the winter period. If during hibernation digestion nonetheless ensued, the fine particles reached the brain only as vapours, *vapores*, where they had the effect, according to Borch, of further blocking the transfer of *spiritus*, which was the real bearer of energy, from passing in that direction, and so this process could only result in extending sleep.[28]

---

25 Borch, *Dissertatio de animalibus hyeme sopitis*, p. 312–314, and see in comparison Thomas Willis, *De anima brutorum, quae hominis vitalis ac sensitiva est, exercitationes duae* (Amsterdam: Johan Blaeu, 1674), ex. 1, cap. 5, p. 52–58.
26 Borch, *Dissertatio de animalibus hyeme sopitis*, p. 314–315.
27 Liceti, *De his, qui diu vivunt sine alimento*, II,1, p. 19–20, and see Giovanni Argenterio, *Opera numquam excusa* (Venedig: Iuntas, 1606), *Commentaria in Hippocratis Aphorismata*, Aphorismus II, p. 26.
28 Borch, *Dissertatio de animalibus hyeme sopitis*, p. 315–320.

In a similar way to his Italian predecessors, Borch finished by considering analogies to hibernation in the human sphere. Many of the apparent sleep-virtuosos of the past, not least the famous Seven Sleepers, who in their cave had outslept their martyrdom by 250 years, were probably to be ascribed to the world of legend. But on the other hand numerous medics, such as Johann Schenck and Marcello Donato, had described pathological cataleptic sleeping states, which in their symptoms could certainly be compared to hibernation.[29] The best known was the near-ecstatic state that befell the philosopher Duns Scotus and which had ultimately led to the *doctor subtilis*, so tradition told, being buried alive.[30] Of similar character at the physiological level were the cases of complete abstinence that were recorded in the annals of medical history, such as the unfortunate Jeanne Balant from Confolent whom François Citois had described, or the case of a certain Eva Flegenia, who had got through 12 years without eating, or Apollonia Schreiner, who had done it for 29 years, if one chose to believe the medic Wilhelm Fabritius. In all these cases the digestive process had neutralised itself in its central functions and had come to a halt due to an inner equilibrium. The air alone had sufficed as a basic form of nutrition.[31]

4 An early Modern Encyclopedia of Animal Sleep: Carl August von Bergen

How sharply the treatment of the phenomenon could change is shown by a work written on the topic a few decades after Ole Borch by the Frankfurt medic Karl August von Bergen, who had studied under Herman Boerhave in Leiden and Johann Salzmann in Strasbourg.[32] Step by step, as will be seen, both the Galenic doctrine of humours and the Paracelsist theory would be pushed into the back-

---

[29] Marcellus Donatus, *De historia medica mirabili libri sex* (Frankfurt: Johann Jacob Porsius, 1613), II,7, p. 181–187, and Johann Schenck of Grafenberg, *Observationum medicarum rariorum libri VII, in quibus nova, abdita, admirabilia, monstrosaque exempla proponuntur* (Frankfurt: Johannes Beyer, 1655), p. 66–70.
[30] The story of John Duns Scotus is reported e.g. in Marcus Antonius Sabellicus, *De omnium gentium omniumque seculorum insignibus memoriaque dignis factis et dictis exempla* (Basel: Petri, 1541), VII,4, p. 421–422, und Petrus Crinitus, *De honesta disciplina libri XXV* (Lyon: Gryphius, 1543), XXIV,11, p. 364–365.
[31] Borch, *Dissertatio de animalibus hyeme sopitis*, p. 320–333, and on comparable stories Simone Porzio, *De puella germanica, quae fere biennium vixerat sine cibo, potuque, ad Paulum III. Pontificem Maximum disputatio* (Florenz: Laurentius Torrentinus, 1551), passim.
[32] For a short hint on the life of Karl August von Bergen see Hirsch (1875), p. 367–368.

ground. The central focus was placed more firmly instead on comparative anatomy, though that still did not permit a convincing new hypothesis to be developed. At the same time, there had been substantial advances in the observation of animals.[33] The marmot, as had been shown by Johann Georg Altmann in his natural history of the Swiss glaciers, slept from September to May. Fourteen days before hibernation, it completely stopped its consumption of food, in order not to further burden its intestinal tract. By spring the fat that the animal had put on had vanished. The animals slept so deeply that farmers could carry them off without waking them.[34] Other naturalists too had made efforts to get a more exact description and confirmation of the processes at work, including Gabriel Rzączyński,[35] who had written a natural history of Poland, and Roland Sibbald, who had done the same for his homeland of Scotland. Badgers, so Sibbald had shown, furnished their winter quarters with great effort, and carefully cushioned their sleeping place; indeed, the beasts gave the impression that they were walling themselves in.[36] Von Bergen surveyed a series of reports like this, which corrected some of the claims of the ancient and early modern zoologists. A clear distinction must be made, so von Bergen, between two variants of hibernation: on the one hand a total lethargy, a *torpor*, which in its total stupor must preclude any consumption of food, and on the other hand a winter rest that still at least partially permitted the search for foodstuffs to continue. Given this observation, it was doubtful whether the hibernation of many animals really possessed such rigorous traits as many authors had claimed. Had Jonstonius really been able to stab a knife into his unfortunate dormouse without waking the creature from its stupor? Martin Lister, an eminent English zoologist,[37] had kept a dormouse himself and the beast had repeatedly, so Lister had shown, returned to consciousness during its period of sleep and had even been able to consume foodstuffs.[38] This still left open the question, so von Bergen, of whether observations made on animals in captivity could really be transferred to life in the wild.

---

33 Karl von Bergen – Franciscus Heyn (resp.), *Disputatio physica circularis de animalibus hieme sopitis* (Frankfurt/Oder: Johannes Christian Winter, 1752), § 1–12, p. 3–10.
34 Johann Georg Altmann, *Versuch einer historischen und physischen Beschreibung der helvetischen Eisberge* (Zürich: Heidegger, 1753), p. 203–209.
35 Gabriel Rzączyński, *Historia naturalis curiosa regni Poloniae in tractatus XX divisa* (Sandomir: Collegium Societatis Jesu, 1721), tract. 8, sectio 1, § 21, p. 225–228.
36 Robert Sibbald, *Scotia illustrata sive Prodromus historiae naturalis, in quo regionis natura, incolarum ingenia & mores, morbi iisque medendi methodus, et medicina indigena accuratè explicantur* (Edinburgh: 1683, 3 vol.), vol. I, lib. III, sectio 2, cap. 4, p. 9–12.
37 On the life and work of Lister Roos (2011), there esp. 375–420.
38 Martin Lister, *Apicii De obsoniis et condimentis sive arte coquinaria libri decem* (Amsterdam: Janssonius, 1709), VIII, p. 244.

As regards the birds, too, many suggestions about the winter rest of swallows and storks had likewise been shown to be false. No swallow passed the winter by spending months underwater, as Olaus Magnus had still maintained.[39]

Von Bergen believed that he was able to generalise on the basis of these various observations; the question had to be posed once again of what the phenomenon of hibernation meant in organic terms. There was a comparatively unified picture of animals increasing their consumption of food some time before hibernation; but also, so von Bergen, they largely ceased looking for food prior to hibernation. Digestive activities during hibernation and a peristalsis of the intestinal tract were thus to be prevented. Bears and badgers set great importance on insulating their winter dens. Access to fresh air existed, but did not seem to be a priority. Boyle had shown that all organisms depended on an air supply; at the same time, however, Borch had been incorrect when he declared air to be a basis of nutrition. Why then would the animals not have ensured better air supply? The creatures' most important concern was, so experience showed, to keep their heat loss minimal. Their straw beds, their caves and burrows in the ground, but also their rolled-up posture ensured a minimal loss of the body's own heat, even when body temperature sank. Although, as has been seen, at least for some animals some active phases were possible during hibernation, they seemed not to make any further demands on their digestive tract. At least some zoologists, so von Bergen, had reported the amount of body fat that most of the animals put on before the onset of winter. It must therefore be of some service to the winter rest, but evidently without causing heavy digestive activity.[40]

To follow up the phenomenon further, von Bergen pursued anatomy. Gerhard Leonard Blasius, Daniel Nebel, as well as Johann Jacob Scheuchzer and Martin Lister had dissected bears, badgers and hedgehogs, investigating primarily their intestinal tract. Neither hedgehogs nor badgers had an intestine that visibly distinguished rectum, colon and caecum.[41] However, it was striking, as at least Scheuchzer had confirmed, that the so-called *foramen ovale*, the connection between the auricles of the heart, remained open, and so the pulmonary

---

39 Bergen – Heyn, *Disputatio physica circularis de animalibus hieme sopitis*, § 13–17, p. 10–15.
40 Bergen – Heyn, *Disputatio physica circularis de animalibus hieme sopitis*, § 22–30, p. 17–21, and see for von Bergen Johann Nicolas Pechlin, *De aeris et alimenti defectu, et vita sub aquis meditatio* (Kiel: Schultze, 1676), 5, p. 64–78, and Caspar Bartholin, *Exercitationes miscellaneae varii argumenti inprimis anatomici* (Leiden: Haak, 1675), ex. 4, p. 63–79.
41 Gerardus Leonardus Blasius, *Anatome animalium, terrestrium variorum, volatilium, aquatilium, serpentum, insectorum, ovorumque, structuram naturalem ex veterum, recentiorum, propriisque observationibus proponens, figuris variis illustrate* (Amsterdam: A Someren, 1681), there e.g. on the *glires* 20, p. 77.

and blood circulation, as we would today say, remained linked to each other.⁴²
But what would that mean for hibernation? That the animals' body formed a closed system during the winter, as the thinkers of the early seventeenth century had claimed? With a certain frustration the Frankfurt doctor was forced to conclude that comparative anatomy was not yet advanced enough to contribute specialist information useful for the question.⁴³

Nonetheless, some first results could be recorded, even though they were not directly aimed at physiology. Some first pointers were offered by the ordinary phenomenon of sleep. The blood circulation of animals slowed detectibly, and also almost all other *actus vitales* were largely curbed in their functions; there were no excreta, no peristalsis of the intestine, no *chylopoiesis*, that is, production of gastric juices. It remained unclear, so von Bergen, whether all animals actually continued to breathe during hibernation and so whether regular air supply really was indispensable. The absolute necessity of *respiratio*, on which Boyle had set such weight, was thus cast in doubt. Von Bergen does not risk a response to him, but like his predecessors he tends towards the model of a largely self-contained bodily system, the dynamic forces of which had for the most part neutralised each other. The *humores*, the humours, were largely blocked from fluctuating during sleep by the large drop in body temperature; also, so von Bergen, fat too must ensure that the circulation of blood was largely checked. In their equilibrium heat and *humor* had, as it were, each put the other out of action. But how far did this neutralisation go? Did it result in a total standstill?⁴⁴

William Harvey suggested that the blood circulation of the sleeping animal was entirely suspended; its heartbeat stopped.⁴⁵ Other scholars, including Johann Nikolas Pechlin, shared this view.⁴⁶ Nicolas Hartsoeker in his *Discours physiques*, which included a large treatise on digestion, had in 1708 made another proposal. The moisture and salinity of the air led to the thickening of the blood, but thus also to the large-scale blockage of the blood circulation and, as already seen, of all digestive processes.⁴⁷ Gabriel Rzączyński had even suggest-

---

42 Johann Jacob Scheuchzer, "Anatome taxi suilii maris", in: *Acta physico-medica Academiae Caesareae Leopoldino-Carolinae Naturae Curiosorum*, vol. III (Nürnberg: Endter, 1733), observatio 43, p. 127–133.
43 Bergen – Heyn, *Disputatio physica circularis de animalibus hieme sopitis*, § 30–31, p. 21–22.
44 Bergen – Heyn, *Disputatio physica circularis de animalibus hieme sopitis*, § 32–35, p. 22–24.
45 William Harvey, *Exercitationes de generatione animalium* (Amsterdam: Elzevier, 1651), ex. 50, p. 282–298.
46 Pechlin, *De aeris et alimenti defectu, et vita sub aquis meditatio*, 5, p. 64–78.
47 Nicolas Hartsoecker, *Conjectures physiques* (Amsterdam: Desbordes, 1706–12, 3 vol.), vol. II, Troisième discours, p. 22–53.

ed that the same blockage, an insulation that was reinforced by the body fat, must also close up the pores of the animal's body entirely.[48] Thus, so the consensus, there was no metabolism at all during hibernation. Von Bergen did not venture a final decision, but he sympathised with the theory of the vacuum-insulated body that his predecessors had postulated. At least indirectly, Borch's, and so also Liceti's, theories had been shown to be viable. At the same time, for von Bergen in 1752 one thing was also clear: More studies were needed to get to grips with the topic.[49]

5 Concluding Remarks

The extent to which hibernation continued to engage later zoologists and physiologists is shown by the many treatises that, as was to be expected, were still written on the topic even directly after von Bergen's treatment. More large studies specifically on hibernation can be noted just in the first decade of the nineteenth century, and more would follow in the rest of the nineteenth century.[50] Many observations, such as the reduced pulse or the absence of digestion in the sleeping animals, were subsequently confirmed by well-known anatomists such as Lazzaro Spallanzani or John Hunter in Glasgow, but now with a very much better understanding of the material.[51] Not least, Albrecht von Haller had demonstrated that the layers of fat in the animal body during the period of sleep did not, as had been believed, serve as insulation of the body, but in fact as nourishment. The metabolism of the animals thus did not in fact come to a stop.[52] Lavoisier had further been able to show the role that must be played by oxygen in this con-

---

[48] Rzączyński, *Historia naturalis curiosa regni Poloniae*, tract. 8, Sectio 1, § 21, p. 225–228.
[49] Bergen – Heyn, *Disputatio physica circularis de animalibus hieme sopitis*, § 36-37, p. 24.
[50] As examples see George Baird – Henry Reeve, *De animalibus hieme sopitis* (Edinburgh: C. Stewart 1804), or enlarged Henry Reeve, *An essay on the torpidity of animals* (London: Longman and Hurst, 1809), Giuseppe Mangili, *Saggio d'osservazioni per servire all'storia dei mammiferi sogetti a periodico-letargo* (Milano: Reale Stamperia, 1807), id., *Dei mammiferi soggetti a periodico letargo. Memoria* (Pavia: Bissoni, 1818), Jean-Antoine Saissy, *Recherches expérimentales, anatomiques, chimiques etc. sur la physique des Animaux mammifères hybernans, notamment les Marmottes* (Paris: H. Nicolle, 1808), Adolph Wilhelm Otto, *De animalium quorundam per hyemem dormientium, vasis cephalicis et aure interna* (Breslau: 1825), and later e.g. Hans Karl Leopold Barkow, *Der Winterschlaf nach seinen Erscheinungen im Thierreich* (Berlin: Hirschwald, 1846).
[51] Lazzaro Spallanzani, *Tracts on the nature of animals and vegetables* (first as *Opuscoli di fisica animale e vegetabile*) (Edinburgh: William Creech, 1799), p. 251–324, John Hunter, *Observations on Certain Parts of Animal Economy* (London: Castle Street, 1786), p. 87–114.
[52] On Haller's debate of hunger and animal sleep see already Hintzsche (1963), p. 33–46.

text; further, it had become clear that the shortage of it could promote sleep itself. How the image of hibernation would change in the following decades goes beyond the topic of this study. It should by now have become clear that a model which actually took its cue from Albertus Magnus, using a terminology still dominated by Aristotelianism, was still capable of carrying the debate far into the eighteenth century.

Bibliography

Fink Jensen, Morten (2000): "Ole Borch mellem naturlig magi og moderne naturvidenskab". In: *Historisk tidsskrift* 100, p. 35–68.

Fink Jensen, Morten (2006): "Ole Borch – et biografisk rids". In: Riis Larsen, Børge (Ed.), *Ole Borch (1626–90) – En dansk renæssancekemiker*. Copenhagen: Nyt Teknisk Forl., p. 11–20.

Hintzsche, Erich (1963): "Der Hunger in physiologischen Lehrbüchern von Haller bis Valentin". In: *Gesnerus* 20, p. 33–46.

Hirai, Hiro (2011a): "Earth's Soul and Spontaneous Generation: Fortunio Liceti's Criticism against Marsilio Ficino's Ideas on the Origin of Life". In: Clucas, Stephen (Ed.): *Laus Platonici Philosophi: Marsilio Ficino and his Influence*. Leiden: Brill, p. 273–300.

Hirai, Hiro (2011b): *Medical Humanism and Natural Philosophy. Renaissance Debates on Matter, Life and the Soul*. Leiden: Brill, p. 123–150.

Hirsch, August (1875): s.v. "Karl August von Bergen". In: *Allgemeine Deutsche Biographie*. Vol. 2. Leipzig: Duncker & Humblot, p. 367–368.

Manuppella, Giancinto (1967): *Estêvão Rodrigo de Castro: Obras Poéticas em português, castelhano, latim, italiano*. Coimbra: Universidade.

Ongaro, Giuseppe (1964): "L'opera medica di Fortunio Liceti (nota preliminare)". In: *Atti del XX° congresso nazionale di storia della medicina*. Rome: Arti grafiche E. Cossidente, p. 235–244.

Roling, Bernd (2016): "*De asitia:* Fortunio Liceti, Estêvão Rodrigues de Castro und die universitäre Aufarbeitung der Magersucht im 17. Jahrhundert". In: de Boer, Jan-Henryk– Füssel, Marian –Schütte, Jana Madlen (Eds.): *Zwischen Konflikt und Kooperation. Praktiken der europäischen Gelehrtenkultur (12.–17. Jahrhundert)*. Berlin: Duncker & Humblot, p. 191–211.

Roos, Anna Maria Eleanor (2011): *Web of Nature. Martin Lister (1639–1712). The first Arachnologist*. Leiden: Brill.

Andreas Blank
Antonio Ponce de Santacruz on Nutrition and the Question of Emergence

Abstract: Theories of emergent properties are build around the idea that, once material composites have reached some level of complexity, causal powers arise that cannot be reduced to the powers of the constituents. This idea can be traced back to ancient Aristotelian and Galenic views, but seems to be absent from early modern natural philosophy. The present article argues that emergentist intuitions play a role in the discussion of nutrition in the early seventeenth-century commentary on the Hippocratic *Aphorismi* by Antonio Ponce de Santacruz, royal physician to the Spanish king Philip IV. Santacruz understands the new causal powers of the substantial forms that arise from nutrition in close analogy with the new causal powers that he ascribes to the substantial forms of mixtures and the substantial forms of elements. Thereby, he complements a theory of material upward causation through a theory of formal downward causation – a kind of causation that modifies the material basis from which new causal powers have emerged. As Santacruz conjectures, these new causal powers involve emanative causation – the type of causation that brings about an effect without undergoing a change in the cause.

1 Introduction

Applying the notion of emergence to any philosophical position before the advent of the British Emergentists in the nineteenth century may seem to be threatened by anachronism. In recent work in the philosophy of science, one certainly finds numerous ways to analyze this notion that have no close parallels in the history of philosophy. For instance, some philosophers understand it as a purely epistemological notion that describes situations where, due to the exceedingly high complexity of a physical system or a mathematical model, we are unable to predict future outcomes.[1] Also, there are formulations of the notion of emergence that essentially involve the analysis of the relation of the laws of different

---

Note: Work on this article was supported by a research position at the Alpen-Adria Universität Klagenfurt, funded by the Austrian Science Fund (P-33429).

1 See, e.g., Bedau (1997).

sciences.² Evidently, the technical sophistication of the analysis of relations between scientific laws of different domains has no parallels in anything written before the 1970s or so. Things stand differently, however, with approaches that are based on the ontological concept of emergent properties. At the heart of these approaches is the view that, once material composites have reached some level of complexity, causal powers arise that cannot be reduced to the powers of the constituents.³ Emergent properties thus are distinguished from properties of complexes that are nothing other than the summation of the causal powers of the constituents. The concept of non-reduceable causal powers is still widely discussed in the contemporary literature,⁴ and it is this concept that could be instructively applied to earlier periods in the history of philosophy.

As Victor Caston has argued, the concept of emergent properties was clearly articulated by some ancient thinkers, including Aristotle, Galen and Alexander of Aphrodisias (Caston 1997). Also, Richard Sorabji (2010) and Jonardon Ganeri (2011) have analyzed a version of this view in John Philoponus. Furthermore, Olaf Pluta (2007) has brought to light that this view left traces in medieval and Renaissance thought, often complicated by theories of celestial causation. These ancient and medieval versions of emergentism deviate from most contemporary accounts of emergent properties in that they regarded new causal powers to have been derived from substantial forms that emerged from the potencies of matter. Still, the view that new causal powers come into being through the occurrence of new substance can be found in recent work by Trenton Merricks (2001 – although, of course, Merricks would not express this idea in terms of the Aristotelian concept of substantial form). The early moderns used the term *eductio* and its cognates to designate the relation between properties of the constituents of a composite, the newly generated substantial form of the composite, and the causal powers deriving from this substantial form.

Emergentism remained a viable option in sixteenth- and early seventeenth-century natural philosophy. There is a wealth of relevant early modern sources to support this view, and I have explored some of them elsewhere Blank (2014, 2016, 2017 and 2018). In the present article, I would like to add a significant piece of evidence to this overall picture by arguing that emergentist intuitions play a role in the discussion of nutrition in the natural philosophy of Antonio Ponce de Santacruz, royal physician to the Spanish king Philip IV. In particular, I will explore how Santacruz in his commentary on the first part of Avicenna's

---

2 See, e.g., Batterman (2002).
3 See, e.g., Broad (1925).
4 For an overview, see Macdonald/Macdonald (2010).

Canon (1624) uses the analogy between nutrition and two other natural processes: the process that takes place when genuine mixtures come into being, and the process that takes place when elements gain their independence after having been part of a genuine mixture. Santacruz analyses the occurrence of genuine mixtures (in contrast to the occurrence of mere aggregates) as involving the coming into being of a substantial form of a composite. Similarly, he analyses the occurrence of independent elements as involving the emergence of elementary forms from qualities that persisted in the mixture. Likewise, Santacruz understands nutrition as a process in which nutriments acquire a new form – the substantial form of a living being. What is more, he understands the new causal powers of the substantial forms that arise from nutrition in close analogy with the new causal powers that he ascribes to the substantial forms of mixtures and the substantial forms of elements – as he conjectures, these new causal powers all involve emanative causation.

To set the stage, I will outline the emergentist aspects of Santacruz's account of the origin of the forms of elements and the forms of mixtures (section 2). Subsequently, I will investigate the account he gives of the material causes that give rise to the substantial form of a living being in nutrition (section 3). Finally, I will examine how he uses the theory of emanative causation to account for how the substantial form of a living being in turn is capable of modifying the material causes from which it emerges (section 4).

2 Forms of Elements and Forms of Mixtures

The concept of emergence occurs in Santacruz's discussion of what happens to elementary forms when composites dissolve. Santacruz is a fierce critic of the "syndiacritic" hypothesis, according to which all that happens in laboratory processes is the putting-together and separating of corpuscles that retain their identity throughout these processes. As William Newman has documented, the most important empirical support for this hypothesis came from the possibility of the so-called reduction to the pristine state (*reductio ad pristinum statum*) that was observed in laboratory processes during which chemical substances first undergo chemical reactions that completely change their physical appearance, while at the end undergoing chemical reactions that restore the physical appearance of the initial ingredients (Newman 2006, p. 41–43, 98–100, 112–115). If at the end the same substances can be retrieved, the conjecture was plausible that these substances were there all the time. Santacruz contests that this is the only plausible conjecture. Contrary to the syndiacritic hypothesis, he maintains that the elementary forms cease to exist in genuine mixtures.⁵ As he suggests, an alternative explanation of the observable *reductio ad pristinum statum* is that what happens in the dissolution of genuine mixtures is that forms of elements "emerge from the matter of the mixture".⁶

Santacruz presents this idea not only as an alternative to the syndiacritic hypothesis but also as an alternative to Scotus's theory of eminent containment: according to Scotus, elementary forms are contained in the form of the mixture as the vegetative soul is contained in the sensitive soul.⁷ As it was understood in late Scholastic thought, a cause contains its effect eminently when the cause possesses a property that is found in the effect in a more perfect way, which is why the cause is capable of transferring this property in a diminished way to the effect.⁸ Evidently, analyzing the relation between the form of a mixture and the forms of elements in terms of eminent containment is opposed to the view that the potentialities of matter could cause something that is *more* perfect than the potentialities of matter are. Eminent containment thus is incompatible with emergentism.

This is why Santacruz is careful to refute the view that elementary forms are contained eminently in the forms of mixtures. As he points out that, according to the proponent of this theory, eminent containment leads to a more perfect realization of the inferior effects. For instance, the sensitive soul leads to more perfect vegetative operations. However, he objects that the form of the mixture does not lead to more perfect operations of the elements; on the contrary, these operations are weakened.⁹ Moreover, he notes that the idea of eminent containment cannot be applied to cases where it would lead to contradictions. For instance, the rational soul cannot contain the form of a lion, and the sensitive soul cannot contain the form of fire because predicating "lion" of the rational soul and "fire" of the sensitive soul leads to contradictions. Likewise, "it is contrary to the form of the mixture to contain four contrary forms, or even a single elementary form, because here, too, the predication leads to a contradiction."¹⁰ The contradiction

---

5 Antonio Ponce de Santacruz, *Opuscula in Primam Primi Avicennae pro lectionibus primariis. Opuscula medica et philosophica vol. 4* (Madrid: Junta 1624), p. 8.
6 Santacruz, *Opuscula in Primam Primi Avicennae*, p. 12.
7 See Ioannes Duns Scotus, *In primum et secundum Sententiarum quaestiones subtilissimae* (Antwerpen: Ioannes Keerbergius 1620) 2. Sent., dist. 25, q. 1, n. 13–15.
8 See, e.g., Francisco Suárez, *Disputationes metaphysicae. Opera omnia vol. 25–26* (Paris: Vives 1866), disp. 29, sect. 2, nu. 16. On this concept and its reception in Descartes, see Gorham (2003).
9 Santacruz, *Opuscula in Primam Primi Avicennae*, p. 10.
10 Santacruz, *Opuscula in Primam Primi Avicennae*, p. 10: *repugnat naturae sensitivae [animae] continere formam ignis, quia & ipsa praedicatio repugnat: repugnat formae mixti continere quatuor formas contrarias, aut unam solam elementarem, quia repugnat etiam praedicatio.*

that Santacruz has in mind derives from the contrariness of substantial forms: As he argues, fire by itself is not ordained towards the form of the mixture but rather destroys it.

By contrast, contrariness is absent from the relation between the temperament of qualities and the forms of the mixture that emerges from it: "The temperament of qualities alone [...] is ordained as a material disposition towards the form of the mixture because through such a temperament the form is conserved in matter."[11] Moreover, the relation between the temperament and substantial form is not restricted to conservation dependence. This is so because "by the same dispositions through which a form is conserved in matter, another similar form can be introduced into matter."[12] Santacruz conjectures that mixtures decay out of themselves, through their internal contraries, even if no extrinsic cause acts upon them.[13] Moreover, he holds that, when mixtures dissolve, celestial causation can increase the intensity of these qualities that were lost in the temperament.[14] When the qualities that previously gave rise to the substantial form of the mixture are reinforced by external influences, they can give rise to the substantial forms of elements: As he puts it, "as far as their entity is concerned, these qualities precisely do not act for the sake of forms other than the forms that are connatural to them, with which they have a connatural connection; and consequently, they prepare matter for them."[15]

Of course, the notion of temperament itself raises intricate questions concerning emergence. As Santacruz notes, Avicenna understands temperament as "a quality that derives from the mutual action and passion of contrary qualities that are found in elements".[16] Likewise, he paraphrases what Alexander of Aphrodisias maintains in *de Mixtione* as follows: "The fight between elements is carried as far as, once the excesses of contrary qualities have been abolished, through which they were different from each together, they generate a single

---

11 Santacruz, *Opuscula in Primam Primi Avicennae*, p. 10: *Sola [...] temperies qualitatum oriatur ut dispositio materialis ad forma mixti, quia per talem temperiem conservatur forma in materia.*
12 Santacruz, *Opuscula in Primam Primi Avicennae*, p. 228: [E]isdem dispositionibus quibus una forma conservatur in materia, potest alia similis forma introduci de novo in sua materia.
13 Santacruz, *Opuscula in Primam Primi Avicennae*, p. 12.
14 Santacruz, *Opuscula in Primam Primi Avicennae*, p. 228.
15 Santacruz, *Opuscula in Primam Primi Avicennae*, p. 12: praedictae qualitates, quantum attinet ad suam entitatem, praecise non agunt propter alias formas, nisi propter suas connaturales, cum quibus dicunt naturalem connexionem; & ex consequenti disponunt materiam ad illas.
16 Santacruz, *Opuscula in Primam Primi Avicennae*, p. 14: qualitas, quae provenit ex mutua actione & passione contrariarum qualitatum in elementis inventarum. See Avicenna, *Canon medicinae: quo universa medendi scientia pulcherrima et brevi methodo planissime explicatur. Ex Gerardi Cremonensis versione* (Venice: Giunti, 1595), liber 1, sent. 1, doctr. 3, cap. 1.

quality out of all potencies."[17] Santacruz rejects such a view and adopts Aquinas's understanding of temperament as an "intermediary quality" that takes up the nature of all contraries.[18] Much of what Santacruz says about the nature of the temperament can be understood as an explication of this enigmatic notion.

To begin with, Santacruz is clear that he accepts a compositional analysis of temperament, according to which the temperament "contains" or is "composed of" primary qualities (hot, cold, wet, dry).[19] However, against Fernel's view that the temperament is an aggregate of modified primary qualities that has no new causal powers (Fernel 2005, p. 404), Santacruz claims that the temperament is not an accidental being like a heap of stones.[20] As he argues, this is for two reasons: (1) The temperament possesses unity with respect to the dependence relations between the primary qualities: Each primary quality is modified by the other primary qualities to such a degree that it could not be what it is independently of the other primary qualities. He holds that, in this sense, the primary qualities that compose the temperament constitute a single, accidental form.[21] (2) The temperament possesses unity with respect to disposition: The primary qualities that compose the temperament constitute a single disposition toward a single form of the mixture.[22] In the case of animals, this "single total disposition" toward the animal soul is called "natural heat".[23] And a quotation from Aquinas makes clear that what Santacruz has in mind here is substantial form: "in this way, the quality of the mixture is the proper disposition for the substantial form of the mixture, for instance of a stone or of whatever soul."[24]

On first sight, analyzing the unity of the temperament in terms of a unitary disposition seems to be in tension with the idea that, when a mixture dissolves, the elements emerge again. For if there is a single uniform disposition, it is not clear where the diversity of dispositions comes from that can be observed in the emergence of elements.[25] Santacruz responds to this worry that qualitatively dif-

---

17 See Todd (1976), p. 158 (*De mixtione* 233,2–5).
18 See Aquinas, *ST* I, q. 76, art. 4, ad 4 (*Pars prima Summae Theologiae: a quaestione 50 ad quaestionem 119 cum commentariis Thomae de Vio Caietano*, Rome: ex typ. Polyglotta, 1889).
19 Santacruz, *Opuscula in Primam Primi Avicennae*, p. 17.
20 Santacruz, *Opuscula in Primam Primi Avicennae*, p. 19.
21 Santacruz, *Opuscula in Primam Primi Avicennae*, p. 20.
22 Santacruz, *Opuscula in Primam Primi Avicennae*, p. 20.
23 Santacruz, *Opuscula in Primam Primi Avicennae*, p. 20.
24 Santacruz, *Opuscula in Primam Primi Avicennae*, p. 20, cites Aquinas, *ST* I, q. 76, art. 4, ad 4: *huiusmodi qualitas mixtionis est propria dispositio ad formam substantialem mixti, puta lapidis, vel animae cuiuscumque.*
25 Santacruz, *Opuscula in Primam Primi Avicennae*, p. 22.

ferent regions in the mixture correspond to what remains of the elements in mixture.[26] This view is closely connected with his analysis of nutrition since nutrition involves a separation of different parts. Therefore, Santacruz concludes that there must have remained something in the mixture that corresponds to this variety of parts.[27] He also points out that these qualitative differences are essential for explaining the properties of mixtures (such as coagulation, fluidity, flexibility, fragility, viscosity, inflammability).[28] What is more, he offers an argument for why different regions in mixtures necessarily must be qualitatively different. He argues that, in order for remission of qualities to take place, elements must come close to each other; but since elements cannot be equally close to each other, also qualitative change is unequal. Hence, there are different primary qualities and different secondary qualities such as tenuity and density throughout any mixture.[29]

In Santacruz's view, qualitatively different regions are essential for the emergence of plants, zoophytes and animals: "as the other forms of plants and animals require a diversity of parts and as such a form is not educed unless such a varied disposition is given, the same has to be said in mixtures."[30] In Santacruz's view, primary qualities are thus the basis from which substantial forms of elements and the substantial forms of inanimate mixtures and the substantial forms of living beings emerge. Santacruz clearly does not regard the causal powers of substantial forms as mere combinations of the powers of primary qualities since he maintains that substantial forms perform actions by means of which something comes into existence anew.[31] Moreover, he characterizes the function of the new causal powers of substantial forms in their tendency of keeping the entities that they inform in their most perfect state.[32] What kind of causation can fulfill this function?

At this juncture, Santacruz adopts the Platonic notion of emanative causation. As it was generally understood, a being that operates by emanative causation brings forth an effect that possesses the same essence as the cause but realizes this essence in a less perfect way; moreover, by bringing about an effect in

---

26 Santacruz, *Opuscula in Primam Primi Avicennae*, p. 23.
27 Santacruz, *Opuscula in Primam Primi Avicennae*, p. 23.
28 Santacruz, *Opuscula in Primam Primi Avicennae*, p. 24.
29 Santacruz, *Opuscula in Primam Primi Avicennae*, p. 24.
30 Santacruz, *Opuscula in Primam Primi Avicennae*, p. 25: *ut reliquae formae plantarum & animalium requirunt diversitatem partium, & non educitur talis forma, nisi supposita tali dispositione varia: ita dicendum in mixto.*
31 Santacruz, *Opuscula in Primam Primi Avicennae*, p. 196.
32 Santacruz, *Opuscula in Primam Primi Avicennae*, p. 195.

an emanative way, the cause does not undergo change itself.[33] Santacruz uses this concept when he conjectures that the actions of elementary forms could be regarded as instances of emanative causation.[34] One example that he gives for the new causal powers of elementary substantial forms is their alleged capacity to reduce water to its natural coldness.[35] Similarly, actions of the substantial forms of complex mixtures such as body parts include directing the temperament toward an optimal state, in this case the healthy state of a body part: "The natural temperament alone is [...] by itself preserved by the substantial form of the body part, and if such a temperament is lost (as happens in diseases), it can emanate from an internal principle, once what is an obstacle for it is removed."[36] Thus, the substantial form of a body part is described both as emerging from the temperament and as having a causal influence on the temperament. The causation relevant for the first aspect could be characterized as upward causation – the causation that leads from suitably prepared matter to a substantial form with new causal powers. The causation relevant for the second aspect could be characterized as downward causation – the causation that leads from powers that emanate from the substantial form to a modification of the temperament according to the goals defined by the substantial form. As we shall presently see, Santacruz understands the relation between nutrition and animal souls in an analogous way as a kind of causal circle.

3 Nutrition and Upward Causation

As in the case of inanimate mixtures, material upward causation in the case of the nutrition of living beings starts with the temperament. Santacruz maintains that it is not necessary to stipulate the existence of natural powers in addition to the causal role of the temperament (which itself it not an emergent property).[37] He offers two arguments that draw analogies from what happens in inanimate mixture to what happens in animate bodies: (1) The forms of elements are introduced and preserved by the temperament, which shows that the temperament is

---

33 See, e.g., Francesco Piccolomini, *Universa philosophia de moribus* (Frankfurt: Nicolaus Hoffmann, 1611), p. 447.
34 Santacruz, *Opuscula in Primam Primi Avicennae*, p. 195.
35 Santacruz, *Opuscula in Primam Primi Avicennae*, p. 196.
36 Santacruz, *Opuscula in Primam Primi Avicennae*, p. 10: *Sola [...] temperies naturalis conservatur a forma substantiali membri per se: & si amittitur talis temperies (ut per morbos fit) potest a principio interno emanare, semoto prohibente.*
37 Santacruz, *Opuscula in Primam Primi Avicennae*, p. 228.

sufficient for producing substantial forms;[38] (2) If the temperament produces substance, as it does in the case of elements and mixtures, then there is no additional task for any natural powers distinct from the temperament.[39] Consequently, "the soul is conserved in its matter only through the temperament of primary and secondary qualities; hence, the soul is introduced into the matter of nutriment only through the temperament through nutritive action."[40] More precisely, Santacruz holds that the temperament together with the influence of other body parts suffices to generate a determinate substance.[41]

To clarify how nutrition relates to other bodily processes that have an influence on it, Santacruz draws a distinction between two kinds of power in the body of living beings: supported power (*facultas ministrata*) and supporting power (*facultas ministrans*). A power of the former kind is understood as a power that can bring forth a form and therefore presupposes some previous preparation of matter.[42] A power of the latter kind is understood as a power that only prepares matter for the workings of a supported power.[43] As examples of supporting powers, Santacruz mentions attraction, retention and expulsion, which he regards as powers resulting from the temperament of fibers and the influence that body fluids such as humors have on it.[44]

In Santacruz's view, what nutrition and growth have in common with generation is that they all belong to the category of supported power:

> [A]mong the supported powers, there is a difference, for some make the form of the being in which they are, such as nutrition and growth; for they make form in its own subject, by uniting new matter, or new quantity. But generative power makes form in another subject; for the power of the seed introduces a form into the menstrual blood ... Hence, the nutritive power makes numerically the same form, but generating power makes form of the same kind.[45]

---

38 Santacruz, *Opuscula in Primam Primi Avicennae*, p. 228.
39 Santacruz, *Opuscula in Primam Primi Avicennae*, p. 229.
40 Santacruz, *Opuscula in Primam Primi Avicennae*, p. 228: *anima conservatur in sua materia per solam temperiem primarum, & secundarum qualitatum; ergo per solam temperiem inducetur in materia nutrimenti per actionem nutritivam.*
41 Santacruz, *Opuscula in Primam Primi Avicennae*, p. 229.
42 Santacruz, *Opuscula in Primam Primi Avicennae*, p. 229.
43 Santacruz, *Opuscula in Primam Primi Avicennae*, p. 227.
44 Santacruz, *Opuscula in Primam Primi Avicennae*, p. 239–240.
45 Santacruz, *Opuscula in Primam Primi Avicennae*, p. 227: *[I]nter ministratas [facultates] est differentia, nam aliquae sunt factivae formae eius in quo est, quales sunt nutritiva, & crescitiva: istae enim faciunt formam in proprio subiecto, uniendo novam materiam, vel novam quantitatem. At vero generativa est factiva formae in alio: nam facultas seminis in menstruo sanguine inducit formam*

Consequently, what nutrition and generation have in common is that they "induce ultimate form into matter". This raises the question of whether nutrition and generation are distinct powers at all. In many places in Aristotle's writings, nutrition and generation are treated as a single power, but it turns out to be difficult to find an explication of the sense in which they are one.[46] Still, Santacruz disagrees with thinkers such as Francisco Valles (1524–1592),[47] who believes that there is only a single power that comprises both generation and nutrition. Valles uses the phenomenon of regeneration to argue for the view that quantity is not produced by an action distinct from substantial generation. If so, the only difference between generation and nutrition would be a difference between more and less.[48] Since Valles, at least with a view to the souls of non-human animals and the substantial forms of human body parts, accepts the theory of eduction of forms from the potencies of matter,[49] this is a challenge that Santacruz takes very seriously.

In his response to Valles, Santacruz draws a distinction between two different kinds of instrumental causes. As he puts it, generation is brought about by a "separate instrument", which does not receive any continuous causal influence from the genitor, whereas nutrition and regeneration are brought about through the continuous influence of the other parts of a living being.[50] To explicate the different kinds of instrumental causation, Santacruz makes use of the technical notion of an "instrument of direction" that he derives from Gentile da Foligno (d. 1348).[51] Gentile develops this notion in his discussion of supporting powers, which is why it is highly relevant for Santacruz's analysis of the relation between nutrition and the powers that support it. In particular, the concept of an instrument of direction stems from Gentile's discussion of animal spirits in the living organism:

[...] *Unde facultas nutritiva est factiva formae eiusdem numeri: at facultas generativa est factiva formae eiusdem speciei.*
46 See James G. Lennox's contribution to this volume.
47 On Valles's natural philosophy, see Martin (2002).
48 Francisco Valles, *Controversiarum medicarum et philosophicarum* [...] *libri X* (Hanau: Typis Wechelianis, 1564, ⁴1606), p. 74–75.
49 Francisco Valles, *De sacra philosophia, sive de iis qui physice scripta sunt in libris sacris* (Frankfurt: Romanus Beatus, 1600) p. 16, 17, 45, 69, 221. This does not imply that Valles is an emergentist about inanimate mixtures. On the reductionist aspects of his pharmacological views, see Blank (2018).
50 Santacruz, *Opuscula in Primam Primi Avicennae*, p. 239.
51 On Gentile, see French (2001).

[S]pirit transports potency as an instrument changing the primarily moving motion; but spirit is of this kind for it is the instrument of the soul [...] But you will say, what is the mode of this change of direction? It has to be said that if the spirit transports potency, it does not transport potency itself; since potency and the soul give to the spirit its form. But because the potency gives to this spirit a mode of motion, and when such a spirit reaches the members, the members acquire formal potencies in their operations; and it is as we see that the art that is in the soul of the smith does not give its form to the fire and the hammer; but surely it gives to them a mode of motion by means of which they can draw out the form of a small knife from the potentiality that lies in the matter of iron to actuality.[52]

In this passage, spirit is characterized as an instrument of the soul – not in the sense that it transmits motions originating from the soul, for in this case, spirit could not fulfil a causal function when separated from the soul – but rather in the sense that it can modify motions that originate from other causes, external to both the spirit and the soul. In particular, the modification of motion is described as a change of direction. What is more, the potency of the spirit to bring about changes of direction is ascribed to the form of the spirit. In this sense, by conferring to the spirit the capacity of influencing the mode of motion of other bodies, the soul confers form to the spirit. This form differs from the form of the entire organism; on the contrary, it is brought forth from the potencies inherent in the matter of the spirit.

In one important respect, Santacruz departs from Gentile since he does not regard animal spirits as instruments of the *soul* but rather as instruments of the principal organs – the heart and the brain. But as to the role of animal spirits in modifying the motions of other body parts, Santacruz's view is close to Gentile's: "as the spirit receives its mode of motion from the heart and the brain, the same spirit afterwards directs the body parts toward motion. In the generation of spirit, its matter receives a form to move, or motion itself, and subsequently in the body part, it communicates this motion."[53] This raises the question of whether animal spirits are conjoined or separate instruments:

---

[52] Gentile da Foligno, *Primus Avicenne Canon. Avicenne* [...] *Canonum liber* [...] *una cum lucidissima Gentilis Fulgi. expositione* [...] (ed. Bartolomeo Tantuccio, Venice: apud heredes O. Scoti, 1520–1522), fol. 83v: [D]*ico quod spiritus defert virtutem sicut instrumentum motum defert principale movens: sed spiritus est huiusmodi, est enim animae instrumentum* [...] *Sed tu dices quis modus delationis est iste? Dicendum quod si spiritus defert virtutem, non defert ipsam: quia virtus & anima spiritui dat suam formam: sed quia virtus dat illi spiritui modum motus per quem spiritus ad membra perveniens possunt membra formaliter in eorum operationes: et est sicut videmus quod ars que est in anima fabri non dat suam formam igni & malleo; sed bene dat illi modum motus per quem possunt ad actum deducere formam cultelli de potentia materie ferri.*

[53] Santacruz, *Opuscula in Primam Primi Avicennae*, p. 285: *sicut spiritus recepit modum motus a corde & cerebro, ita ipse spiritus postea active dirigit ipsa membra ad motum. In generatione spi-*

If conjoined, as Gentile's words seem to suggest, and as the example of the artisan declares, there arise great difficulties. For if spirit acquires its motion in the brain, as the hammer receives its motion from the smith, how can Gentile contend with Averroes that this sprit articulates itself [in the body part] in such a way that it develops its own motions? If this were the case, one would have to say that it is not a conjoined instrument of the brain or the heart, for it does not operate by means of the same form that it received but by means of another, partial form.[54]

The worry articulated here seems to be that the pattern of motion that spirit receives in the brain cannot remain unchanged while it travels from its origin to other body parts. But if spirit initiates motion in body parts by means only of a part of the pattern of internal motions that it received initially, then it would act as a principal agent, not as an instrument of a primary agent. Santacruz argues that Gentile could reply that what matters is not the preservation of motion but rather the "preservation of the mode of motion". This, of course, sounds enigmatic. What does the concept of the preservation of the mode of motion mean? Santacruz offers the following analogy:

> When I receive the influence of the heavens, I do not destroy it but rather receive it according to the mode of the recipient. Similarly, spirit receives this quality from the brain and the heart, namely, motion with a purpose. Again, when it enters single parts, it does not lose this mode but rather confers dispositions to body parts themselves to move in such-and-such a way, although this motion is determined by the temperament and the form of the body part.[55]

This analogy suggests that spirits are not active principles that initiate the motion of other body parts but rather active principles that change the dispositions of other body parts; and these dispositions determine how these body parts move. What has to remain constant, accordingly, is not particular motions of spi-

---

ritus illius materia suscepit formam ad movendum, seu ipsam motionem, postea vero in membris ipsis communicavit illam.

54 Santacruz, *Opuscula in Primam Primi Avicennae*, p. 285: *si coniunctum, ut verba Gentilis explicare videntur, & exempla artificis declarant, magnae oriuntur difficultates: nam si spiritus in cerebro acquirit istam motionem, ut malleus a fabro, quo pacto ipsemet Gentilis cum Avicenna contendit spiritum in partibus dearticulari ad proprios motus obeundos? Ita enim dicendum est, non esse instrumentum cerebri coniunctum, vel cordis, siquidem non operatur per illamet formam, quam recepit, sed per aliam partialem.*

55 Santacruz, *Opuscula in Primam Primi Avicennae*, p. 285: *ego, qui nunc recipio influxum caeli, non illum destruo, sed ad modum recipientis recipio. Ita similiter spiritus qualitatem istam recipit a cerebro, & corde, motionem, scilicet ad opera. Rursus cum in partes singulares inciderit, non amittit modum illum, sed in talem motionem membra ipsa disponit; quantumvis determinetur a temperie, & forma partis.*

rits but rather the capacity of spirit to change in other body parts their disposition for motion.

Santacruz gives the following analysis of the relation between efficient and material causation in the eduction of substantial forms:

> Even if the agent introduces dispositions into matter through efficient causation, and the qualities that are produced belong to the accidental forms, which are dispositions preceding the existence of form in matter: nevertheless, the whole matter with its dispositions behaves as a subject and purely materially with respect to the principal form. Hence, we concede readily that what produces dispositions produces in the way of efficient causation. Again, the produced dispositions concur as forms of some kind to render matter more perfect. Finally, however, with respect to the action to be brought forth, all these qualities behave as material causes with respect to this form.[56]

For instance, animal spirits modify the dispositions of motion of the body parts involved in nutrition, which in turn give rise to the substantial form of a living being. This is why Santacruz understands the natural qualities of spirits as qualities that do not belong to the efficient aspect of the instrument but rather to its material aspect.[57] These qualities "concur instrumentally, not by bringing about action through efficient causation but by formally perfecting a principle that brings forth actions."[58]

Animal spirits differ from animal seeds, however, because they require a continuous causal influence from the primary agent. This is so because the mode of motion conferred by them "does not consist in those qualities that can persevere in the absence of an agent."[59] Santacruz adduced experimental evidence that shows that local motion of body parts ceases as soon as nerves are obstructed. In such an obstruction, two things take place: animal spirits no longer can flow through the nerves, and no mode of motion can be propagated by means of animal spirit. However, the propagation of modes of motion does not coincide with the flow of animal spirits. Spirits flow into body parts after an

---

[56] Santacruz, *Opuscula in Primam Primi Avicennae*, p. 290: *Licet enim agens disponat materiam efficienter, & qualitates productae sint quaedam formae accidentales, quae disponunt praevie ad esse formae in materia: tota tamen materia disposita habet se subiective, & mere materialiter respectu formae principalis: Itaque concedimus libenter, quod producens dispositiones, producit efficienter: Rursus dispositiones productae, ut formae quaedam concurrunt ad ornandam materiam. Ultimo tamen respectu actionis eliciendae, respectu ipsius formae materialiter se habent omnes istae qualitates.*
[57] Santacruz, *Opuscula in Primam Primi Avicennae*, p. 291.
[58] Santacruz, *Opuscula in Primam Primi Avicennae*, p. 291: *possit dici instrumentaliter concurrere qualitates, non efficiendo actionem, sed formaliter perficiendo principium elicitivum actionis.*
[59] Santacruz, *Opuscula in Primam Primi Avicennae*, p. 286.

obstruction has been removed; this does not happen instantaneously but rather in time. This is why what moved instantaneously are not spirits but rather "a quality that is diffused throughout spirits and nerves in a moment, as if it lacks a contrary."[60]

Santacruz compares the propagation of modes of motion in spirits with the behavior of sensible species in a medium.[61] As he describes them, sensible species are information-carrying structures that are not corrupted by other sensible species and, in this respect, do not possess contraries; however, they cease to exist in the moment in which their source ceases to exist.[62] Santacruz's use of the analogy between how sensible species transport information through a medium and how sprits propagate modes of motion from the principal organs to another body indicates that his view is not that a portion of spirit receives a pattern of motion in the brain and subsequently travels through the nerves to a body part; in this case, it would in fact be difficult to imagine how the original pattern of motion could remain intact all the way along. Rather, he takes spirits to form a continuum that forms a medium through which patterns of motion originating from the brain and the heart are propagated; what remains intact is not the structure of a particular portion of spirit that moves from one place in the body to another place but rather the patterns of motion that are propagated in a continuous medium. It is these patterns of motion that function as "modes" of motion because they are capable of modifying dispositions inherent in body parts. The motions of spirits, one could say, are efficient causes of the modification of the dispositions inherent in body parts. And these dispositions, together with the motions of surrounding bodies, are the efficient causes of motions of body parts.

Surprisingly, as to the nature of instrumental causation involved in the workings of spirits there is an analogy with how Santacruz believes that animal seeds operate. He holds that the structure of the material parts of the seed modifies the motions of particles external to the seed.[63] This is so because the capacity of modifying the motions of other particles in this way does not presuppose any continued influence of the parent's soul on the seed; all that is needed is that the parent's soul previously conferred a certain structure to the constituents of the seed.[64] What is more, Santacruz expands the concept of an instrument of direction in a way that goes beyond Gentile's analysis of spirit by including the notion of qualitative change into his analysis of instrumental causes. The primary

---

60 Santacruz, *Opuscula in Primam Primi Avicennae*, p. 293.
61 On the medieval background of this theory, see Maier (1963).
62 Santacruz, *Opuscula in Primam Primi Avicennae*, p. 293.
63 Santacruz, *Opuscula in Primam Primi Avicennae*, p. 72, 284–285.
64 Santacruz, *Opuscula in Primam Primi Avicennae*, p. 71–72, 285.

qualities of the material parts of the seed modify the primary qualities of the parts that are added during the process of nutrition and growth.[65] Again, the causal powers of primary qualities are independent of a continued causal influence from the parent's soul;[66] all that is needed is that the parent's soul has previously tempered these qualities in a suitable way.[67] Moreover, since the potencies of local and qualitative modification inherent in seeds are the outcome of the agency of the parents' souls and since these potencies realize the reproductive goals of the parents, seeds function as instruments of the parents.[68] Thus, both animal spirits and animal seeds operate as instrumental causes because they modify the motions of bodies with which they come into contact according to the goals of their primary agents. In this way, what nutrition and generation have in common with all other body functions that involve supporting natural powers is the role that instrumental causes play in the emergence of substantial forms – be it the forms of body parts or the forms of living beings.

4 Nutrition and Downward Causation

Santacruz's theory of instrumental causes thus offers an analysis of the role that material causation plays in the emergence of the substantial forms of living beings and at the same time serves to explicate both the analogies and the differences between nutrition and generation. Still, this raises the question of what the new causal powers of the substantial forms of living beings are supposed to consist in. As in the case of the substantial forms of complex mixtures such as body parts, Santacruz understands animal souls to be embedded in a causal circle: "[A]s the ultimate disposition is preserved by the form, so is the vegetative grade preserved by the sensitive grade, [...] in the same way as a single form preserves its disposition and conversely is preserved by it, albeit in different kinds of causation."[69] As will become clear presently, the two kinds of causation relevant here are the emergence of an animal soul from the potencies of matter and the emanation of faculties from the animal soul, which direct the animal toward its optimal state.

---

65 Santacruz, *Opuscula in Primam Primi Avicennae*, p. 72.
66 Santacruz, *Opuscula in Primam Primi Avicennae*, p. 69.
67 Santacruz, *Opuscula in Primam Primi Avicennae*, p. 74, 290.
68 Santacruz, *Opuscula in Primam Primi Avicennae*, p. 284.
69 Santacruz, *Opuscula in Primam Primi Avicennae*, p. 10: *Et sicut ultima dispositio conservatur a forma, ita gradus vegetativus a sensitivo [...] ut una forma conservat suam dispositionem; & e converso conservatur ab illo in diverso genere causae.*

> The quality that emanates from the soul, by means of potency, is something superadded to the temperament in the mode of a more perfect actuality and form. And it is not necessary that it is of the same specific nature as the first qualities, but stands in a more eminent relation that regards the temperament united as such, and preserves and confirms this unity.[70]

Does it make sense to ascribe to qualities emanating from the soul a function in preserving the unity of the temperament? Recall that Santacruz accepts Avicenna's view that mixtures, due to the contrary nature of the qualities of their constituents, have an internal tendency toward decay. In this sense, preventing this internal tendency is a genuine task that the qualities emanating from the soul could fulfil. Moreover, it is evidently a task that cannot be fulfilled by the primary qualities themselves since the tendency toward decay results from their contrariness. This is why Santacruz takes the qualities emanating from the soul to be "something added to the temperament itself by means of a more perfect actuality and form."[71]

This is the crucial point where Santacruz diverges from Galen. As Santacruz diagnoses it, Galen did not know anything about "the emanation of faculties from the soul".[72] Unlike Galen, Santacruz understands a process such as nutrition to be a single total effect of the two distinct agents – the temperament of various body parts and the vegetative powers of the soul – without distinguishable aspects in the effect.[73] Still, the question remains why one should accept the reality of vegetative powers emanating from the form. Santacruz offers the following argument:

> [M]ore is required to change matter than to preserve form in matter. For to preserve the form, no passive resistance has to be overcome; but to change matter, there intervenes some passive resistance; and hence a greater power is required for transforming matter. And this is the reason why inanimate bodies, such as stones, do not transmute their prox-

---

[70] Santacruz, *Opuscula in Primam Primi Avicennae*, p. 231: *Illa autem qualitas, quae dimanat ab anima, media potentia, est aliquid superadditum ipsi temperamento per modum actus & formae perfectioris, Neque est necessarium, ut sit eiusdem specificae naturae cum primis, sed eminentialis cuiusdam rationis, quae respicat temperamentum unitum ut sic, & illam unionem conservet & confirmet.*

[71] Santacruz, *Opuscula in Primam Primi Avicennae*, p. 231: *aliquid superadditum ipsi temperamento per modum actus & formae perfectioris* [...].

[72] Santacruz, *Opuscula in Primam Primi Avicennae*, p. 231.

[73] Santacruz, *Opuscula in Primam Primi Avicennae*, p. 231.

imate matter, because they have power limited to self-preservation only. Therefore, living bodies have faculties in order to augment the power of their primary qualities.[74]

Hence, it is the capacity for self-change characteristic of living beings that speaks in favor the existence of vegetative powers emanating from souls. While temperament, together with the influence of animal spirits, is capable of bringing forth a substantial form, the development and improvement of organs is something that requires additional powers that do not reduce to combinations of primary qualities. This is why the soul plays a role in downward causation by means of which powers that emanate from the soul change bodily structures.

As Santacruz notes, his conjecture concerning downward causation raises the question of whether the way in which powers "result from" the soul is a kind of efficient causation.[75] The problem is that not all instances of "resulting from" are causal relations, and not all qualities that "result from" something else have causal potencies. For example, relations cannot produce anything new because they do not possess activity by themselves. This is so because they do not add anything to the foundations from which they result (the things or qualities that stand in a relation to each other). Hence, they can be active only in virtue of the activity of their foundations. Likewise, relations cannot cause relations because relations result whenever their foundations are given. By contrast, the soul stands in a dual causal relation to powers. One relation is a relation of final causation:

> those faculties that relate to the whole, insofar as they are ordained towards the good of the entire substance (for body and soul sense, vegetate and generate as a whole of some kind) nevertheless because through these operations and powers the soul perfects its matter, these powers ultimately are directed towards the soul as their goal.[76]

---

74 Santacruz, *Opuscula in Primam Primi Avicennae*, p. 232: *magis [...] requiritur ad conservandam formam in materia. Nam ad conservandam formam nulla advertitur resistentia passi; ad transmutandam vero materiam intervenit aliqua resistentia passi: & sic maior virtus requiritur ad transmutandum. Et haec est ratio quare corpora inanimata, ut lapides, &c. non transmutent materiam sibi proximam, quia limitatam habent virtutem ad se conservanda tantummodo. Quare corpora viventia habent facultates, ut augeantur virtus suarum primarum qualitatum.*
75 Santacruz, *Opuscula in Primam Primi Avicennae*, p. 189.
76 Santacruz, *Opuscula in Primam Primi Avicennae*, p. 192: *illae facultates, quae respiciunt coniunctum, licet in bonum totius suppositi ordinentur (nam corpus & anima sentiunt ut totum quoddam, vegetantur & generant) tamen quia per tales operationes, & facultates anima perficit suam materiam, inde est, quod ultimate, potentiae ordinentur in animam tanquam in finem.*

At the same time, he also maintains that the soul is an efficient cause of the powers that emanate from it, a view that he defends against three objections raised by Domingo de Báñez (1528–1604).[77] First, Báñez argues that the relation of resulting from something cannot be an instance of efficient causation, since otherwise accidents would be produced immediately by substances, not mediated through powers; but only God can produce something immediately.[78] Second, Báñez argues that if the soul acts by means of efficient causation, it would initiate alteration, which consists in motion towards some quality. Such motion would take time, which is contrary to the understanding of emanation as something that takes place instantaneously.[79] Third, Báñez argues that efficient causes precede temporally their effects, while the soul is generated in the same instant as its powers.[80] As to the first objection, Santacruz argues that accidents that follow from the essence of the soul emanate immediately; but accidents that do not follow from the essence of the soul require another accident. In this sense, some powers can emanate immediately, by means of another accident.[81] As to the second objection, Santacruz maintains that not all emanation is alteration but rather an action through which something new arises; but he does not see any problems with the claim that some qualities that arise by means of emanation lead to alteration that takes place in time.[82] As to the third objection, Santacruz denies that it is necessary that the efficient cause precedes temporally. As an example, he mentions the relation between the sun and the light that it produces – a relation that in the Platonic tradition has often been used as an example for emanative causation.[83] This is why he takes neither the supposed immediacy nor the supposed instantaneous character of emanation to be an obstacle for the claim that powers emanating from the soul can fulfill a role as efficient causes of changes in the bodily organs from whose temperament the soul emerges.

---

[77] Santacruz also ascribes these arguments to Cajetan's commentary on *ST* I, q. 54, a. 3, where I have not found them.
[78] Dominico Báñez, *Scholastica commentaria in primam partem angelici doctoris S. Thomae. Tomus I: Usque ad LXIII quaestionem* (Dvaci: ex typographia Petri Dorremans, 1591) ad *Sum. theol.* I, q. 77, a. 6, dub. 3.
[79] Dominico Báñez, *Scholastica commentaria* ad *ST* I, q. 77, a. 6, dub. 3.
[80] Dominico Báñez, *Scholastica commentaria* ad *ST* I, q. 77, a. 6, dub. 3.
[81] Santacruz, *Opuscula in Primam Primi Avicennae*, p. 196.
[82] Santacruz, *Opuscula in Primam Primi Avicennae*, p. 196.
[83] Santacruz, *Opuscula in Primam Primi Avicennae*, p. 196; see Lindberg (1976), p. 96–98.

5 Conclusion

Santacruz's theory of nutrition thus is embedded in a comprehensive emergentist worldview that regards the substantial forms of elements, the substantial forms of mixtures, the substantial forms of body parts and the substantial forms of living beings as emergent phenomena with novel causal powers. The use that Santacruz makes of the notion of an instrument of direction establishes a close analogy between nutrition and generation without, however, confounding the two processes since generation involves instrumental causes that are separate from the principal agents, while nutrition involves instrumental causes that are continuously influenced by the principal agents. What both processes share is the function of instruments of direction: material beings that modify the motion and the qualities of material objects external to them. The emergence of animal souls, in Santacruz's view, is based on such material changes whose structure is common to nutrition and generation. Santacruz analyses the new causal powers of substantial forms in terms of emanative causation. The substantial forms of elements, the substantial forms of mixtures and the souls of living beings have in common that they emanate powers that direct the entity that they inform toward its most perfects state. This holds for the forms of elements (which direct elements toward their natural degrees of primary qualities and their natural places); it holds for the forms of body parts (which direct body parts toward the temperament that defines health); and it holds for animal souls (which govern the development of organs during the life cycle of an animal). Nutrition thereby is embedded in a causal circle: nutritive powers initially arise from complexes of primary qualities that constitute a unitary disposition toward the substantial form of the living being; and the substantial form of the living being is what preserves the temperament of its body parts and modifies it according to the goals defined by the substantial form. In this way, the soul itself modifies the material basis from which it emerges, thereby leading to a dynamic interaction between the natural powers inherent in the temperament of body parts and the powers emanating from the soul.

Bibliography

Batterman, Robert (2002): *The Devil in the Details: Asymptotic Reasoning in Explanation, Reduction, and Emergence*. Oxford: Oxford University Press.

Bedau, Mark (1997): "Weak Emergence". In: Tomberlin, James E. (ed.), *Philosophical Perspectives, 11: Mind, Causation, and World*. Malden, MA: Blackwell, p. 375–399.

Blank, Andreas (2014): "Material Causes and Incomplete Entities in Gallego de la Serna's Theory of Animal Generation". In: Nachtomy, Ohad/Smith, Justin E. H. (eds.), *The Life Sciences in Early Modern Philosophy*. Oxford: Oxford University Press, p. 117–136.

Blank, Andreas (2016): "Daniel Sennert and the Late Aristotelian Controversy over the Natural Origin of Animal Souls". In: idem (ed.), *Animals. New Essays*. Munich: Philosophia, p. 75–99.

Blank, Andreas (2017): "The Question of Emergence in Protestant Natural Philosophy, 1540–1610". In: *Magyar Filozófiai Szemle* 61, p. 7–22.

Blank, Andreas (2018): "Sixteenth-Century Pharmacology and the Controversy between Reductionism and Emergentism". In: *Perspectives on Science* 26/2, p. 157–184.

Broad, Charles Dunbar (1925): *The Mind and Its Place in Nature*. London: Routledge & Kegan Paul.

Caston, Victor (1997): "Epiphenomenalisms, Ancient and Modern". In: *Philosophical Review* 106, p. 309–363.

Fernel (2005): *Jean Fernel's On the Hidden Causes of Things. Forms, Souls and Occult Diseases in Renaissance Medicine*. With an edition and translation of Fernel's *De abditis rerum causis* by J. M. Forrester; introduction and annotations by J. Henry and J. M. Forrester, Leiden and Boston.

French, Roger (2001): *Canonic Medicine. Gentile da Foligno and Scholasticism*, Leiden/Boston: Brill.

Ganeri, Jonardon (2011): "Emergentisms, Ancient and Modern". In: *Mind* 120, 671–703.

Gorham, Geoffrey (2003): "Descartes' Dilemma of Eminent Containment". In: *Dialogue* 42, p. 3–25.

Lindberg, David Charles (1976): *Theories of Vision from Al-Kindi to Kepler*, Chicago/London: University of Chicago Press.

Macdonald, Cynthia/Macdonald, Graham (2010): "Introduction". In: iidem (eds.), *Emergence in Mind*. Oxford: Oxford University Press, p. 1–21.

Maier, Annelise (1963): "Das Problem der 'species sensibiles in medio' und die neue Naturphilosophie des 14. Jahrhunderts". In: *Freiburger Zeitschrift für Philosophie und Theologie* 12, p. 3–32.

Merricks, Trenton (2001): *Objects and Persons*, Oxford: Clarendon Press.

Martin, Craig (2002): "Francisco Valles and the Renaissance Reinterpretation of Aristotle's *Meteorologica IV* as a Medical Text". In: *Early Science and Medicine* 7, p. 1–31.

Newman, William (2006): *Atoms and Alchemy. Chymistry and the Experimental Origins of the Scientific Revolution*, Chicago/London: University of Chicago Press.

Pluta, Olaf (2007): "How Matter Becomes Mind: Late-Medieval Theories of Emergence", In: Lagerlund, Henrik (ed.), *Forming the Mind. Essays on the Internal Senses and the Mind/Body Problem from Avicenna to the Medical Enlightenment*. Dordrecht: Springer, p. 149–167.

Sorabji, Richard (2010): "Introduction". In: idem (ed.), *Philoponus and the Rejection of Aristotelian Science*. London: Duckworth, p. 1–40.

Todd, Robert B. (1976): *Alexander of Aphrodisias on Stoic Physics. A Study of the De Mixtione with preliminary essays, text, translation, and commentary*. Leiden: Brill.

Contributors

Tommaso Alpina is Wissenschaftlicher Mitarbeiter at MUSAPh (Munich School of Ancient Philosophy) of the Ludwig-Maximilians-Universität in Munich. He received his Ph.D. in Islamic Philosophy from the Scuola Normale Superiore of Pisa. His main areas of research are the reception of Aristotelian philosophical psychology and zoology in Arabic philosophy, notably in Avicenna, and the connections between natural philosophy and medicine. His first monograph *"Subject, Definition, Activity. Framing Avicenna's Science of the Soul"* is forthcoming in De Gruyter – Scientia Graeco-Arabica series.

Hynek Bartoš is an Associate Professor in the Faculty of Humanities at the Charles University in Prague. He is the author of *Philosophy and Dietetics in the Hippocratic On Regimen* (Brill 2015) and a range of essays on the history of ancient Greek philosophy and medicine. Most recently, he co-edited (with C. G. King) the volume *Heat, Pneuma, and Soul in Ancient Philosophy and Science* (CUP 2020).

Andreas Blank holds a Lise-Meitner Research Position at the Alpen-Adria Universität Klagenfurt, Austria. He has published some 60 articles on early modern philosophy and science in edited volumes and journals such as "British Journal for the History of Philosophy", "History of Philosophy Quarterly", "Journal of the History of Ideas", "Intellectual History Review", "History of European Ideas", "The Monist", "Studia Leibnitiana", "Perspectives on Science", "Annals of Science", "Science in Context", and "Early Science and Medicine".

Andrea Libero Carbone's research focuses on Aristotle's biology, and particularly on a survey of argumentation practice in Aristotle's biological works aimed at exploring the relationship between visual thinking and topical arguments. Among his main publications are a monograph on Aristotle's use of visual diagrams in his biology, *Aristote illustré* (2011), and several annotated translations. He is currently head of the Book Department of the Cultural Services of the French Embassy in Rome.

Sophia M. Connell is Senior Lecturer in ancient philosophy at Birkbeck College, University of London. She did her MPhil and PhD at the University of Cambridge and is a former Fellow of Selwyn College, Cambridge. Her main research interests are in ancient Greek philosophy, particularly Aristotle's natural science. She also works on ethics and the history of analytic philosophy. She has published *Aristotle on Female Animals* (CUP 2016) and is the editor of *The Cambridge Companion to Aristotle's Biology* (forthcoming).

Mary Louise Gill is David Benedict Professor of Classics and Philosophy at Brown University. She is the author of *Aristotle on Substance: the Paradox of Unity* (PUP 1989) and *Philosophos: Plato's Missing Dialogue* (OUP 2012). She has co-edited three volumes, including *Self-Motion: From Aristotle to Newton*, and *Unity, Identity and Explanation in Aristotle's Metaphysics*. She is currently working with James G. Lennox and Tiberiu Popa on a translation of and commentary on book IV of Aristotle's *Meteorology* for the Clarendon Aristotle Series.

Gweltaz Guyomarc'h is Associate Professor ('Maître de Conférences') at the University of Lyon. His research interests pivot on Aristotle and the Aristotelian tradition, esp. Alexander of Aphrodisias. He has published *L'Unité de la métaphysique selon Alexandre d'Aphrodise* (Vrin 2005); F. Baghdassarian, G. Guyomarc'h (ed.), *Réceptions de la théologie aristotélicienne. D'Aristote à Michel d'Ephèse* (Peeters 2017); L. Lavaud, G. Guyomarc'h, *Alexandre d'Aphrodise. Commentaire à la Métaphysique d'Aristote, Livres Petit Alpha et Beta*, introduction, traduction et notes (Vrin 2020).

R.A.H. King teaches Philosophy in Berne. He is the author of *Aristotle on life and death*, the editor of *The good life and Conceptions of life in early China and Graeco-Roman antiquity*, and he contributed as translator to the Cambridge works of Plotinus.

Martin Klein, Ph.D. (2016), Humboldt-University of Berlin, is a Lecturer in Philosophy at the University of Würzburg, Germany. His research focusses on medieval epistemology and metaphysics, and the history of the philosophy of mind. He is the author of *Philosophie des Geistes im Spätmittelalter* (Brill 2019) and co-editor of *Medieval and Early Modern Epistemology: After Certainty* (PSMLM 17, 2020).

Giouli Korobili is a Marie Skłodowska-Curie Post-doctoral Fellow at the University of Utrecht. She studied Classical Philology and Philosophy at the National and Kapodistrian University of Athens (BA), at the University of Ioannina (MA) and at Humboldt University of Berlin (PhD). She has contributed to a number of collective volumes on Aristotle, ancient medicine and Byzantine Aristotelian commentators. She is currently revising her PhD thesis on Aristotle's *On Youth and Old Age, on Life and Death, on Respiration* for publication with Springer series Studies in the History of Philosophy of Mind.

David Lefebvre is Professor of Ancient Philosophy at Sorbonne University and member of the Centre Léon Robin (CNRS). He has published on Aristotle and the Peripatos – Theophrastus, Strato, Boethos. He is the author of *Dynamis: Sens et genèse de la notion aristotélicienne de puissance* (Vrin 2018). His most recent publications include also a new translation of Aristotle's *Generation of Animals* (2014). He has co-edited *Dunamis: Autour de la puissance chez Aristote* (Peeters 2008), *La Métaphysique de Théophraste: Principes et Apories* (Peeters 2015) and Aristotle's *Generation of Animals*, A Critical Guide (CUP, 2018).

James G. Lennox is Professor Emeritus of History and Philosophy of Science, University of Pittsburgh. His publications include *Aristotle: On the Parts of Animals* (OUP 2001) and *Aristotle's Philosophy of Biology: Essays on the Origins of Life Science* (CUP 2001). He co-edited, *Being, Nature, and Life in Aristotle* (CUP 2010), *Self-Motion from Aristotle to Newton* (PUP 1994), and *Philosophical Issues in Aristotle's Biology* (CUP 1987). He is the author of "Aristotle and the Origins of Zoology" for *The Cambridge History of Science:* Volume 1.

Roberto Lo Presti is Lecturer of Classical Philology and Ancient Philosophy at the Humboldt-Universität zu Berlin. He has mainly worked on ancient medicine, and the relationship between medicine and philosophy in ancient, medieval and early modern times. He is currently working on a project on love and erotic desire in ancient and medieval philosophy. He has recently edited *Human and Animal Cognition in Early Modern Philosophy and Medicine* (Pittsburgh UP 2017, co-edited with S. Buchenau) and *Werner Jaeger: Wissenschaft, Bildung, Politik* (De Gruyter 2017, co-edited with C. Guthrie King).

Robert Mayhew is Professor of Philosophy at Seton Hall University (in New Jersey). His area of specialization is ancient Greek philosophy, with a concentration on Aristotle and other early Peripatetics. His recent publications include *Aristotle's Lost Homeric Problems: Textual Studies* (OUP 2019), *Theophrastus of Eresus: On Winds* (Brill 2018), and *Aristotle: Problems* (Loeb Classical Library–Harvard UP 2011).

Elisabeth Moreau is a Visiting Postdoctoral Research Fellow at Princeton University and a Research Associate at the Université libre de Bruxelles (Belgium). Trained in History and Philosophy of Science, she works on medicine, alchemy and natural philosophy in late Renaissance Europe. Her PhD dissertation (2018) is centered on the emergence of atomistic and corpuscular theories in Galenic physiology between 1567 and 1634. Her current postdoctoral project is focused on the body's assimilation of food and drugs in early modern medicine from the perspective of matter theories. For her doctoral training, she has held fellowships at Science History Institute (Philadelphia) and FNRS (Belgian National Fund for Scientific Research), and has been a visiting researcher at UPenn and the University of Cambridge.

Bernd Roling is Professor for Classical and Medieval Latin at the Institute for Greek and Latin Philology of the Freie Universität Berlin. His research interests include high medieval and early modern Latin poetry, medieval and early modern philosophy, especially philosophy of language; the history of early modern science, university history, with special focus on Scandinavia; and early modern esoteric traditions. Recent monographs are: *Christliche Kabbalah und aristotelische Naturphilosophie im Werk des Paulus Ritius* (Niemeyer 2007); *Locutio angelica. Die Diskussion der Engelsprache im Mittelalter und der Frühen Neuzeit als Antizipation einer Sprechakttheorie* (Brill 2008); *Drachen und Sirenen: Die Aufarbeitung und Abwicklung der Mythologie an den europäischen Universitäten* (Brill 2010); *Physica Sacra: Wunder, Naturwissenschaft und historischer Schriftsinn zwischen Mittelalter und Früher Neuzeit* (Brill 2013). He has just finished a book on the Swedish polymath Olaus Rudbeck and his reception in 18th century Northern Europe, entitled *Odins Imperium*, 2 vols. (Brill, forthcoming).

Christoph Sander is a postdoctoral researcher at the Bibliotheca Hertziana – Max Planck Institute for Art History in Rome. Currently, he investigates the production, typology and use of diagrams in early modern science. He has published widely on the history of early modern philosophy and its institutional embedding. He is the author of *Magnes: Der Magnetstein und der Magnetismus in den Wissenschaften der Frühen Neuzeit* (Brill 2020).

Index locorum

Achilles Tatius
– *Leucippe et Clitophon* [ed. Vilborg]
  – III 2, p. 50   298
Aetius Amidenus
– *De re medica*
  – V 27 1 = DK 31 A 77   XIII
  – VII 39, p. 308 [Basel: Froben, 1535]   298
Agricola, Georgius
– De *natura fossilium* [Basel: Froben, 1546]
  – I, p. 171   299
  – V, p. 251   300
Albertus Magnus
– *De nutrimento et nutribili*   341
– *Parva Naturalia*   341
– *Physica* [ed. Hossfeld]
  – lib. 7 tract. 1 cap. 1, p. 523   290–291
Alcmaeon Crotonensis
  – DK 24 B 4   XIII
Alexander Aphrodisiensis
– *apud Simplicius*, In Aristotelis Physica commentaria 282,21–24   204
– *De anima* [ed. Bruns]   197
  – 2,25–15,29   204
  – 5,4–6   204
  – 5,10   204
  – 5,11–12   204
  – 5,12–18   204
  – 7,14–10,10   204
  – 8,11–13   210
  – 8,12–13   204
  – 8,17–22   204
  – 8,25–9,4   205
  – 10,28–29   204
  – 11,5–13   205
  – 11,6–7   204
  – 15,28–29   216
  – 16,11–12   205
  – 21,24–22,23   204
  – 27,4–8   205
  – 27,5   205
  – 27,8–15   209
  – 28,20–25   211
  – 29,1–3   200
  – 29,3–10   214
  – 29,10   214
  – 29,13   200
  – 29,16–17   211
  – 30,2–3   210
  – 30,5–6   210
  – 30,26–29   199
  – 31,1–6   198
  – 31,2–3   199
  – 31,2–6   215
  – 31,9   214
  – 31,25–32,19   203
  – 31,27–28   203
  – 32,6–11   202
  – 32,11–20   214
  – 32,11–23   202
  – 32,21–22   212
  – 32,23–33,10   208
  – 32,24   203
  – 33,1–2   208
  – 33,8   208
  – 33,12–34,26   210
  – 34,27–35,2   209
  – 35,8   209
  – 35,9–17   207
  – 35,16–17   214
  – 35,17–22   211
  – 35,19–20   212
  – 35,22–23   214
  – 35,26–36,2   210
  – 36,3   208
  – 36,4   213
  – 36,5–9   214
  – 36,13–15   211
  – 36,19–21   107
  – 36,19–37,3   107
  – 37,4–38,11   106
  – 38,12–13   214
  – 69,5–6   213
  – 73,22–23   214
  – 75,2–10   212
  – 75,8   199

- 75,12–13   213
- 75,14–15   213
- 75,24–76,1   211–212
- 75,26–30   205
- 75,31   214
- 75,31–76,6   211
- 76,1–16   215
- 78,5–6   214
- 92,18   214
- 92,19–20   214
- 92,20–21   213
- 94,9–10   199, 200
- 94,11–17   211
- 94,15   199
- 95,31   199
- 96,12   210
- 99,9   199
- 99,10   210
- 99,18   214
- 99,21–22   213
- *De mixtione* [ed. Bruns]
  - XVI 233,2–5   359
  - XVI 235,22 (= 41,21–22 Groisard)   209
  - XVI 236,18–26 (= 43,12–23 Groisard)   209
  - XVI 236,23–24 (= 43,19–20 Groisard)   210
  - XVI 236,24–26 (= 43,21–23 Groisard)   209
- *De providentia* [ed. Ruland]
  - 77,12   205
  - 83,6–87,4   205
- *In de sensu* [ed. Wendland]
  - 87,11–12   239
  - 165,26–166,2   199
- *In Metaphysica* [ed. Hayduck]
  - 249,5–14   216
  - 263,32–33   211
  - 373,25   216
  - 375,10–11   216
  - 424,28–31   215
  - 425,1   216
- *Mantissa* [ed. Bruns]
  - §1 105,29–30   215
  - §3 115,24–25   204
  - §4   203
  - §4 118,16–26   203
  - §4 118,27–29   209
  - §4 118,28–35   206
  - §4 118,35–38   206
  - §6 123,23–34   199
- *Quaestiones* [ed. Bruns]
  - 2 3   205
  - 2 23 73–74 (= 128 Hervet)   289
  - 2 23 74 (= 128 Hervet)   290
  - 2 27   200, 212
  - 3 9   199
  - 3 14   204
  - 77 12   200

Alexander Aphrodisiensis Latinus
- *Problemata*
- *Ex Theodori Gazae versione* [Venice: Aldus, 1504]
  - fol. 256v   300–301
- *Ex Angeli Politiani versione*
  - fol. T4v Aldus Romanus 1498 = fol. 3r Vercellensis 1501   300

Alexander Aphrodisiensis[Ps.]
- *Problemata* [ed. Ideler]
  - 1 19–21, p. 4   300

Al-Hasan ibn Musa al-Nawbakhti
- *Commentary on Aristotle's de Generatione et Corruptione* [ed. Rashed]
  - 168–173   289

Allatius, Leo
- *De magnete* [Vallicellianus MS Allacci LXXVII]
  - fol. 16r–20r   302
  - fol. 18r–v   301

Altmann, Johann Georg
- *Versuch einer historischen und physischen Beschreibung der helvetischen Eisberge* [Zürich: Heidegger, 1753]
  - p. 203–209   350

Anaxagoras
- DK 59 A 46   XII

Anonymus
- *Tractatus utriusque philosophie* [ed. Lohrmann]
  - p. 240   298–299

Argenterio, Giovanni
– *Opera numquam excusa, Commentaria in Hippocratis Aphorismata* [Venice: Giunta, 1606]
  – Aphorismus II, p. 26   348
Aristoteles
– *Analytica Posteriora*
  – I 2   168
  – I 2 71b33–72a5   25
  – I 2 72a29–30   32
  – II 16 98b32–38   166
  – II 17 99a28–29   167
  – II 19 100a12–13   191
– *Categoriae*
  – 7 7b15–8a12   24
– *De anima*   XI, 44
  – I 1   200
  – I 1 402a5–7   45
  – I 1 402a7   53
  – I 1 402a10   53
  – I 1 402b1–3   200
  – I 1 402b9–16   228
  – I 2 403b25–27   47
  – I 2 404a1–2   129
  – I 2 405a7–9   129
  – I 2 405a19–21   297
  – I 4 407a30–34   51
  – I 5   77
  – I 5 410b23   53
  – I 5 411a30   53
  – I 5 411b19–20   77
  – I 5 411b27–28   239
  – I 5 411b27–30   78
  – II   12
  – II 1   54, 58
  – II 1 412a2–10   45
  – II 1 412a7–9   45
  – II 1 412a12–20   46–47
  – II 1 412a13–15   222
  – II 1 412a14–15   46, 108
  – II 1 412a20   46, 75
  – II 1 412a27–28   36
  – II 1 412a28–b6   36
  – II 1 412b1–4   156, 287
  – II 1 412b3–4   160
  – II 1 412b5–7   46
  – II 1 412b6–7   45
  – II 1 412b6–9   43
  – II 1 412b18–22   27
  – II 1 412b29–30   127
  – II 1 413a1–12   58
  – II 1 413a3–7   192
  – II 2   XII, 200, 204
  – II 2 413a11–16   25
  – II 2 413a20   200
  – II 2 413a20–25   236
  – II 2 413a22   253
  – II 2 413a27   108
  – II 2 413a28–30   77
  – II 2 413a31–32   78, 200
  – II 2 413b1–2   222
  – II 2 413b1–10   321
  – II 2 413b10–15   200
  – II 2 413b11–13   217
  – II 2 413b15   200
  – II 2 413b24–29   192
  – II 2 414a19–27   49
  – II 2 414a26–27   55
  – II 2–4   77
  – II 3 414b8–14   142
  – II 3 414b28–33   229
  – II 3 415a1–2   75
  – II 3 415a2–3   78
  – II 4   11, 13, 18, 21–40, 79, 101, 105–110, 112, 122, 149, 198
  – II 4 415a14–16   36
  – II 4 415a14–22   4, 22, 228
  – II 4 415a14–23   198, 202
  – II 4 415a19–20   208
  – II 4 415a20   105
  – II 4 415a20–22   23, 24
  – II 4 415a22–26   23
  – II 4 415a22–b7   6, 81, 90
  – II 4 415a23   78, 79, 82
  – II 4 415a23–25   222, 279
  – II 4 415a23–b7   202
  – II 4 415a24–b7   9
  – II 4 415a25   260
  – II 4 415a25–26   24
  – II 4 415a25–27   4, 5
  – II 4 415a25–28   6
  – II 4 415a25–29   7
  – II 4 415a26   57, 103, 105, 159, 234
  – II 4 415a26–27   91

- II 4 415a26–b7   40, 231
- II 4 415a27   78, 81
- II 4 415a28   6, 105
- II 4 415b6–7   105
- II 4 415b7–28   202
- II 4 415b8–12   35
- II 4 415b8–14   59
- II 4 415b12–14   12
- II 4 415b12–15   35
- II 4 415b13   215
- II 4 415b15–21   36
- II 4 415b25–27   53
- II 4 415b27–28   202
- II 4 415b28–416a2   108
- II 4 415b28–416a9   29
- II 4 415b28–416a18   202, 231
- II 4 415b29–416a2   XII
- II 4 416a1–30   298
- II 4 416a5   156
- II 4 416a6–18   92
- II 4 416a8–9   144
- II 4 416a9–18   29, 34, 144
- II 4 416a10–18   153
- II 4 416a13–16   153
- II 4 416a13–18   108, 110
- II 4 416a18   110
- II 4 416a18–21   202
- II 4 416a19   78, 113
- II 4 416a19–21   8
- II 4 416a19–b9   26
- II 4 416a19–b12   145, 147
- II 4 416a21   202
- II 4 416a21–22   202
- II 4 416a25–27   127
- II 4 416a34–b3   30
- II 4 416b3–7   105
- II 4 416b9–11   222
- II 4 416b9–15   23
- II 4 416b11–12   207
- II 4 416b11–13   105
- II 4 416b11–20   86–87, 88, 109, 203
- II 4 416b12–18   15, 56
- II 4 416b13   80
- II 4 416b13–15   78
- II 4 416b13–16   106
- II 4 416b13–20   34
- II 4 416b14   109
- II 4 416b15   80, 108
- II 4 416b15–16   24, 106
- II 4 416b16–17   106, 112–113
- II 4 416b17–18   15
- II 4 416b17–19   78
- II 4 416b17–20   16
- II 4 416b18–19   7, 106
- II 4 416b20–23   7, 28, 38, 105, 106
- II 4 416b21–32   53
- II 4 416b22   106
- II 4 416b22–25   16
- II 4 416b23–25   7, 18, 40
- II 4 416b25   57
- II 4 416b25–27   38
- II 4 416b25–29   28, 29
- II 4 416b25–30   231
- II 4 416b26   78
- II 4 416b26–28   106
- II 4 416b28–29   160, 240
- II 4 416b30–31   26, 230
- II 4 416b30–32   127
- II 4 417b26   106
- II 5   53
- III 2 425b26–426a1   24
- III 2 427a9–14   199
- III 3 427a17–19   238
- III 5 430a17–19   192
- III 9 432b3–8   XII
- III 9 432b8–11   108
- III 9–10   77
- III 10 433a21–26   XII
- III 10 433b19–21   104
- III 12 434a22–26   53, 108
- III 12 434a24–25   87, 109
- III 13 435b1   161
- *De caelo*
  - I 3 270b20–25   325
  - III 1 298a32   156
  - IV 1 308a29–31   189
  - IV 22 730b25–26   63
- *De generatione animalium*   53, 184
  - I 2 716a1   239
  - I 2 716a5–6   75
  - I 2 716a8   79
  - I 2 716a31–35   104
  - I 6 717b24   71
  - I 6 718a6–7   71

Index locorum

- I 12 719a34 71
- I 12–13 719a29–720a11 66
- I 13 720a9–10 68
- I 16 721a26 104
- I 17 721b7–8 67
- I 17–18 64
- I 18 722b22–23 109
- I 18 723a9–24 112
- I 18 723b29–30 104
- I 18 724a16–19 67
- I 18 724a36–724b8 67
- I 18 724b8 112
- I 18 724b25–26 64
- I 18 724b35–725a2 65
- I 18 725a2–3 65, 67
- I 18 725a3–7 66
- I 18 725a4–5 64
- I 18 725a4–9 68
- I 18 725a11 64
- I 18 725a11–13 67, 213
- I 18 725a14–28 64
- I 18 725a28 65
- I 18 725a34–35 65
- I 18 725b3–4 68
- I 18 725b14–17 68
- I 18 725b18 68
- I 18 725b19–25 67
- I 18 725b25–726a6 109
- I 18 725b31–33 73
- I 18 726a3–7 67
- I 18 726a10–15 68
- I 18 726a14–16 68
- I 18 726a19–20 141
- I 19 726b1–3 26
- I 19 726b2–3 66
- I 19 726b16 80
- I 19 726b17–19 104
- I 19 726b22–24 27
- I 19 726b25–30 68
- I 19 727a3 69
- I 19 727a13–14 70
- I 19 727a16–26 69
- I 19 727a33–37 73
- I 19 727b31–33 26
- I 20 728a15–18 70
- I 20 728a21–25 70
- I 20 728a26–27 67
- I 20 729a9–12 63
- I 20 729a9–14 162
- I 20 729a10 71
- I 20 729a10–14 107
- I 20 729a11 79
- I 20 729a22–34 63
- I 20 729a32–33 141
- I 20 729a33 79
- I 20–21 729a34–b8 104
- I 21 729b5–6 103
- I 21 729b6–8 117
- I 21 730a5–8 74
- I 21 730a14–15 104
- I 21 730a29–33 71
- I 21–22 730b4–32 104
- I 22 730b20 63
- I 22 730b21 104
- I 23 730b34 75
- I 23 731a5–9 166
- I 23 731a24–33 75
- I 23 731a25–32 8
- I 23 731a29–30 239
- II 10, 136
- II 1 110–114, 122
- II 1 731b24–25 13
- II 1 731b24–732a1 11, 90
- II 1 731b24–732a12 13
- II 1 731b26–27 13
- II 1 731b27–28 13
- II 1 731b28–29 13
- II 1 731b29–30 XII
- II 1 731b30–31 13
- II 1 731b31 14
- II 1 731b32–732a1 14
- II 1 732a10 79
- II 1 732a12–14 75
- II 1 732a16–20 29
- II 1 732b31–33 119
- II 1 733b20–21 107
- II 1 734b7–9 80
- II 1 734b19 111
- II 1 734b24–27 27
- II 1 735a12–26 110, 111
- II 1 735a13 111
- II 1 735a13–14 111
- II 1 735a14 112
- II 1 735a16–19 78

- II 1 735a16–21   87
- II 1 735a17–19   112, 120
- II 1 735a19–20   111
- II 1 735a21   112
- II 1–6   38
- II 2 736a1   177
- II 3   121, 205
- II 3 736a32–36   82
- II 3 736a35–36   76
- II 3 736b10–11   76
- II 3 736b21–29   192
- II 3 736b27–28   80
- II 3 736b29–33   205
- II 3 736b33–737a7   29
- II 3 736b34–737a2   325
- II 3 736b35–737a1   104
- II 3 737a18–24   121
- II 3 737a18–30   81
- II 3 737b2   149
- II 3 737b4   149
- II 3 737b4–5   149
- II 4   78, 114–121, 122, 230
- II 4 738a10–15   69
- II 4 738a25–30   69
- II 4 738a27–30   69
- II 4 738a35–38   329
- II 4 738b7–9   63, 73
- II 4 738b12–14   63
- II 4 738b20–26   67
- II 4 738b21   75
- II 4 738b26   67
- II 4 739a3–4   71
- II 4 739a6–10   68
- II 4 739b20–25   154
- II 4 739b20–28   107
- II 4 739b21–24   71
- II 4 739b25–26   116
- II 4 739b33–740a1   114
- II 4 739b33–740a3   119
- II 4 739b33–740a5   111
- II 4 739b34–740a4   76
- II 4 740a8   118
- II 4 740a17–23   114
- II 4 740a20–21   111
- II 4 740a25–27   77
- II 4 740a33–35   160
- II 4 740b2–8   114, 116, 166
- II 4 740b3   111
- II 4 740b13   115
- II 4 740b19–21   17, 63, 79–80
- II 4 740b24–741a3   114
- II 4 740b25–34   104
- II 4 740b25–741a3   5–6, 16
- II 4 740b26–30   80
- II 4 740b29   115
- II 4 740b29–741a2   87
- II 4 740b29–741a3   159
- II 4 740b30   17, 115
- II 4 740b32–33   117
- II 4 740b33   115
- II 4 740b35   103, 116, 117
- II 4 740b36   118, 119
- II 4 740b37   118, 120
- II 4 740b37–741a3   120
- II 5   74–77
- II 5 741a6–30   74–75
- II 5 741a10–12   80
- II 5 741a13–14   75
- II 5 741b7   67
- II 5 741b16–18   73
- II 6   121–123
- II 6 742a14–16   104
- II 6 742a18–b6   37
- II 6 742b4   73
- II 6 742b35–37   73
- II 6 743a8–11   135
- II 6 743a11–26   87
- II 6 743a13   87
- II 6 743a17–26   119
- II 6 743a18   87, 89
- II 6 743a18–21   131
- II 6 743a25–26   104
- II 6 743b5–12   149
- II 6 743b26–32   133
- II 6 744b11–27   122
- II 6 744b12–28   72
- II 6 744b16–27   87
- II 6 744b27–36   72
- II 6 744b27–745a4   121–122
- II 6 744b27–745a10   86, 98
- II 6 744b28   73
- II 6 744b28–37   127
- II 6 744b32–37   146
- II 6 745a4   72, 73

Index locorum — 385

- II 6 745a4–5   88
- II 6 745a4–9   29
- II 6 745a8–9   89
- II 6 745a11   88
- II 6 745a33–35   148
- II 6 745b4–9   110
- II 6 745b19–20   66
- II 6 745b22–29   114, 119
- II 6 745b29   71
- II 6 746b25–29   109
- II 8   78
- II 8 748b20–24   73
- II 8 749a4   97
- II 8 749a5–6   96
- III 1 750a21–26   164
- III 1 751b26–27   71–72
- III 1 752a2   71
- III 2 752a18–23   158
- III 2 753b23–29   160
- III 2 753b26–27   76
- III 5 756b5–11   71
- III 7 757b6–8   74
- III 7 757b13   71
- III 9 762b3–4   80
- III 11   65
- III 11 762a18–21   150
- III 11 762a20   177
- IV 1 765b14–15   75
- IV 1 765b28–35   166
- IV 1 765b35–766a3   67
- IV 2 767a16–24   74
- IV 2 767a27–28   81
- IV 2–3   94
- IV 2–4   121
- IV 3 767b33–768a9   82
- IV 3 768a10–21   82
- IV 3 768b16–769a6   94
- IV 3 768b29–33   94
- IV 3 768b31–32   91
- IV 3–4 771a17–b14   96
- IV 3–4 771a17–772b12   96
- IV 3–4 771a27–b14   166
- IV 4 771a28–30   73
- IV 4 771a31–32   93
- IV 4 771b33–772a2   110
- IV 4 771b33–772a4   92
- IV 4 771b33–772a25   162
- IV 4 772a4–8   96
- IV 4 772a12–25   162
- IV 4 772a28   93
- IV 4 772b15–18   94
- IV 4 773b4–5   93
- IV 5 773b8–25   73
- IV 5 774a35–b4   73
- IV 6 775a15–16   81
- IV 8 776b25–28   69
- IV 8 777a4–8   116
- IV 8 777a7 = DK 31 B 68   XIII
- V 1 778b23   76
- V 1 778b25–779a2   76
- V 1 779a1–2   119
- V 3 783b12–23   166–167
- V 3 783b17–18   164
- V 4 784b25–28   154
- V 8 789a4–7   147–148
- V 8 789a9–13   148
- V 8 789b8–9   104
- V 8 789b19–20   148
- *De generatione et corruptione*   48, 53
- I 5   15, 43, 44, 53, 59, 112, 230
- I 5 320a2–3   267
- I 5 321a19–24   55
- I 5 321b16–22   56
- I 5 321b17   56
- I 5 321b19–22   56
- I 5 321b22–28   54
- I 5 321b22–31   58
- I 5 321b28–34   55
- I 5 321b28–322a28   15
- I 5 322a10–13   80
- I 5 322a10–16   56
- I 5 322a16–28   56
- I 5 322a20–23   80
- I 5 322a20–28   109
- I 5 322a22–28   56
- I 5 322a24–26   207, 209
- I 5 322a28–33   57, 109
- I 5 322a28–34   58
- I 5 322a33   57
- I 7 324a9–11   32
- I 7 324a24–b4   29–30
- I 10 327a30–328b24   326
- II 2 329b29–33   143
- II 8 335a11–14   160

- *De incessu animalium*
  - 4 705a29–b8   160
  - 4 705a32–b1   160
- *De insomniis*
  - 3 461a12–13   76
- *De juventute et senectute, de vita et morte, de respiratione*   153
  - 1–6   153
  - 2 468a21   240
  - 3 468b18–23   158
  - 3 468b23–28   158
  - 3 469a1–2   116
  - 4 469a26–27   87
  - 4 469b6–20   29, 35
  - 4 469b11–17   153
  - 5   53, 161
  - 6 470a19–22   165
  - 6 470a19–32   155, 163
  - 6 470a23–27   161
  - 7(1) 470b6–13   XIII
  - 12(6)   XIV
  - 12(6) 473a3–6   142
  - 13(7) 473a15–17   XIII
  - 14(8) 474a25–29   127
  - 14(8) 474b13–14   57
  - 14(8) 474b15–19   160
  - 17(11) 476a16–17   51
  - 19(13) 477a28   161
  - 20(14)   163, 185
  - 20(14) 477a32–b4   163
  - 20(14) 477b14–17   163
  - 20(14) 477b26–478a7   163
  - 23(17) 478b28   157
  - 23(17) 478b31–33   154
  - 23(17) 478b31–479a1   157
  - 23(17) 478b32–479a1   240
  - 24(18)   52, 53, 59
  - 24(18) 479a29–30   113
  - 24(18) 479a32–33   91
  - 27(21) 480b21–30   149
- *De longitudine et brevitate vitae*
  - 5 466a18–19   127
  - 5 466a18–20   132
  - 5 466a29–b9   164
  - 5 466b9   64
  - 5 466b22–24   127
  - 5 466b28–33   65
  - 6   53
  - 6 467a6–9   167
  - 6 467a22–30   158
  - 6 467a30–b5   160
- *De motu animalium*
  - 5 700a35–b3   106
- *De partibus animalium*   53
  - I 1 639b22–30   149
  - I 1 642a24–31   149
  - I 5   168
  - I 5 644b22–26   11–12
  - I 5 644b29   44
  - II 1 647a25–27   75
  - II 2 647b10–14   26
  - II 2 647b21–28   139
  - II 2 647b26–27   141
  - II 2 648b35–649a17   27
  - II 3   128
  - II 3 649b19–27   27
  - II 3 649b26–29   141
  - II 3 649b28–35   27–28
  - II 3 650a2–7   156
  - II 3 650a2–8   158–159
  - II 3 650a2–14   140–141
  - II 3 650a2–23   240
  - II 3 650a2–31   161
  - II 3 650a3–4   127
  - II 3 650a9   IX
  - II 3 650a14–b13   143
  - II 3 650a20–23   161
  - II 3 650a20–27   119
  - II 3 650a21–23   156
  - II 3 650a30–31   127
  - II 3 650a34–35   26
  - II 3 650b8–10   73
  - II 3 650b8–11   127
  - II 4 651a7–8   140
  - II 4 651a10–12   140
  - II 4 651a11–13   141
  - II 4 651a13   116
  - II 4 651a13–15   26
  - II 4 651a14–15   141
  - II 6 652a21–24   144
  - II 7   129, 177
  - II 7 652a27–28   130
  - II 7 652a33–36   130
  - II 7 652b7   130

- II 7 652b7–15   130
- II 7 652b15–26   131
- II 7 652b33–35   132
- II 7 653a31–32   90
- II 9 654a33   90
- II 9 654b13   90
- II 9 654b29–32   89
- II 10 655b29–32   142
- II 10 655b32–36   156
- II 10 655b34–36   160
- II 10 655b35–37   139
- II 10 656a2   156
- II 16   95
- II 16 658b32–35   95
- II 16 659a15   96
- III 1   66
- III 2   97
- III 4 665a32   97
- III 4 665b27–666a3   157
- III 5 668a13–27   26
- III 5 668a19–22   160
- III 6 668b33–669a1   132
- III 8   66
- III 11 673b9–11   133
- III 14 674a9–21   143
- III 14 674a19–22   127
- III 14 674b10   IX
- III 19   IX
- IV 2 677a27   66
- IV 4 678a9–15   160
- IV 4 678a11–15   161
- IV 4 678a17–20   127
- IV 5 680b31   143
- IV 5 682a22–27   147
- IV 5 682a23–25   153
- IV 6   97
- IV 7 683b17–24   156
- IV 9   95
- IV 10   97
- IV 10 685b12–15   95
- IV 10 686a5–10   133
- IV 12   95
- *De philosophia*   128
- *De sensu et sensibilibus*   165, 193
  - 1 436a1–6   156
  - 1 436a11   53
  - 1 436a17–b3   149
  - 4 441a11–20   165
  - 4 442a6–8   160
- *De somno et vigilia*   174
  - 1   174
  - 1 454a11–17   175
  - 1 454a11–19   78
  - 1 454b9–11   175, 189
  - 1 454b11   188
  - 2 455a4–5   175
  - 2 455b10   175
  - 2 455b16–25   175
  - 2 455b18   188
  - 2 455b28–30   175
  - 2 455b34–456a6   75
  - 2 456a4–11   175
  - 2 456a24–29   175
  - 3   176
  - 3 456a32–b5   161
  - 3 456b5–6   230
  - 3 456b17–26   176
  - 3 457a8–14   177
  - 3 457a9–25   176
  - 3 457a27–29   185, 187
  - 3 457a31   185
  - 3 457a33–b2   176
  - 3 457b20–23   176, 177
  - 3 457b26–31   177
  - 3 458a25–28   176, 177
- *Ethica Eudemia*
  - II 3 1219b38–39   87
  - VIII 3 1249b10–24   37
- *Ethica Nicomachea*
  - I 6 1097b34–1098a4   87
  - I 10 1100a3   77
  - I 13 1102a32–b2   117
  - I 13 1102b2   117
- *Historia Animalium*   53
  - I 2–3 489a3–8   66
  - I 4 489a20–21   127
  - I 5 489b8–10   146
  - I 16 494b25–29   131
  - III 2 511b11–12   141
  - III 15   66
  - III 18 520b7–9   89–90
  - IV 1 524a4–11   3
  - IV 1 524b2–4   131
  - IV 1 524b32–34   131

- IV 2 527a8   66
- IV 2 529a14   66
- V 1 539a21   239
- V 1 539a23   65
- V 5 541a3–11   66
- V 6 541b8–12   3
- V 12 544a12–13   3
- V 12 544a13–15   3
- V 19 551a4   65
- V 19 552a16–18   65
- V 31 556b26   65
- VI 2   71
- VII(VIII) 1 581a14–16   164
- VII(VIII) 1 588b24–30   7
- VII(VIII) 1 589a2–7   8
- VII(VIII) 12 597a5   97
- VII(VIII) 13–17 599a–600b   340
- VII(VIII) 28   97
- VII(VIII) 28 606a8–b10   93
- VIII(IX) 1 608b4–7   69
- IX(VII) 1 581b29–33   68
- IX(VII) 2   67
- IX(VII) 11 587b30–31   70
– Metaphysica   44, 57–60
  - A 3 984a8–9   129
  - A 8 989a16   57
  - A 9   75
  - Δ 5 1015a20   57
  - Δ 11 1019a1–4   24
  - Δ 11 1019a4–14   24
  - Δ 12 1019a15–b15   32
  - Δ 12 1019b35–1020a6   32
  - Δ 25   215, 216
  - Δ 25 1023b20   215
  - Δ 25 1023b23–25   216
  - Δ 28 1024a29–31   14
  - E 1 1026a10–13   59
  - Z   31
  - Z 3 1029b3–12   25
  - Z 7 1032a24–25   32
  - Z 7 1032a26–32   57
  - Z 7 1032a32–b1   32
  - Z 7 1032b9–14   30
  - Z 7 1032b21–23   30, 32
  - Z 8 1033b19–23   59
  - Z 8 1033b26–32   59
  - Z 8 1033b27–32   17
  - Z 8 1034a2–5   17
  - Z 8 1034a4–6   59
  - Z 8 1034a6–9   60
  - Z 9 1034a21–24   32
  - Z 10 1035a4   216
  - Z 10 1035a21   216
  - Z 10 1035b14–21   216
  - Z 10 1035b24–25   27
  - Z 11 1036b2–7   60
  - Z 11 1036b23   59
  - Z 11 1036b26–32   58
  - Z 16 1040b8–9   57
  - Z 17   36
  - Z 17 1041b14   57
  - Z 17 1041b21   57
  - Z 17 1041b25–32   59
  - Z 17 1041b26   57
  - H 2   36
  - H 4 1044a32–b3   26
  - H 6   36
  - Θ   31, 206, 208
  - Θ 1 1045b33–1046a2   31
  - Θ 1 1046a10–11   32
  - Θ 1 1046a11–13   32
  - Θ 2 1046a36–b7   237
  - Θ 2 1046a36–b29   206
  - Θ 6 1048a25–30   31
  - Θ 6 1048a30–b9   33
  - Θ 6 1048a35–b9   48
  - Θ 6 1048b18–36   33
  - Θ 7 1048b37–1049a3   26
  - Θ 7 1049a8–12   26
  - Θ 7 1049a15–16   51
  - Θ 8   207
  - Θ 8 1049b4–10   58
  - Θ 8 1049b4–12   33
  - Θ 8 1049b8–9   33
  - Θ 8 1049b12–17   22
  - Θ 8 1049b27–29   32
  - Θ 8 1050a4–b34   39
  - Θ 8 1050a34–b4   39
  - Θ 8 1050b6–28   35
  - Λ 3 1070a4–8   32
  - Λ 3 1070a7–8   32
  - Λ 4 1070b28–35   30
  - Λ 7 1072b1–4   37
  - M 4 1078b19–21   129

– *Meteorologica*   334
  – I 1   53
  – I 1 338a25–339a5   53
  – I 1 339a5–8   53
  – I 1 339a5–10   239
  – I 4 341b6–12   176
  – II 2 355a5   127, 142, 147
  – II 4 360b30–32   155
  – II 4 362a5–7   155
  – I–III   244
  – IV   136
  – IV 1 378b10–379b9   153
  – IV 1 379a23–26   127, 147
  – IV 2 379b18–25   164
  – IV 2 379b18–26   80
  – IV 2 379b30–32   139
  – IV 3 380a11–12   164
  – IV 3 380a11–381b22   331
  – IV 3 380a20–22   166
  – IV 3 380a21–22   139
  – IV 3 380a23–26   165
  – IV 3 380b34   332
  – IV 3 381b12–14   139
  – IV 12   27
  – IV 12 389b28–390b2   27
  – IV 12 389b29–390a2   27
  – IV 12 390a15–16   27, 29
  – IV 12 390b3–14   138
  – IV 12 390b14–22   27
– *Parva Naturalia*   X, 12, 53
  See also De sensu et sensibilibus, De somno et vigilia, De insomniis, De longitudine et brevitate vitae, De juventute et senectute, de vita et morte, de respiratione
– *Physica*
  – I   46
  – I 1   168
  – I 1 184a10–b14   25
  – I 1 184a21–25   53
  – I 7   46, 53
  – I 7 190a13–21   24
  – I 9   46, 53
  – I 9 192a31–33   267
  – II 1   58
  – II 1 192b20–27   32
  – II 2 194a27–b8   37

  – II 2 194a27–36   38
  – II 2 194a35–36   37
  – II 3 194b16–23   53
  – II 3 195b21–25   30
  – II 7 198a22–27   37
  – II 8 198b23–28   147
  – II 8 199b10   156
  – III 1 201a10–11   32
  – III 2 201b31–32   32
  – III 2 202a9–12   32
  – III 3 202a15–21   24
  – III 3 202b5–16   24
  – III 3 202b5–29   32–33
  – IV 4 212a24–26   189
  – VII 2 243a   291
  – VII 3 246b4–6   161
  – VIII 5 257b9–10   31
– *Politica*
  – I 1 1252a26–34   10
  – V 3 1302b33–38   94
  – V 9 1309b23–31   96
  – VII 4 1326a35–b2   93
  – VII 7   185
– *Topica*
  – VI 10 148a29–31   253
Aristoteles[Pseudo]
– *Problemata physica*
  – IV   187
  – XIV   173, 185
  – XIV 8   184
  – XIV 8 909b9–13   184
  – XIV 15   184, 185
  – XIV 16   184
  – XIV 16 910a38–b3   184
  – XVIII   174
  – XVIII 1   174, 181
  – XVIII 1 916b2   181
  – XVIII 1 916b2–4   181
  – XVIII 1 916b4–7   183
  – XVIII 1 916b7–12   186
  – XVIII 1 916b8   186
  – XVIII 1 916b9   192
  – XVIII 1 916b10–11   186
  – XVIII 1 916b16   192
  – XVIII 1 916b18   192
  – XVIII 4–6   192
  – XVIII 7   173, 174, 177, 180–192

- XVIII 7 917a18–20   181–182
- XVIII 7 917a19   181
- XVIII 7 917a20–23   182–185
- XVIII 7 917a23   182, 187
- XVIII 7 917a23–28   186–188
- XVIII 7 917a24   186
- XVIII 7 917a25   187
- XVIII 7 917a25–26   186
- XVIII 7 917a26   187
- XVIII 7 917a27   186
- XVIII 7 917a28–36   188–191
- XVIII 7 917a29   187
- XVIII 7 917a30   192
- XVIII 7 917a31–32   191
- XVIII 7 917a35   187
- XVIII 7 917a37–b3   188, 191–192
- XVIII 7 917a38   187
- XVIII 7 917a39   187
- XXVI 8 941a11–13   191
- XXVII   173
- XXX 1   173, 185
- XXX 1 954a11–26   185

Auctoritates Aristotelis [ed. Hamesse]   266
- 4 7, p. 167   266
- 4 9, p. 168   267
- 4 10, p. 168   276

Averroes Latinus (for the Arabic, see Ibn Rushd, Muḥammad ibn Aḥmad al-Ḥafīd)
- *Colliget* [Venice: Giunta, 1574]
  - fol. 95vb   325
- *Commentaria in libros de Physico auditu* [Venice: Giunta, 1562–1574]
  - lib. 7 summa tertia, vol. IV, fol. 315r   292
  - lib. 8 cap. 3, vol. IV, fol. 374v   292

Avicenna Latinus (for the Arabic, see Ibn Sīnā, Abū ʿAlī)
- *De anima* [ed. van Riet]   222
  - praef. 9.12–10.15   238
  - praef. 10.16–21   245
  - praef. 10.21–11.24   247
  - praef. 11.27–12.43   246
  - praef. 12.44   245
  - praef. 12.45   245
  - praef. 12.46–48   245
  - praef. 13.59–62   238
  - praef. 13.60–61   246
- I 1 14.71–15.78   236
- I 1 15.77–78   236
- I 1 16.87–18.10   238
- I 1 30.73–75   251
- I 1 31.85–33.5   243
- I 3 62.82–64.12   243
- I 3 31.11–32.14   247
- I 5 79.3–4   228
- I 5 79.4–5   228
- I 5 79.5   233
- I 5 79.5–80.10   229
- I 5 80.17–81.28   228
- I 5 81.29–82.32   229
- I 5 81.29–82.39   229
- I 5 87.19–90.60   226
- II 1 99.79–102.15   232
- II 1 101.6–7   232
- II 1 101.10   232
- II 1 103.4–5   230
- II 1 103.5–7   230
- II 1 103.5–105.30   230
- II 1 103.8   230
- II 1 104.13–14   231
- II 1 104.17–18   230
- II 1 104.22   231
- II 1 105.30 108.77   231
- II 1 108.78–86   231
- II 1 108.87–110.4   231
- II 1 110.5–10   232
- II 3 132.7–133.24   238
- II 4 146.21   223
- III 7 265.78   225
- III 8 275.60   223
- IV 4 67.70   223
- V 8 174.36   239
- V 8 175.49–51   239
- V 8 175.54–55   240
- V 8 176.64–70   239
- V 8 176.70–71   240
- V 8 176.71   240
- V 8 176.71–72   240
- V 8 177.95   241
- V 8 179.27   241
- V 8 185.26   241
- *De generatione et corruptione* [ed. van Riet]
  - 8   230

- 8 84.31–87.91   230
- *Liber canonis*
- *Ex Gentilis Fulginei versione* [Venice: Scotus, 1520]
  - fol. 37v   328
  - fol. 83v   365
- *Ex Gerardi Cremonensis versione* [Venice: Giunta, 1507]   325, 334
  - I 1 i 2, fol. 1v, a41–55   224
  - I 1 iii 1   359
  - I 1 vi 1, fol. 23r, b12   233
  - I 1 vi 1, fol. 23r, b13   233
  - I 1 vi 1, fol. 23r, b15–16   233
  - I 1 vi 1, fol. 23r, b17   233
  - I 1 vi 1, fol. 23r, a40–53   225
  - I 1 vi 4, fol. 24r, b30–54   241
  - I 1 vi 4, fol. 24v, a9–12   242
  - I 1 vi 4, fol. 24v, a16–22   242
  - I 1 vi 4, fol. 24v, a51–55   241
  - I 1 vi 5, fol. 24v, b41–25r   226
  - I 1 vi 5, fol. 24v, b26   226
  - I 1 vi 5, fol. 25r, a42–b4   226
- *Ex Gerardi Cremonensis versione* [Venice: Giunta, 1595]
  - I, p. 20b–21a   329
  - I, p. 21b   333
  - I, p. 73a   326
  - I, p. 111b   326
- *Philosophia prima sive Scientia divina* [ed. van Riet]
  - I 2 15.74–79   225
  - I 9 5   278
  - V 4 255.70–256.78   246
  - VIII 7 430.36–37   253
  - IX 3 464.92–93   225

Baird, George – Reeve, Henry
- *De animalibus hieme sopitis*   353

Banes, Dominicus
- *Scholastica commentaria*
  - ad *ST* I q. 77 a. 6 dub. 3   372

Barkow, Hans Karl Leopold
- *Der Winterschlaf nach seinen Erscheinungen im Thierreich*   353

Bartholin, Thomas
- *Historiae anatomicae* [Copenhagen: Petrus Hauboldt, 1654]
  - centuria II, historia 47, p. 245–248   346

Bartholin, Caspar
- *Exercitationes miscellaneae varii argumenti inprimis anatomici* [Leiden: Haak, 1675]
  - ex. 4, p. 63–79   351

Basilius Valentinus OSB
- *Basilius Innovatus* [Hamburg: Heyl, 1717]
  - p. 678   303
- *Geheime Bücher oder letztes Testament* [Straßburg: Dietzel, 1645]
  - p. 132   303
- *Letztes Testament und Offenbahrung* [Jena: Eyring, 1626]
  - p. 225   303

Beeckman, Isaac
- *Jounal tenu par Isaac Beeckman de 1604 à 1634* [ed. De Waard]
  - III, p. 124   285

Blasius, Gerardus Leonardus [Amsterdam: A Someren, 1681]
- *Anatome animalium*
  - 20, p. 77   351

Borch, Ole
- *Dissertatio de animalibus hyeme sopitis* [Copenhagen: Ioannes Sebastianus Martini, 1715, vol. I]
  - p. 297–299   344
  - p. 297–351   344
  - p. 299–300   345
  - p. 300–304   345
  - p. 304–307   345
  - p. 307–308   346
  - p. 308–309   346
  - p. 309–311   346
  - p. 311–312   347
  - p. 312–314   348
  - p. 314–315   348
  - p. 315–320   348
  - p. 320–333   349

Campanella, Thomas
- *De sensu rerum et magia* [Frankfurt: Egenolff Emmel/Gottfried Tampach, 1620]
  - I 8, p. 27   302

– *Disputationes in quatuor partes suae Philosophiae realis* [Paris: D. Houssaye, 1637]
– q. 29 art. 7, p. 273   303
Canepari, Petrus Maria
– *De atramentis* [Venice: Evangelista Deuchino, 1619]
– descr. 1 cap. 6, p. 26   302
Cardanus, Hieronymus
– *Contradicentium medicorum* [Lyon: Gryphius, 1548]
– lib. II tract. 1 contr. 3, p. 18–21   300
– lib. II tract. 1 contr. 3, p. 20 (= vol. VI, p. 442 Huguetan/Ravaud)   300
– lib. II tract. 1 contr. 3, p. 21   300
– lib. II tract. 2 contr. 7, p. 101   301
– lib. II tract. 5 contr. 9, p. 277 (= vol. VI, p. 442 Huguetan/Ravaud)   300
– lib. II tract. 5 contr. 9, p. 277 (= vol. VI, p. 566 Huguetan/Ravaud)   300
– lib. II tract. 6 contr. 17, p. 434 (= vol. VI, p. 641 Huguetan/Ravaud)   300
– *De subtilitate* [Basel: Petrina, 1560]   300
– V, p. 345   299
– V, p. 355–356   299
– V, p. 360   299
– V, p. 361   301
– VII, p. 441   301
– VII, p. 494 (= fol. 158v Fezandat/Granion 1550)   301
– *In calumniatorem librorum de subtilitate* [Opera omnia vol. III, Paris, Huguetan/Ravaud, 1559]
– p. 693   302
Celsus
– *De medicina*
– I praef. 20   128
Citois, François
– *Abstinens confolentanea* [Montpellier: Chouët, 1602]
– p. 1–12   342
– *Abstinentia puellae confolentaneae ad Israeli Harveti vindicata*   342
Claudianus [ed. Platnauer]
– 29(48) vol. II, p. 236   298

Cornarius, Ianus
– *Pedanii Dioscoridis De materia medica* [Basel 1557]
– V 111, p. 477   298
Costaeus, Ioannes
– *De universali stirpium natura* [Turin: Nicolai Bevilaqua, 1578]
– p. 268   305
– *Disquisitiones physiologicae in primam primi* Canonis Avic. Sect. VI,3 [Bologna: Giovanni Rossi, 1589]
– p. 515   305
Crinitus, Petrus
– *De honesta disciplina* [Lyon: Gryphius, 1543]
– XXIV 11, p. 364–365   349
Cyrano de Bergerac, Savinien
– *Les Estats et Empires de la Lune et du soleil* [ed. Alcover]
– p. 295   304
de Boodt, Anselmus
– *Gemmarum et lapidum historia* [Hanau: Wechel, 1609]
– II,244   301
de Castro, Estêvão Rodrigues
De asitia tractatus [Firenze: Pignoni, 1630]
– 13–14, p. 41–49   343
– 17–18, p. 55–58   343
de Clave, Étienne
– *Paradoxes* [Paris: La veufue Pierre Chevalier, 1635]
– II 27 p. 474–475   303
Democritus
– DK 68 B 149   XIII
de Porta, Ioannes Baptista
– *Magia naturalis* [Naples: Salvian, 1589]
– VII 50, p. 146   301
Diogenes Laertius
– V 23   173
– V 26   173
– V 59   177
– IX 47   129
– IX 81   199
Donatus, Marcellus
– *De historia medica mirabili* [Frankfurt: Johann Jakob Porß, 1613]
– II 7, p. 181–187   349

Eckius, Ioannes
- *Aristotelis Stagyritae Acroases Physicae* [Augsburg: Grimm/Wirsung, 1518]
    - fol. 91r 301
Empedocles
- DK 31 a 77 XIII
Ennemoser, Joseph
- *Der Magnetismus* [Leipzig: Brockhaus, 1819]
    - p. 20-23 309
Euripides
- *Ion*
    - 1170 XII
Eustachius, Bartholomaeus
- *De rerum structura, officio & administratione* [ed. Pini, Venice: Vicenzo Luchina, 1564]
    - 32, p. 106-109 294
Faber, Petrus Ioannes
- *Panchymici seu Anatomiae totius universi opus* [Tolouse: Pierre Bosc, 1646]
    - IV 26, p. 242 304
Falloppius, Gabriel [Frankfurt: Wechel, 1584]
- *De simplicibus medicamentis purgantibus*
    - 3, p. 27 294
- *De tumoribus praeter naturam*
    - 3, p. 703 294
- *Observationes anatomicae*
    - p. 442 294
Fernelius, Ioannes
- de *abditis rerum causis* [ed. Forrester] 320, 323, 325
    - p. 478-497 325
    - p. 498-503 325
- *Physiologia* [ed. Forrester] 323, 334
    - V-VI 321
    - p. 208-213 326
    - p. 210-212 328
    - p. 256-263 325
    - p. 304 321
    - p. 310-321 321
    - p. 312 322
    - p. 320 322
    - p. 320-321 324
    - p. 402 324
    - p. 404-407 324, 326
    - p. 406-407 331
    - p. 414-415 327
    - p. 420-421 328
    - p. 420-423 331-332
    - p. 424-425 332
    - p. 426-428 330
    - p. 428-429 330
    - p. 434-435 329
    - p. 438-441 322
    - p. 442-443 329
    - p. 444-447 330
    - p. 454-455 331
    - p. 458-459 332
    - p. 462-463 333
    - p. 464-465 333
    - p. 464-467 333
- *Universa medicina* 320
Flud alias de Fluctibus, Robertus
- *Philosophia moysaica* [Gouda: Peter Rammazeyn, 1638]
    - sect. 2 lib. II membr. 2 cap. 2, fol. 97v 305
    - sect. 2 lib. II membr. 2 cap. 2, fol. 111r 304
    - sect. 2 lib. III membr. 1 cap. 1, fol. 126v 304
Franciscus Suárez
- *Disputationes metaphysicae*
    - disp. 29 sect. 2 n. 16 358
Franciscus de Marcis
- *Quaestiones in secundum librum Sententiarum* (Reportatio) [ed. Suarez-Nani et al.]
    - q. 38, p. 121-122 268
Galenus [ed. Kühn]
- *Ars medica*
    - 4 [I,316-317] 331
- *De alimentorum facultatibus* 341
    - I 1 [VI,463] 333
- *De atra bile*
    - I 1 [V,104-148] 333
- *De elementis ex hippocratis sententia*
    - I 1 [I,413] 328
    - I 9 [I,489] 328
    - II 330
    - II 5 [I,507-508] 325

– *De facultatibus naturalibus*   321, 322, 323, 332
  – I 1 [II,2]   288
  – I 4–5   288
  – I 5 [II,10]   IX
  – I 5 [II,10,13–11,5]   101–102
  – I 6 [II,15,12–13]   101
  – I 11 [II,24]   329
  – I 15 [II,60]   288
  – II 3 [II,86,7–9] (=SVF II 462)   102
  – II 7 [II,106–107]   325
  – II 9 [II,135–140]   333
  – III 4–8   232
  – III 8 [II,173–174]   XIII
– *De marcore*   285, 286
  – 3 [VII,674]   286
  – 5 [VII,683]   286
  – 9 [VII,700]   286
– *De placitis Hippocratis et Platonis*
  – VI 3 [V,519–532]   321
– *De sanitate tuenda*   286
– *De semine*
  – II 5 [IV,642,1–3]   101
– *De simplicium medicamentorum temperamentis et facultatibus*
  – III 15 [XI,612]   289
– *De temperamentis*
  – III 1 [I,654–655]   325
– *De theriaca ad Pisonem*
  – 3 [XIV,224–225]   325
– *De usu partium*
  – IV 3 [III,269–270]   332
– *Definitiones medicae*
  – 288 [XIX,426]   68
Galenus[Pseudo]
– *Quod qualitates incorporea sint*
  – 3 [XIX,469,15–472,2]   199
Garzoni, Leonardus
– *Trattati della calamità* [ed. Ugaglia]
  – p. 119   302
  – p. 221–222   302
Gilbert, William
– *De magnete, magneticisque corporibus, et de magno magnete tellure* [London: Short, 1600]
  – I 1, p. 3   305
  – II 39, p. 112   295

Giraldus Odonis
– *in secundum* librum *Sententiarum*
  – d. 18 q. unica   267
Guillelmus de Ockham
– *Quaestiones in secundum librum Sententiarum* (Reportatio) [ed. Brown/Gál]
  – q. 19, p. 413–414   264
  – q. 19, p. 420–421   264
– *Quaestiones in librum tertium Sententiarum* (Reportatio) [ed. Kelley/Etzkorn]
  – q. 4, p. 135   263
  – q. 6, p. 163   264
– *Quodlibeta*
  – II q. 10–11   263
  – II q. 11   264
– *Scriptum in librum primum Sententiarum – Ordinatio* [ed. Brown/Gál]
  – dist. II q. 10, p. 348   264
Hartsoecker, Nicolas
– *Conjectures physiques* [Amsterdam: Desbordes, 1706–1712, vol. II]
  – troisième discours, p. 22–53   352
Harvey, William
– *Exercitationes de generatione animalium*
  – ex. 50, p. 282–298   352
Heraclitus
  – DK 22 B 117   XII
  – DK 22 B 118   XII
Hierocles
– *Elementa Ethica*
  – 1a 9   102
Hippocrates [ed. Littré]
– *Aphorismi*   132
  – I 14   132
  – I 15   132
  – V 18   133
  – V 62 [IV,554–556]   132
– *Coacae praenotiones*
  – 489   133
– *De aere aquis locis*   132
– *De alimento*
  – 8   329
  – 29   139
– *De arte*
  – 12   132
– *De carnibus*
  – 1   129, 150

– 1 132,1–10 Potter   134
– 2   135, 150
– 2 132,12–14 Potter   134
– 3   135, 136, 138
– 3 136,21–27 Potter   135
– 4   132, 135
– 5   132, 135
– 5 1   149
– 6   XIV, 136, 138
– 6 142,7–13 Potter   136
– 7   135
– 8   135
– 8 144,10–12 Potter   140
– 9   135
– 9 144,25–146,4 Potter   135
– 9 146,5–20 Potter   140
– 10   135
– 12   135
– 12 148,8–22 Potter   137
– 12–13   137
– 13   135
– 13 148,23–152,6 Potter   137–138
– 14   XIV, 135
– 15 152,23–25 Potter   133
– 15–18   135
– 16 154,6–8 Potter   133
– 17   133, 135
– 18 158,1–4 Potter   148
– *De diaeta*   128–150
– I 1   129
– I 3   130, 132, 144
– I 3 [VI,472]   XIV
– I 4   137
– I 6   131
– I 7   131
– I 8   133
– I 9   136, 137
– I 10   131
– I 13   137
– I 16   131
– I 33   132
– II 49   133
– II 49 172,7–10 Joly-Byl   132
– *De flatibus*   149
– 3   137
– 15 252,13–15 Jones   145

– *De genitura*   65, 134
– I 2   71
– III 1 [VII,474–5]   65
– *De humoribus*
– 11   155
– *De liquidorum usu*
– 2   132
– 2 1   133
– *De locis in homine*   68
– 2   133
– *De morbis*
– I 11   132
– IV   65, 134
– IV 1 [VII,542]   XIV
– IV 2 [VII,544]   XIII
– IV 47   137
– *De morbo sacro*   132, 134
– 16–17   133
– *De natura hominis*   132
– 5 [VI,40–44]   330
– 6 [VI,44–46]   331
– 12   132
– *De natura pueri*   65, 134
– 1 = 12 [VII,486]   137
– 11 = 22 [VII,514]   137
– 24–46   155
– *De prisca medicina*
– 11 [I,594]   332
– *De septimanis*   149
– *De vetere medicina*
– 3,4 [I,576]   XIII
Hohenheim, Theophrastus von (Paracelsus)
– "*Herbarius*" *von den Heilswirkungen der Niswurz der Persicaria, des Salzes, der Engel-Distel, der Korallen und des Magneten* [ed. Sudhoff]
– p. 52   297
– p. 53   297
– p. 56   297
– *Astronomia Magna oder Philosophia sagax*
Ed. Palthenius
– p. 99   308
Ed. Sudhoff
– p. 39   296
– p. 163–164   296, 308
– *de natura rerum* [ed. Sudhoff]
– 11, p. 331   300

- *de sensu et instrumentis* [ed. Goldammer]
  - p. 91   297
- *Die Bücher von den unsichtbaren Krankheiten* [ed. Sudhoff]
  - p. 311   296
  - p. 316   296
  - p. 363   296
- *Von der Wassersucht*   295
- *Von der Wassersucht* [Andere Redaktion] [ed. Sudhoff]
  - p. 12–13   296
- *Von Fasten und Casteyen* [ed. Goldammer]
  - p. 435–436   296

Hunter, John
- *Observations on Certain Parts of Animal Economy* [London: Castle Street, 1786]
  - p. 87–114   353

Iamblichus
- *de anima*   198
  - apud Stobaeus I,17(18 H.), 368,12–20 (SVF II,826; LS 53K)   198, 207

Ibn Sīnā, Abū ʿAlī (Avicenna)
- *Aqsām al-ʿulūm al-ʿaqliyya (On the Divisions of the Intellectual Sciences)* [ed. Cairo: Dar al-ʿarab, 1980]   224
  - 110.7–10   224
- *Adwiya Qalbiyya (On Cardiac Remedies)*   223, 239
  - 2–9   242–243
- *Afʿāl wa-Infiʿālāt (Activities and affections)*
  - II 1–2   249
- *Biography*
  - 54.4–56.1   224
  - 58.7   224
  - 65.5–67.4   223
  - 68.6–72.8   251
- *Burhān (Demonstration)* [ed. Strobino]
  - II 7 163.14–20   225
- *Ḥayawān (Animals)* [ed. Rahman]
  - III 1   226, 234, 241
  - XII 2   240
  - XII 8   241
  - XIII 3   240
  - XV 1   241
- *Ilāhiyyāt (Divine things)* [ed. Rahman]
  - I 1–2   224
  - I 2 14.18–15.3   225
  - V 4 220.13–18   246
  - VIII 7 368.4–6   254
  - IX 3 393.16–17   225
- *Kawn wa-Fasād (Generation and Corruption)* [ed. Rahman]
  - 6–7   249
  - 8   230
  - 8 144.11–146.9   230
- *Maʿādin wa-Āṯār ʿulwiyya (Minerals and Upper Signs)* [ed. Rahman]   244
- *Nabāt (Plants)* [ed. Rahman]
  - 1   246, 253
  - 1 3.4–5   249
  - 1 3.13–15   250
  - 1 3.15–19   250
  - 1 3.19–20   250
  - 1 4.1–4   250–251
  - 1 7.12   251
  - 1 7.12–20   251
  - 1 7.14   251
  - 1 7.16   251
  - 1 7.19   251
  - 3   246, 247
  - 3 13.17–14.5   247
  - 7   247, 248
  - 7 33.5–12   248
  - 7 33.16–17   248–249
  - 7 34.9–16   249
  - 7 34.11   247
  - 7 38.4–5   249
- *Nafs (On the Soul)* [ed. Rahman]
  - praef.   247
  - praef. 1.9–11   238
  - praef. 1.11–2.1   245
  - praef. 2.1–3   247
  - praef. 2.5–17   246
  - praef. 2.18   245
  - praef. 3.1   245
  - praef. 3.2   245
  - praef. 3.9–11   238
  - praef. 3.10   246
  - I 1   237, 238, 253
  - I 1 4.5–10   236
  - I 1 4.9–10   236
  - I 1 5.3–6.1   238
  - I 1 12.15–16   251
  - I 1 13.8–14.8   243

– I 3 30.5–31.11   243
– I 3 31.11–32.14   247
– I 5   228, 230, 232, 233, 238, 243, 252
– I 5 39.13–14   228
– I 5 39.14   228, 229, 233
– I 5 39.15–18   229
– I 5 40.4–13   228
– I 5 40.14–16   229
– I 5 40.14–41.3   229
– I 5 44.3–45.16   226
– II 1   228, 232, 238, 240, 243, 247, 252
– II 1 50.13–51.16   232
– II 1 51.10   232
– II 1 51.12   232
– II 1 52.4–5   230
– II 1 52.5   230
– II 1 52.5–53.2   230
– II 1 52.6   230
– II 1 52.9–10   231
– II 1 52.12–13   230
– II 1 52.16   231
– II 1 53.2–54.18   231
– II 1 54.18–55.5   231
– II 1 55.5–19   231
– II 1 55.19–56.4   232
– II 3 68.6–19   238
– II 4 76.20   223
– III 7 149.5   225
– III 8 156.15   223
– IV 1   226
– IV 2–3   226
– IV 4 201.13   223
– IV–V   243
– V 8   226, 234, 241
– V 8 262.18   239
– V 8 263.9–10   239
– V 8 263.13   240
– V 8 263.20–264.3   239
– V 8 264.4   240
– V 8 264.4–5   240
– V 8 265.1   241
– V 8 266.4   241
– V 8 269.14–15   241
– *Qānūn fī l-ṭibb (Canon of Medicine)* [ed. Institute of History of Medicine and Medical Research, Ǧāmiʿa Hamdard, New Delhi]
– I   221, 223–224
– I 1   233
– I 1 i 1 33.8–9   224
– I 1 i 2 36.8–14   224
– I 1 vi 1 122.18   233
– I 1 vi 1 122.19   233
– I 1 vi 1 122.22   233
– I 1 vi 1 123.6–11   225
– I 1 vi 3   252
– I 1 vi 3 126.6–8   251
– I 1 vi 4 126.19–28   241
– I 1 vi 4 127.4–5   242
– I 1 vi 4 127.7–9   242
– I 1 vi 4 127.26–27   241
– I 1 vi 5 128.17–129.30   226
– I 1 vi 5 129.13–20   226
– II   224
– II–V   223
– III 1 v 1   223
– III 6 i 2   223
– III–IV   223
– *Fī Naqḍ Risālat Ibn al-Ṭayyib fī l-quwà l-ṭabīʿiyya (Refutation of Ibn al-Ṭayyib's Treatise On the Natural Faculties)*   232
– *Samāʿ ṭabīʿī (Physics)* [ed. Rahman]
– I 5   236–237
– IV 9   237
– *Šifāʾ (Book of the Healing)*
– see Nafs, Ḥayawān, Nabāt, Afʿāl, Ilāhiyyāt, Samāʿ, Maʿādin wa-Āṯār ʿulwiyya
– *The Physics of the Healing*
– II 8   292

Ibn Rushd, Muḥammad ibn Aḥmad al-Ḥafīd (Averroes)
– *Long Commentary on the Physics of Aristotle*   291, 292
– *Middle Commentary on the Physics of Aristotle*   291, 292

Ioannes Buridanus
– *Expositio et Quaestiones in Aristotelis De caelo et mundo* [ed. Patar]
– II tract. 3 cap. 2, p. 148–151   281
– *Expositio in librum De motibus (motu) animalium* [ed. Scott/Shapiro]
– 5   275

- *Quaestiones de secretis mulierum* [ed. Beneduce/Bakker]
  - q. 3–7   278
- *Quaestiones in Ethica Nicomachea* [Paris: Minerva, 1489]
  - VI q. 4   261
- *Quaestiones super De Anima (secundum tertiam lecturam)* [ed. Klima et al.]
  - II q. 2 n. 2   260
  - II q. 2 n. 24   276
  - II q. 4   261
  - II q. 5 n. 20   263
  - II q. 5 n. 24   263
  - II q. 8 n. 8   266
  - II q. 8 n. 19   266
  - II q. 9 n. 31   276
  - III q. 6   260
  - III q. 6 n. 10   272
  - III q. 6 n. 18–19   272
  - III q. 17   261, 277
  - III q. 17 n. 3   265
  - III q. 17 n. 4   265
  - III q. 17 n. 9   277–278
  - III q. 17 n. 16   264
  - III q. 17 n. 18   268, 277
  - III q. 17 n. 23   278
  - III q. 20 n. 13   274
  - III q. 20 n. 18   275
- *Quaestiones super libros De generatione et corruptione Aristotelis* [ed. Streijger et al.]
  - I q. 7, p. 74–79   266
  - I q. 7, p. 75   271
  - I q. 8   261
  - I q. 8, p. 85   264
  - I q. 10, p. 98   266, 275
  - I q. 14, p. 118   274
  - I q. 16, p. 126   265
  - I q. 16, p. 126–127   273
  - I q. 16, p. 127   274
  - I q. 17, p. 132   265
  - I q. 17, p. 133   265
  - I q. 17, p. 134   266, 274, 275
  - I q. 18 ad 4, p. 141   291
  - II q. 7, p. 226   276
  - II q. 12, p. 251–253   278
- *Quaestiones super octo Physicorum (secundum ultimam lecturam)* [ed. Streijger et al.]
  - I q. 8, p. 87–90   273
  - I q. 17, p. 171   267
  - I q. 17, p. 170–172   269
  - I q. 19, p. 199   270
  - I q. 20, p. 202–205   276
  - I q. 20, p. 204   267
  - II q. 5, p. 274–276   278
  - II q. 13, p. 339–342   278
  - III q. 5, p. 49–50   274
  - III q. 9, p. 97   274
  - IV q. 2, p. 214   276
  - IV q. 11, p. 300–303   273
- *Summulae de Dialectica* [ed. Klima]
  - tract. 1 cap. 3,2   271
  - tract. 3 cap. 6,1   271
  - tract. 7 cap. 4,2   271

Ioannes Duns Scotus
- *In primum et secundum Sententiarum quaestiones subtilissimae*
  - sent. 2 dist. 25 q. 1 n. 13–15   358

Ioannes Tzetza
- *Historiarum variorum chiliades* [ed. Kiessling]
  - chiliada 4 epistula, p. 145   298

Israeli, Isaac
- *Liber Pantegni* [in: *Opera omnia*. Lyon: Barthélemy Trot, 1515]
  - fol. 1r–144r   329

Jonstonius, Johannes
- *Historiae naturalis de quadrupedibus* [Amsterdam: Johann Jacob Schipper, 1657]
  - I 16, p. 114–115   346

Joubert, Laurent
- *Paradoxa* [Lyon: Senneton, 1582, vol. I]
  - decas I paradoxum II, p. 11–19   342

Kircherus, Athanasius
- *Magnes*
  - lib. III pars 5 cap. 7, p. 721 Scheus   294–295
  - lib. III pars 7 cap. 2, p. 548–552 Zanobius   346
  - lib. III pars 7 cap. 5, p. 817 Scheus   295
  - lib. III pars 7 cap. 5, p. 818 Scheus   296

Kyranides
- *Ed. Delatte*
  - p. 48   298
  - p. 86   298
  - p. 87   298, 301
  - p. 97   298
- *Ed. Rivinus*
  - p. 30   298
  - p. 61   298
  - p. 69   298
  - p. 82   298
  - p. 123   298
- *Ed. Kaimakēs*
  - I 7 63–67   298
  - I 21 90–103   298
  - I 21 103   301
  - I 24 97   298
  - II 20 21   298

LaGalla, Iulius Caesar
- *Disputatio de sympathia et antipathia*
  - Vallicellianus MS Allacci XXX.4, fol. 61r   302

Lemosius, Ludovicus
- *In tres libros Galeni de Naturalibus facultatibus commentarii* [Salamanca: G. Foquel, 1591]
  - ad I 14, p. 92   301
  - ad I 14, p. 94   293

Libavius, Andreas
- *Antigramania* [Frankfurt: Kopff, 1594]
  - p. 606–607   303
- *Res Chymicae* [Frankfurt: Kopff, 1599]
  - lib. III ep. 13, p. 121   308
- *Singularium, Pars tertia* [Frankfurt: Kopff, 1601]
  - I 10, p. 86   295
  - I 10, p. 99   303
  - I 10, p. 101   303

Libavius, Andreas/Riolanus, Ioannes
- *Alchymia triumphans* [Frankfurt: Ioannes Saur/Peter Kopff, 1607]
  - 63, p. 646–654   303

Licetus, Fortunius
- *De feriis altricis animae Nemeseticae Disputationes* [Padua: Variscianus, 1631]
  - disputatio IX, p. 82–86   344

- *De his, qui diu vivunt sine alimento* [Padua: Pietro Bertelli, 1612]
  - II 1, p. 19–20   348
  - III 42–48, p. 30–33   343
  - III 70–71, p. 45   343
  - III 78–79, p. 48–49   343
- *De spontaneo viventium ortu* [Vicenza: Francesco Bolzetta, 1618]
  - I 116, p. 115   294
- *Litheosphorus, sive, De lapide Bononiensi* [Udine: Nicolo Schiratti, 1640]
  - 42, p. 184–187   294

Lister, Martin
- *Apicii De obsoniis et condimentis sive arte coquinaria* [Amsterdam: Janssonius, 1709]
  - VIII, p. 244   350

Luiz, Antonius
- *De occultis proprietatibus* [Lisboa: Luis Rodrigues, 1540]
  - II praef., fol. 16r   293

Magirus, Petrus
- *Antilogia inutilis futilisque* [Linz: 1639]
  - p. 151   302

Mangili, Giuseppe
- *Dei mammiferi, sogetti a periodico letargo. Memoria*   353
- *Saggio d'osservazioni per servire all'storia dei mammiferi sogetti a periodico-letargo*   353

Marsilius Ficinus
- *De vita* [ed. Kaske/Clark]
  - III 3, p. 257   299

Mersenius, Marinus
- *Quaestiones celeberrimae in Genesim* [Paris: Carmoisy, 1623]
  - ad Gen 1,13 art. 3, p. 948   301, 302

Michael Ephesius
- *In libros de Generatione Animalium commentarium*
  - 47,3   68

Montanus, Ioannes Baptista
- *Medicina universa* [Frankfurt: Wechel, 1587]
  - pars 2, p. 375   292
  - pars 2, p. 375–376   294
  - pars 2, p. 383   294

– pars 2, p. 387   294
Nemesius Emesinus
– *De natura hominis* [ed. Morani]
   – 1, p. 3   298
Nicolaus Cusanus
– *Idiota de sapientia* [ed. Baur/Steiger]
   – I 16, p. 34   291
– *Sermones* [ed. Pauli]
   – CLVIII 7, p. 175–176   291
Nicolaus Damascenus
– *De plantis* [ed. Drossaart Lulofs/Poortman]
   – I   244
   – I 1 1   250
   – I 1 10   250
   – II   244
   – II 2 150   244
Nicolaus Oresmius
– *Questiones super physicam* [ed. Caroti]
   – lib. 7 q. 3 ad 4, p. 733   291
Nierembergius, Johannes Eusebius
– *Historia naturae maxime peregrinae* [Antwerpen: Balthasar Moret, 1635]
   – X 1, p. 206   347
Nolle, Henricus
– *Naturae Sanctuarium* [Frankfurt: Rosa, 1619]
   – XI 2, p. 653   302–303
Olaus Magnus
– *Historia de gentibus septentrionalibus* [Rome: 1555]   340
   – XVIII 17–23, p. 614–620   340
   – XVIII 38–40, p. 635–637   340
– *Orphei Lithica*
   – p. 24–25 Abel   298
Otto, Adolph Wilhelm
– *De animalium quorundam per hyemem dormientium, vasis cephalicis et aure interna*   353
Fabius Pacius
– *Commentarius in Galeni libros methodi medendi* [Vicenza: Greco, 1597]
   – ad I 6, p. 170–171   294
– *Papyri graecae magicae* [ed. Preisendanz/Henrichs]
   – II 18   298
   – IV 1721   298
   – IV 2627   298
   – IV 2875   298
Pechlin, Johann Nicolas
– *De aeris et alimenti defectu, et vita sub aquis meditatio* [Kiel: Schultze, 1676]
   – 5, p. 64–78   351, 352
Petrus Hispanus
– *Comentario al "De anima" de Aristóteles* [ed. Alonso]
   – lib. I lect. 13, p. 442–444   298
Piccolomineus, Franciscus
– *Universa philosophia de moribus* [Frankfurt: Nicolaus Hoffmann, 1611]
   – p. 447   362
Pineda, Ioannes de
– *In salomonem commentarios Salomon praevius* [Venice: Thomas Ballionus, 1611]
   – p. 37   285–286
Philoponus[Pseudo]
– *In de anima* [ed. Hayduck]
   – 9,35–38   217
   – 237,11–23   217
   – 287,17–288,5   28
   – 287,25–26   28
   – 289,2–7   230
Plato
– *Phaedo*
   – 64d   XII
   – 96b2–8   133
   – 96c3–d5   146
– *Philebus*
   – 26d8–9   57
– *Res publica*
   – IV 434–441c   XII
– *Symposium*
   – 207c–d   15
   – 207c8–d5   10
   – 207d3–e2   15
– *Timaeus*   44
   – 27d–28b   57
   – 66b   332
   – 70d–71a   XII, XIII
   – 80d–e   140
   – 81b   209
Plinius Maior
– *Naturalis historia*
   – VIII 126–135   340

– X 1   324
Ponce de Santacruz, Antonio
– *Opuscula* in *Primam Primi Avicennae* [Madrid: Junta, 1624]   356
  – p. 8   358
  – p. 10   358, 359, 362, 369
  – p. 12   358, 359
  – p. 14   359
  – p. 17   360
  – p. 19   360
  – p. 20   360
  – p. 22   360
  – p. 23   361
  – p. 24   361
  – p. 25   361
  – p. 27   363–364
  – p. 69   369
  – p. 71–72   368
  – p. 72   368, 369
  – p. 74   369
  – p. 189   371
  – p. 192   371
  – p. 195   361, 362
  – p. 196   361, 362, 372
  – p. 227   363
  – p. 228   359, 362, 363
  – p. 229   363
  – p. 231   370
  – p. 232   370–371
  – p. 239   364
  – p. 239–240   363
  – p. 284   369
  – p. 284–285   368
  – p. 285   365–366, 368
  – p. 286   367
  – p. 290   367, 369
  – p. 291   367
  – p. 293   368
Porphyrius
– *De abstinentia* [ed. Nauck]
  – IV 20   297–298
Porzio, Simone
– *De puella germanica, quae fere biennium vixerat sine cibo, potuque*   349
Philoponus[Pseudo]
– *In de anima*
  – 571,11–13   199

– 571,14–16   199
Rzączyński, Gabriel
– *Historia naturalis curiosa regni Poloniae* [Sandomir: Collegium Societatis Jesu, 1721]
  – tract. 8 sect. 1 § 21, p. 225–228   350, 353
Reeve, Henry
– *An essay on the torpidity of animals*   353
– *See also* Baird, Henry
Renodaeus, Ioannes
– *De materia medica* [Paris, De la Nouë, 1608]
  – II 9, p. 205   301
– *Institutiones pharmaceuticarum*
  – IV 12, p. 95 [Hanau: David Aubrius, 1631]   303
  – IV 12, p. 128 [Paris: De la Nouë, 1608]   301
Riolanus, Ioannes
See Libavius, Andreas
Robertus Grosseteste
– *Summa philosophiae* [ed. Baur]
  – tract. 17 cap. 13, vol. II p. 613–614   291
Sabellicus, Marcus Antonius
– *De omnium gentium omniumque seculorum insignibus memoriaque dignis factis et dictis exempla* [Basel: Petri, 1541]
  – VII 4, p. 421–422   349
Saissy, Jean-Antoine
– *Recherches expérimentales, anatomiques, chimiques etc.*   353
Scaliger, Iulius Caesar
– *Exotericarum exercitationum liber quintus decimus, De subtilitate, ad Hieronymum Cardanum* [Paris: Michael de Vascosan, 1557]
  – ex. 101 15–18, fol. 149v–152v   302
  – ex. 101 18, fol. 151v   302
  – ex. 102 5, fol. 155v   302
  – ex. 102 6, fol. 156v–157r   294
Scheuchzer, Johann Jacob
– *Anatome texi suilii maris* [Nürnberg: Endter, 1733]
  – observatio 43, p. 127–133   352

Schwenckfeld, Kaspar
– *Stirpium & fossilium Silesiae catalogus* [Leipzig: David Albertus, 1600]
  – III, p. 384   301
Sextus Empiricus
– *Adversus Mathematicos*
  – VII 103   199
  – VII 234   207
– *Pyrrhonicae hypotyposes*
  – I 94–97   199
  – I 99   199
Sibbald, Robert
– *Scotia illustrata sive Prodromus historiae naturalis* [Edinburgh: 1683]
  – vol. I lib. III sect. 2 cap. 4, p. 9–12   350
Simplicius [ed. Hayduck]
– *In Aristotelis Physica commentaria*
  – 265,1–3   205
  – 371,33 = DK 31 B 61   XIII
Simplicius[Pseudo]
– *In de anima* [ed. Hayduck]
  – 115,29–116,8   28
  – 116,16–17   230
Spallanzani, Lazzaro
– *Tracts on the nature of animals and vegetables* [Edinburgh: William Creech, 1799]
  – p. 251–324   353
Strato Lampsacenus
– *De somno* [ed. Sharples]   174, 177
  – fr. 57   178, 180
  – fr. 61   193
  – fr. 62   193
  – fr. 66   178, 180
  – fr. 66–69   178
  – fr. 67   178, 179–180
  – fr. 66–69   178
– *Stoicorum veterum Fragmenta*
  – II 462   102
Tanckius, Joachim
– *Schatzkammer der Natur*
  – Kassel UB MS chem. 99, fol. 20r–21v   303
Tenzelius, Andreas [Jena: Johann Birckner, 1629]
– *Medicina diastatica*
  – 4, p. 58–67   304

Themistius
– *In de anima paraphrasis* [ed. Heinze]
  – 37,21–23   199
  – 117,3–4   199
Theodorus Gaza, see Alexander Aphrodisiensis Latinus
Theophrastus
  – fr. 307 A FHS&G   193
  – fr. 330 FHS&G   193
– *De lassitudine*   193
– *De vertigine* [ed. Sharples]
  – 1 1–5   176
  – 9 69–70   190
  – 9 75–76   190
Theophylaktos Simokates
– *Epistulae* [ed. Zanetto]
  – p. 15   298
Thomas Aquinas
– *In librum primum Aristotelis De generatione et corruptione* [ed. Leonina]
  – cap. V lect. 14, p. 313a   266
– *In physicam* [ed. Maggiòlo]
  – lib. 7 lect. 3 n. 7, p. 461   291
– *Quaestiones disputatae de anima*
  – art. 12 co.   262
– *Summa Theologiae*
  – I q. 76 art. 3 arg. 1   262
  – I q. 76 art. 3 ad 1   262
  – I q. 76 art. 4 ad 4   360
  – I q. 77 art. 2 co.   262
  – I 1. 77 art. 6 co.   262
Thomas de Vio Caietanus
– *In Summam Theologiae*
  – ad I q. 54 art. 3   372
UB München 2° Cod. Ms 578
  – fol 26r   295
Vallesius, Franciscus
– *Controversiae medicae et philosophicae* [Hanau: Wechel, 1564, [4]1606]
  – p. 74–75   364
– *De sacra philosophia* [Frankfurt: Romanus Beatus, 1600]
  – p. 16   364
  – p. 17   364
  – p. 45   364
  – p. 69   364
  – p. 221   364

van Helmont, Ioannes Baptista
- *De magnetica vulnerum* curatione [Amsterdam: Louis Elsevier 1652]
  - p. 597   297
van Linschoten, Jan
- *Reizen van Jan Huyghen van Linschoten*
  345
von Bergen, Karl – Heyn, Franciscus (resp.)
- *Disputatio physica circularis de animalibus hieme sopitis* [Frankfurt/Oder: Johannes Christian Winter, 1752]
  - § 1–12, p. 3–10   350
  - § 13–17, p. 10–15   351
  - § 22–30, p. 17–21   351
  - § 30–31, p. 21–22   352
  - § 32–35, p. 22–24   352
  - § 36–37, p. 24   353
von Hohenburg, Hans Friedrich Herwart [Ingolstadt: Eder, 1623]
- *Admiranda Ethnicae Theologiae mysteria propalata*
  - 54, p. 189–190   298

von Grafenberg, Johann Schenck
- *Observationum medicarum rariorum* [Frankfurt: Johannes Beyer, 1655]
  - p. 66–70   349
Walterius Burlaeus
- *super Aristotelis libros De physica auscultatione lucidissima commentaria* [Venice: Bernia, 1589]
  - ad VII 10, p. 865–867   291
  - ad VIII 35, p. 1006   291
Wilbrand, Johann Bernhard
- *Darstellung des thierischen Magnetismus* [Frankfurt: Sauerländer, 1824]
  - p. 36   309
Willis, Thomas
- *De anima brutorum* [Amsterdam: Johan Blaeu, 1674]
  - ex. 1 cap. 5, p. 52–58   348

Index rerum

abstinence 296, 341–344, 349
actuality 17, 35, 36, 44–46, 49, 58, 76, 80, 81, 104, 112, 114, 121, 168, 238, 247, 276
   See also potentiality; ἐνέργεια; ἐντελέχεια
adhesion 234
aether 325
agency, agent XII, XV, 30–34, 37, 39, 53, 74, 75, 80, 92, 114, 131, 134, 135, 159, 160, 263, 272, 277–279, 289, 366, 367, 369, 370, 373
air XI, XII, XIV, 27, 47, 91, 98, 103, 114, 115, 117, 120, 129–143, 145–148, 150, 153, 155, 159–161, 163–165, 177, 183–192, 232, 235, 251, 289, 302, 320, 326, 327, 330–332, 339, 341, 342, 345, 347–349, 351, 352, 360, 362
   See also cooling
alchemy 285, 302–306, 309, 320, 332, 335
analogy 149, 286, 307
– between blood and water 27
– between brain and celestial influence 366
– between partial substantial change and complete substantial change 277
– between digestion of chyle and wine fermentation 331, 332
– between embryo and plant 119
– between growth and generation 115, 117
– between animal hibernation and human abstinence 342, 349
– between the instruments of soul and hand/rudder 28, 30
– between animal/human form and the measure to mark a river mile 43, 55
– between magnetism and nutrition 287, 289, 290, 293–295, 297, 306, 307, 325
– between nutrition and coming into being of mixtures 356, 357, 362, 363
– between nutrition and elements gaining their independence from genuine mixtures 356, 357

– between nutrition and generation 369, 373
– between parts of brain and soldiers 190, 191
– between plants and animals 7, 74, 77, 79, 90, 119, 156, 157, 159, 160, 161, 166, 168, 239, 331, 332
– between powers of soul and parts of animate body 212
– between the road Athens-Thebes and the road Thebes-Athens 24
– between sensible species and spirits 368
– between soul and apple 198, 199
– between soul and craftsman 104, 105, 115, 117, 123, 131
– between dignity and nature of souls 205
– between soul and forms of elements 204
– clepshydra XIII
– disanalogy 290
– functional 239
– structural 223
   See also like to like (principle); similitude
anatomy 66, 132, 144, 146, 149, 225, 233, 240, 243, 321, 328, 351, 353
– comparative 328, 352
animals
– elements 174
– faculties 154, 253, 369, 370
   See also faculty
– form 43, 63, 75, 79, 80, 93, 94, 361
– generation 8, 10, 14, 15, 38, 63, 74, 75, 102, 112, 268, 275, 289, 307
– matter 17, 28, 93, 260, 277
– principle 71, 237
anorexia 342, 343
assimilation
– of blood 329
– of elements 320
– of food/nutriment IX, XIV, XV, 57, 127, 160, 208, 209, 234, 235, 240, 290, 320, 323
– of patient to agent 32

– of secondary humors 323, 329
    See also like to like (principle)
athanor 335
atomism 288, 305, 307, 353, 389
attainment 89–91, 98, 234
attraction, action at a distance 147, 155, 286–309, 323, 324, 326, 363
    See also faculty, attractive
augmentation 217, 229, 236, 260, 266, 269, 273–276, 280, 371
    See also faculty, growth-promoting; growth; αὔξησις; τὸ αὐξητικόν

badger 340, 344, 350, 351
balance XIV, 74, 91, 97, 98, 130–134, 144, 146, 147, 161, 163, 184, 321, 326–328, 330, 343
bear 339–341, 344–346, 351
being
– cause 12, 13, 36, 38, 59, 216
– principle 57
    See τὸ εἶναι; οὐσία
beings, independent X, XIII, 6, 9–18, 21, 23, 52, 57, 58, 59, 65, 78–80, 82, 102, 105–120, 123, 144, 153–157, 159, 166, 168, 201, 203, 205, 206, 210, 212, 214, 222, 226, 233, 236, 237, 240, 241, 243, 248, 251, 253, 254, 259, 260–266, 268, 269, 271, 273, 275, 277–280, 287, 321, 357, 360–364, 367, 369, 371, 373
– generation 279
– human see human
– parts of living being 27, 80, 108, 110, 112, 119–121, 364
beneficiary 37, 40
    See also cause, final
bile XIV, 66, 330, 333
– black 183, 185, 320, 328, 331, 332, 333
– green 333
– preternatural 333
– yellow 320, 328, 331, 332, 333
    See also humor
bird 71, 75, 93, 95, 340, 346, 347, 351
blood XIII, XIV, 9, 15, 16, 17, 18, 26–28, 33, 66, 69–71, 73, 74, 79, 80, 93, 97, 102, 112, 116, 119, 120, 122, 130–132, 139, 140, 141, 143, 144, 146, 147, 157, 175–177, 184, 230, 265, 288, 320, 327–332, 345, 352
– circulation 231, 234, 240, 307, 345–348, 352
– concocted see blood, as final food
– as constituent matter 26–28, 116, 118, 141
– conversion 9, 26, 176, 322, 327–329, 332, 334, 335, 347
– decomposed XIII, 116
– bloodletting 330
– as final food/nourishment 18, 26, 27, 67, 71, 73, 79, 102, 116, 119, 140, 141, 143, 144, 146, 147, 166, 176, 177, 184, 230, 265, 328, 329, 332
    See also food
– form 184
– menstrual 66, 69, 73, 74, 122, 141, 363
– preexisting see blood, residual
– residual 16, 17, 26, 27, 66
– vessels XIII, 69, 70, 135, 138–141, 143, 144, 146, 149
    See also vein
    See also humor
body
– elements 179–190.
– living XI, XIV, 25–28, 35, 36, 38, 39, 43, 44, 49–53, 57, 66, 67, 70, 79, 105, 106, 118, 140, 161, 204, 277, 285, 286, 288, 289, 325, 363
– organic 29, 210, 229, 235, 252
– human XIII, XIV, 132, 135, 140, 185, 224, 225, 247, 253, 267, 272–276, 279, 280, 343, 364
– principle 224
    See also soul
– matter 26, 27, 44, 46, 51, 52, 60, 72, 88, 89, 141, 245, 248, 260, 276, 277, 334, 371
– simple 145, 203–205, 249
– body-soul-relationship X, XI, 13, 28, 38, 44–50, 52, 58, 104, 106, 145, 149, 150, 159, 173, 175, 222, 259–261, 263, 264, 266–268, 275, 276, 278, 280, 321, 371

bones   XIII, 9, 15, 43, 56, 60, 72, 85–90, 94, 98, 112, 121, 122, 131, 136–139, 141, 146, 148, 213, 273, 274
– generation   273, 274
– principle   90
botany   223, 224, 226–228, 238, 241, 244–249
brain   X, 129, 130–134, 147, 176, 177, 180, 190, 193, 233, 234, 331, 345, 348, 365, 366, 368

calefaction   *See* heating
capacity   *See* faculty; δύναμις
cardiocentrism   233, 240
causality   202, 308
– causal agency   278, 279
– causal circle   362, 369, 373
– causal explanation   169.
– causal influence   275, 362, 364, 367
– causal powers/potencies   356, 357, 360–362, 369, 371, 373
– causal relation   294, 305, 308, 371
causation
– celestial   356, 359
– downward   362, 369–372
– emanative   357, 361, 362, 372, 373
– order   294
– per se   299, 361, 369
– transmission theory of   32
– upward   362–369
– material   362; *see also* cause, material
cause
– absolute   145
– co-cause   *see* cause, secondary
– concurrent   *see* cause, secondary
– efficient (ὅθεν ἡ κίνησις)   XI, 18, 21, 25, 28, 29, 32, 35, 37, 38, 53, 63, 90, 111, 113, 115, 117, 144, 146, 148, 175, 187, 274, 303, 367, 368, 371, 372
– first moving   29, 30, 37, 38, 39
  – inner   38
  – intermediate moving   29, 38, 39
  – last moving   29, 38, 39
  – main moving   147
– external, extrinsic   109, 237, 359, 365
– external moving   112, 113

– final (οὗ ἕνεκα)   6, 7, 11–16, 18, 21, 25, 33, 35–40, 57, 70, 80, 82, 88, 91, 116, 131, 147, 148, 150, 168, 175, 207, 208, 210, 211, 213, 214, 215, 216, 218, 369, 371, 373
  See also teleology; τέλος.
– formal   17, 21, 25, 35, 36, 37, 117, 150
– helping   *see* cause, secondary
– hidden, "occult"   320, 325
– inherent   294, 299
  See also causation, per se
– instrumental   364, 368, 369, 373
– internal   112
– internal moving   118
– main   150
– material   63, 73, 89, 90, 146, 148, 150, 175, 357, 367, 369
– first material   141
– natural   58
– partial   263
– potential   66
– primary   29
– secondary   29, 34, 38, 144, 145, 153
– simple   29
– subordinate   147
  See also causation, order
– unnatural   175
  See also αἰτία, αἴτιον, μεταίτιον, συναίτιον
celestial   *See* divine
change
– external   111, 112
– internal   111, 112
– of living things   44, 58
– material   280, 370, 373
– non-essential   51
– of non-living things   44, 58
– non-substantial   33
– per se   24, 31–33, 44, 46, 53, 56, 57, 371
– principle   32, 75, 112
– qualitative   XIII, 161, 164, 165, 361, 368
– quantitative   56, 109, 112, 272, 273, 274, 275, 276, 277
  see also growth
– self-change   *see* change, per se

- substantial 63, 265–272, 274, 275, 277, 278, 280
  - complete 277
  - partial 260, 266, 267, 269, 272, 275, 279, 280
  See also κίνησις; μεταβολή
chyle 322–332
climate 93, 173, 185
coagulation 320, 321, 329, 334, 335, 361
cold/coldness See air
color 23, 248, 299, 320, 330, 332, 334
combustion 330–333, 334
concoction 18, 26–28, 57, 66, 67, 69–73, 79, 80, 106, 116, 127, 128, 138, 139, 141–144, 146, 147, 153, 159–161, 164, 165, 176, 177, 184, 187, 252, 319, 321–334, 342, 347
  - cause 142
  See also digestion; food; πέψις
continuity
  - of being 11, 158
  - of individuals see individual, persistence
cooling 6, 17, 18, 29, 53, 74, 91, 132, 135, 137, 140, 155, 157, 161, 175–177, 180, 184–188, 275
  - as instrument/intermediate mover 6, 17, 18, 29, 235
copy 59, 243, 304
corporeality 176, 198, 205, 210, 213, 236, 264, 267, 278
correlative object 17, 21–25, 35, 105, 106, 116, 228
corruption XI, 21, 34, 39, 91, 108, 139, 175, 250, 251, 260, 262, 266–268, 270–280, 333, 359, 370
  See also food, corrupted
courage 184

death X, 3, 12, 134, 137, 154, 157, 178, 179, 206, 251, 272, 276, 301
decay See corruption
desire XII, 10, 25, 40, 131, 132, 260, 287, 289, 290, 306
dialectics 45, 48, 50, 254
diaphragm 143, 324
dietetics IX, XIII, 130, 132, 319

digestion X, XII–XIV, 26, 66, 68, 73, 77, 104, 105, 127–150, 161, 208, 209, 231, 235, 240, 251, 252, 259–281, 319–335, 339–343, 347–349, 351–353
  See also concoction; food; faculty, digestive; πέψις
disease XIII, 65, 66, 68, 69, 70, 134, 145, 154, 185, 221, 233, 235, 319, 320, 330, 362
  - cause 66, 69, 145, 224, 235.
divinity/divine 6, 9–13, 18, 39–40, 128, 203, 205, 237, 243, 254, 268, 320, 325, 326, 334
  - sphere 278
  See also life
drinks XII, XIV, 130, 137–139, 142, 146, 157, 176, 184, 296, 305
drugs 247, 289, 307, 319, 324, 325, 330, 331, 380
drunkenness 178, 179
dry/dryness See earth
dualism 259, 280

earth
  - as element XII, XIV, 29, 47, 66, 87, 88, 92, 97, 108, 131–133, 135, 136, 141–144, 150, 153, 157, 159, 161, 163–165, 176, 204, 231, 232, 235, 251, 320, 326, 327, 330–332, 347, 360
  - as digestive organ of plants 77, 240
  - in the sense of ground 155, 176, 178, 304
  - heat of earth 143, 155, 165
  - as nutriment of plants 136, 143, 159
  - planet earth 155, 304
elements XIV, XVI, 11, 29, 97, 132, 136–138, 144, 147, 149, 161, 204, 205, 211, 224, 232, 235, 278, 286, 293, 319–335, 355–363, 373
  - form 204, 357–362, 373
  - generation 238
  - potencies 360
  See also air; earth; heat; water; magnet, elemental
embryo XIII, 38, 48, 63–82, 88, 89, 92–94, 96, 101–123, 134–137, 162, 212, 278, 322
  - cause 107

– generation 92, 93, 101, 111, 118, 120
– matter 67, 70, 74, 93, 107, 115–118, 120, 121
– principle 120
Empedocles XII, XIII, 28, 29, 163, 202, 231, 250, 287
emotion 242, 252
end See cause, final; see also teleology; τέλος
environment XII, 94, 97, 102, 136, 140, 147, 155, 161, 163–168
essence 49, 51, 67, 82, 85, 92–98, 129, 131, 141, 149, 157, 198, 202, 207, 210, 241, 245, 250, 252, 253, 260, 262, 266, 268, 271, 277, 291, 297, 319, 323–325, 329, 334, 361, 372
essentialism 246, 248
evaporation X, 87, 147, 161, 162
exhalation 176, 177, 185, 187, 190, 192
experience/experiment XI, 34, 68, 133, 140, 150, 214, 240, 260, 295, 299, 301, 304, 325, 340, 344, 345, 348, 351, 357
eyes 33, 72, 133, 158, 190
eyebrows 178–180

faculty, capacity, potency 34–36, 46, 70, 78, 104
– in relationship to activity/actuality 22, 343
– attractive 232, 235, 286–289, 293, 300, 307, 322, 323, 325–327, 334
– auxiliary 322, 323, 324
– causal 371
– cognitive 246
  – internal 226
– concocting 322
– cooling 362
– digestive 231, 232, 235
– emanation of 369, 370
– expulsive (*expellens*) 232, 235, 322, 327
– formal 365
– formative 102, 103, 105, 106, 108, 234
– generative 16, 63, 64, 79, 88, 102, 119, 120, 145, 159, 162, 202, 203, 229, 231, 233, 234, 247, 364
– organogenetic 120
– gennetic 102, 103, 105

– growth-promoting 57, 229, 231, 234, 322, 371
– higher 21
– human 253
– imagery/form-bearing 226, 234
– imaginative/cogitative 226
– informative 102
  See also faculty, formative
– intellective 4, 22, 206, 262–265, 267
– intermediate 241
– material 268, 272, 279, 356, 364, 365, 369
– of influencing the mode of motion of other bodies 365, 367, 368
– of setting in motion 229
– natural 102, 227, 231–234, 241, 242, 253, 289, 321, 322
– nutri-generative 64, 159
– nutritive (*vis nutritiva, altrix*) X–XII, XV, XVI, 4, 6–9, 14–18, 21–40, 59, 63–82, 75, 77–81, 101–123, 127, 145, 149, 150, 153, 154, 159, 161, 162, 168, 173–194, 197–218, 221, 222, 229–235, 240, 242, 247, 250, 253, 279, 321–323, 364
– perceptive 4, 5, 21, 22, 180, 189, 193, 206, 229, 238
– pluripotency 74
– preservative 103
– procreative 322
– productive 12, 16, 114–121
– rational 21, 22, 154, 190
– recollective 226
– reproductive 4, 6–8, 16, 18, 78, 81, 102, 103, 120, 200, 214, 234, 247, 322
– repulsive 286
– retentive/retaining (*continens*) 232, 235, 322
– sensory 74–76, 88, 192, 206, 212, 213, 215, 262–265, 267, 358
– one-capacity one-function thesis 207
– served 231, 234, 363
– serving 231, 232, 234, 363
– of soul in general 4–8, 14, 15, 18, 23–25, 198, 221, 224, 226, 228, 232, 234, 238–241, 253, 370
– subordinate 232, 235, 253

– subservient 232
– threptic *see* faculty, nutritive
– transformative 235
– vegetative 102, 234, 242
– vital 214, 227, 231, 233, 240–242, 252, 253
  See also potentiality; τὸ αἰσθητικόν; τὸ αὐξητικόν; τὸ γεννητικόν; τὸ διανοητικόν; δύναμις; τὸ θρεπτικόν; τὸ κινητικόν; τὸ νοητικόν; τὸ ποιητικόν
fat 67, 73, 89, 90, 109, 135, 137, 140, 286, 340, 341, 350–353
fatigue 69, 187, 189, 190
fear 173, 185, 241, 275
female 3, 8, 10, 11, 16–18, 26, 38, 63–82, 88, 105, 107, 111, 113–115, 121, 122, 162, 184, 244, 278, 300, 340
– matter 74, 81, 104, 114
ferment/fermentation 320, 321, 330–335
fetation 71, 76, 77
  superfetation 73
fibers 324, 363
fig-juice 162
filtration 327
fire *see* heat
fish 144, 340, 346
flavor 198, 199, 223, 320, 330
food 25, 26
– concocted *see* food, digested
– corrupted 272–277
– digested 105, 138, 139, 141, 145, 159–161, 184, 208, 269, 272, 277, 280, 323, 327–329, 331, 334
– elements in XIV, 323–327
– external *see* food, first
– form 79, 272
– final 26, 67, 73, 79, 80, 102, 106, 116, 119
– first 26, 67, 73, 77, 79, 80, 88, 102, 116, 122, 143, 208
– incoming *see* food, ingested
– ingested XIV, 143, 155, 160, 161, 163, 166, 322, 323, 326, 330, 331, 334
– as instrument 23, 25, 28–30, 34, 38
– internal *see* food, final
– as last mover 30, 38
– matter 53, 334

– qualities of 87
– raw *see* food, first
– undigested 143, 145
  See also concoction; digestion; faculty, nutritive; πέψις; τροφή
flesh XII, 9, 15, 26, 27, 43, 54–57, 60, 72, 74, 80, 87–89, 93, 109, 112, 122, 135, 139, 146, 209, 235, 273, 274, 328.
– generation 89, 273, 274
flowering 159, 164
form 31, 51
– accidental 275, 276, 360, 367
– complex 218
  – nutritive soul as complex form 201–206
– composition 266
– conception, general 97
– corruption 271, 272, 280
– final 26, 27, 70
– first 145
– form of forms 204, 210, 325
– form-giver (*dator formarum*) 278, 325, 326, 359, 364, 365
– form-matter-relationship 35, 43, 45, 46, 48, 49, 50, 54–59, 215, 245, 266, 269, 334
– gradation 204
– growth 54–57, 91, 94
– immaterial 265, 267, 280
  – substantial 272–273
  See also indivisibility
– last 145
– latent 80
– living 58, 60, 79, 110, 357, 369, 373
– material 264, 267
– substantial 279
– in mind 37
– organic 210, 245
– origin of 325, 363
– partial 366
– parts of form 215–218, 267, 268
– perpetuation 63, 78, 80, 81
– persistence 39, 59, 60
– preservation 40, 43, 359, 370
– principal 367
– production 271
– reception 234, 270, 290, 357, 365, 366

– replication 38
– separation 58
– of soul 30, 79, 204, 205, 217, 245, 268, 269
– soul as form 44–47, 50, 52, 58, 75, 92, 115, 280
  See also body-soul-relationship
– soul principle 34
– substantial 36, 78, 259–261, 264–269, 271, 275–280, 319, 323, 325–327, 334, 355–357, 359–364, 366, 369, 371, 373
– eduction of 367
– specific 326, 334
– superior 326
– transmission of 32, 290
– unity of 268
  See also cause, formal; faculty; εἶδος; λόγος
fruit 83, 139, 156, 162, 164–166, 244, 331

gallbladder 331
generation IX, 6, 24, 31, 33, 55, 57, 60, 64, 79, 80, 102, 113, 135, 363–364, 369, 373
– of another like itself 40
  See also similarity
– artificial 158
– cause 106, 113
– efficient cause of 18
– elementary 238
– female contribution to 16
– as function 7–9
– instrument of 364
– natural 58, 112, 135, 268, 272
– partial 276
– of parts 112
– principle 75, 110, 111, 112, 119, 244, 260
– of pneuma 240
– *secundum quid* 270
– sexual 112
– *simpliciter* 270
– spontaneous 65
– substantial 31, 75, 82, 259, 267, 269–271, 275, 277–280, 364
  – complete 279
  – partial 259, 265, 266, 268, 273, 274, 279

See also faculty, generative; γένεσις; γέννησις
goal See cause, final
growth
– cause 23, 92, 108, 114, 123, 127, 144, 147, 159
– limited 34, 85–98, 144
– principle 111–113
– unlimited 29, 144, 153
– See also augmentation; faculty, growth-promoting; αὔξησις; τὸ αὐξητικόν
gullet 302

head 177, 180, 187, 189, 190
health IX, XIII, XIV, 30–32, 37, 44, 66, 68, 69, 134, 143, 145, 161, 224, 225, 253, 319, 320, 321, 326, 330–332, 362, 373
  cause 224
heart XIV, 5, 16, 26, 34, 38, 66, 73, 74, 76, 77, 90, 101, 110–121, 123, 131–133, 136, 139, 140, 142, 146, 157, 160–163, 165, 175–177, 180, 184, 187, 190, 192, 193, 233, 234, 239, 240, 265, 324, 328, 351, 356, 365, 366, 368
– generation 101, 110–113, 117, 118, 223, 241
– heat XIV, 131–133, 136, 140, 142, 157, 177, 184, 265
  See also heat, body heat
– matter 94
  See also cardiocentrism
heat
– body heat XII, 51, 98, 132, 133, 139, 140, 346, 347, 351
– connate *see* heat, internal
– concentration of 176, 177
– elemental quality XI, XII, XIV, 15, 27–29, 31, 34, 47, 56, 65, 74, 91, 92, 103, 108, 110, 113, 115, 117–120, 127–148, 150, 153–155, 157, 159–163, 165–167, 176, 177, 185, 187, 188, 190, 191, 204, 231, 232, 235, 251, 320, 324–327, 330, 339, 341, 345, 346, 358–360, 365
– external 27, 28, 155, 156, 161, 163–165, 167, 168
– foreign *see* heat, external
– in generation 29

- growth-promoting 91, 147
- hungry 141, 144, 153
- innate see heat, internal
- internal X, 27, 28, 105, 119, 127, 128, 130, 131, 133, 137, 139, 140, 142, 146, 147, 150, 153–158, 160, 161, 163–165, 167, 168, 177, 231, 240, 246, 251, 252, 264, 320, 322, 324–326, 331, 333, 334, 342, 343, 345, 352
- heat as instrument/intermediate mover see heating
- lack of 154, 157
- life-bearing see heat, internal
- loss of 35, 180, 251, 346, 351
- natural 90, 97, 142, 154, 155–162, 164–166, 184, 265, 360
- principle of 90, 134, 135, 154, 155, 157, 158, 160, 161, 168
- soul heat 71, 161, 177
- vital see heat, internal
    See also heating; τὸ θερμόν
heating 17, 18, 29, 53, 135, 275
- as instrument/intermediate mover 6, 17, 18, 28–30, 38, 53, 56, 104, 115, 120, 127, 131, 145, 150, 159, 168, 235
- overheating 74
    See also heat
helping-cause see cause, secondary; συναίτιον
Hippocratic XVI, 32, 65, 130–133, 137, 155
    See also in the Index locorum
homeostasis 161
homonymy 27, 28, 32, 111, 112
hot/hotness See heat
human
- being XIII, 14, 17, 21, 110, 114, 166, 201, 210, 226, 233, 240, 241, 243, 248, 253, 260–266, 268, 269, 271, 273, 275, 277, 280
- form 43, 58, 60, 138
- generation 14, 59, 135, 278
- matter 58, 266
- parts 268
    See also body, human
humor 65, 183, 184, 224, 234, 265, 319–335, 341, 343, 345, 347, 348, 352, 363

See also bile; blood; melancholy; phlegm
hunger 142, 290, 297, 341, 343, 347

incorporeality XI, 198
individual
- existence 102, 231, 234
- persistence 12–15, 33, 40, 43, 60, 78, 102, 231, 250, 253, 322
    See also self-maintenance
indivisibility 59, 96, 97, 199, 216, 265
- of immaterial form 263, 267, 272
- local 199
- numerical 199
instrument
- conjoined 366
- of direction 364, 368, 373
- material 104, 121, 123
- separate 364, 365
    See also faculty, instrumental; power, instrumental
instrumentalism 103
intestines 70, 76, 137–140, 143, 144, 146, 322, 324, 327, 333, 341, 350–352
iron 27, 285–309, 324, 348, 365
    See also magnet/magnetism

kidneys 288, 323

leaf-shedding 159, 164, 166–167, 169
lethargy 340, 348
life 21, 23, 39, 46–52, 58, 60, 76, 118, 127, 222, 250–254, 260, 320
- animal 7, 8, 77, 108, 238, 241–243, 253, 254
- aura vitalis 346
- cause of 43, 59, 127, 189
- celestial 254
- conditions of X, 136, 201, 205, 206, 214, 221–254
- cycle XI, 52, 59, 373
- explanation of 52
- life-like 300
- metal life 300, 301
- organic see life, sublunary
- potential 36, 47, 51

- principle of   9, 12, 158, 160, 235, 243, 246, 247, 320, 324
- sublunary   222, 238, 239, 243–245, 248, 252
- sustaining   9, 158, 163, 214, 341
- style (βίος)   76
- vegetative   76, 108, 150, 153–169, 214, 215, 218, 226, 227, 239, 241–243, 246, 247, 249–254

like to like (principle)   XIV, 79, 139, 141
limbs   94, 140, 245, 268, 273–275, 277, 289
liquor vitalis   345
liver   XIII, 66, 73, 76, 104, 135, 184, 233–235, 295, 321–324, 327–331, 333, 341, 347
locomotion   7, 9, 11, 44, 58, 77, 97, 143, 238, 242, 250–252, 254, 273–275, 298
   See also motion

magnet/magnetism   285–309, 324–326
- cause   293, 304
- generation   304
- elemental magnet   290, 296
male   3, 8, 10, 17, 18, 37, 38, 63–82, 94, 101, 103–107, 113–115, 120, 121, 123, 162, 177, 244, 278, 300, 340
matter   12, 356, 358, 359, 362, 364, 369
- aptitude   278
- change of   43, 275, 277, 280
- composite   270
- composition   261
- constituent   26–28
- dispositions   163, 234, 280, 367
- incorruptible   265
- information   17, 105, 115, 232, 272, 274–277, 279, 364, 365, 367, 370, 371
- informed   277
- living   43, 44, 50, 53, 58–60, 70, 115, 204, 358
- new   362
- pre-existing   26–28
- preparation   362
- prime   141
- proper   55, 163
- replacement   16, 58–60
- seminal   67, 92

- simple   204
- of substance   43
- types of   55, 73, 163
- unchanged   267
   See also form-matter-relationship; ὕλη
measure   43, 54, 55
melancholy   174, 176, 183–185, 187, 188, 320, 328, 330–333
   See also bile, black
menstruation   64, 66, 67, 69, 70, 73, 74, 81, 104–106, 108, 111, 113, 116, 118, 120–123, 141, 363
   See also blood; seed
mercury   335
metabolism   IX, 139, 353
metaphor   XIII, 77, 89, 175, 298, 306–309, 330, 332, 334
   See also analogy
metaphysics   4, 9–13, 17, 44, 58, 59, 223, 224, 254, 260, 261, 264, 280
meteorology   176, 191, 244
milk   XIII, 116, 137, 147, 154, 162, 331
minima   327–329, 333, 334
mixture   299, 319
- complex   362, 369
- of elements   149, 323, 327, 330, 334, 355, 361
- of hot and cold   185
- forms of   357–362, 373
- genuine   356–358
- of humors   341
- inanimate   361, 362, 364
- nutritive   163
- primary   249
- properties of   361
- rule of   323, 326, 327
- secondary   249
- theory   326–328
   See also temperament
moist/moisture   *See* water
motion
- constitutive   120, 121
- internal   366
- mode of   365–369, 373
- principle   XIV, 30, 33, 115, 204, 237, 297
- with a purpose   366, 367
- self-motion   39, 80, 287, 294, 306, 308

- simple   205
- source of   35
- vital   306
- voluntary   236–238, 241, 250
- without volition   237
    See also locomotion
mouth   66, 104, 119, 142, 143, 287, 324, 344
mucus   342, 343, 347, 348

Natural philosophy   XII, XV, 37, 129, 149, 153, 156, 221–254, 261, 266, 273, 285–309, 320, 321, 323, 326, 344, 355, 356, 364
nutrition
- cause   92, 144, 231, 293
- instrument of nutrition   231, 232, 299
- limit   53
- matter   6, 16, 53, 56, 88, 116–118, 141, 160, 268, 269, 271, 275, 276, 279, 280, 321, 334, 362
- nutrition as principle   222
- principle of nutrition   XIV, 106, 110, 205, 244, 247
    See also faculty, nutritive

Oesophagus   143, 302
Organon   45, 49, 50, 53

Paracelsianism   285, 320, 335, 344, 349.
particle   332, 348, 368
parts
- animal   16, 17, 63, 74, 75, 80, 86, 95, 106, 115, 122, 133, 149, 174, 213, 275
- elemental   327, 328, 334
- essential   266, 271
- material   27, 369, 276
- natural   66
- non-uniform/instrumental   27, 55, 56, 118, 139, 149
- part-whole relationship   XI
- uniform   15, 26, 27, 55, 56, 87, 88, 108, 118, 138, 139, 141, 146, 147
- substantial   271, 326
- vital   113
    See also body, living
patient   30–34, 37, 39, 330, 343

perception   X, 4, 5, 7, 11, 22–25, 47, 53, 54, 74, 76, 78, 130, 133–135, 168, 175–177, 182, 190, 192, 193, 206, 212, 214, 229, 238, 242, 250, 251, 254, 290, 300, 306
    See also faculty, perceptive
pericarp   158, 164, 165
peristalsis   351, 352
perspiration   323
pharmacology   247, 249, 289, 325, 330, 364
phlegm   XIV, 166, 320, 328, 330–333, 341, 342
    See also humor
physiology   IX, X, XII, 128, 154, 156, 173, 182, 221, 225, 228, 286–288, 290, 295, 298, 307, 309, 319–321, 325, 331, 334, 342, 345, 349, 352, 353, 379
plants
- cause   248
- form   361
- generation   136, 244
- heat   154–158, 160–168
    See also heat
- parts   75, 108, 157, 158, 160, 161, 164–168, 244, 294
- principle   161, 168, 239, 244, 247
- properties   248
pore   273, 274, 323, 329, 342, 353
potency   See faculty
potentiality   XVI, 4, 17, 26, 31, 35, 36, 43, 47, 51, 55, 57, 63, 65–67, 73–77, 80, 81, 87, 93, 109, 110, 115, 117, 121, 130, 158, 168, 229, 309, 358, 365
potentiality-actuality model(s)   31, 33, 34, 39
    See also actuality; faculty; δύναμις
powers
instrumental   263, 278
    See also faculty
preservation   See faculty, preservative; self-maintenance
preternatural   331–333
principle
- active   32, 33, 366
- chemical   138
- common   134, 135
    See also heat
- controlling   72

– external   101, 111, 113
– first moving   30, 113, 149, 182
– formal   245
– immaterial   289
– internal   101, 112, 362
– male   71, 75
– natural   160
– passive   32
– principle-matter-relationship   92
– overarching   XIII, 179, 248
    See also ἀρχή
priority
– in account (λόγῳ)   22
– of activity   208, 4
– of the actual   16, 35
– Double Priority Principle   4, 17
– of the end   214
– existential   24
– of form   245.
– of function   16, 18, 35
– logical   218
– of the object   4, 5, 22
– of reproduction   7
prioritization   7, 8, 16, 18
procreation   See faculty, procreative; generation
product
– waste products   66, 69
psychology   XI, XV, 85, 101, 103, 104, 154, 194, 221, 223, 224, 227–230, 232, 238–243, 245, 247, 253
pulse   353
purgation   323, 324, 331, 332
putrefaction   XIII, 65, 330–335
pylorus   327

qualities   IX, XIV
– *qualitas acquisita*   298
– balanced, imbalanced   321, 326
– contrary   139, 145, 359, 368, 370
– corporeal   198
– elementary   XIV, 135, 137–139, 141, 147, 232, 235, 327, 357
    See also air; earth; heat; water
– corresponding   137–139, 289
– generative   65
– hidden, "occult"   320, 321, 323, 324

– *qualitas impressa*   291
– incorporeal   198
– *qualitas inducta*   291
– intermediary   360
– natural   367
– nutritive   132
– passive   164
– primary   244, 320, 326, 330, 334, 335, 360, 361, 369, 370, 371, 373
– proper   272
– proportion   331
– secondary   330, 361
– sensible, sensory   199, 332, 334
    See also change, qualitative
quantification   260, 273–276
quantity   14, 23, 34, 53, 56, 86, 87, 91, 93–95, 105, 109, 118, 136, 145, 147, 162, 166–168, 177, 185, 187, 209, 277, 280, 294, 330, 342, 345, 363, 364
    See also change, quantitative

regeneration   260, 264
relative(s)   23, 24, 30, 48, 54, 55, 91
    See also correlative
replication   14, 31
– formal   9, 10, 37, 38
reproduction   X, XII, XIII, 3–18, 21, 23–26, 30, 31, 34, 37–40, 66, 78, 81, 90, 91, 101–103, 105, 106, 108, 109, 112, 113, 117, 120, 159, 184, 200–203, 205, 208, 210–214, 222, 232, 234, 236, 239, 244, 247, 249, 253, 260, 300, 322, 325, 369
– principle   244, 247
    See also faculty, reproductive
residue   X, 26, 63–74, 79, 80–82, 85–89, 98, 104, 106, 108, 109, 111, 113, 115–118, 120–122, 139, 142–144, 156, 164, 183, 184, 186, 187, 189, 208, 322, 323
    See also food; περίττωμα
ripening   159, 164–166, 331, 334
    See also πέπανσις
rodents   340, 343, 344, 346

saliva   324
salt   333, 335, 348
sciences, subordination of   221, 224, 225, 227, 248, 254

seed 3, 7, 8, 51, 64–73, 76–82, 101–104, 106, 107, 109, 113, 115, 120, 121, 123, 139, 157, 158, 164–167, 184, 202, 213, 244, 247, 249, 278, 289, 293, 299, 320, 321, 329, 363, 367, 368, 369
– principle 76, 158
    See also σπέρμα
self-maintenance 7, 10, 21, 25, 30, 31, 34–35, 38–39, 40, 102, 103, 113, 319
self-motion see motion
sensation 76, 77, 180, 213, 215, 236–238, 241, 242, 250, 252, 290, 366
    See also faculty, sensory
shape 6, 32, 54, 72, 80, 85, 104, 105, 107, 110, 114, 162, 198, 199, 232, 268, 269, 273, 277
similitude, similarity
– of agent and patient 289
– of corpse and ensouled body 276
– of relief and the removal of harmful fluids 68
– of form 37, 290, 359
– functional 156
– growth/generation by adding/reproducing what is similar 80, 137, 138, 139, 202, 209, 229, 231, 234, 236, 276, 287
    See also like to like (principle)
– of human and animal 268
– of magnet and the object attracted 292, 293, 295, 304, 305, 309
– between forms of movement 306
– of must and chyle 332
– of self-motion and magnetism 287
– of sleep and epilepsy 177
– of animals' sleep and the state of plants 76
– of soul/body and fire/water 150
– of spermatic secretions and menstrual discharge 69
– structural 156
– substantial 324
– See also analogy; assimilation
size 29, 34, 35, 56, 69, 85, 86, 89–98, 102, 109, 110, 118, 153, 162, 167, 168, 207, 214
– cause 93
– limit 86, 90–93, 97, 162, 168

skin 136, 140, 149, 222, 274, 323, 342, 346
sleep X, 33, 76, 131, 173–194, 212, 338–354
– cause 177, 181, 182, 184, 191
sortal 59
soul 320, 321
– animal 50, 73, 75, 130, 131, 159, 201, 213, 228, 229, 237, 238, 248, 277, 278, 360, 362, 369, 373
– appetitive XII, XVI
– celestial 237
– corruptible 262
– faculties See faculty
– generation 278, 372
– human 201, 211, 228, 229, 233, 238, 248, 260–263, 268, 272, 273, 276–280, 302
– immaterial 260–269, 272–275, 277–280
– impulsive 213
– incorruptible 262, 272
– instrument 25, 36, 38, 239, 334, 365
– irrational 217
– material 260, 262, 263, 269, 276–280
– natural 321–323
– non-animal 159
– non-human 272, 278, 364
– parts X–XII, 6, 78, 81, 88, 111, 112, 119, 153, 155, 159, 173–176, 178–180, 182, 186, 188–191, 193, 194, 197–218, 228, 233, 247, 268
– potential 75
– primary/first 7, 16, 18, 28, 40, 106, 112, 247
– principle of soul 16, 34, 77, 86, 202
– soul as principle 34, 35, 52, 76, 81, 103, 211, 216, 218, 229, 233, 235–240, 243, 246–248, 252–254, 262, 289, 341
– rational XII, XVI, 21, 174, 182, 188–191, 193, 194, 229, 262, 358
– vegetative 199–201, 204, 206–215, 218, 228–230, 232, 237, 238, 246, 248, 262–264, 267, 279, 358
– world-soul 325
    See also body-soul-relationship; faculty; form; ψυχή

spirit (*spiritus*) 239, 291, 296, 297, 300,
    303–305, 324, 341, 345, 347, 348,
    364–369, 371
- potency 365
- generation 365
    See also πνεῦμα
spleen 66, 73, 76, 324, 331
stomach XIII, 26, 66, 71, 73, 76, 77, 104,
    119, 137–140, 143, 144, 146, 184, 295,
    297, 308, 319, 321–329, 334, 335
storks 346, 351
stupor 345, 350
subject
- logical 50
- matter 46, 47, 49, 50, 53
- passive 32
- underlying 212–214
subjecthood 44, 58
substance IX, 12, 17, 23, 31, 34, 35, 39, 43,
    45–52, 57, 58, 82, 86–88, 105, 106,
    109, 110, 136, 141, 149, 159, 199, 216,
    231, 234, 251, 265, 266, 276–278, 286,
    290, 292, 295, 299, 304, 306, 320–
    329, 334, 335, 343, 356, 357, 363, 371,
    372
- celestial 243
- composite 245
- material 260
- natural 261
- sublunary 204
    See also being; change; substantial;
    form, substantial; generation, substan-
    tial; οὐσία
sulfur 335
swallows 51, 346, 351
sweat 323, 343
synonymy See causation, transmission
    theory of causation

taste 25, 165, 332, 334
teeth XIII, 104, 137–139, 141, 147, 148
teleology 8, 27, 70, 73, 89–91, 95, 97, 98,
    110, 130, 133, 148, 150, 197, 201, 208,
    210, 211, 214–216
    See also cause, final; τέλος
temperament 173, 185, 224, 231, 235, 242,
    244, 247, 248, 250, 251, 319, 326, 329,
    330, 331, 334, 345, 359, 360, 362, 363,
    366, 370–373
- principle 244
    See also mixture
therapeutics 319, 320
transmutation 292, 335
transplantation 158, 224, 304

urine 68, 288, 319, 323
user 37–39, 159
uterus 63, 68–71, 81, 113, 121, 136–138

vapor 176, 323, 329, 342, 345, 348
vein XIV, 69, 140, 141, 160, 175–177, 224,
    273, 274, 289, 290, 299, 309, 322, 323,
    327–329, 332, 334, 347
    See also blood

wake 33, 76, 174–176, 179–181, 187, 189,
    191, 192, 339, 346, 350
- cause 191, 192
warmth, warm See heat
water, wet XII, XIV, 27, 43, 47, 54–56, 127,
    130–133, 135–138, 140–147, 150, 155,
    157, 159, 162, 163, 165–167, 176, 177,
    185, 188, 232, 235, 246, 251, 252, 289,
    290, 302, 320, 326–331, 341, 343,
    346–348, 352, 362
- generation 302
weight loss 348
wind egg 71, 72, 74, 75, 82
wine 30, 55, 56, 176, 327, 328, 331, 332,
    334
wisdom 87, 185

zoology 128, 223, 224, 238, 239, 241, 243,
    245–247, 340, 343, 350, 351, 353

τὸ αἰσθητικόν (to aisthêtikon) 22, 175, 178,
    217
αἰτία, αἴτιον (aitia, aition) 12, 29, 35, 89,
    107, 142, 145, 160, 167, 189, 217
ἀρχή (archê) 12, 16, 30, 32, 33, 35, 46,
    107, 114, 115, 118, 134, 135, 141, 157,
    158, 160, 164, 217, 222, 239
αὔξησις (auxêsis) 46, 54, 86, 88–92, 107–
    109, 114

τὸ αὐξητικόν (to auxêtikon)   23, 24, 56, 57,
    72, 86, 105, 109, 121, 122
βίος (bios)   76
γένεσις (genesis)   5, 8, 31–33, 57, 65, 107,
    108, 112, 113
γέννησις (gennêsis)   5, 8, 10, 79, 108
τὸ γεννητικόν (to gennêtikon)   16, 105, 108,
    111, 113, 120
διάνοια (dianoia)   182, 186, 187, 189, 192,
    193
τὸ διανοητικόν (to dianoêtikon)   217
δύναμις (dynamis)   5, 6, 16, 17, 22, 31–37,
    39, 40, 46, 48, 57, 82, 93, 95, 101,
    114–116, 118, 158–162, 200, 209, 222,
    232, 241, 288, 298
εἶδος (eidos)   39, 40, 46, 54, 55, 139, 204–
    206, 216, 254
τὸ εἶναι (to einai)   12, 23, 35, 107, 208, 210
ἐνέργεια (energeia)   IX, 22, 30–33, 35, 39,
    48, 80, 114, 209
ἐντελέχεια (entelecheia)   32, 36, 46, 49, 55
ἔργον (ergon)   8, 39
τὸ θερμόν (to thermon)   113, 128, 129, 135,
    140, 155, 157, 160–164, 166, 167, 176,
    177, 184
τὸ θρεπτικόν (to threptikon)   7, 16, 22, 72,
    86, 107, 111, 113, 114, 121, 122, 217, 218
κίνησις (kinêsis)   31, 35, 114, 121, 175, 183,
    186, 217, 238
τὸ κινητικόν (to kinêtikon)   33, 297
λόγος (logos)   6, 22, 29, 33, 35, 36, 92,
    114, 117, 159, 163, 193, 200, 202, 203,
    206, 208, 230
μεταβολή (metabolê)   160, 163

μεταίτια (metaitia)   145
τὸ νοητικόν (to noêtikon)   22
νοῦς (nous)   XI, 187, 189, 192, 193
ὄργανον (organon)   36, 104, 114, 159
οὐσία (ousia)   8, 12, 15, 23, 31, 33–35, 39,
    45, 46, 49, 57, 59, 106, 109, 110, 149,
    150, 157, 199, 204, 208
πέπανσις (pepansis)   164–166
περίττωμα (perittôma)   64, 66, 162, 176, 183
πέψις (pepsis)   XIII, 57, 79, 142, 155, 160,
    161, 164, 166, 252, 331
πνεῦμα (pneuma)   XIV, 102–104, 113, 115,
    117, 123, 136, 137, 165, 176–180, 183,
    184, 186–193, 203, 224, 227, 233,
    239–242, 298, 345
τὸ ποιητικόν (to poiêtikon)   24, 75, 153, 213
σπέρμα (sperma)   38, 64–66, 68, 70–73,
    77, 79–82, 162, 164, 167
συναίτιον (synaition)   29, 38, 145, 153, 167
τέλος (telos)   16, 36–38, 40, 162, 166, 175,
    211
τροφή (trophê)   IX, 4, 5, 16, 22–24, 26, 44,
    46, 48, 64, 79, 86, 102, 105–110, 114,
    122, 136, 137, 139, 142–146, 153, 159–
    161, 164, 166, 176, 202, 209, 230, 234
ὕλη (hylê)   26, 46, 54, 55, 92, 114, 163, 164
φύσις (physis)   11, 33, 36, 37, 58, 65, 92,
    114, 141, 161, 163, 165, 183–185, 189,
    203, 204, 288, 290
ψυχή (psychê)   XI, 7, 12, 16, 17, 46, 75, 92,
    107, 113, 114, 130, 150, 156, 175, 189,
    191, 200, 203, 209, 217, 238, 239, 288,
    297, 298

www.ingramcontent.com/pod-product-compliance
Lightning Source LLC
Chambersburg PA
CBHW031749220426
43662CB00007B/332